ENVIRONMENT AND NATION

By GRIFFITH TAYLOR

A. GEOGRAPHY

Australia, Physiographic and Economic (4th edition),	Oxford, 1928.
A Geography of Australasia - - -	Oxford, 1914.
*New South Wales - - - -	Melbourne, 1912.
*The Geographical Laboratory - - -	Sydney, 1925.
*Wall-Atlas of Australian Maps - - -	Oxford, 1929.
Australia, a Descriptive Text - - -	Chicago, 1931.
Europe, a Descriptive Text - - -	(In the Press).

B. METEOROLOGY

*Climate and Weather of Australia - -	Melbourne, 1913.
Australian Environment (Government Printer),	Melbourne, 1918.
Australian Meteorology - - - -	Oxford, 1920.
*Koeppen's World Climatology, Vol. IV - -	Berlin, 1932.

C. ANTARCTICA

With Scott—The Silver Lining - - -	London, 1916.
Physiography of MacMurdo Sound, (Harrison)	London, 1922.
*Hints to Scientific Travellers, Vol. IV - -	The Hague, 1926.
Antarctic Adventure and Research - -	New York, 1930.

D. GEOLOGY

The Archeocyathinae (Cambrian Corals) -	Adelaide, 1910.
*David's Geology of Australia, Vol. I - -	(In the Press).

E. ETHNOLOGY AND ENVIRONMENTAL CONTROL

Environment and Race, Oxford, 1927.	(Japanese edition, 1930).
Atlas of Environment and Race - -	Chicago, 1933.
Environment and Nation - - -	Toronto, 1936.

*Joint Author.

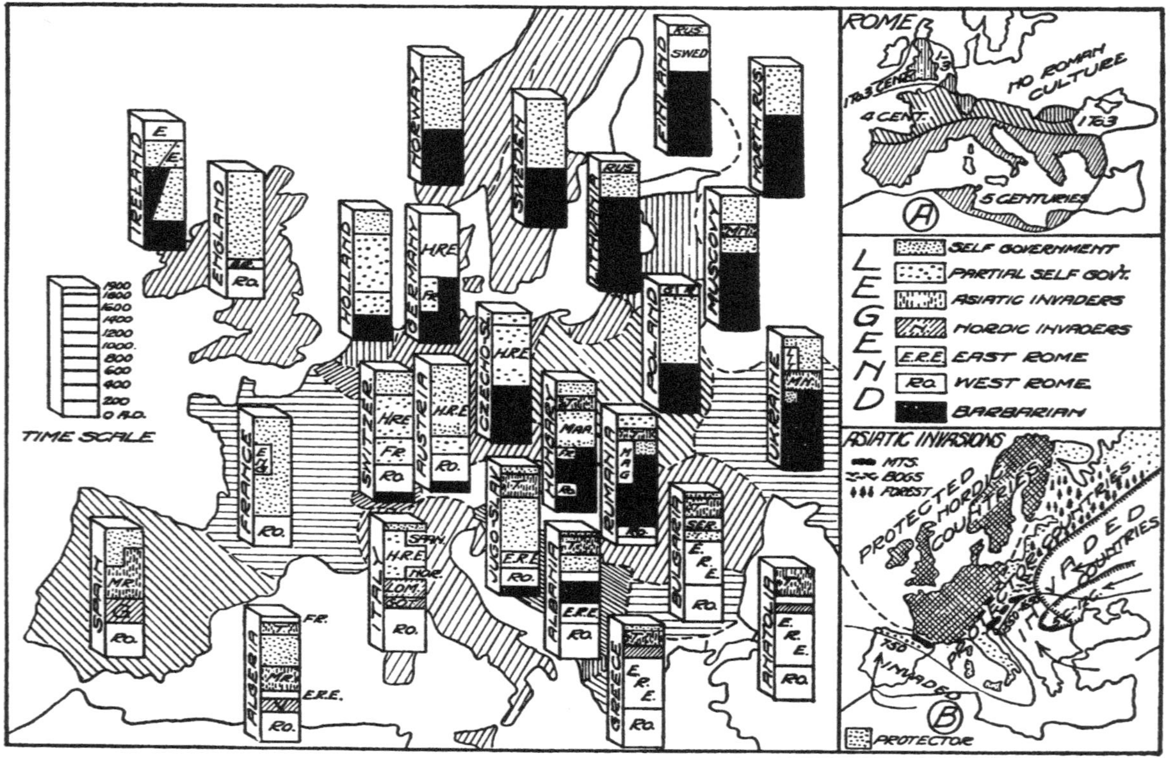

A Time-Space Chart showing Historical Strata for the Chief European Nations. Time-Scale at the left. Barbarian, Roman, Nordic and Asiatic Invaders and Self-Government (*i.e.*, by a ruler of the same nation) are the chief features indicated to scale in the columns of Strata. Insets show the effect of Roman Control, and of Asiatic Invasions. (See p. 26.)

ENVIRONMENT AND NATION

GEOGRAPHICAL FACTORS IN THE CULTURAL AND POLITICAL HISTORY OF EUROPE

BY

GRIFFITH TAYLOR

D.Sc., B.E. (Mining), B.A. (Research, Cambridge)

Professor of Geography, The University of Toronto

TORONTO

THE UNIVERSITY OF TORONTO PRESS

1936

SCHOLARLY REPRINT SERIES
ISBN 0-8020-7046-9
ISBN 978-1-4875-8250-0 (paper)
LC 37-22776

Reprinted in 2018
BY
THE UNIVERSITY OF TORONTO PRESS

PRINTED IN CANADA

PREFACE

THIS book is a sequel to my "Environment and Race," published by the Oxford Press in 1927. In that book the effect of Environment on the world-wide groups of Man known as Races was studied. In the present work a smaller unit, "the Nation," is discussed in the same fashion. A third book, "Environment and Settlement," dealing with the still smaller units of towns and settlements is in preparation.

In a sense the book is a protest against narrowing the fields of science unduly. "We live by advancing and changing and discovering, not by defining and hedging about" (Bowman). In America more than in Britain or on the Continent professional Geographers of to-day are concerned almost wholly with the economic aspects of man's habitat. Cultural and topographic aspects are not so generally discussed. I have no doubt that this latter side of Geography will become more "fashionable" in the near future. To the writer it is the most fascinating of all, perhaps because it is concerned with a field to a large extent unclaimed by the Geographer, Geologist, Historian, Ethnologist or Sociologist, and yet owing much to all these disciplines.

I believe that the application of a somewhat novel geographical technique to historical problems will be found interesting by readers. The 200 maps and diagrams have almost all been specially drawn for the work. I know of no similar study which describes categorically the background and chief events of each of the countries of Europe. I hope that it will serve perhaps as a reference book for that large section of readers who are much more interested in the cultural than the economic aspects of the European Scene.

To that group of researchers which sees no merit in "generalisations" or in "objective treatment" it is possible that this book will not make much appeal. It is fair to point out that any writer who tries to discuss such a large field as that treated in this volume must necessarily deal only with the more salient facts. There is no generalisation without its exceptions. But it is the writer's experience that young students are greatly helped if they can identify certain leading features in the mass of details comprising European History. Furthermore, it has always been the writer's practice, where possible, to chart every fact with which his study is concerned, and it should be unnecessary to state that a scientist tries to do this impartially. He asks certain historians to restrain their criticism until they themselves have tried this "objective approach."

The author places above all other considerations the preservation of World Peace; so that he believes that all educators should keep ever in mind the saying "In times of Peace, make Peace secure." It is his belief that no one can properly understand the European Political Complex without knowing the fundamentals of the cultural and topographic background. These interact, and it is precisely because the geographies omit the "Culture" and the histories omit the "Topography" that the writer ventures to hope that this book will find a useful place in the classes of both subjects.

Needless to say, many books have been consulted and the chief of these are listed in the footnotes. But there are half a dozen which should be in the library of every student in this field, and I am glad to acknowledge my indebtedness to them here.

Isaiah Bowman: *New World*, 1928, New York.

Cambridge Medieval History, Cambridge University Press.

H. F. Helmolt: *History of the World* (Vols. IV to VII), 1907, London.

E. de Martonne: *Europe Centrale* (2 vols.), 1931, Paris.

M. Newbigin: *Southern Europe*, 1932, London.

W. Z. Ripley: *Races of Europe*, 1900, London.

W. R. Shepherd: *Historical Atlas*, 1921, New York.

I have also to express my debt for innumerable incidental dates and facts to that compendium of knowledge, "The Encyclopedia Britannica."

The writer has been lecturing on the lines of this book, *i.e.*, the geographical interpretation of European History, since 1928; admittedly only a short time in which to gain a comprehensive idea of the complicated trends which go to form European History. He has, however, in the aggregate spent several years wandering about Europe, and has a personal knowledge of all the main countries except Russia. This background has been supplemented by the kindness of a number of his colleagues while he was a teacher at the University of Chicago. Among the 900 members of the Faculty it is possible to find experts in most of the branches of human knowledge. Thus Parts I and II of the book have been read by Professor Arthur Scott, and I am greatly indebted to him for his very detailed comments. The chapter on the Spread of Christianity has been carefully revised by Professor John McNeill. Professor Joranson has read Part III and Professors Harper and Noe have criticised much of Parts IV and V. I take this opportunity to give them my hearty thanks for the considerable time they have spent in improving the book.

GRIFFITH TAYLOR

Department of Geography,
The University of Toronto.

CONTENTS

PART I

Physical Factors in European History

PART II

Cultural Factors in European History

PART III

The Protestant Nations of Northwest Europe

PART IV

The Catholic Nations of the Southwest and Centre

PART V

The Eastern Nations of Europe. Mainly Greek Church

LIST OF ILLUSTRATIONS

CHAPTER VI—LANGUAGE

CHAPTER VII—EARLY HISTORY

CHAPTER VIII—GREECE AND ROME

CHAPTER IX—FOLK-WANDERING

CHAPTER X—RELIGION

CHAPTER XI—RENAISSANCE

CHAPTER XII—BRITAIN

CHAPTER XIII—SCANDINAVIA

CHAPTER XIV—GERMANY

CHAPTER XV—NETHERLANDS

CHAPTER XVI—SWITZERLAND

CHAPTER XVII—FRANCE

CHAPTER XVIII—SPAIN

CHAPTER XIX—ITALY

CHAPTER XX—AUSTRIA

CHAPTER XXI—BOHEMIA AND HUNGARY

CHAPTER XXII—POLAND

CHAPTER XXIII—RUMANIA

CHAPTER XXIV—BALKANS

CHAPTER XXV—RUSSIA

CHAPTER XXVI—EAST BALTIC

CHAPTER XXVII—CONCLUSIONS

PART I

PHYSICAL FACTORS IN EUROPEAN HISTORY

CHAPTER I

CHARACTERISTICS OF THE NATIONS

A. Introduction

There appears to be in recent years a rather well-defined change of approach in the modern treatment of history. The earliest studies of history were often based mainly on the contemporary biographies of notable leaders who were generally sovereigns. *Personality* was greatly stressed, so that at times it seemed perhaps the main factor in history. Later, say a few decades ago, it was generally realised that the *People* were even more interesting than the Princes, and so we find historical treatment largely concerned with *social studies*. But there is another side of history which has been but little developed hitherto—at any rate as a major topic. It is a commonplace that while men like Charlemagne and Napoleon have "altered the map of Europe," some factor seems to wipe out their extended boundaries within a decade or so of their deaths. In general, the national areas revert to something of their condition before the great conqueror took charge. The main reason for this reversion is that there is a large number of *environmental factors* which only act very slowly on national development, but in the long run outweigh the control exercised by personality, however vigorous it may seem to be for a time. It is these slow-working environmental factors—whether *topographic*, *climatic*, or in a less particular sense *cultural*—which will be investigated more and more fully in the future. The present book is an attempt to show how much history has been influenced by these non-personal environmental factors. It is in a sense a liaison-study of a field lying between geography and history, but belonging to both.

Yet it is not correct to class the present study with a number of works which have appeared under some such title as "Historical Geography." The latter studies for the most part were concerned mainly with the fluctuating boundaries of various nations or kingdoms. The geography there discussed was almost entirely *Political* Geography. The *Where* rather than the *Why* was the portion contributed to such an "Historical Geography" by the geographical data. But the modern geographer is—or should be—as much concerned with causes as with places. In the words of Francis Bacon, "Vere scire est per *causas* scire." And so in our study we shall throughout attempt to show not only how various tribes and nations expanded in this or that direction, but also what part was played in their history by such factors as mountain barriers and mountain passes, by broad rivers, by open plains, by arid plateaux, by unkindly or unfamiliar climates, and by mutual attractions between allied races and languages. Naturally, few historians in the past have been adequately acquainted with that branch of geography which some have termed Environmental Control, while with others there is a tendency to name it Human Ecology. There have been, however, a number of geographers who have studied historical problems along these lines. Among these may be mentioned Newbigin and Fleure. Both of these, however, have confined their attention to relatively small areas. Nor have they perhaps kept definitely in view the effect of environment on *national* development, so much as the part it played in determining changing boundaries in the areas concerned.

For reasons which will be obvious enough, the writer has confined his studies of national development to the lands included in Europe and the immediate vicinity. Here is a sufficient diversity to satisfy any writer, and the historical literature is get-at-able and abundant. The present book is in a sense a sequel to the earlier book, "Environment and Race," which was published in 1927. In that book the part played by the ever-present environmental factors in determining the evolution and migration of the peoples of the world was studied in regard to the more salient features. It described an interesting chapter in human history, covering approximately the period between the age of Paleolithic Man and the Folk-Wandering (A.D. 500).

The present work carries on the study from about this date, when several of the modern nations of Europe had their birth, up to the present. Whatever be the evils linked with the Treaty of Versailles it at least resulted in a map of Europe in which for the first time the political boundaries agreed approximately with those defining national cultures. In other words, the nations have reached some sort of equilibrium, based, in the writer's opinion, on the total effect of all those diverse environmental factors referred to earlier in this chapter. It is an interesting speculation to consider what will be the condition of the world a century hence. Surely the parochial divisions of to-day—the nations with their petty but disastrous differences—will then have given place to something approaching a World State. To coin a phrase "May Internationalism intèr Nationalism"! Howbeit the present is the "Period of the Nation," which is no less interesting because it is but a temporary stage in the development of the Human Race.

B. Characteristics of a Nation

It is unfortunate that to most laymen the distinction between Race and Nation is not at all clear. In fact, most of them use the terms as synonymous. This is perhaps natural enough as long as we find in serious literature such terms as "European race," "French race," "the varied races of the old Austrian Empire," still used quite freely, though few if any anthropologists would countenance such terms. As regards lesser known parts of the world, such as Melanesia and Polynesia, it is still by no means rare for these *purely geographical* names to be used in ethnological literature as if they had a definite *racial* meaning. Furthermore, the old practice of separating human groups according to the *languages* they speak, is still used for many outlying parts of the world, probably because language is the sole attribute of the various groups of which we have much knowledge. This criterion, of course, has no real relationship to *Race*, which belongs purely to the *biological* side of Anthropology. The concept of a *Nation* is unfortunately not much more easily understood.

In the course of a dozen years the writer has often looked for some logical discussion of the geographical factors which determine the evolution of a distinct nation. Webster gives two definitions of Nation; as follows: "(1) (Ethnology) A part or division of the people of the earth, distinguished from the rest by common descent, language or institutions; a race; a stock. (2) *The body of inhabitants of a country united under an independent government of their own.*" The present writer strongly objects, as stated, to the word *nation* being taken as synonymous with *race* or *stock*. Often enough, common ancestry, common language, and common institutions are involved in a given nation, but exceptions to one or other of these factors are extremely common. On the other hand, the criterion of an "independent government" is not so much open to objection. To be sure, the Finns never had an independent government (until 1918), but they have formed a fairly definite nation throughout historic times.

It is clear, therefore, that we are dealing with a very complex problem, so complex that there is no simple answer to the question "What constitutes a nation?" But it is an important problem lying on the borders of Geography, History and Sociology; and as such seems to me to offer an interesting field for the use of what I am sometimes tempted to call the "New Geography."

Thus we find that many factors have to be considered in arriving at a concept of a nation. Language, Race, Tradition and Independent Government have been mentioned. But a survey of history shows that a well-defined environment, especially if it have a *natural centre*, is perhaps of as much importance.

We may arrive at a clearer idea of the relative value of these factors if we show in a table how such characters are distributed among some of the chief nations of Europe during the past ten centuries.

TABLE A

AN ATTEMPT TO EVALUATE FACTORS IN A NATION[1]

	France	England	Spain	Switzerland	Italy	Belgium	Germany	Russia	Rumania
Language	Unit	Unit	¾ Unit	4 Langs.	Unit	2 Langs.	Unit	Unit	Unit
Religion	Unit	Unit	Unit	2 creeds	Unit	Unit	2	Unit	Unit
Race	3 races	2 races	Unit	Unit	2 races	2 races	2 races	Unit	Unit
Tradition	Strong	Strong	Strong	Feeble	Discontinuous	Feeble	Discontinuous	Fair	Fair
Independent Government	10 centuries	9	8	5	10?	1	10?	6	1
Well-defined environment	Unit	Unit	Unit	Fair	Poor	Poor	Fair	Fair	Poor
Natural centre	Very good	Good	Poor	Fair	Poor	Fair	Fair	Good	Poor

Assuming that these seven characteristics include most of the main ingredients of nationhood, we see that France possesses all of them in a notable degree, except Race. England is nearly as strong. Spain resembles England closely, though Madrid is a poor substitute for London as a capital. Switzerland has no strong (*i.e.*, clear-cut) characters except Race. In all other features it is diverse, yet the Swiss Nation seems to be fairly harmonious and to present a united

[1]This table is admittedly tentative, but the conclusions summarised here will be justified—the writer hopes—in later chapters. Professor Scott stresses long-continued political association and defence against common enemies (as tending to produce a common consciousness) as further important factors.

front to the world. Italy has the advantages of a single language and religion, yet in all other respects it is built of diverse strains and has not a satisfactory environment to promote complete unity. Belgium has little but unity of religion to cement its people into any kind of nation. Germany resembles Italy in many respects, but Berlin is perhaps a better centre than Rome. Russia is strong in human links but weaker in environmental links than most of the other areas, and the same is true of Rumania.

In dealing with such a complex concept as a Nation we cannot hope to arrive at very acceptable quantitative conclusions. It is obvious that France is in a strong position as regards permanent nationhood, while Belgium's condition is much more precarious. We may appeal to the actual record of history, which shows that *Linguistic divisions* agree more closely with the modern map of the Nations of Europe than do the divisions of Religion, Race, lengthy independent government or even Environment. So strong is the language bond that sometimes purely linguistic maps are erroneously labelled "Racial" maps![2]

C. A Classification of European Nations

For many years the writer has been experimenting with graphs in three dimensions. As used for Topography these graphs (Block Diagrams) are unexcelled. They render almost unnecessary the pages of description (not to mention contour maps) with which many topographic papers are still filled. But they can, of course, be used for three independent variables of any kind. In the present discussion of factors affecting European Nationality we can construct a three-dimension graph which enables us at a glance to pick out affinities among the thirty European nations of to-day.

Our variables are Race, Language and Religion. There are in each case three dominant classes, which may be indicated in Table B.

These three types seem to centre about three "nuclei" around Moscow, Rome and northwest Germany.

However, almost all the remaining regions in Europe depart in some fashion from this simple grouping of types. In the Three-dimension Graph (Fig. 1) the nine variations (three each in Language, Religion and Race) are arranged somewhat like nine "Sets of Apart-

[2]For example the "Racial" Map of Europe published by the National Geographic Society (of Washington) in 1918. It resembles Fig. 122 rather than Fig. 22.

ments" in intersecting rows. Since the Mediterranean Race was the first to enter Europe we may represent nations of this type as living on the "ground floor." The Nordics may be shown as occupying the next floor, while the Alpine Race (which seems to be slowly submerging the other two all over the world)[3] may be represented as occupying the top floor in each "apartment."

TABLE B

MAJOR TYPES OF LANGUAGE, RELIGION AND RACE

Culture—Type	Language	Religion	Race	National Example
I Slav	Slav	Greek	Alpine	Russia, Bulgaria
II Roman	Romance	Roman	Mediterranean	Spain, Portugal
III German	Teutonic	Protestant	Nordic	Denmark, Holland

Let us divide our area (which may be taken to represent Europe) into three "Parks," and name them in accord with the *Languages*, "Teutonic Park," "Romance Park," and "Slav Park." The cross avenues (in an east-west direction) indicate the Religions of the nations occupying the respective "apartments."

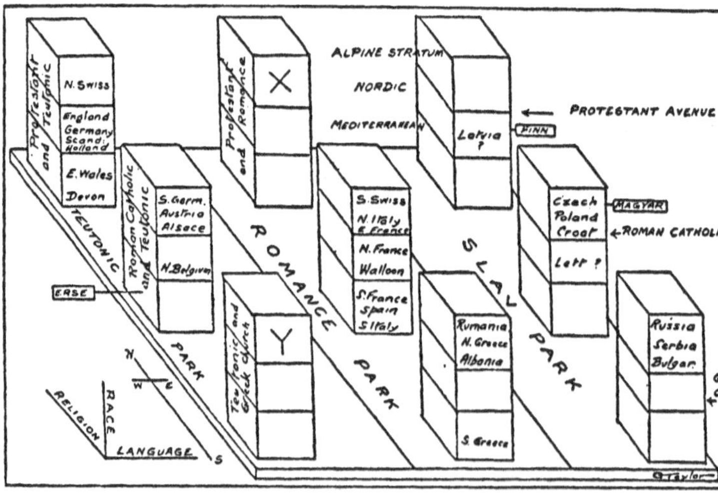

Fig. 1.—A Block Diagram correlating National Variations in Race, Religion and Language.

---

[3]See the writer's paper, "Nordic and Alpine Races and Their Kin," *Journal of Sociology*, Chicago, 1931.

To give an example: the Swiss, for the most part, speak German (a Teutonic language); hence they have a place in the western "Park." They are mostly Protestant; hence they live in the northern "Avenue." They belong to the Alpine Race; hence they can be shown in the top "flat." Here the word SWISS will be found in the Block Diagram (or Model). To give another example: Poland is Slav (Eastern Park), Catholic (Middle Avenue), and Alpine (Top Flat). France is so diverse in Race that we must place various divisions in each of the three kinds of "flats." But it is dominantly Catholic and Romance-speaking, so that all go into the single central building.

To accommodate a few less important nations, characterised by special *languages*, I have attached labels to the nearest nations which agree with them in Race and Religion. Thus the Magyar nation is fairly close to the Poles, and the Finn is fairly close to the Swede. In Western Ireland, Erse (Irish) is spoken by the primitive "Mediterranean Catholic." This small national group is remote from any other.

This diagram (Fig. 1) therefore enables us to find the larger groups into which the nations might be expected to fall. Thus England is closest to North Germany, Scandinavia and Holland (which are all in the same flat). Its people have much in common with the Swiss. No other Protestant groups are very close. The Catholic groups of Ireland and Belgium are allied by similarity of race and language.

An allied language almost invariably implies a considerable body of common ideals and a somewhat similar national background. In fact, the writer believes it is much the most important "cement" in welding a nation. This is why in the "Russification" or "Prussianising" of a nation it is decreed that the language taught in the schools must be that of the dominating power. Nearly all the so-called "Ethnic Maps" of Europe, which have been published in such profusion during this century, should more properly be labelled *Linguistic Maps*. Moreover, there are only a few exceptions to the general agreement of language and nation. Switzerland and Alsace spring to one's mind.

Returning to the model (Fig. 1) we find the well-defined Romance group of France, Spain and Italy all occupying one building. Religious allies lie to east and west. Linguistic allies are found in the front of our diagram in Rumania and (to some extent) in Greece. Two groups of Slav nations are indicated—Russia, Serbia and Bulgaria in one building, and Poland, Czecho-slovakia and Croatia in the other.

Perhaps of most interest are the two empty buildings X and Y. What do they mean? Surely we learn from X that Protestant religions do not appeal to folk with a background of Romance culture; and from Y that the Greek orthodox religion has not attracted the Nordic race in northwest Europe, which of course was christianised from Rome. By studying the nations in the "ground floors" we may learn how the Mediterranean Race has varied. In fact, most of the essential differences and similarities of European nations are shown in this somewhat novel "three-dimension" diagram.

D. Patterns of National Development

The salient features in the development of the main nations of Europe can be arranged in several fairly definite patterns. All start as communities of *Barbarians* and all have arrived at a fairly satisfactory measure of *self-government*,[4] in many cases as the result of the Great War of 1914-18. In the frontispiece to this book an attempt is made to summarise on one map the salient features of the history of the chief European nations. In each country is placed a *column*, which represents to scale the period from A.D. 0 to A.D. 1900. This column is divided into sections corresponding to the rulers of the country in the period indicated. The initials indicate various conquerors or forms of government. Among others are Barbarian (black), Roman (R), Holy Roman Empire (H), East Roman Empire (ERE), Turk (T), Goth (G), Vandal (V), Moor (M), Mongol (MN), etc. The period in which each country experienced self-government (either by tyrants, princes, or people derived from the nation itself) is indicated by *dots*. An analysis of the histories of the chief countries, reduced to the simplest terms, is also given in Table C.

Perhaps the most striking events in the history of the twelve nations there summarised are the *Rule of Rome* and *Eastern Invasions* (Fig. 2). Eight of the twelve nations benefited largely from the *Pax Romana*, five of them suffered from Eastern Invasions. (See maps inset in Frontispiece.) Since the Western Roman Empire fell before the onslaught of Barbarians the general evolution even of the western nations was not on the whole very different from those unaffected by Rome.

In essence, therefore, each of the European nations passes through

[4]By this term I do not mean *democracy*, but "government by a ruler (often a tyrant) of similar nationality to the governed."

TABLE C

Approximate Summary of National Developments

	France	England	Spain	Switzerland	Italy	Austria
1918 1600	Republic Strong Ministers	Limited Monarchy Strong Ministers	Republic Weak rule	 Republic	Strong Ministers Foreign aggression	Republic Strong Rule by Hapsburgs
1200	Strong Personal Rulers	Strong Personal Rulers	Strong Personal Rulers	Military Republic	Local Republics and Weak Kings	
800	Strong King	Strong Kings	Eastern Invaders	Imperial Union	Imperial Union	Imperial Union
400	Nordic Invaders	Nordic Invaders Weak British	Nordic Invaders	Nordic Invaders	Nordic Invaders	Nordic Invaders
0	Roman Empire	Roman Empire Barbarian	Roman Empire	Roman Empire	Roman Empire 'Golden Age'	Roman Empire
400 B.C.	Barbarian	Barbarian	Barbarian	Barbarian	Roman Rule	Barbarian

	Germany	Poland	Russia	Rumania	Yugo-Slavia	Greece
1918 1600	Republic Rise of Hohenzollerns	Republic Foreign aggression	Republic Weak rule Strong Personal Rule	Kingdom Balance of Power saves from Turkish aggression	Kingdom	Republic
1200	Many weak kingdoms	Rule by weak nobles and foreign kings	Foreign Invaders	Foreign domination usual	Turks Some Strong Kings	Turks Weak Control by Eastern Roman Empire
800	Imperial Union	Strong Rulers	Several Weak Kingdoms	Continuous Asiatic Invasions		
400	Barbarian	Barbarian	Barbarian		Barbarian	
0	Barbarian	Barbarian	Barbarian	Barbarian Roman	Barbarian	Roman Empire
400 B.C.	Barbarian	Barbarian	Barbarian	Barbarian	Barbarian	Rome Weak Rulers 'Golden Age'

(N.B.—Refer also to the two Time-Space Charts given on pages 126 and 127).

the following stages. First there arises an assemblage of tribes—mostly of similar racial origin and speaking cognate languages. A strong leader settles in some satisfactory "national nucleus" (here comes in the geographic factor), and gradually builds up a union of associated groups generally with feudal status. These units are often of the nature of *marches*, *i.e.*, regions capable of defence against outlying enemies—and again here is introduced the geographic factor. If the ruler has vigorous descendants the nation develops in what we may call a normal fashion, through control by a dynasty or dynasties of strong personal rulers. Then as the nation becomes too large for

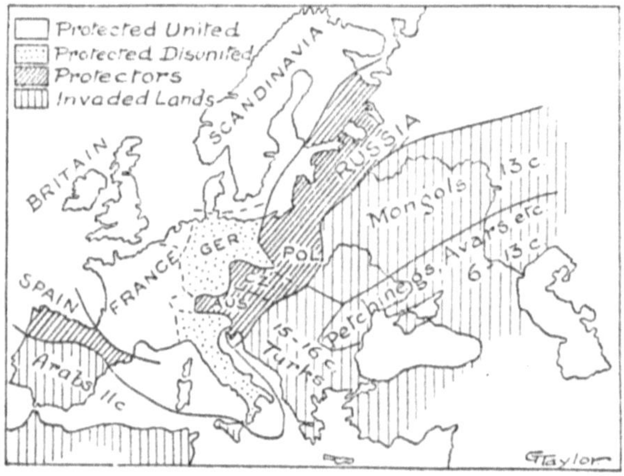

Fig. 2.—A Classification of Europe into four divisions based on freedom from Asiatic Invasions. Note the favourable position of the Nordics.

such rule, we usually find the development of strong ministers (sometimes deteriorating into dictators), and finally of parliamentary rule, either associated with democratic monarchs or with the republican form of control.

Among the nations of Europe, France perhaps best exemplifies this typical progression. It is true that it had a period of some five centuries of Roman Culture between the *barbarian times* of Vercingetorix and those, much less rude in character, of Clovis the Frank. Such rulers as Louis IX and Richelieu illustrate the later developments of our pattern. England has had almost exactly the same national evolution, luckily without the oppression of the peasants which led to the dramatic episode of the French Revolution. The Scandinavian development has been similar. Of the remaining countries Austria (in its present limited region) seems to have most closely agreed with France and England. We shall find it instructive to discuss later why this *central* region paralleled the characteristic development of Northwest Europe. It seems to the writer to be due to geographic control.

A marked variation in the pattern is offered by Germany and Italy. In the case of these countries there was no obvious *national nucleus*, for both regions are complex in their environment. This directly led to the development of a number of independent states, the whole series at first linked together in the high-sounding title of the Holy Roman Empire. The Alps separated the German and Italian cultures, and as regards actual practical politics they fell apart after the battle of Legnano (1176). Of course many *political* factors also delayed their independence, as will be seen in later chapters. Hence, not till the second half of the 19th century were these two great nations able to realise their respective unities. Switzerland also in some sort exhibits this historical pattern. Here a series of isolated "republics" takes the place of the larger independent units of Germany and Italy. Perhaps the pattern of Switzerland with its very varied warp and woof—without a *Swiss* cultural basis—is rather a sample of what may develop in the *future* World State, than a close analogue of Germany and Italy; with which, however, it had a parallel development till the 16th century.

There are some interesting parallels to be drawn between the histories of Spain and Russia—though their environments have not at first glance anything in common. Both, however, are marginal countries and have suffered the infliction of Eastern Invasions. In Spain the Moors dominated from 700 to 1300, while in Russia the Mongols controlled affairs from 1200 to 1400. It is to be noted that the Moors, though their homeland was for the most part Barbary in Africa, had an essentially Arab (*i.e.*, *Eastern*) Culture, based as much upon Asia as was that of the hordes of Jenghiz Khan. After the invaders were driven out by the rise of strong personal rulers (both in Spain and Russia) we see the strength of the dynasties—whether Hapsburg or Romanov—slowly waning, and leading to revolutions and republics.

Poland seems to have a special "pattern" due to the folly of her aristocracy. At first the Poles were Barbarians, civilised about 900 by Christianity rather than by the *Pax Romana*. In this they agreed with Germany and Russia. Then followed a period of strong personal rule, so that Poland in the 15th and 16th centuries was one of the strongest nations in Europe, hardly rivalled by the Hapsburg realms. But the members of the Slachta (a diet of selfish nobles) cut their own throats by oppressing the peasants and by appointing puppet kings whom they would not obey. Complete dominance by neighbouring European powers for over a century (after such a proud record) was an unfortunate experience hardly exhibited on such a scale by any other of our patterns.

Somewhat alike in pattern are Greece, Yugo-Slavia and Rumania, the three remaining examples in our table. Here in all three the dominance of the Turks has been the leading "motif" in the national pattern. To-day Rumania is the strongest nation of the three, yet it has the most extraordinary history of any large nation in Europe. Its development has been the reverse of normal—accepting our earlier definition as a criterion. About A.D. 100 it received some Roman culture, but Roman authority endured hardly a century. Thereafter with only the briefest intervals it was either being traversed by eastern hordes or conquered by the Crescent, right up to modern times. The Rumanian nation seems to the writer to have persisted primarily for two reasons. Firstly, by virtue of the mountain nucleus of their nation—the Carpathian Plateaux. Secondly, by reason of their astonishing powers of "peaceful penetration" which seem to exceed those of any other European group. A third factor of special importance in southeastern Europe in later times is the "Balance of Power." This has pitted the Russian against the Austrian, and both against the Turk. The emergence during the 19th century of Greece and Yugo-Slavia is also definitely due in large part to this rivalry.

Somewhat akin to Spain and Russia in her development is Yugo-Slavia. Like the former she had a valuable period of Roman culture and in all three nations there followed a period of strong personal rulers. All three, moreover, were swept by barbaric invaders for several centuries. But in Yugo-Slavia the Eastern dominance persisted until the 18th century, so that the native Serbian culture has had no chance to grow to the maturity of that exhibited in Spain or Russia. Moreover, there is less cohesion in Yugo-Slavia to-day with its strong religious cleavage than in Spain or Russia. Nor does the Serbian environment lend itself to unity in the same fashion as the four square peninsula of Spain or the illimitable plains of Russia. One may therefore predict troublous times for Yugo-Slavia before the south Slavs settle down to a united nationality.

The Greek nation has had perhaps the most diverse history of all—and it is not too much to say that its attempt to control two distant and diverse environments has been a large factor in determining its long period of negligible importance. Starting off about B.C. 500 with a history glorious alike in war and culture it soon reached its nadir a few centuries later. Then Greek culture reached a second peak—but now transplanted far to the east in the setting of the large imperial city of Constantinople. Here the rulers of the East Roman Empire were largely Greek or dominated by Greek culture. But the population of Thrace was by no means all Greek. Bulgars, Slavs and Asiatics were perhaps as important. Meanwhile the old homeland of Hellas with its centre of Athens merely changed from an unimportant province of Rome to an equally unimportant province of Constantinople. Thus there was a complete divorce between the new centre of Greek culture and the national home of the Greeks. This break in the evolution of the Greek nation was rendered even more complete by the Turkish dominance over both Greek centres which endured right into the 19th century. We may perhaps state that the main factor which ultimately restored the Greek nation in its old home of Hellas was that intangible something called "National Consciousness." This was founded on the Greek pride in their "Golden Age"—but no other nation in Europe has any parallel to offer for this long lapse from nationhood, which in Greece endured from B.C. 146 to A.D. 1821.

There are thus many diverse controls woven into the patterns exhibited by the nations of Europe. In the above sections we have rapidly traversed some of the main historical factors. Standing out above other controls are the *Pax Romana* (see Inset A in Frontispiece) and the Asiatic Invasions. These latter in turn were controlled by the Environment to so great a degree that we can usefully consider the Evolution of European Nations from a point of view based on "Protection from Invasion." Thus some countries were retarded by actual *Invasion* and some retarded by *resisting* invasion. Other happier areas largely occupied by the Nordic Race progressed far beyond their neighbours because they were *protected* from invasion (see Inset B in Frontispiece). This aspect of our problem will be developed in later chapters.

In a more limited sense Nationality did not develop until much later dates than those given in the frontispiece. J. H. Rose discusses this point admirably in his book "Nationality in Modern History" (1916, New York). In his opinion the English nation was not properly developed until the Hundred Years' War cemented Normans and Saxons. The same period saw France welded together by Joan of Arc, though it was reserved for Rousseau with his "Social Contract" in 1762 to crystallise the idea of French Nationality. Rose expresses the stages of evolution of the European Nations in the following phrases: (1) Imperial Unity of Rome, (2) Tribal Chaos, (3) Painful jostlings, (4) Assorting process under Monarchs, (5) Groupings of related tribes, (6) Homogeneous "national blocks" of to-day.

CHAPTER II

ENVIRONMENT AND NATION IN BRITAIN

A. Environment in England and Wales

The liaison between geography and history which is studied in this book may well be illustrated in a preliminary chapter dealing specifically with Great Britain. (See also Chapter XII.) It will be convenient to explain certain rather novel methods of approach before the general sections dealing with the major historical factors are entered upon. In the first place, it is futile to attempt to show the relation of landscape (or topography) to history unless the reader understands in some degree the simple principles which control the topography of a land. This involves the discussion of a few elementary facts in geology, and for such a discussion Britain is particularly well suited, for it was in Britain that the "Build" of a country was first investigated. The "type" names were given here—so that, in fact, rock formations all over the world bear such names as Devonian and Cambrian from their early association with Devon and Wales (Cambria). In the second place, the whole sequence of geological history is represented in what is perhaps a simpler and more complete fashion in England than in any other part of the world of such small size. A traveller from Snowdon (in North Wales) to London passes over a wonderful series of rocks, which epitomise the world's history for 500 million years. We are not in this book directly concerned with events which happened before man appeared on the earth, but indirectly the topography due to the rocks laid down in the last 500 million years has affected the life of every person who has ever lived.

In Fig. 3 (p. 35) is given a map of England and Wales according to a somewhat new plan. It is neither a *geological map* (showing the rocks of various ages), nor a *block diagram* (which represents topography in relief), but a combination of both, to which I have given the name "Mantle-Map." In most parts of the world the rock-formations (or strata) are laid one above the other, so that the younger formations appear somewhat like layers covering or "mantling" the older rocks. These younger rocks appear to be flung over the flanks of the young mountains, and to cover the relics of the *old* mountains. For our purpose it is usually sufficient to know if the rocks are older than the

TABLE D

SHOWING RELATION OF GEOLOGY TO TOPOGRAPHY IN EUROPE

Millions of years ago	Age	Formation	Topography, etc.
Last 40,000 years	Recent	Recent	Loss of much land in North Sea, Baltic, etc.
1	Tertiary	Pleistocene	Glacial Till and Moraines
5		Pliocene	Soft beds in England
15		Miocene	Age of "Alpine Storm"
30		Eocene	Soft beds in England
50	Mesozoic	Cretaceous	Chalk over Central Europe
70		Jurassic	Hard "Cuestas" in England
100		Triassic	Soft "Vales" in England
150	Paleozoic	Permian	Armorican mountain-building
200		Carboniferous	Great Coal-Plant Swamps
300		Devonian	Caledonian mountain-building
370		Silurian	Hard Rocks in Wales
450		Cambrian	End of a Period of Great Mountain-building

chief "Age of Coal" (*i.e.*, Pre-Carboniferous) or are a little newer *i.e.*, Mesozoic, or belong to the Tertiary or Recent epochs.

The oldest rocks usually contain the ores of the metals, which are not common in rocks younger than Carboniferous. In general, large areas of *agricultural* land belong to the younger formations, and are especially common in the alluvial soils belonging to the *Recent* period. The great Ice Ages occurred during the Pleistocene, within the last million years. It seems likely that this striking change in the environment was responsible for the evolution of man. During this period many mammals died out, but one plastic type—the Apes—gave rise to a new mammal particularly suited to cope with the new conditions. It seems certain that the slow changes in environment *which still occur* are potent in the evolution of man—which must still be going on. (This has always seemed to the writer an important reason why such studies as the present should appeal to every intelligent reader.)

Returning once more to the "Mantle-Map" in Fig. 3 it is easy to see the main geological features at a glance. The east of England consists of the *younger* layers—which lie *over* the western layers. In the west are the "lowest" layers of rocks, the oldest of all appearing in North Wales. Indeed, here are the original *Cambrian* rocks, the oldest in the above table. Most of Wales, however, consists of Silurian rocks, so named from the "Silures"—a British tribe who dwelt here in ancient days. In Devon are the original "Devonian" rocks which also occur in South Wales. Many of these rocks were greatly folded and much altered in the periods of early mountain-building mentioned in the above table.

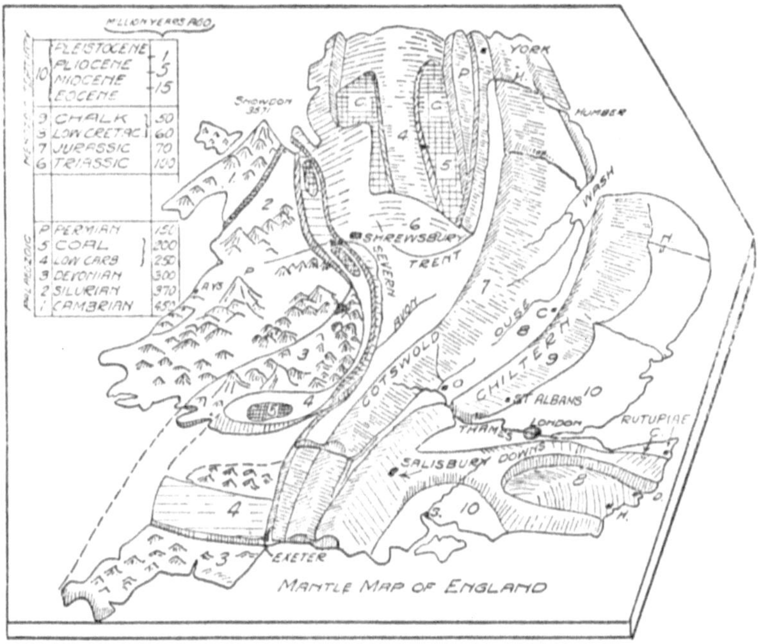

Fig. 3.—A Mantle-Map of England showing the control of Topography by Geology. Numbers 1 to 5 constitute the Older-Mass and 6 to 10 the Younger-Mass. Coalfields determining present population are shown at 5.

The chief coalfields of England are to be found where the next "mantle" or formation occurs. This is the Carboniferous (labelled 4 and 5 in the map). Thus we see at a glance why the industrial region of England is confined to the north and west. It is because the coal-bearing rocks are deeply covered by younger rocks in the east and south of England. These four formations constitute the "Older-Mass" of England.

The "Younger-Mass" of England consists of a series of alternating hard and soft formations running from the North Sea to the Channel.

The reddish Triassic rocks are soft and have weathered rather readily. Hence they form the Midland Plain, watered by the Trent, Dee and Avon. All the "mantles" of the Younger-Mass slope (or dip) to the southeast. The Jurassic limestones were deposited a little later than the Triassic beds, and now stand up as a long scarp (or *cuesta*) which forms hills like the Cotswolds, Edgehill or the York Moors (south of Middlesborough). The next formations belong to the Cretaceous, which occurs as lower clay beds and upper limestone beds. The latter are built of the tiny marine shells which constitute chalk (Latin *creta*), and this forms a cuesta parallel to the Jurassic cuesta. The chalk cuesta builds up the Chiltern Hills and the North and South

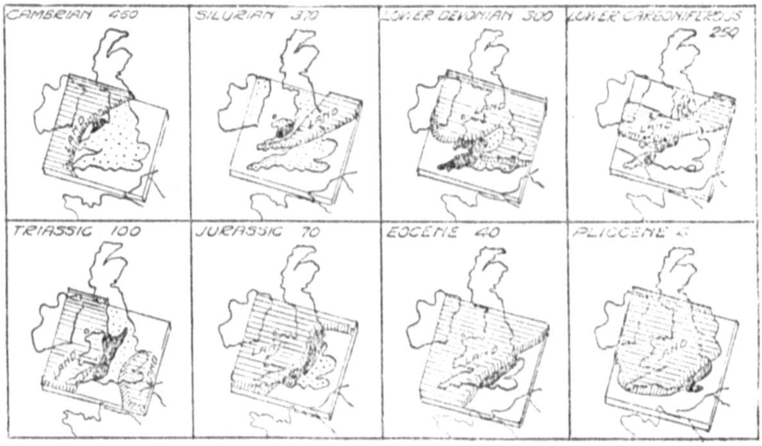

Fig. 4.—The Evolution of England shown in eight diagrams at various geological Epochs. Figures represent Millions of Years ago. The dotted areas are covered by *later* sediments, the black areas show where the sediments are *now* visible as outcrops.

Downs. Two small basins around London and Southampton contain the youngest (Tertiary) beds in England.

It is worth while to study for a moment the actual evolution of a land area such as England. This is shown in the series of eight small maps in Fig. 4. The Age of a rock is the time when it was laid down, and is measured generally by its fossils. In England most of the beds were laid down in the sea—as is shown by the *marine* character of the fossils. We see that in Cambrian times (about 450 million years ago) there was a land covering the northwest of Britain, while sea occupied the southeast. Along the coast were laid down beds of

Cambrian sediments—which to-day appear as the hard Cambrian Rocks of North Wales. Most of the Cambrian deposits in the east of England are, of course, still covered deeply by younger beds, and so do not appear on the surface or affect the topography.

In Devonian times the sea was chiefly in the south. The chief Devonian rocks occur in Devon and South Wales, those further to the east being deeply covered. In Carboniferous times shallow lakes developed both north and south of the central land area. In these grew peat-plants like those in the Florida Everglades to-day. Slowly the lake floors subsided, the plants continuing to grow on the peaty debris. In this way many feet of peat were produced which ultimately became the rich coalfields of Britain. In Triassic times the land lay chiefly in the west, and a series of red deposits—at times marine, perhaps at times freshwater or even due to wind erosion—covered the centre and east of England. In Jurassic times the sea retreated to the east. Marine animals produced shells of lime and these built up the hard limestone deposits of the Jurassic Cuesta shown in Fig. 3. The last phase of the Evolution of Britain is indicated in the Eocene and Pliocene maps, where the deposits of Tertiary Age are confined to the southeast corner of England.

In relatively late geological times the "mantles" of the Younger-Mass in Fig. 3 have been gently buckled to form a broad *ridge* in the Pennine area, and a *Dome* in Kent (in the southeast). It is an axiom in the Evolution of a Topography that *uplifted rocks are quickly eroded away*, while depressed rocks are protected from erosion and survive. Hence the top of the Pennine ridge and the top of the Kent Dome have both been eroded (by rain, rivers, wind, frost, etc.), and older rocks are here exposed. In the map we see the old gritstones (4) exposed in the centre of the Pennines, while the younger actual coal-bearing beds (5) are exposed at the sides. So also the soft beds (8) of the Cretaceous appear in the centre of the Kent Dome (the sides of which consist of younger rocks), just as the fruit appears in an apple-pie when the top of the crust is sliced off.

The coastline of England clearly is in agreement with the "build" so far elucidated. Thus the Wash is a broad flat estuary where the waves have succeeded in breaching the Chalk Cuesta (9). The Thames estuary is due to similar action in the soft Tertiary Rocks (10). South Foreland and Beachy Head are the ends of the harder *chalk* limestones. Spithead and the Solent are drowned Tertiary areas exactly similar to the Thames estuary, and giving rise in Southampton to a port second only to London. The large bay southeast of Exeter is where the soft Triassic vale (6) reaches the coast, and the estuary of the Severn is partly due to the same soft rocks.

The rivers also conform to the build as we should expect. The Severn flows along the junction of the "Older-Mass" and "Younger-Mass" chiefly on the edge of the soft Triassic rocks. The Trent and its tributaries are almost confined to these soft rocks. The soft vale (8), between the two cuestas from the Wash to the Channel, includes the rivers Witham, Welland, Nen, Bedford-Ouse, upper Thames, and the upper Bristol-Avon. The highest mountain in the area depicted is Snowdon (3,571 feet). This is also the oldest and hardest rock in the area, but it must not be supposed that its elevation is due to its age. The age of a rock is the age when it was *deposited;* in this case about 500 million years ago. But the age of a mountain is when it was *uplifted,* which is an entirely different phenomenon. Erosion is relatively rapid, and few mountains would last for 50 million years. Hence the Cambrian rocks of Snowdon have had time to be worn down to sea level ten times since they were first deposited. We deduce then that the west of Britain is highest because it has been uplifted fairly lately in Tertiary times, *i.e.*, within the last few million years. Since it consists of *ancient hard* rocks, it has resisted erosion much more than the "Younger-Mass" to the east—which may or may not have been uplifted to the same extent. We have now considered how a *simple succession* of hard ancient rocks and softer younger rocks has given rise to diverse topography as is clear in Figs. 3 and 4. This has greatly affected history. In a later section we shall see how a different phenomenon, the *subsidence* of a large portion of the Earth's Crust, has dominated the topography (and history) of Scotland.

B. *Environment and Nation in England*

England has in many ways a remarkable environment. It lies in the uniform-rain zone in Europe (p. 72), so that if it had been a low plain it would in all probability have been covered with a uniformly dense forest of oak, beech, etc. This, however, would have been very unsuitable for early settlement, as we shall see when we consider where the Neolithic people of Britain made their homes. Right from the beginning the topography determined the distribution of population. But a word or two first may be given to the situation of Britain. It is an island large enough to be self-contained in early historical days, yet small enough to be controlled by a powerful monarch in those crucial times when nations were developing. Its island position has been a help to it right up to the present. The "narrow seas" are wide enough to prevent aggression, but not too wide to break the lines of commerce with the continent. Its seas are notably shallow, and rich in fish, thus giving rise to a flourishing fishing industry, which in turn led to the relatively early development of a navy and materially helped to keep England independent. It is suited for wool-growing, almost the most valued industry in the middle ages, and has coal resources only excelled by a few other countries. This latter endowment gave it a long lead at the dawn of the Industrial Age—while its island position enabled it to keep aloof from many continental wars. Its position facing the mouths of the Rhine gave it a direct hold on the commerce carried by that river, while its situation in the zone of southwest winds led American commerce later to its shores.

We know little of the distribution of population among the pre-Roman tribes of Britain. Undoubtedly their chief monument and one of their most important meeting places was Stonehenge some 40 miles northwest of Southampton. We may profitably examine the conditions which made this a Neolithic centre of population. From the mantle-map (Fig. 3) we see that four ridges meet here. These are the North and South Downs, the Chiltern Hills and the Dorset Heights. Here also the Jurassic Cuesta at the Cotswolds makes its nearest approach to the chalk cuesta. Stonehenge is also near to the Welsh and Cornish moors whose rugged summits rise high above the forested lowlands. Now we can understand the situation of Stonehenge; for Neolithic man could not cope with the trees and beasts of the forests and so lived for the most part on the open grassy "downs" of the chalk or limestone cuestas, and on the sparsely covered moors of the west. This is indicated on the map given in Fig. 60 based on research by Fleure and James.

It seems likely that Britain was inhabited by rather dark, narrow-headed (or *dokeph*) aborigines, almost wholly of Mediterranean Race, until about B.C. 1000. It is almost certain that they spoke a tongue entirely different from any Aryan language. From racial analogy we should expect it to be akin to the language of the less disturbed primitive folk of similar race in the Atlas Mountains of Africa. This is Hamitic, and the Welsh philologists Rhys and Jones show that many of the non-Aryan features in the Welsh and Irish speech are common in early Hamitic languages, such as that used by the early Pharaohs, and inscribed on their monuments (Fig. 25).

About B.C. 1000 it is supposed that the first wave of Aryan-speakers reached Britain. They spoke Goidelic, which is an early form of Gaelic. They came from central Europe, as the place-names show (p. 120), and were possibly few in number and mainly of Alpine race. This language slowly displaced the old language everywhere except in the far north and west of Britain. Perhaps about 200 B.C. a second wave of Aryan-speakers arrived by the same route.

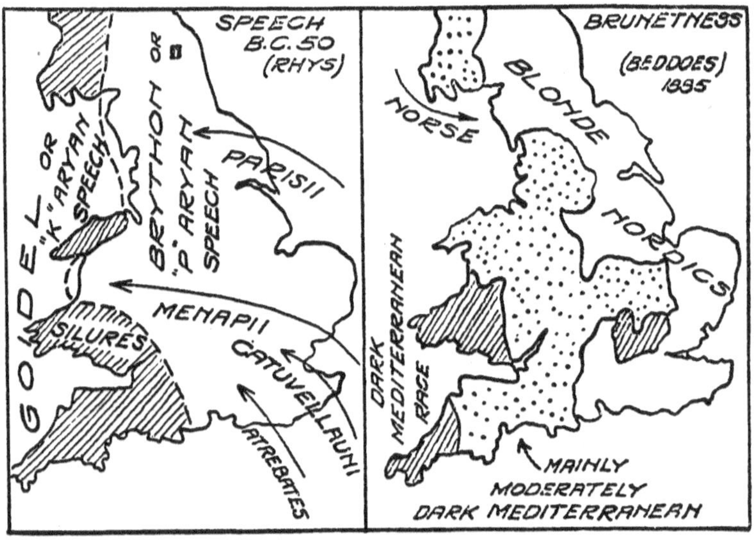

Fig. 4a.—Two maps of England showing Invasions of Britain by Welsh-Speakers (in part Belgae tribes) a few centuries before Christ; and (on the right) the present ethnic classes.

They introduced Brythonic, an early form of Welsh. This in turn displaced Gaelic to the west, and was the language of England, Wales and South Scotland when the Roman conquest occurred about A.D. 43. These Britons had begun to settle in the lower lands, and had cleared some of the forests. Iron had become fairly common in Britain a century or so before the Roman occupation, but the primitive British tools had not made much impression on the great forests which still occupied the ill-drained flat floors of the vales shown in Fig. 3. Many of the Brythonic tribes were merely sections of the larger tribes in France. The British tribes of Catavelaunus, Menapii, Parisii and

Atrebates are older forms of the familiar French names Chalons, Menin, Paris and Arras (Fig. 4a).

The Róman Conquest of Britain. The Romans occupied Britain for about 350 years, but it was too far away from their home environment to become a Roman Province in the sense that the South of France developed into *the* "Provence." The occupation was mainly one of military domination, and this in turn was very largely determined by the environment. It would be not much of an exaggeration to say that the Romans settled rather sparsely in the "Younger-Mass;" that they scattered military posts through the "Older-Mass;" and that they only attempted to control the region north of Hadrian's Wall (see Fig. 5) for a short time in any fashion.

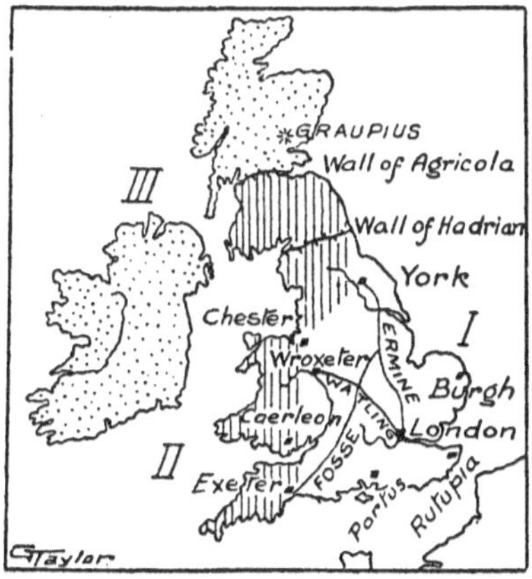

Fig. 5.—Britain in Roman times. The Younger-Mass (I) contained the settled areas. The Older-Mass (II) had sparse military occupation. The rest (III) was practically unaffected by Rome.

Their main line of defence extended from York to Chester, Caerleon and Exeter. This is almost the line bounding the Triassic Plain on the west. All through North England, Wales and Cornwall the Brythons still maintained their independence in the rugged hills, though small Roman strongposts were maintained at intervals.

The Roman generals took advantage of the two narrow and relatively low areas in the north to build actual walls against the wild *Picts* of the north. These fierce highlanders were at that date essentially like many of the aborigines in the Atlas Mountains to-day, and probably spoke a cognate language. In A.D. 80 the Romans advanced to the Tay and fought the battle of Graupius—a name which seems to be a modification of Grampians. The wall of Antonine was built from Glasgow to Falkirk about A.D. 140—the Romans thus taking advantage of the crustal subsidence described on page 46. Further south a much larger work, consisting of a trench and a strong wall 10 feet high, extended from the Solway Firth up the Pennines to the head of the South Tyne, and so eastward to "Wallsend" near Newcastle.

Rome soon adopted the little fishing port of Lynn-dun as the centre of her colony (p. 220). It is no accident that this is also practically the centre of the Younger-Mass, where the youngest and uppermost beds of rock are to be found. In a general fashion the two cuestas, the vales and the Tertiary beds around London (see Fig. 3) are linked to form the "London Basin," in the same way as the rocks of the same age and character form the "Paris Basin" (p. 314). The tributaries of the Thames, the main river of southeast England, flow towards the centre of the Basin; for the Cretaceous cuesta is pierced by the Thames at Goring Gap. The North Downs has also been pierced by its tributaries, the Mole and Medway. Thus in a limited sense, in southeast England "All roads lead to London."

It is usual to describe the famous Roman Roads as crossing hill and dale "as straight as a dart;" but as a matter of fact, although they ignored contours to some extent, they were compelled to deviate from the direct route by forests and fens in much the same way as roads in later times. As we have seen, the really rugged country of Wales and Scotland was almost ignored by the Romans.

The main port of ingress of the Romans was Rutupia (Richborough) where the North Downs reach the sea near Dover. The great Anderida Forest filled the bowl of the Weald and was not much traversed by the Romans. An important road ran north of the Downs to Silchester (near Reading) and so to Winchester and Exeter. Watling Street went north to Verulam (St. Albans) and so to Viroconum (Wroxeter) and Chester. It had many lonely stretches near the Forest of Arden, north of the Avon. Ermine Street also passed northward and skirted the great Fens of Cambridgeshire on its way to Lincoln and York (Eboracum). An important by-road (the Fosse) linked Lincoln with Exeter, and in part followed the old Neolithic track along the Jurassic Cuesta (Fig. 5).

In their earlier days in Britain the chief enemies of the Romans were the Britons of Cornwall (*i.e.*, Horn of *Wales*), Cambria (*Cymry* = Wales) and *Cumber*land. In the later years of their domination the almost universal law of European invasion began to operate again, and the Romans in turn were *attacked by foes on their eastern flank.* Hence they appointed a "Count of the Saxon Shore" to curb the forays of the Nordic pirates. These began to make their raids long before the first important Saxon settlements in Britain, which were probably about 450 A.D. A series of forts was built from the Wash to the Isle of Wight, of which Brancaster, Burgh, Othona, Lympne and Portus were some of the chief (Fig. 5). But the Roman Empire began to crumble after 400 A.D. Fierce enemies attacked her bulwarks along the Rhine and Danube, and about 410 the Roman legions left Britain to defend the Empire nearer its heart. We know very

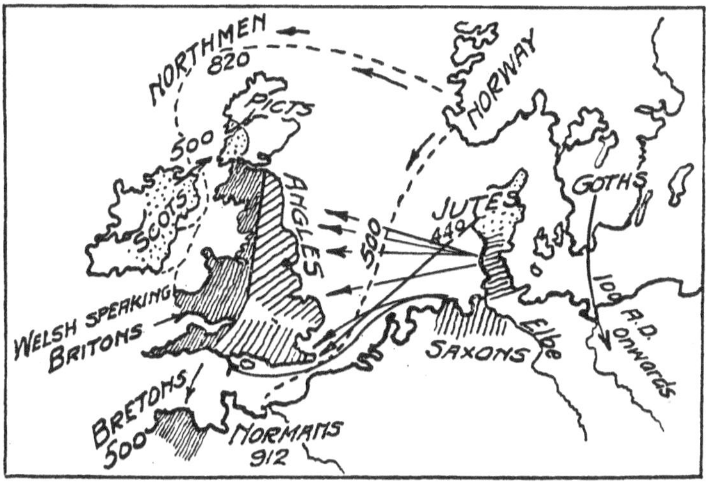

Fig. 6.—Movements of the Nordic Invaders from A.D. 100 to A.D. 900.

little about British history in the next century or two. The British tried with very little success to copy Roman methods of defence. A leader called Arthur won some victories, which have doubtless been wildly exaggerated in the Arthurian Legends based on his exploits. As early as 440 two Saxon leaders, called Hengist and Horsa, are said to have made forays and settlements in Kent, and by 500 much of eastern England was under Saxon control.

Perhaps the outstanding geographical aspect of this Nordic invasion was that it took place at many places under different leaders all round the east and south coast. The small ships of the Saxons, Angles, and especially of the Danes, easily ascended the rivers flowing across the Younger-Mass, and so penetrated some distance into the low eastern country. The map (Fig. 6) illustrates the original homes of these newcomers—the first of the English. For it must not be forgotten that before their settlement Britain was certainly not English—but merely an extension of Wales.

As is often the case, islands were the sites first occupied by the van of the invasion. Thanet—then isolated from Kent—and the Isle of Wight were both occupied by Jutes—who almost certainly came from Jutland in Denmark. But their name and characteristics were lost among the much greater numbers of Saxons and Angles. The homes of the latter lay south and north of the lower Elbe; and between them there were only slight cultural differences in dialect and customs. With the increasing slackening of a united defence by the Britons these two tribes poured in large numbers into England. The Saxons occupied the North Downs, and then the South Downs (South Saxons or "Sussex"). The East Saxons soon settled Essex, and later spread into Surrey (perhaps the "Sud-reich" or south-realm). The West Saxons ascended the Solent and spread rapidly through Wessex.

The Angles also moved westward from their homeland and naturally reached areas north of the Saxons. The "North-folk" and "South-folk" colonised Norfolk and Suffolk, while ships sailing up the Humber settled Lincoln to the south and "North-Humber-Land" to the north. Not long after 500 A.D. England presented a picture which still preserved the ancient topographic divisions of 400 years earlier—though with different nations in control. The Romanised Britons were either submerged by the Nordic invaders in the Eastern Younger-Mass or had fled to the Western Older-Mass. Many Britons fled to France from Wessex about 500 A.D. and founded Brittany or Little Britain. It is possible that these latter refugees were not peasants, but largely the Alpine (?) *aristocrats* who introduced Welsh into Britain. This may explain why to-day the people of Brittany are essentially broad-headed Alpines—while the Welsh-speakers in Britain are mostly of the narrow-headed Mediterranean Race.

It is important to note that the division of the British nation into Blonde "Nordics" in the east and Brunette "Mediterraneans" in the west has not altered in essentials from the early days of the Saxon invasion to the present—as a reference to Beddoes' map (Fig. 4a) will show. (A further discussion of the control on English History exercised by geographical factors is given in Chapter XII.)

C. *Environment in Scotland*

We are accustomed to think of earthquakes as agents of devastation, yet it is no exaggeration to say that faulting—which is the geologist's term for an earthquake—is responsible for a Topography which has almost entirely determined the prosperity of the Scottish nation. Since the Scottish topography is different from anything we have so far considered—and yet resembles what is found in many other parts of Europe—a few paragraphs may be devoted to its discussion.

Fig. 7.—Map of Scotland showing the control of population by the Rift Valley or Graben. (Figures per square mile.)

In Fig. 7 the map clearly shows that the population of Scotland is negligible except in the narrow "waist" of Scotland lying between Ayr and Dundee. All to the north has a very sparse population—generally below 30 per square mile. It is also sparse in the southern uplands; but in the Ayr-Dundee belt of lowlands most of the country has more than 60 to the square mile; while several large districts rise to 500 per square mile. It is by no means sufficient to say that this is due to the coalfields near Glasgow and Edinburgh.

Let us imagine ourselves in an aeroplane over the Firth of Clyde, observing the land lying to the northeast. We shall readily be able to distinguish three major divisions in the landscape before us. To the north are the rather rugged granite mountains forming the Grampians (Fig. 8). They rise to 4000 feet in Ben Lawers, but end rather

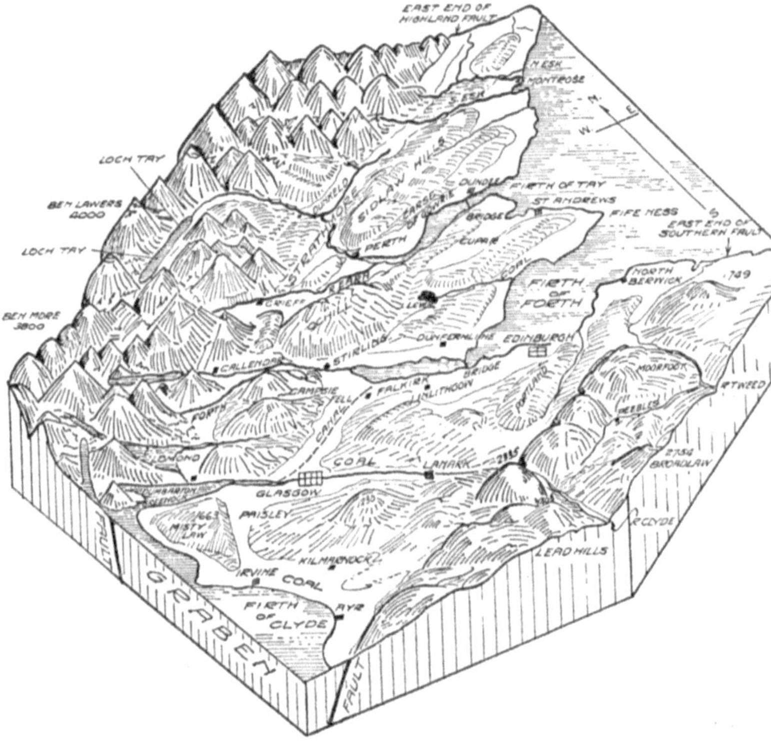

Fig. 8.—A Block Diagram of the Central Valley (or Graben) of Scotland. The hills from Misty Law to Sidlaw are volcanic. The Faults are shown in front of the Block.

abruptly along a line from Loch Lomond to Aberdeen. To the south of the Grampians is a broad belt of lowlands about 40 miles wide separated again by a fairly well-marked line (from Ayr to Edinburgh) from the rounded hills of the southern uplands. These latter rise to 2,754 feet at Broadlaw.

This topography is due to the presence of a very characteristic area of subsidence, termed by the geologist a "Graben" (a word allied to "grave"). A graben is a rectangular block of the earth's crust bounded by earthquake-cracks (or faults), which has subsided relatively to the adjacent crust. This depressed area consists, as usual, of *younger* rocks than have been preserved in the adjacent mountains. The topographic changes are made clearer by reference to the vertical sections shown in Fig. 9. In the left hand section Scotland is shown in the Carboniferous period long before the faulting developed. The section is quite hypothetical, of course, as are all such reconstructions. A layer of coal-peat is shown at 4 which developed in swamps on the surface (p. 35). Two great cracks developed across this landscape constituting the Faults mentioned previously. Relatively to the areas north and south of these faults the central area now sank many

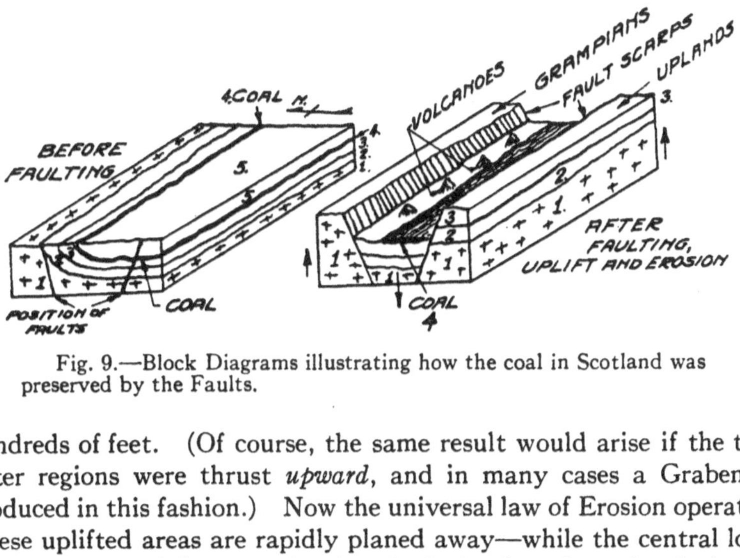

Fig. 9.—Block Diagrams illustrating how the coal in Scotland was preserved by the Faults.

hundreds of feet. (Of course, the same result would arise if the two outer regions were thrust *upward*, and in many cases a Graben is produced in this fashion.) Now the universal law of Erosion operates. These uplifted areas are rapidly planed away—while the central lowland is protected from erosion by the lack of grade and velocity in its rivers, and often by a covering of silt deposited in the low areas. This stage is also shown in Fig. 9. In the Grampians we assume that all the layers which possibly once covered the granite have vanished. In the southern uplands all the coal beds (4) have been eroded, though 3, 2, and 1 (of older Paleozoic age) are shown as still present. In the Graben the complete series (1 to 5) is still present. Usually some volcanic action accompanies widespread subsidence of this type. Accordingly we find a series of small volcanic hills—such as the Ochill Hills—along the north side of the Graben. These are also easily seen in the Block Diagram (Fig. 8). We shall find many similar graben with a similar evolution in the rest of Europe, such as the Rhine Graben south of Mainz and the Rhone Graben south of Lyons.

The control exercised upon human affairs by this graben may now be briefly referred to. The alluvial soils are naturally deeper in this broad area of lowland than in the remaining rather mountainous districts of Scotland. Hence agriculture is in a more flourishing state. Communications such as roads, canals and railways, are much more numerous here than elsewhere. The very ancient rocks north and south of the Graben do not in general weather to a rich soil—which is in any case nowhere deep or widespread. The coal seams, which have been protected from erosion—though so near the surface—by their humble position at sea level are, however, the main reason for the prosperity and dense settlement in the Graben. In this small region, only $\frac{1}{5}$ of that of Scotland, live $\frac{3}{4}$ of the Scottish nation.

D. The Early History of Scotland

We have seen that the Angles who founded Angle-land (England) derived from Schleswig-Holstein, north of the Elbe, about 500 A.D. So also the word Scotland is of very similar origin, for the Scots did not reach Scotland till about the same date. Just as the English are "foreigners" from Germany, so the Scottish came from Ireland. But the stocks building the final Scottish nation are probably more complex than those of England. It is likely that the aborigines of the Scottish Highlands—the Picts—were more primitive than those of England and Wales, since they were further from the continent. Some early inscriptions cut on the edges of stones in the Ogam letters do not belong to Aryan languages. They may turn out to be Hamitic, as Rhys and Jones have suggested for the early speech further south.

In Roman times the northern mountains of Scotland were in Pictish hands, but the Brythons held the lowlands. This explains why many of the legends of King Arthur—the Brythonic Saga—deal with the *borders of Scotland* and England. Here also many of the place-names are Welsh rather than Gaelic. Thus the border mountains are *Pen* and the river-mouths are *Aber*, while in the Highlands to the north the Gaelic names are *Ben* and *Inver*. In the fourth century Picts raided all the Scottish Lowlands, and after 410 penetrated right into England. Meanwhile the Scottish tribes from north Ireland were raiding Argyle. They were fairly closely allied in race to the Picts—but spoke an early Gaelic speech akin to Erse. About A.D.

500 they began to settle north of the Firth of Clyde in the district called Dalriada. After 500 the Angles also rapidly extended their power northward from "North-humber-land" and occupied the "Lothians" around Edinburgh. They displaced the Brythons to the West, but there was no marked topographic feature in Scotland (as there was in England) separating their regions.

While the Nordics were gradually conquering England, the Brythons everywhere retreating to the west, the process was more complex in Scotland. Here the Irish-Scottish were apparently the most energetic stock. They were also relatively more civilised than the Angles of Lothian, for they had been Christians for a century. About 860 the Scottish chiefs completely conquered the Picts and according to legend massacred most of them. A little later an alliance was made between the Scottish and the Brythons south of the Clyde which was directed against the Angles of Lothian. Moreover, after the Danes had conquered the Angles in England, the Angles of Lothian in Scotland could fairly claim to be more "English" even than the people of England. Early in the eleventh century Malcolm the Scot conquered the Angles of Lothian, and his realm agreed fairly well with the Scotland of to-day.

Thus the history of Scotland is quite different from that of England. In the latter the conquerors came ever from the east. Gaelic-speakers conquered the aborigines and then in turn *moved west* before the Brythonic-speakers, the Romans, the Angles and the Danes. The rugged Older-Mass preserved many of the refugees and their culture. But in Scotland not only the Brythons of the Lowlands were finally dominated by an *eastward* wave of Gaelic-speakers (the Scots); but also the Angles of Nordic strain. No doubt merging occurred along the line of contact; but the blonde character of the people along the east coast of Scotland shows that the Angles did not lose their individuality there.

Right through the later history of Scotland the Faultscarp north of the Graben (Fig. 8) has separated the "Highlander," mainly of Scots-Irish ancestry (with some Pictish strain perhaps), from the Lowlander. The latter is essentially Nordic in the east and what one may call "Mediterannean of a Welsh type" in the southwest. Many cultural features also change at the Highland boundary. For instance, the Lowlanders are essentially Protestant, and according to popular belief, cautious and persevering; while the Highlanders were for centuries Catholics of a fiery and impulsive temperament.

The islands and firths which characterise Northwest Scotland are almost identical in age and structure with those of Norway. It is not strange therefore that the Vikings made raids and settlements in the home-like region extending from the Orkneys to the Hebrides, from the 9th century onwards. Indeed about 1100 the Norse gained the Suder-eys (*i.e.*, the Southern Isles) off the Firth of Clyde. Haco's Norse Armada was destroyed in 1263, and most of the Western Isles became Scottish. It was not, however, till the end of the 15th century that the Northern Isles were handed over to James III as his wife's dowry. The relations of Scotland and England may now be briefly considered.

The Stewart line of Kings began in the 14th century, and dominated Scotland for several centuries. In 1502 James IV married Margaret, daughter of Henry VII of England, but in spite of this, alliances with France dominated Scottish policies, and James V twice married French women. His daughter Mary Stewart was educated in France, and in 1558 she married the French Dauphin. Gradually dislike of the French and of the Catholic Party helped to turn the Scotch towards Calvinism. In 1567 Mary was deposed and her young son James VI was made king. In 1603 he succeeded Elizabeth on the English throne, by virtue of his descent from Henry VII. The attempts of Charles I, Charles II and James II to put down the Reformed Religion led to constant struggles in Scotland (p. 231). William III found bitter opponents among the Highland Clans, but the Union of the two Parliaments in 1707 gradually brought about peace between England and Scotland. (A further discussion of British history is given in Chapter XII.)

CHAPTER III

THE BUILD OF EUROPE AND ITS GENERAL EFFECT ON HISTORY

A. The Shields and Ancient Folds

The continent of Europe has at first glance a somewhat complicated "build," which is, however, largely an extension of what we have studied in some detail in the last chapter on the Build of Britain. There we saw that Britain could be considered best as an "Older Mass" and a "Younger Mass," each of which had its special topography, which in turn reacted on the dwellers therein. Generalising a little, we saw that the oldest rocks occurred in Wales and Scotland while the rocks became younger and softer and the topography flatter as we journeyed to the east. These features are repeated on a grander scale in Europe. In the north is an eastern extension of those ancient *Caledonian Folds* which build up the Grampians. Across the centre of Europe run two zones, consisting of the *Ancient Coal Trough* (whose western end is found in Wales and Scotland), and the *Relics of the Permian Folds* which are also represented in Britain, especially in Cornwall and South Ireland. These represent the Older-Mass. But the two major features of the continent have no adequate representative in Britain. These are the *Great Plains* of the east and the *Young Mountains* of the south. It is true that the latter are formed largely of the same strata as the Younger-Mass of Britain—but they are so folded and crumpled that they have entirely lost any resemblance to their quiescent "relatives" in Britain. The crumpling has altered the very substance of the shales, sandstones and limestone; while the folding has elevated them to 20,000 feet above the sea, though later erosion has removed more than the upper half of these gigantic folds (Fig. 10).

The time is at hand when no geography will be considered satisfactory which merely describes the position and height of mountains without any reference to the build or age. The chapter on England, it is hoped, has demonstrated how much more valuable it is to know somewhat of the origin of mountains. Moreover, all mountains are due to essentially simple processes which can readily be grasped by the reader, even if he has little or no knowledge of geology.

If we are told that Mont Blanc is 15,730 feet high and lies on the border of France, while Galdhopig rises to 8,385 feet in the south of Norway—we have no useful idea of the environment concerned. But if we learn that Mont Blanc is the summit of a series of earth waves built up relatively lately, this means that it belongs to the class of young mountains, which have many characteristics in common. Considering the other example, Galdhopig is the rounded summit of an

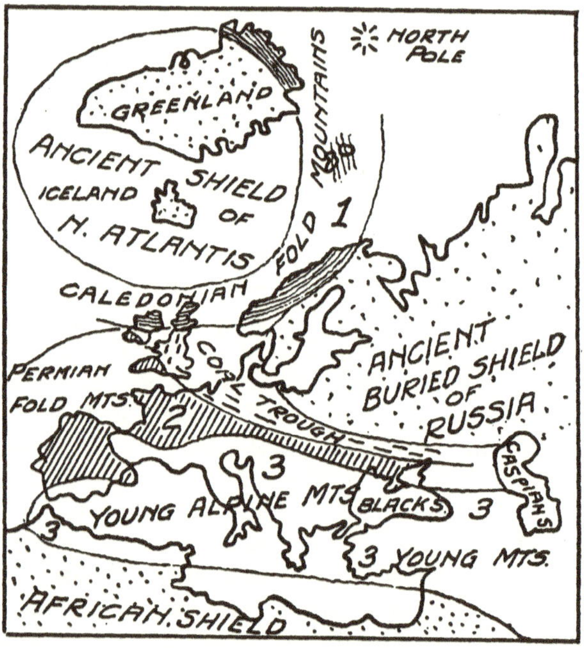

Fig. 10.—A Generalised map of Europe showing the three Resistant Shields, the Coal Trough and the Weak Rocks which were crumpled to form the three series of Young Mountain Ranges labelled 1, 2 and 3.

elevated peneplain (p. 60), which in turn is the worn-down relic of a very ancient mountain-system. Its topography is utterly unlike that of Mont Blanc, though both are the highest points in the regions concerned.

The basic units in the Build of Europe are three large and very resistant areas called Shields (Fig. 10). The essential characteristic of a shield is that it *resists folding;* though it may tilt somewhat or become wholly or in part submerged by soil, rock, or water. One of these shields contains Greenland and Iceland. Another underlies the plains of Russia. A third dominates North Africa. If we like to think of Europe as ground between "upper and nether millstones" formed of these shields, we shall have a useful analogy to the processes involved in the growth of Europe. Somewhere about 300 million years ago, about the beginning of Devonian times (p. 34), the earth entered a phase of unrest. It is believed that the weak rocks in northern Europe yielded to strain, between the "millstone" of Greenland and that of Russia. The two shields compressed the crust between into a series of gigantic folds—producing no doubt mountains very like those of the Himalayas or Andes of to-day. These "Caledonian Ranges" seem to have formed a ring around Greenland as shown in Fig. 10.

Then for over a hundred million years the earth's crust seems to have been quiescent. The "Caledonian" mountain ranges were relatively soon eroded by the action of rain, rivers, frost, wind, etc. (For the hills, so far from being "unchanging," as the poets would have us believe, are almost the least permanent features of the crust.) Gradually they wore down to flattened ridges and finally to undulating rocky platforms not much above sea level. This last stage is the *peneplain*, which is the natural end of a mountainrange (Fig. 12).

We may suppose that this corrugating of the crust in Devonian times stiffened it, and when the next great phase of unrest occurred (in Permian times, about 150 million years ago) the new mountain ranges developed somewhat to the south of the Caledonian Ranges. They seem to have been "ground" between the Russian Shield and the African Shield. In Europe, this range is called the Armorican or Hercynian range, from Armorica (Brittany) and Hercynia (Harz Mts. in Germany) which were traversed by them. (In America the similar period of mountain-building is called the Appalachian Period.) This great range or series of ranges extended across Europe and its relics are found to-day in Spain, Cornwall, France, Bohemia and South Russia. Like the relics of the Caledonian Mountain Period, the Armorican Mountains appear to-day as peneplains or as relic-stumps of the ancient high mountains.

There remains one ancient unit, the Coal Trough. This was a region of level swampy lands in Carboniferous times, some 200 million years ago. As it slowly subsided the ferns and trees of that period grew up on the peaty debris in the swamps. In this fashion were formed layers many feet thick of peat, which in later ages were converted into coal. This great Coal Trough did not participate in any of the major foldings which followed, or we may be sure the coal would have been washed away. Such a waste probably occurred in the uplands of Scotland (p. 47). Possibly the Coal Trough was firmly buttressed by the Russian Shield. At any rate, the Permian Mountains did not destroy its general basin-like structure; nor did the "Alpine Storm," still later. To-day the densest population of Europe extends along the Coal Trough, from Glasgow through Belgium to Saxony, Poland and South Russia (Fig. 14). It is hardly an exaggeration to say that these millions of Europeans owe their existence to topographic conditions of 200 million years ago.

B. The Alpine Storm and Associated Areas

There now ensued another long period, about one hundred million years, which was free from world-wide unrest. But in early Tertiary times—say twenty or thirty million years ago—the crust again began to yield to accumulated strains. The mountains produced in this age of unrest are so young that they still dominate the world's landscapes. The mountain-building culminated in Europe in Miocene times, some ten million years ago, but the Himalayas were probably somewhat later, while the region down the west of the Pacific is undoubtedly still in the throes of earth-movement. This last of the crustal periods of movement is often called the "Alpine Storm," for the crust in south Europe was buckled up into waves and breakers not unlike those of the ocean in a period of storm. The Alps, then, are the product of the Alpine Storm, but in the last half dozen million years or so the upper part of the crustal "waves" has been eroded, leaving us mountains ten to fifteen thousand feet high, which are certainly much less magnificent than they were at the close of Miocene times.

We must examine a little more closely the plan of the young mountains in the remainder of Europe. The bold arc of the Alps is repeated in the other areas affected. The Carpathian arc is even more striking, and the young folds to the east curve equally sharply to the Black Sea (see inset in Fig. 13). These belong to the northern series of fold mountains, the Alpine arc proper. But a southern series of folds of the same age and character (but "splashed" *southward* as noted later) runs parallel to the Alpine arcs. It constitutes the Dinaric arcs—named from the Dinaric Mountains in Yugo-Slavia. They form a very striking "horse shoe" around the Adriatic (see inset Fig. 13) and continue westward to Africa (as the Atlas Mountains) and eastward across the Aegean Sea to build the Taurus Mountains of Anatolia. As Marion Newbigin points out, the western Mediterranean Sea lies between these two series of arcs, but the eastern Mediterranean is well to south of both of them, while the Black Sea corresponds better to the Western Basin. The topographic factors which characterise the more familiar western area are therefore repeated again and again as we journey eastward.

This last phase of mountain-building affected most strongly the weak young sediments which had accumulated in the "Tethys Sea" south of the earlier mountains. These marine sediments were lifted up thousands of feet and flung into Switzerland *over* the quieter crustal regions to the north (see section, Fig. 11). But such catastrophic crumplings in the south certainly caused "uneasiness" in the northern parts of Europe. It seems likely that the Paris Basin, including its northern "rim" in England, was squeezed somewhat, perhaps helping to produce the domes of the Weald and the Pennines already described (p. 37). The Armorican relics were too resistant to fold, but they were in places elevated *en masse*, as in Spain. The ancient peneplains resulting from the wearing down of the Caledonian Ranges, also experienced considerable uplift. Hence the Fells of Norway, some 4,000 feet above the sea, have only recently occupied that position, and during middle Tertiary times were probably at sea level in the same condition as the rocky peneplain of Finland is to-day.

We can therefore summarise the topography of Europe in a table, which also indicates some of the human responses to such topography.

In the previous chapter some account was given of the way in which a series of cuestas could develop from various hard and soft layers deposited on the shores of an eroding land-mass. Exactly the same processes occurred in North France to produce the Paris Basin (see Fig. 11). Indeed, the south of England is structurally part of the Paris Basin. The Channel tunnel, when constructed, will extend from Dover to France and pass through unbroken formations under the Channel. The latter is partly due to subsidence, and partly to wave-erosion, but there seems little doubt that there was dry land at the Straits of Dover only a few thousand years ago.

C. *The Relic-Blocks and Their Effect on the Alpine Storm*

The Armorican Ranges may now be briefly described. Probably 150 million years ago they formed a series of arcuate ridges—as do

TABLE E

MAJOR TOPOGRAPHIC UNITS IN OR NEAR EUROPE (Fig. 15)

Class	Locality	Features	Human Use
Shields			
(*a*) North Atlantic	Greenland, etc.	Level, mostly low; resists folding	Too cold
(*b*) Russian	Russian Plains	Level, often buried; resists folding	Forest and farms
(*c*) African	North Africa	Moderately level and low; resists folding	Now too hot and dry
Fold Mountains			
(*a*) Caledonian	Scotland, Norway	Now peneplain, considerably uplifted	Soils poor or absent
(*b*) Armorican	Spain, France, etc.	Now peneplain, somewhat elevated	Soil thin and poor
(*c*) Alpine Storm	Around Mediterranean	Lately folded, still great mountains	Very rugged
Troughs			
(*a*) Coal Trough	Scotland to S. Russia	200 million years old, preserves coal deposits	Thickly populated
(*b*) Alfold Type	Hungary, etc.	Late Fold-basins filled with silt	Rich soils
(*c*) Mediterranean	Several basins	Fold-basin* filled with sea water	Navigation and fish
(*d*) Caspian Type	Caspian and Black Seas	Fold-basin filled with brackish water	Navigation and fish

*The folding is often accompanied by marked faulting.

the Alps to-day. But nowadays they are represented merely by solid blocks of resistant material of which five are indicated in Fig. 11. In Brittany is one such block, no longer a mountain, but a low yet somewhat rugged complex of rocky hills, which represent the stumps of part of the ancient range. Nearby is the Cevennes, which is considerably higher, but, like the Armorican area, is built up of ancient folded rocks. No doubt it unites with Armorica under the mantle of younger rocks now separating them. Indeed, we may safely picture these old rocks as fairly continuous under the English Channel and reaching their allies in Cornwall and South Ireland. The Cevennes, owing to their greater uplift, are in a more "dangerous" position than the Armorican Relic. Rivers are rapidly wearing away the slopes, and in a relatively short time the Cevennes will be worn down to a peneplain again. The Spanish Meseta (*i.e.*, table-land) is of similar structure, but much larger. Its edges are rapidly being trenched by rivers, especially on its western, wetter border. The Black Forest and Bohemian region exhibit similar topographic features.

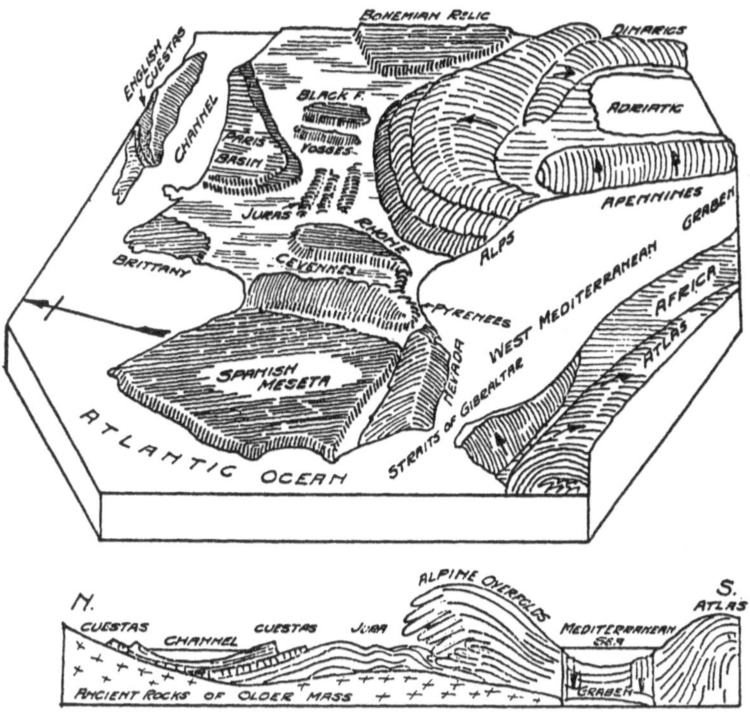

Fig. 11.—Diagram showing the Salient Features of the Build of Western Europe (much simplified). The folds (moving away from the Mediterranean) build up the Alps, Pyrenees, Atlas and the Apennines. Note the Armorican Relics and the Paris Basin. (Looking to the East.)

But the presence of these Armorican relics has had a great effect on the distribution of the young mountains of the "Alpine Storm." The mechanics of mountain-building is a little too complicated for our present discussion; but as the weak strata were "gripped" between the huge resistant shields (as described earlier) they yielded and crumpled readily. The Armorican relics were too resistant to fold. They seem to have been elevated somewhat; but in addition they appear to have determined the position of the young folds, for we see in Fig. 11 that the Pyrenees have been squeezed in between the Spanish Meseta and the Cevennes Relic. So also the Alps have been overfolded into the space between the Bohemian Relic and the Cevennes. The Jura folds are but small, simple affairs, but they fit in snugly between the "Black Forest" relic and the Cevennes.

On the south of the Mediterranean the Atlas Mountains are due to movements in the same Alpine Storm. Here, however, the overfolds are to the south (see Fig. 11), while the Mediterranean Sea between occupies an area in which the crust has sunk. It is, in effect, a gigantic graben. All through the belt of young mountains in Europe and Asia we see similar phenomena. A central "plug" sinks deeply and has the effect of squeezing the weak strata at each side into overfolds progressing *away from* the plug. We shall see later that the Alfold plain in Hungary is a similar "plug," which has sunk deeply and (in effect) squeezed the Carpathians over to the north and east, while the Serbian Folds move to the west. (See inset in Fig. 13.)

The section below Fig. 11 illustrates these folds, overfolds and subsidences. In the north (in Cornwall) the ancient rocks of the Older Mass appear at the surface. Just to the south are the slightly compressed sides of the Paris Basin. Again to the south are the folds of the Jura Mountains, due to the simple crumpling of surface rocks.

Now we enter the complicated region of the Swiss Alps. Here all the surface rocks which once lay peacefully horizontal in "Tethys" (the ancestor of the Mediterranean Sea), are now violently folded and *overfolded*, and flung far over the quiescent surface near the Jura ranges. To the south of the Alps is the graben of the present Mediterranean. No one imagines that it was the sinking of the floor of the Mediterranean which produced these tremendous crumplings of the crust. Yet to the south in Africa we see the Atlas Mountains made up of what we might term "splashes of crust" also moving away from the Mediterranean. However, enough has been written to give the reader an idea of the build of Western Europe. Further details of the mechanism involved belong to the province of the geologist.

D. The Evolution of the Landscapes of Europe

It will be time well spent if we follow through the complete history of the erosion of a mountain range (Fig. 12). We shall then have the key to the origin of almost every scene in the European landscape, and it is unnecessary to dwell on the intimate relation between hill and valley on the one hand, and man and his activities on the other. We may picture the crust after a very long period of quiescence appearing something like the sketch at *A* (which may be taken to represent many areas in Russia). Here river-soils form a plain covering four layers

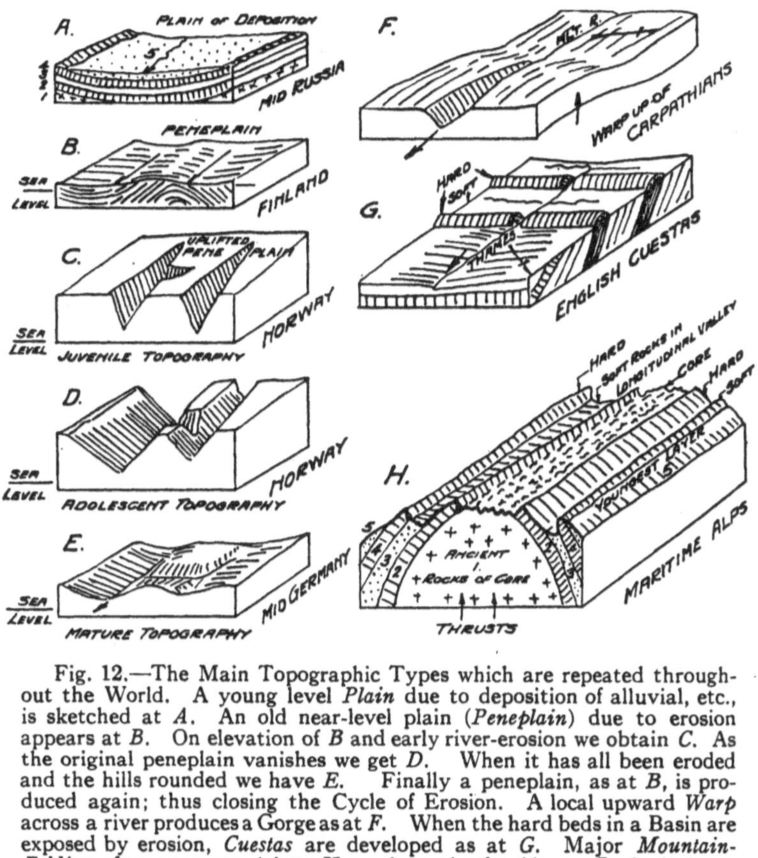

Fig. 12.—The Main Topographic Types which are repeated throughout the World. A young level *Plain* due to deposition of alluvial, etc., is sketched at *A*. An old near-level plain (*Peneplain*) due to erosion appears at *B*. On elevation of *B* and early river-erosion we obtain *C*. As the original peneplain vanishes we get *D*. When it has all been eroded and the hills rounded we have *E*. Finally a peneplain, as at *B*, is produced again; thus closing the Cycle of Erosion. A local upward *Warp* across a river produces a Gorge as at *F*. When the hard beds in a Basin are exposed by erosion, *Cuestas* are developed as at *G*. Major *Mountain-Folding* often converts *A* into *H*, as shown in the Alps. Geological sections are shown at the front of *A*, *B*, *G* and *H*.

of fairly horizontal weak rocks. Now let us suppose that mountain-building affects this area. The folding produces some such structure as that shown at *H*, which represents approximately parts of the Western Alps as they appear to-day. The top of the folds has been eroded. The hard core of the Older-Mass now appears as the summit of the range, while younger layers (2-5) mantle the slopes, being eroded according as they are more or less resistant. We may imagine further that erosion continues for many million (perhaps a hundred million) years. The mountain range is worn down to a flat rocky peneplain (shown at *B*). Notice that the *peneplain* consists of *ancient* rocks cut to a level, while the *true plain* is built up of layers of recent soils, etc. (as at *A*). This sort of a landscape (*B*) is common in Finland, or in the "Barren Grounds" of Canada.

Now let us imagine our peneplain (*B*) elevated to some 3,000 feet, as has happened in Norway. At once the rivers trench deeply into the edges of the elevated peneplain. They are marked by many waterfalls, and gradually cut canyons deeply into the plateau. In Greenland such canyons penetrate nearly 200 miles into the plateau. For long ages there is still much of the flat peneplain surface left—and in this condition the topography including the river valleys is said to be *juvenile* (*C*). Further erosion broadens the valleys, which are called *adolescent* when the last of the flat peneplain is vanishing (*D*). Then comes *maturity*, shown at *E*, which is the common type of landscape in much of England and Germany. Finally, and very slowly, the rivers wear away the rounded hills until the peneplain stage (at *B*) is reached again. Broad valleys often develop, filled with silt under certain conditions, in which rivers meander lazily about with very little velocity. Such rivers are said to be *senile*. We have now completed the "topographic cycle," and it should be realised that in scenery, unlike the human subject, the younger it is the more interesting it is. Waterfalls, canyons, high crags, snow mountains and everything the tourist most admires are the direct result of fairly recent uplift. The dreary plains of Finland or Northern Canada represent the end of the cycle and the old age of scenery.

One or two other types of landscape are sufficiently common to merit a little attention. It is rather usual for a small portion of the crust to be warped up, often right across the bed of a river (see Fig. 12 at *F*). If it is a small river, it is usually blocked and a lake results. Such warps in part account for the great lakes of the Saint Lawrence. But if it is a large river, often enough the river can cut down its bed as quickly as the warp rises across it. We get as a result an *antecedent gorge*. Some of the best in the world occur on the Danube River, notably near the Iron Gates in Rumania. Here the young fold of the Carpathians has risen across the Danube. For hundreds of miles each side of this locality the Danube winds through flat plains in a *senile* condition. But just above the Iron Gates (in the Kazan Gorge) it flows through huge precipitous granite crags which are less than a quarter of a mile apart. The popular idea of an earthquake *crack* enabling the river to pass the mountain is far from the truth. We learn from such scenery that the river is far older than the mountain. The Alt River, a much smaller stream than the Danube, also cuts through the Carpathians about 100 miles to the northeast.

A final example is illustrated at *G* in Fig. 12. Here is a pair of cuestas of the type we have discussed in regard to the England midlands (p. 36). It is fairly obvious why the river can only cut a narrow gorge through a hard cuesta—while it has eroded a wide valley as it crosses the vale. But how did it make its way through the cuesta in the first place? In many cases these rivers are very old, and flowed (as now) towards the centre of the Basin, *before* the differences between the cuesta and the vale were etched out so sharply. Hence the river, as in the case of the antecedent gorge (at *F*), really originated in a topography unlike that seen to-day.

E. *The Barrier between the Two Main European Habitats*

If we ignore the lesser divisions of the continent, it is clear that we have to do essentially with two main habitats of the European peoples. Along the coasts of the Mediterranean are an abundance of fertile if restricted areas with a special set of products which include not only wheat, wine, olives, etc., but also in the early days civilisation itself. Far to the north are the great plains with their cold, raw climates and with entirely different products. In the early days amber, hides, wool, fur and slaves were usual commodities. But separating these two contrasted regions is that same belt of *Young Mountains* with which this chapter has been mainly concerned. From the Pillars of Hercules at Gibraltar to the shores of Palestine—a distance of 2,400 miles in a direct line—nature has erected an almost continuous barrier, for the most part several thousand feet high. In early days the movements of man were tremendously restricted by these young mountains, and if to-day we can burrow under them and fly over them, the natural gaps are still of paramount importance.

Ignoring the Straits of Gibraltar at the extreme western end, there are only six low gaps in this giant wall. Luckily, as Ellen Semple[1]

[1] "Geography of the Mediterranean Region", 1931, New York.

has pointed out, they are fairly well spaced along the northern Mediterranean coasts. Of these six only one is at sea level. This is the Bosporus by which the Mediterranean trader could reach the Black Sea. Along the north coast of this sea the Young Folds have been drowned; so that the Russian Plains can easily be reached each side of the Crimea, which contains a fragment of the Young Folds (Fig. 129). Troy owed some of its importance to its command of the western end of this narrow pass; where storms are so common that in early times small ships were often hauled overland. Constantinople, the capital of the Eastern Roman Empire, was placed on the Bosporus mainly

Fig. 13.—The Topographic Barrier north of the Mediterranean. The six Main Routes in the early days were (1) Bosporus at sea level, (2) Saloniki-Morava to the Danube, (3) Pear-Tree from the Adriatic to the Sava and Danube, (4) Brenner over the Alps (used later), (5) Rhone Graben, (6) Narbonne to the Atlantic. Inset is a sketch (from Kober) showing the two lines of Mountain Arcs with the "Plugs" or "Median Masses" between (shown by dots).

to control this artery of ancient commerce. The Greeks, of course, at the dawn of history had linked up the Black Sea trade with that of the Mediterranean. Indeed, it is supposed that the Voyage of the Argonauts is a legend based on Greek traders who voyaged into the Black Sea and then perhaps ascended the Danube and Sava in their eagerness to trade with the barbarians of the north. Possibly they then carried their ships over the "Pear Tree" Pass (see below) and so reached the Adriatic.

The second gate is the combined valley of the Vardar-Morava (Fig. 13). We shall see later that this depression from Saloniki to

Belgrade is in part due to fold topography. The two rivers rise near Uskub, where a gap only 1,500 feet high connects the two. But since this route is over 300 miles long and (until lately) was exposed to brigands on the mountains which flank it nearly the whole way, it has not become a general route until modern times. Indeed, even to-day the main railway runs from Belgrade to Istanbul (Constantinople or Byzantium) rather than along the lower and quicker route to the Mediterranean at Saloniki. However, many important migrations of Slavs and Goths not to mention Romans, Serbs and Turks have used this route.

The third gate is little known to-day. It is usually called the Pear Tree Pass. It leads from the head of the Adriatic Sea to the Danube Valley at Laibach. At its western end was the famous old Roman town of Aquilea, whose destruction in A.D. 450 probably led to the founding of Venice. Twenty miles north is Gorizia, the scene of bitter fighting for the command of this gate in the Great War. Then a wide gap leads eastwards to the head of the Sava at Laibach, about 50 miles away. The pass is only 2,900 feet above the sea, and is hence 1,500 feet lower than the famous Brenner Pass. Down the Pear Tree Pass poured the barbarians, eager to ravage the rich lands of Italy. It was used by the Cimbri in the centuries before Christ. Theodosius of the Eastern Empire advanced to attack Rome by this route at the close of the 4th century A.D. The Visigoths, the Huns and the Ostrogoths all used it. In 600 it saw the Avars and later the Magyars. Indeed, the March of Istria was established by the Holy Roman Empire specifically to hold back barbarian invaders during the 9th and 10th centuries. In later times its importance dwindled; partly because its hinterland was held for centuries by the ignorant and unprogressive Turk. But this wide gap, the most attractive on the west of the Balkans, must inevitably be largely utilised in the future—as, indeed, the growth of Trieste near its western end is some indication. (The railway, however, crosses the divide a little south, near Zirknitz, the wettest place in Europe.)

Some 120 miles to the westward is the Brenner Pass—the outstanding pass in the Alps. It is only 4,495 feet above sea level, whereas most of the others are close to 7,000 feet (see Fig. 101). It was one of the main causes of the growth of Venice and Genoa in the south, and of Innsbruck, Munich and Nuremburg in the north. Since, however, its structure will be described in the section on the Alps, and its trade in another chapter, it is not necessary to emphasize further its importance here. Although it was often used by barbarian invaders of Italy and by the "Holy Roman" Emperors later in their journeys from Germany to Rome, no carriage-road was made across it till 1772.

The fifth of the great gates is the Rhone Graben—often called the "Way of Light." The civilisation of France, Britain, and to a large extent of Germany, came by way of this route; which connects the Mediterranean with the Seine basin across a divide about 1,600 feet high near Langres. Its use by pioneers and merchants goes far back of history. Marseilles, as we have seen, dates back to B.C. 600. Greek trading posts extended from the Pyrenees to Genoa long before the rise of Rome. Hannibal crossed the Rhone in his march to Rome in 201 B.C., but though Spain soon became a Roman province the Rhone region was not incorporated (and administered from Narbonne) till 118 B.C. Much traffic reached the Rhone by road from Spain over the Col de Perthus and from the Italian Littoral, as well as the greater volumes which came by sea to Massilia and other ports. The early invaders of Rome used this route. The Cimbri advanced south a few years after the Province was established and defeated Rome in B.C. 105. The Teutons were crushed at its mouth (Aix) in a decisive battle in B.C. 101. Later Caesar made his main line of attack on Gaul up the Rhone Valley, seizing the Burgundian Gate (Belfort) at its northern end in B.C. 57. Christianity advanced up the Rhone Graben as early as the second century A.D., and it is believed reached the middle Rhine about this date also. It is unnecessary to elaborate further its importance—for this is brought out fully in the discussion of the geographic aspects of the Renaissance (p. 180).

The sixth gate in our series is of much less importance, though probably it was one of the most used at the dawn of history. Neolithic traders crossed from the Mediterranean coast at Narbonne to Brittany and the northwestern islands by the gap between the Cevennes Massif and the Pyrenees (Fig. 13). This is the Gate of Narbonne or Carçassonne, which is now traversed by the Canal du Midi. The Neolithic folk, however, bore away to the north, and probably kept to the open western slopes of the Cevennes and Auvergne. The coasts of the open ocean had little attraction in early days nor did the basin of Aquitaine ever rival that of Paris as a centre of human interest. Yet Toulouse (Tolosa) at the west end of the Gate was a place of some importance in Roman times; and became the capital of the Visigoths during the 4th and 5th centuries. Being clearly off the main routes of the continent it does not rank with the other five, which may perhaps be arranged in the following order of importance; first the Bosporus, then the Rhone Gate, the Brenner, the Pear Tree Pass, and the Vardar-Morava Gate, and lastly the Narbonne Gate.

F. General Conditions of Topographic Control

From the preceding paragraphs it should be clear that the suitability of a country for human occupation depends very largely on its topographic history. The fundamental topographic reasons which have led to the growth of huge populations along the coal belt have been referred to (Fig. 15). Metal resources do not as a rule attract

Fig. 14.—The Great Coal Trough across Europe. Figures show 100 million tons of resources. Nations without much coal are shaded. (See Fig. 147.)

such dense groups of peoples, for it is usual to carry metalliferous ores *to the coal* for treatment. However, it may be here noted that metal mines usually occur in *ancient* rocks, for in them there has been more chance for the ores to be deposited. The latter are derived in general from hot gases or heated waters from the *deeper* parts of the crust; and shattered older rocks are more likely to have collected these than young unaltered level-bedded rocks. Hence we find mines more usually in ancient peneplain surfaces than elsewhere—because here the "roots of the crust" are exposed. The gold and silver fields of the Canadian Barren Grounds are cases in point; so are the tin mines of Cornwall, the rich mines of the Spanish Meseta, of the Bohemian "Ore Mountains and of the peneplain in Sweden (Fig. 16).

But turning to agriculture—which is a more certain and lasting resource than mining—we can make even clearer correlations. The young mountains are obviously too rugged, too subject to the erosion of rapid streams, to be good sites for the deposition of large areas of soil. So also uplifted peneplains are bad; here also the surface is likely to lose soil rather than to conserve it. As the "erosion cycle" progresses (see p. 59) the conditions become more and more favourable. In the "mature" stage of erosion the debris carried by the rivers begins to be deposited—for now the river grades are become flatter. It is indeed in such topographies—in the later stages of

Fig. 15.—Map of Europe showing the location of the chief characteristic Topographic Types.

maturity and in the stages of senility (see p. 59)—that some of the most favourable agricultural regions are to be found. Of this type are the smiling farmlands of England, France and Germany. Here the country is undulating—the river valleys are broad, but are well defined by rounded hills. The last stage is shown by flat regions in which the rivers wind about in senile fashion. Such a topography is often met with in the case of hollows formed, not by erosion, but by warping or faulting. The best examples in Europe are the Lombard Plain and the Alfold. In both cases down-warp basins have accompanied the upward buckling of the Alps or Carpathians around them. In both cases the hollow basin has in process of time been largely filled by the debris brought down by rivers from the surrounding hills. This rich deposit of alluvial is utilised for wheat and mixed farming in both areas—where hundreds of square miles without any rocky outcrop are to be seen.

A few words must be added about the Russian Plains. Here a shield seems to be present—but it does not in general show at the surface, save on its northwest border. Deposit after deposit for many millions of years seems to have been laid down on this shield, which clearly must have been in general below sea level. But though it seems to have sunk more or less steadily and been covered with many rock layers—especially in the south, and even by glacial debris and rich "Black Soils" as we shall see later—it has resisted folding forces. This has had satisfactory results as far as man's occupation of it is concerned. It is surely better to have the southern half of Russia (see Fig. 130) still covered with rich black soil, because of the resistant character of the sub-surface shield, than to have the land buckling up with the loss of most of the surface layers. This sort of thing has happened elsewhere, as in southern Italy—where very late formations have been buckled up to form highlands, much later than has been the case with most of south Europe. This Italian topography is doubtless much more scenic and attractive to the tourist, but the stable, low, monotonous plains are the land-forms which give most subsistence to mankind (Fig. 15).

These general remarks on topographic control apply best, of course, in modern times. Obviously, however, in the very early days when man lived in isolated groups, protection and refuge from enemies bulked very large. Rugged topography, it is true, repelled early man, but so did vast open plains, which he has only relatively recently learnt to utilise adequately. And so we find that intermediate types of landscape appealed most to our early forefathers. The sunny nooks of the Mediterranean, with their convenient crags to carry the fort or acropolis of the settlement, attracted them much more than the thousands of acres of the "Black Soils" of southeast Europe. The undulating lands of France and Germany seem to have carried more population in the early days than the plains to the north and north-east. In the north of Europe, as we shall see, the temperature control dominates man. In the far southeast drought is the master; but in the greater part of the continent topography has always profoundly affected man.

CHAPTER IV

CLIMATE AND HISTORY IN EUROPE

A. Climatic Regions in Europe

However much we may admire the accomplishments of the ancients of Asia and Africa, there can be no doubt that in the last 500 years it is the inhabitants of Europe who have impressed themselves on the rest of the world. Only about one-third of the world is inhabited essentially by people of European descent, but they govern ninety per cent. of the land area (J. W. Gregory). Many factors have contributed to this significant result. It is not as some think due mainly to Nordic racial supremacy, for South America and Mexico were conquered by folk of the Mediterranean race. Many of the British adventurers were also of that race, though perhaps the Nordics predominated. We are, moreover, apt to forget the vast Asiatic conquests of the Russians, who are dominantly Alpine in race, while the Rumanians, Poles, Magyars, Czechs, etc., who are thronging in millions to the New World, are also Alpine. Material resources have helped to put Europe ahead of other regions in the race for World Power, but the feature in which Europe differs most from other large land areas is the *climate* with which it is endowed. This must assuredly have played an important part in her rise to greatness.

It has been stated that an astronomer in other worlds (if such there be) peering at our planet would be most struck by one peculiar phenomenon. He would, of course, note the two ice-caps, especially obvious in the northern winter. But running obliquely into this zone of ice he would observe a long "Black Gulf," extending in January almost from latitude 40° N. to latitude 70° N. This would, of course, represent the warm dark waters of the Northeast Atlantic, contrasted with the floating ice and snow-covered lands of the northern hemisphere. No similar feature occurs elsewhere in our planet. In other words, there is something anomalous about the climate of the most important part of Europe.[1] The western half is abnormally free from the cold winters of the latitudes involved. Such *deviations from the average temperatures* of the various latitudes can be plotted on a map.

[1]A map showing this Anomaly is given in Bartholomew's large Meteorological Atlas, Plate II.

We find that the Lofoten Islands off Norway have a (warm) deviation of 40°, which, in fact, is the most striking "anomaly" on the earth's surface. Throughout Europe only the eastern fringe of Russia is as cold as the *average* conditions in the latitudes concerned. Hence Europe as a whole stands out with a most remarkable climate—in scientific language it has a *positive* anomaly rising to 40° F. in the northwest. Much of North America and Asia suffer from a strong

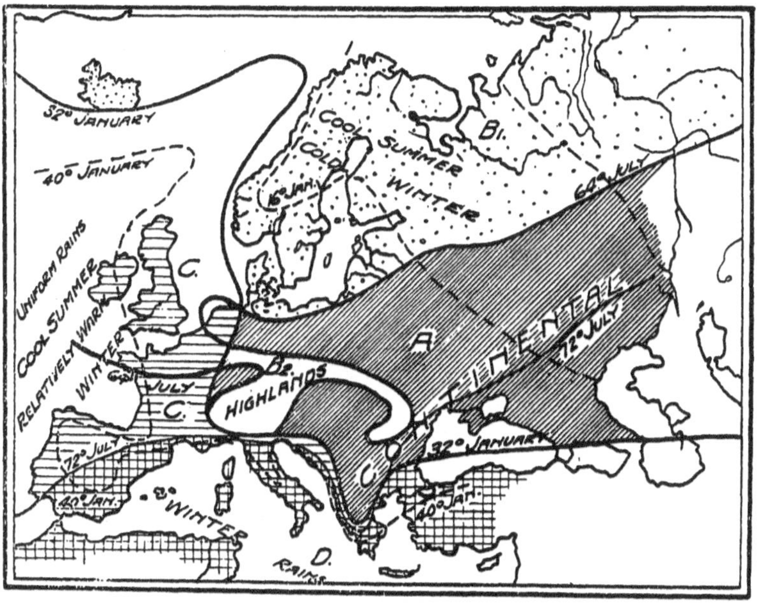

Fig. 16.—The four main types of Climate in Europe. At *A* is the *Continental* Climate (over 64° in July, under 32° in January), At *B* is the *Northern* type with light summer rain. *B2* is similar; being cold owing to elevation. *C* is the *Marine* equable type with uniform rain. *D* is the *Mediterranean* climate with winter rains.

negative anomaly. Chicago, for instance, is 15° F. colder than the average for places on latitude 42° N. in January. Churchill (on Hudson Bay) is about 30° colder than the average for the latitude—and so is much like central Siberia in this respect.[2] It is clearly necessary to consider this factor of climate in some detail, since it has

[2]Chicago has a stimulating climate—and its abundant coal supplies render it satisfactory for man in winter. But it is clearly not so suitable for medieval settlement as most of Europe.

probably played a part greater than, for instance, that of Race in the world's history.

This is not the place to explain in detail the reasons for the unusual climate of Europe. Many factors have collaborated. The general circulation of air over the oceans produces westerly winds to the west of Europe. In part owing to the trend of the coasts around the Caribbean seas, a vast body of water is warmed here and then driven north and northeast into the Atlantic. This current actually flows against the Trade Winds—at any rate in winter—but is picked

Fig. 17.—Rainfall, Storm-Tracks, Winds and Pressure in January. Some Storms cross the Mediterranean. Figures are inches of rain.

up later by the westerlies and bathes the west coasts of Europe. These warm winds and warm waters are the chief factors in producing relatively mild winters as far north as Spitsbergen (80° N.). Their most obvious effect is the extraordinary deflection of the isotherms in January in western Europe (see Fig. 16). In January the Lofoten Islands in latitude 70° N. have the same temperature as central Greece in latitude 40°. In other words, a traveller could journey through 30 degrees of latitude (*i.e.*, 2,100 miles) from Greece towards the pole (by way of Trieste, Munich and Bergen) in this month, and find the climate becoming no colder! In the same way it is often pointed out in England that Cornwall has the same January temperatures as the Riviera or Athens, although it is so much farther north. On the temperature map in Fig. 16 the January isotherm for 32° is shown running north-south as long as it is affected by the warm westerly winds, but curving rather abruptly to the east as it reaches the Black Sea. Similar directions obtain in the warmer more western isotherms in this month; but as we move east over Asia—away from the westerlies—the isotherms tend more and more to resume their normal east-west direction.

Turning now to the summer conditions (in July), the temperature anomaly has vanished in most of Europe (Fig. 16). Only the east

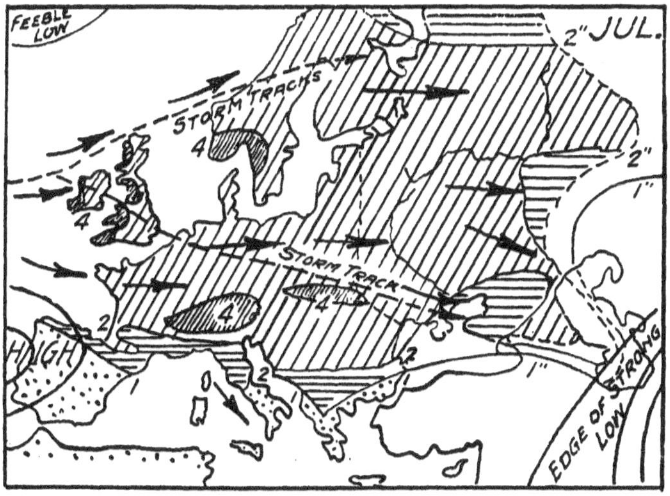

Fig. 18.—Rainfall, Storm Tracks, Winds and Pressure in July. Storms chiefly over centre and north of Europe. Figures are inches of rain.

of Russia is a few degrees colder than might be expected from its latitude. We find the isotherms running more or less parallel to the equator, as is normal. The lands and seas do not differ much in temperature. Lofoten is about 50° F. while Greece is now about 72° F., agreeing with the difference in latitude. Of course, elevation introduces some complications into the problem. A climb of 1000 feet reduces temperature about 3° F., while a journey of some 200 miles towards the Pole is necessary to bring about the same fall of temperature in a horizontal plane. Hence such regions as the Alps or Carpathians act climatically very much as if they were "fragments," say of Finland, placed in the centre of the continent. These cold central regions are labelled *B*2 in Fig. 16.

Equally important with winter and summer temperatures are the *ranges* of temperature. Climates are classified broadly into two main divisions: those with *continental* climates and those with *marine* climates. These are essentially functions of the range of temperature. The reader will have gathered already that in the west of Europe the temperature does not change greatly in the year. In other words, it has a *marine* type of climate. On the other hand, the east—remote from the favouring westerlies—has a very cold winter, as we have seen. Moreover, the ocean also absorbs heat but slowly in summer and so helps to produce mild summers in its neighbourhood. On the other hand, the lands warm rapidly, and so those remote from the sea have more extreme weather both in summer and winter. In the map (Fig. 16) those areas which are below *32* in January and above *64* in July have a range of *32* degrees (or more). They experience a *continental* climate, and this is confined in Europe to the southeast part of the continent.

We can now usefully divide Europe into four temperature regions, which agree fairly with the rainfall regions considered in the next paragraph (Fig. 16). *A* is the continental type with a range of 32 degrees. Its rains are light and chiefly in summer. *B* has a cold, dry winter and cool, wet summer (*B*2 is much the same). *C* has a moderate winter and cool summer with uniform rains, while *D* has somewhat the same marine conditions but the rains are lighter and fall in winter.

Of equal importance with temperature as regards human habitation is the rainfall of a country. Europe is blessed with the best distribution of rainfall of any continent. Only negligible areas in the far north and near the Caspian Sea have less than ten inches a year. East Spain and South Russia are the only *warm* countries with less than 20 inches a year; and in cool countries this supply is fairly adequate for man. Again only in a few regions, such as Croatia and the Scottish Highlands, is the rainfall excessive. By far the larger part of Europe receives the very satisfactory rainfall of 15 to 45 inches. But the *season* of the rainfall varies and is very important. It will now engage our attention.

In most textbooks, it seems to the writer, the reader is led to suppose that rains are mainly due to steady *onshore* winds; which rise over the land, are chilled, reach their point of saturation, and so deposit bountiful rains. This is, of course, fairly correct. But a better picture, nearer to the truth, is to think of the rains being produced in large measure by wandering low-pressure-systems, the so-called *Cyclones*, Storms, or Lows. Where these are abundant there are generally plentiful rains; where they are absent the country is in general dry. These wandering storms move (almost invariably in the European area) from west to east. But they have favourite tracks, depending essentially on the relation of the sun to the continent.

If we study the tracks of these favourable rain-producers[3] we find that there are two zones of storms—a northern and southern, of which the former is much the most important and operates all the year round (see Fig. 17). The *northern track* of storms in winter passes from Iceland across to Lapland and then as summer approaches this track becomes less definite and in June it runs across Scandinavia through Bergen. (Of course the individual cyclones, which are about 500 miles wide, move steadily from west to east about an average of 400 miles a day—and carry the rains across Europe at this rate.) In the second half of the year, as the sun moves south the storms become more numerous in the north, the most frequent occurring there from October through January. Meanwhile the *southern track* is unimportant in summer, but as the sun moves south the storms become quite frequent here in November, passing from the Bay of Biscay across Italy to the Black Sea (Fig. 17). In December and January they move along much the same path, but their frequency diminishes during spring.

From the foregoing studies two salient facts emerge. Firstly, the rains of Europe should be more abundant in the west than the east, since the storms rapidly lose their moisture as they leave the ocean. Secondly, the northwest experiences storms all the year round, and so must have a uniform rainfall; while the south receives more rains in winter, and is relatively dry in summer. The actual rain records justify these conclusions, and enable us to divide Europe into four Rain Regions: *Uniform* and heavy in the Northwest and Centre; *Summer* and light in the Northeast; *Winter* and light in the Southwest; *Arid* in the Southeast.

The actual distribution of rainfall in January and July is given in Fig. 17 and Fig. 18. The winds are in accord with the main areas of low and high pressure. They blow in general from the west, but are intensified in the north in winter by the proximity of a semi-permanent Low or Cyclone (over Iceland) which vanishes in the summer time.

[3]See "Atlas of Meteorology", Bartholomew and Herbertson, Plate 29.

B. The Forests and Grasslands of Europe

If a geographer were limited to one single map on which to base his deductions as to the suitability of the lands for human occupation he would probably choose the map showing natural vegetation. For this map shows us how plant life has responded to temperature, to rainfall (total and seasonal), to soils, to elevation, etc., etc. Man and beast live on the vegetation, so that (except for mining) most human

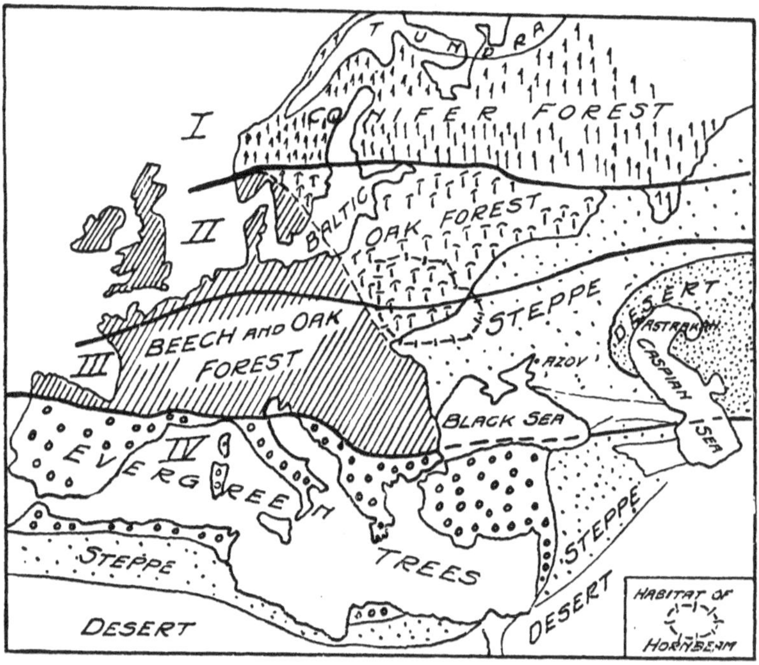

Fig. 19.—A Generalised Vegetation Map of Europe. Belts I to III are Jonasson's Agricultural Belts. I, Forest with some Barley. II, Oats and Rye. III, Wheat and Maize. IV, Olives and Subtropical Grain.

resources can be deduced from a detailed vegetation map. Hence a map showing the natural forests, grasslands, etc., of Europe is of great assistance in understanding past migrations and settlements as well as indicating future densities of population.

The natural vegetation of Europe may be divided into six main types: Evergreen forest, Broad-leaf forest, Coniferous forest, Steppe, Desert and Tundra. These agree fairly closely with our climatic regions, but a few words may here be given concerning *edaphic* (*soil*) *control.* Even under similar climatic conditions different plants will spring up if the character of the soil varies. Sands, wet clays and limestones produce different types of vegetation. Forests prefer wet clays, grass is found on limestones, poor sandy soils encourage heath and other moor plants, all under somewhat similar climatic conditions.

True forests develop where the trees can obtain adequate moisture all through the growing season. The best locality is clearly the uniform-rainfall region in the northwest and centre. Here accordingly are the forests of broad-leaf trees, such as oak, beech, elm, birch, etc. In the north the summer is rainy and this is adequate for the coniferous forests which are adapted to grow very quickly during the summer. In the intense winters of the north their structures can resist frost in a manner impossible to the broad-leaf trees. The conifers do not shed their leaves and so can start functioning in spring much more quickly than the deciduous broad-leaf trees. The critical temperature appears to be about 43° F. If a region has only 5 months of this temperature then conifers will be dominant. If more than 5 months then oaks, etc., seem to be most abundant. These figures hold good right along the southern boundary of the coniferous forest in the Northern Hemisphere (A. Miller, "Climatology", 1931).

The evergreen forest flourishes where the temperature never falls below 43° F. In Europe this isotherm in January runs approximately from the Pyrenees to the Bosporus. Hence the Mediterranean lands are characterised by evergreen forest. But as we have seen, their rainfall is not high and is markedly seasonal, mostly falling in winter. For this reason the evergreen forest is scattered, and, indeed, only in places are the trees close enough together to merit the title of *forest.* Many devices have been developed (such as spiny or waxy leaves) to resist prolonged drought in summer, and shrubs, tussock grass, herbs and tubers are perhaps more characteristic of the Mediterranean region than are evergreen trees of any size.

The grasslands are often called Steppes in cool countries and Savanas in warm countries. We are dealing with steppes exclusively in Europe. Here the rains are limited in season and extent. Trees do not flourish, but grasses are well suited by a spring and early summer rainfall. In the autumn and winter they die, to develop afresh from seeds in the following spring. Hence they are the best adapted to flourish in the southeast of Europe where the summer rainfall is rather low and the temperature in winter too low for evergreen trees. However, there is never any sharp line of division between the various vegetation regions. Thus "corridor forests" often run along streams from the true forest region far into the steppe. So also "corridor grasslands" border the intermittent streams on the desert margins. Koeppen has defined the European *boundary between desert and steppe* (with its light summer rain) as the locus of places where the average temperature (Centigrade) plus 14 equals the annual rainfall in centimetres. Thus Astrakhan (on the Caspian) has a temperature of 10° C. Its rainfall should be 24 centimetres (*i.e.*, 10+14) to be on the boundary, but is only 15 centimetres. Hence it is in the desert, not the steppe. Azov has a temperature of 9° C. Its rainfall to be on the boundary should be 23 centimetres; but is about 43 centimetres. Hence it is well into the steppe region.

The tundra region is of no significance in European history. It is confined to the north coasts of Europe, but extends far into Norway along the broad elevated plateaux or Fells. Mosses and lichens are the chief plants—which are often frozen during eight or nine months in the year.

It must be understood that Fig. 19 shows the vegetation as we imagine it would be if man had not interfered with it. Almost the only forests left undisturbed are to be found in the northern part of the Coniferous forest. Almost all the Beech and Oak forest of the west and centre is altered out of all recognition—where it has not been entirely cut down. On the coasts of the Mediterranean the evergreen copses of oaks and allied trees have been ruthlessly destroyed. The result has been the washing away of the soil-cover on the slopes of all the hills, and in consequence a tremendous decrease in the value of the land for agriculture. There is a general belief that this single fact—the destruction of the forest and soil-cover—has had much to do with the decline in power and population of many Mediterranean peoples since classical times. France and Italy and to a lesser degree Spain and Greece are spending vast sums to try and replace these lost forests, and so in times to come render valuable the areas made desolate by man's short-sighted policy in the past.

The intimate relation between forest, grasslands and man will, of course, be emphasised many times in later chapters. Some of the broader correlations may, however, here be considered briefly. The Phoenician, Greek and Roman cultures grew up exclusively in regions characterised by the evergreen trees of the Mediterranean shores. These folk were accustomed to heathy headlands and scattered clumps of trees, perhaps clustering in the frequent valleys; but dense forests, such as covered central Europe, awakened awe rather than admiration in the southern Europeans. The Romans expanded considerably into the broad-leaf forests, especially on their western margins in Gaul. But their terrible defeat in the Teutoberg Forest early in the first century taught them not to penetrate too far into Germany. Thereafter they adopted the Rhine-Danube line as their main line of defence against the Forest Barbarians, and though Dacia (West Rumania) was conquered by Trajan it was soon relinquished by his successors.

The Nordic Conquerors were eminently men of the forests. All through the early Middle Ages the chief power in Europe was in the hands of men of the forest, *i.e.*, Franks, Germans, British, Poles and Czechs. It is true that in the 16th century Spain (a Mediterranean land) seemed to reach its zenith. But it depended to a large degree on American mines and Austrian armies for its power (see Chapter XVIII), and soon lost ground when these were removed from Spanish control.

Of late years the Steppes are coming into their own. The rich soils of the Cherno-ziom,[4] extending from Kiev well into central Siberia, were ignored as important wheatlands until this last century (Fig. 130). For in Russia the *podsol*[5] regions around Moscow with their open forests were more attractive to the early Slavs originating in the forests to the west. Moreover, the open steppes were often held by the Asiatic invaders, who poured into South Europe almost continuously from the 5th to the 13th century. These nomads, however, made no attempt to use the steppes for croplands. During the recent social troubles in Russia the Ukrainians on their rich wheat farms on the Cherno-ziom have naturally been in a much happier position than the folk of North Russia who live on sterile *podsols*—poor in plant foods. The southern parts of Russia are growing rapidly in population for exactly the same reason as have the lands around Winnipeg in Canada. Both consist largely of rich wheat soils. Thus even to-day the nations are modifying their cultures and habitats in response to environmental factors.

C. *Changing Climate and Its Effect on History*

Many years ago there was considerable doubt as to the reality of the Ice Ages, but nowadays everyone realises that there were four

[4]*Cherno-ziom* means Black Soil.

[5]*Podsols* are soils which have been rendered rather sterile by constant rains washing away soluble plant-foods.

major oscillations of climate during the Pleistocene Age, say within the last 800,000 years. The date of the last of these major ages (the Wurm Ice Age) is placed at various periods from 60,000 to 30,000 years ago. Of course the Ice Age conditions have not yet vanished in Norway. Visitors to Finse (120 miles N.W. of Oslo) may see a small Icecap (Hardanger Jokul) lying inert on the plateau at 5000 feet, and only a mile or so from the railway. The controversy among scientists has now shifted to the climate of the period succeeding the Wurm Ice Age, and especially to the question as to whether there has

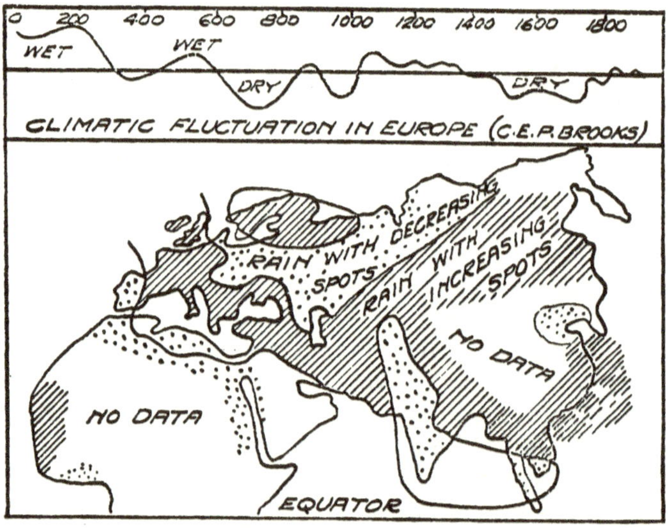

Fig. 19a.—A map showing how the rainfall in adjacent regions responds differently to changes in Solar Energy. The graph above shows fluctuations in rainfall from A.D. 0 to 1900. (From Huntington, "Earth and Sun", 1923).

been any marked change of climate in archeological and historic times, say in the last 8000 years. The writer has no doubt in the matter, but since several notable archeologists and a few geologists are not convinced that there has been any noteworthy change, it is well to examine some of the evidence.[6]

The solution is difficult because it seems that adjacent regions react differently (Fig. 19a) to variation in solar energy; which is generally assumed as the main factor in these recent changes. For instance, the writer about 1924 found a very close relation between

[6]See C. E. P. Brooks, "Climate Through the Ages," for an excellent summary.

droughts and solar variation at Bourke (in east Australia). But Powells Creek, in north central Australia, showed no sign of such a relation. Later work by Kidson and others shows that in eastern Australia greater rainfall follows increased solar energy, but in central Australia more rains occur with less solar energy. Huntington, Walker, and others have mapped out the world into various rain regions (Fig. 19a), some of which react in a positive and others in a negative manner with regard to solar energy. To the writer, this seems the most fruitful advance in our studies of climatic changes which has taken place for many decades.

It has been usual for opponents of the theory of climatic change to say that crops in Palestine grow in much the same localities and have similar harvest times as at the dawn of history. Such a statement is much too general to be of value. No one suggests that the rainfall or temperature changes have been of the order of those marking the Great Ice Ages. Only changes of a few degrees in temperature or an inch or two in rainfall are to be expected. In a diversified region like Palestine the transfer of a crop up or down a hillside might accommodate it to a real temperature change—or up or down a valley to a real change in water supply, without any difference in the locality which could be deduced from early records. Moreover, the rather small climatic changes involved would have far greater effect in *inland* semi-desert areas than in Palestine. However evidence of a physical nature, such as changes in lake areas, or in bogs, or in general of trees,[7] or in the advance of glaciers, is preferable to reports coming down from early times.

In the adjacent table, taken chiefly from data given by Brooks, some of the chief evidence is summarised. I have grouped the material in four geographic regions: A, Scandinavia, Iceland and Greenland; B, Britain and Central Europe; C, the Mediterranean; D, the Caspian area. As far as our studies in national evolution are concerned it is not necessary to go back beyond the time of Christ, but I have added a few earlier dates which link up with the beginnings of Greece and Rome. It is obvious that in the two regions in which we are chiefly interested, climatic changes might operate in rather different directions. Thus, the Mediterranean is on the whole rather hot and dry, so that *heavy rains* would be noticeable. Its most

[7]The trees in the forests of N.W. Europe have completely changed their character several times in the last few thousand years. (See Brooks, *passim.*)

TABLE F

CLIMATIC CHANGES IN OLD WORLD
(Generalised, but mainly after Brooks)

Date B.C.	Scandinavia, etc.	Central Europe	Mediterranean	Caspian Sea
1300		Floods Hungary		
1000	Traffic to Britain	Traffic over Alps Dry period		
800 700	Twilight of Gods (very cold)	Hallstatt wet Alps empty	Mykenae good	
400	Europe wet	Good crops, Rome		Level 150 feet higher
200	-do-	Peat in Frisia	Some malaria, Rome	
A.D. 0		Wet in N. Europe	Rome good	
1st C.		Bad in N. Europe	-do-	
2nd C.	Goths leave Sweden	-do-	Pestilence, Rome	Near present level
3rd C.		-do-	Wetter, Rome Rome deteriorating	
4th C.		German swamps dry	Rome very dry	Aboskum dry. 15 feet lower
5th C.		Ireland flourishes		
6th C.			Arabian invasions	
7th C.		Droughts in Britain	Driest in Greece	
8th C.		Alps traffic revives		
9th C.	Iceland and Greenland flourish,			Derbent level 29 feet up
10th C.	Ice retreats north	Dry, generally		
11th C.	N. route to Greenland	Wet in Britain	Moors flourish in Spain	Baku levels 14 feet lower
12th C.	N. voyages end	Very stormy Holland	Generally wet	
13th C.	Ice moves south	Wet in Britain	Venice flourishes	Sea 37 ft. higher
14th C.	Greenlanders die	Dry, generally	-do-	
15th C.		Dry in Belgium		Caspian still high
17th C.	1658 Baltic thick ice	Glaciers increase	Rainy in Italy	1638 fifteen feet higher
18th C.	1760-1810 Lund wet	1670-1730 Britain dry		
19th C.	1811-1890 Lund dry	1730-1830 Britain dry		1815 eight feet higher

marked changes would be very different from those in Northwest Europe, where the climate is in general cool and moist; and *droughts* are a rarity. The point of view of the historian is naturally bound to be affected by what is his usual environment. We are told that the Egyptians—used to the bounty of the Nile—greatly pitied the Greeks whose water-supply depended on erratic falls from the sky!

Archeological evidence tells us that at the close of the age of Bronze and at the dawn of the Iron Age there was a good deal of traffic between the British Isles and Scandinavia. Storms were rare and civilisation flourished in Ireland—where conditions were unusually dry. At this time snow and ice seemed to be less abundant in the Alps and there was much traffic and settlement therein. During the later Iron Age snow and cold possibly account for what the Norse legends call the "Twilight of the Gods," while it became much wetter at the Hallstatt Iron site in the Tyrol. Argos in Greece was too swampy for settlement, if we may use Homer as a guide. Mykenae, on the other hand, which is not far distant flourished at this time.

In the Roman area the conditions seem to have been wetter than usual around B.C. 400, so that agriculture flourished. But gradually the rains decreased, streams ceased to run and malaria (due to mosquitoes) depopulated part of Italy. During this period the Caspian Sea was subsiding greatly, from a high stage (said to be 150 feet higher than at present) about B.C. 438 to conditions near the present level about A.D. 200. The decline of Rome is directly related to a deteriorating climate by Ellsworth Huntington. The early centuries of our era experienced adverse climates in North Europe, leading to the migrations of the Goths, but apparently the Mediterranean had a good environment until about A.D. 250 when prolonged dry periods set in (Fig. 66). At this time the German tribes were settling on former swamps. Around the 5th century Ireland had become so much drier and more flourishing that Christianity and culture generally advanced there rapidly and missionaries were sent out to all parts of Europe. Several sites, like Aboskum on the Caspian, were occupied about 450, which are drowned by the sea to-day.

The 7th and 8th centuries also seem to have been dry throughout most of our region—and even as far as Arabia. This long period of drought contributed to set in motion the Moslem migrations and greatly helped the spread of the new religion. However, snow and ice being rather rare, traffic over the Alps seems to have increased about the time of Charlemagne. Iceland and Greenland were colonised just before 1000 A.D. There is no question that their climates were much superior to what is now experienced there. Crops were grown in Iceland where now are fields of ice. The voyagers used a northern route to Greenland for two centuries which is now blocked by ice. In the early days of settlement the seals and Eskimo lived near the pack ice, far north of the two Norse villages in Greenland. During this period (1000 to 1200) most of Europe was much wetter than usual, so that the Mediterranean shores flourished. The Moorish culture in Spain may in part be ascribed to this improvement. The Caspian seems to have risen about the same time, and in 1300 it was 37 feet above the present level. The stormiest weather in western Europe occurred in the 13th century, and the coasts of Holland were entirely altered in outline about this time (p. 285). During the 14th century the Norse settlers in Greenland found the ice advancing to the south, and by 1500 the whole population seems to have succumbed to the unfavourable climate. Only shallow graves in frozen soil, occupied by meagre and disease-stricken bodies, remained when the next voyager, John Davis, reached Greenland in 1585.

Since 1400 the climatic fluctuations seem to have been less striking. As stated in the first part of this chapter different regions vary in their response to solar changes. In general, the region of storms (cyclones) and general low pressure reacts in one way, while the belt nearer the equator of high pressure reacts in the other way. It is considered by some scientists that increase in solar energy increases the intensity of the storm-belt, but also increases the strength of the anticyclones (high-pressure systems) to the south. Hence the storm areas are wetter when there is greater solar energy (*i.e.*, a positive effect). But since the anticyclones (which rarely lead to rain) are also stronger, their habitats are drier (*i.e.*, a negative effect). Possibly when solar energy is weakened the boundaries of both cyclone and anticyclone areas are less clearly defined. The storms are less numerous and give less rain in the positive areas, but often wander into the negative (anticyclonic) areas, and so considerably increase the meagre rainfall of the latter. Whatever be the "mechanics" of climatic changes, no impartial student can deny that there have been marked changes in Europe during historical times, which in turn have been of considerable moment in the history of the nations.

PART II

CULTURAL FACTORS IN EUROPEAN HISTORY

CHAPTER V

RACE, POPULATION AND POWER

A. The Importance of Race

To-day the most obvious distinctions between the peoples of Europe are those based on nationality, which in turn depend on the governments which control them, on the languages they speak, and possibly on a dozen other factors which will be discussed in later chapters. Race, in Europe at any rate, is really of minor importance to-day. It is true that the terms "Jewish race," "Aryan race," "European race," etc., are common in literature, but they are all misnomers due to a confusion of ideas. Thus we can talk of a Jewish religion—but not of a Jewish race; or of Aryan-speaking peoples—but not of an Aryan race; or of European culture, but not of a European race. Race is a *biological* concept, and is quite apart from culture, religion, language or nationality. It is to be measured in terms of head index, hair texture, jaw length, etc., to the same end that the breeds of dogs are considered, *i.e.*, to determine the various divisions of the genus.

One aspect of Race is of considerable importance in the study of European nations. The relative status of each of the various races is ascertainable by means of a technique based on ecological data. Indeed the writer has made this problem a major study since 1919, and his book "Environment and Race" (1927) deals fully with this matter. Two important deductions follow from this research. First, there seems to be little evidence for the current division of the peoples of Eurasia (by a north-south line across the main corridors of Central Asia) into the classes of Caucasian and Mongolian*. Rather should we divide them into a marginal zone, the Mediterranean Race, and a central zone, the Alpine Race. Secondly, from biological analogy, the marginal races *furthest from the centre of the biological zones*, are the most primitive. This means that the marginal "Narrowheads" (Dokephs) in southern and western Europe are racially akin to the similar types in southern Asia. So also the "Broad-heads" (Brakephs) who occupy the central region from the Alps to Korea are all Alpines, and are a later racial development than the other folk in Eurasia.

The series of maps given in Fig. 20 is based on a lengthy discussion published in "Ecology" (Chicago) in July, 1934. Biologists will see

*See author's recent paper on the subject in "Human Biology," vol. 8, no. 3, Baltimore, 1936.

in the relative distribution of the five races (A to E) a complete parallel to the zones used by W. D. Matthew in his classic deductions as to the evolution, status and migrations of such higher animals as

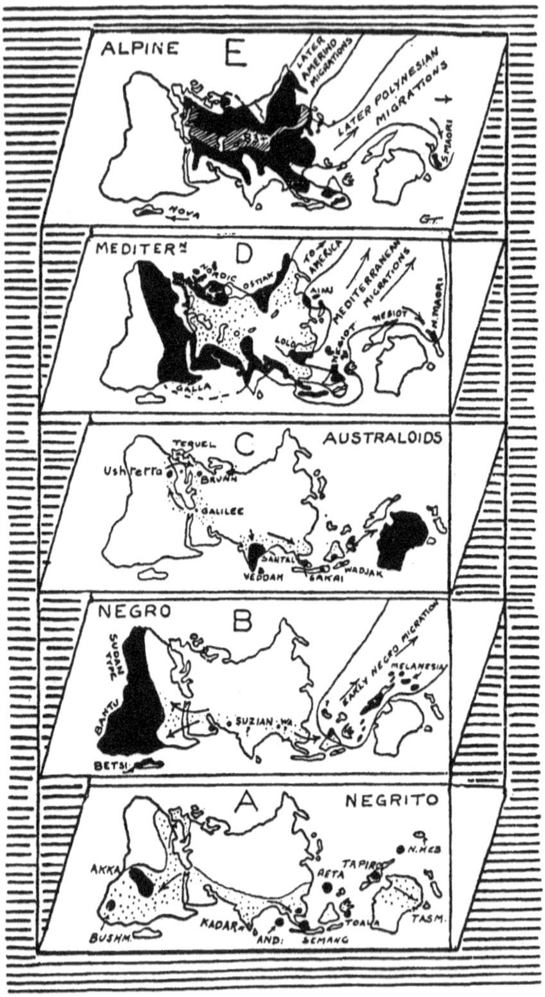

Fig. 20.—A Stage-Diagram showing the distribution of the Five Major Races of Man in approximately concentric zones about central Asia, the common Cradle-land.

The Negrito (A) was first evolved, and was pushed to the margin by the races developing later. The Alpines (E) are the latest race to evolve.

Black areas show present distributions.

Dotted areas are former habitats.

The Alpine and Mediterranean races in Europe are seen to be merely extensions of similar races in Asia.

the Even-toed Mammals (Antelope, Cattle, Sheep) and the Odd-toed Mammals (Tapir, Rhinoceros, Horse). He shows that the *marginal* genus (*i.e.*, Antelope or Tapir) is an earlier type than the genus living in the *centre* of the zones (*i.e.*, Sheep or Horse). So also the Negrito Zone in A has been pushed southwest and southeast by later evolved races. It now survives only at the margins (shown in black). But if we consider archeological and fossil indications (shown dotted) also, this race covers the greatest area of all five.

A major principle of Ecology tells us that the Negrito was therefore the earliest to develop of the five races under discussion. The Negro (B) has spread nearly as far, and is also nowadays found only in the southwest and the southeast of the Old World. The Australoid (C) occupies the third zone, and has clearly not moved so far from the common centre of all the zones in Central Eurasia. The Mediterranean zone (D) still forms a fairly complete ring round the great Eurasian land-mass. There is much evidence (in the areas dotted in D) that this dokeph race once occupied a large part of central Eurasia also. Finally, the Alpine Race still occupies the centre of the zones—the common Cradleland, as Ecology teaches. It would seem to the writer entirely to be expected that the centre of evolution of the higher mammals (due to the special factors for evolution obtaining there) should continue its "function" and later produce the highest mammal of all, *i.e.*, Man. Since the Alpines occupy the centre of these zones Ecology teaches us that they are the latest developments of the Human Race.

The writer briefly introduces these conclusions[1] because they should lead to a less arrogant attitude on the part of the European nations towards the original stocks of Asia, North Africa and America. If one believes, as does the writer, that the Egyptians, Abyssinians and Northern Indians are closely akin to the similar slight dark stocks in Spain and Western Britain, there is little ground for the common European feeling of racial superiority towards these non-European peoples. If, as seems likely, the very broad-headed people in Korea and in much of Japan and North China, have only lately migrated from the general cradleland, then they are further removed biologically from the primitive ape-man than almost all the peoples living in Europe. Once more the writer emphasises his belief that international difficulties are actually due to the clash between human-made cultures rather than between biologically different races. But if the real problems of race can be simplified and minimised as the above brief discussion indicates, this is certainly something gained in our progress to a World at Peace.

[1]The proof will be found in a number of the writer's recent works, *i.e.*, "Environment and Race," University of Oxford Press, 1927. Also see "Human Biology," Baltimore, 1930, pp. 34-62; and Matthew's "Climate and Evolution," New York Academy of Sciences, 1915.

Only a scientist can accurately determine to what race a given European belongs, and it is as well that it is so. There is little animosity between Nordic and Alpine or Mediterranean races (as such), precisely because the man in the street cannot distinguish them at all clearly. On the other hand, in certain other parts of the world there are clearly separable groups of people who differ altogether in culture and often differ in race, and here "race prejudice" has full swing. In the southeast states of U.S.A. the two groups of negroes and poor whites are widely separated in race, but not greatly separated in culture to-day. Here also race feeling is very acute. In India, on the other hand, we have large groups of Indians who differ greatly in culture from the ruling European class. But in race it is clear that both white and "coloured" are often of the Mediterranean Race, since members of the latter stock inhabit half of the British Isles. Moreover, the present writer sees no reason to doubt that many of the Sikhs and allied northern nations of India are very close to the Nordics in race. If this be correct, the peoples of north India form a composite group close indeed in race to that found in Britain, yet here also there is much racial feeling. "Race prejudice" could in most cases be described as "a condition of social friction based on ethnological ignorance."

In Europe, however, there is little "race prejudice," though a great deal of "cultural prejudice" and an almost universal "national prejudice." It is the purpose of the present chapter to show what are the racial differences in various parts of Europe, and how they affect national problems. While there is perhaps no complete unanimity among anthropologists as regards the racial divisions in Europe, almost all would agree to the statement that three such divisions are to be recognized. Some would split these up further. Others, like the writer, would reduce them to two major divisions, with two important subdivisions,[2] giving three classes as in the following table. (Deniker's five subdivisions follow beneath.)

[2]The Nordics differ from the Mediterranean in having fair hair and generally gray or blue eyes. This is a much more important difference than change in skin colour. The writer, however, believes that both originated in a common race in Asia and differentiated through natural selection (of favourable types) in response to very different environments during and after their migrations from central Asia. (See "Environment and Race," p. 143 and p. 158.)

TABLE G

RACES OF EUROPE

BROADHEADS (Brakephs)			NARROWHEADS (Dokephs)		
Alpine Race			Fair Nordics	Dark Mediterraneans	
Eastern (Fair)	Cevenole (Short, dark)	Dinaric (Tall, dark)		Iberian (Short, dark)	Littoral (Tall, dark)

Since to many ethnologists, including the writer, stature and skin colour are of relatively little importance in a primary classification, it is clear that, in their opinion, the major groups, Alpine, Nordic and Mediterranean, constitute the three main races in Europe.

In the writer's opinion, skull-shape is the best single criterion of race, though hair texture is of nearly equal importance. But in Europe there is little difference in hair texture. The breadth of the head therefore assumes first importance. The breadth (or head) Index is the greatest breadth of the head expressed as a percentage of the length. Thus a head with a width of 160 millimetres and a length of 185 millimetres would have a head index of 87. This person would almost certainly belong to the Alpine Race—and if from Europe his folk would come originally from near the Alps or some region to the east. On the other hand, a head 140 mm. wide and 190 mm. long would give an index of 73. The European measured would almost certainly belong to the Mediterranean Race, and his ancestors would be marginal people in Europe. (The Index 81 is approximately the boundary between the two types, though, of course, many individuals overlap.) If he were rather tall and generally blonde he would be a Nordic; if rather short and dark he would be of true Mediterranean race in all probability. In general, the hair of the Alpine Race is wavy to straight, while that of the Nordic and Mediterranean Races is wavy to curly.[3]

The best way to gain a clear idea of racial distribution in Europe is to study the way in which Europe has been peopled by the living races of to-day. The major migrations with their dates and environments are now fairly well understood by archeologists and anthropol-

[3]The various racial problems raised in this chapter are much more fully discussed in the writer's companion volume "Environment and Race," Chapters 3, 4, 10-13.

ogists. It seems likely that folk of very primitive culture inhabited western and southern Europe until about 10,000 B.C. The last major Ice Age (Wurm) had passed away about 10,000 years before, but several minor ice stages (Buhl, Daun, etc.) had shown that the climate of Central Europe had still not reached the more stable and warmer type of to-day. The northern plains were not attractive during most of this time, probably partly tundra (p. 74) and partly coniferous forest. After B.C. 10,000 there was a steady march northwards of these cold environments, and with the growth of oak and beech trees in the northern plains (of Germany, etc.) a later type of man advanced

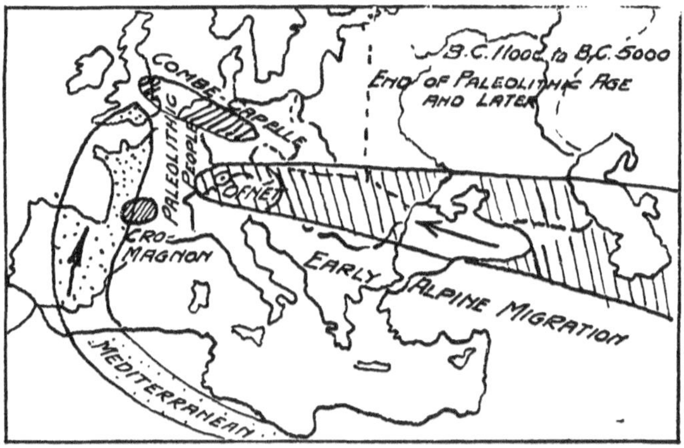

Fig. 20a.—The arrival of the early Mediterranean and Alpine tribes in Europe between 11,000 B.C. and 5,000 B.C.

in large numbers from Africa into southern Europe. These were the first "Mediterraneans."

The Cro-Magnon and Combe-Capelle types of Paleolithic days were either overwhelmed in their earlier habitats of France and Spain or moved northward with the environment to which they were accustomed. Fairly soon the central parts of Europe and the Alpine valleys became much more suited for human occupation. Now it became easy to migrate up the Danube Valley from the west of Asia. Perhaps about 7000 B.C. small parties of a new race migrated into Europe. They were Alpine Broad-heads who had arrived from Asia by a much shorter route. (For there is a good deal of evidence that the Mediterranean Race also dispersed from some Asiatic centre, not far from Mesopotamia, and moved not only west to Africa and Europe but east to India and south China, etc.) These Alpines seem to have been adepts at peaceful penetration, as some of their racial relatives, the Rumanians and Chinese, have been in later years. The state of migration during Neolithic times is shown in a simplified fashion in Fig. 20a.

It is not quite certain where the Nordic Peoples had their habitat while the Mediterranean and Alpine peoples were entering Europe. Northern Europe was not very attractive, nor we may believe was

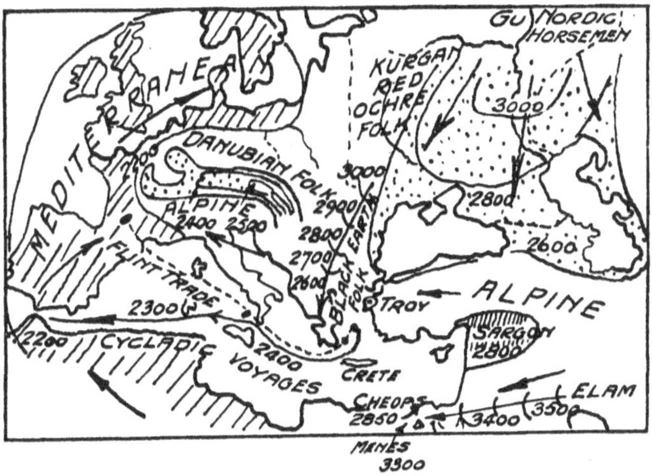

Fig. 21.—Movement of the later Alpine folk and their allies from B.C. 3500 to 2200. The Nordic tribes of the Russian Steppes invade Anatolia, Greece and the neighbouring regions. Historic kings now lived in Syria and Egypt. (Based on Fleure and Peake, "Corridors of Time," 1927.)

northern Asia. There are some racial relatives of the Nordics—such as the Ostiaks of the Yenesei Basin and the Ainu of North Japan—who also were driven from their homelands we may imagine by pressure of folk a little to the south. Some common *central* position is indicated from which these three Nordic groups could migrate, and we may accept the region northeast of the Caspian as perhaps the most likely. At any rate somewhere about B.C. 3000 there was great unrest in south Russia. A group of folk with a typical culture (called the "Black-Earth Culture" by Fleure and Peake) were driven down the west coast of the Black Sea and settled in Greece about B.C. 2600 (Fig. 21). About the same time fierce horsemen (apparently known as Gu tribes) who came from the north appeared around the Black Sea.

Ancient tumuli or burial mounds called Kurgans are scattered through South Russia, and they contain skeletons of narrow-headed folk, who used Red Ochre and are generally accepted as Proto-Nordic peoples. Perhaps they were kin to the Gu horsemen.

More and more broad-heads continued to advance up the Danube Valley, especially from 2500 to 2200 B.C. We may place the early trading voyages of the Greeks from their ports in the islands of the Cyclades about this time. No doubt some sort of interchange before this period had brought flint from France to the southeast, and sent obsidian from Greece and Italy in the opposite direction. About the same time lived Sargon, King of Kings, an historic person of whose life and environment much is known from written records. He flourished in Mesopotamia about B.C. 2800. We may imagine that the Mediterranean Race continued to live all round the Mediterranean and that it spread slowly north into the British Isles, into Scandinavia and even as far east as the south shores of the Baltic. But it never seems to have had much of a footing in the Alps and vicinity, for probably these were occupied by the Alpine folk as soon as the warming of the climate (after the last minor Ice Ages) made settlement possible to any large degree. Thus by 2200 B.C. the three races of Europe were already in much the same positions as we find them to-day. The chief racial movements since that time have been at the expense of the Mediterraneans, who seem to have been displaced from the Baltic by Nordic folk before historic times in those lands. So also the wedge of Alpine folk perhaps descended into France and approached the Pyrenees soon after the dates (rather vaguely estimated) given in Fig. 21.

We may be reasonably certain that these tall fair narrow-headed (dokeph) Nordics of the north, stocky Alpine broad-heads (brakephs) of the centre, and short dark Mediterranean dokephs of the south and west all looked with suspicion and hostility upon each other. Had any prophet suggested that a time would come when no one would care which race a European belonged to, he would have received the attention given to most truthful prophets. To-day there is a real antagonism, based, however, entirely on cultural (*i.e.*, human-made) not on racial (biological) differences. Since sources of friction made by man himself can readily and quickly be cured by man, it is important to realise what these so-called "racial" differences amount to. Since many Germans have glorified the Nordic Race there is a belief in many quarters that Germans are essentially Nordic. This is not correct, for the southern half of Germany, with an unrivalled record in advances in civilisation, is Alpine. Hitler's persecution of Jews on "racial" grounds is absurd, since they also are Alpine like the South Germans; while Yiddish is a German (*i.e.*, Aryan) dialect,[4] written in an unfamiliar script.

The southern parts of Italy have suffered from famine, malaria and other troubles leading to extensive emigration. This leads foreigners to imagine that most Italians are like the southern emigrants and are therefore Mediterranean in Race. In actual fact, the most progressive part of Italy (the north) is almost purely Alpine. One could quote many similar examples, all tending to show that Race and Nation have so little in common that national problems in Europe can safely be considered entirely apart from Race. Yet while this is true to-day, it was certainly not true in the early days of European history. It is for this latter reason that we must give a good deal of attention to racial problems.

B. European Races To-day

The composition of the racial map as it appears to-day may now well occupy us for a time. It is shown in Fig. 22 in a slightly generalised form. Here, as mentioned, an index of about 81 is adopted as bounding the Alpine race. It lies as a great wedge across Europe, with its apex in southwest France—broadening to the east so that it covers the whole continent save the extreme north of Russia and south of Greece. It, of course, extends without break into Asia, and in the writer's opinion extends to the Pacific Ocean. For the Koreans and many other Asiatics differ in no important fashion from the Alpines of Central Europe. The Nordics cluster around the North Sea in an amazingly small area considering the place they have taken in the world to-day. These small Nordic groups have not only dominated the history of most of Europe in medieval times, but have built up large European settlements in Australia and much of South Africa and North America. (It was not till the middle of the 19th century that the huge migrations of Alpines entered North America.)

It is, of course, a little difficult to draw racial boundaries exactly, for a great deal of mixing has occurred on the fringes. For instance, in Poland there are large numbers of folk with rather intermediate

[4]See note on p. 418.

heads, who are rather dark. Some ethnologists would say that these represent a merging of fair Nordics with darker Alpines, others would say they represent a very early migration of Mediterraneans now slightly mixed with Alpines. We note also that many of the eastern Bulgars have rather narrow heads. This is explained as being due to a large early migration of Nordics (coming from east Russia) who

Fig. 22.—The three Races in Modern Europe and their relation to the linguistic and National groups. (From "Amer. Geog. Review," 1921.)

were allied in race and language with the Finns. Sicily offers another example of a group of brakephs among Mediterranean peoples. But Sicily has had such a varied history of migration and conquest that it is almost impossible to interpret its racial evolution.

It is worth while to give in a table the main racial stocks which build up the nations of Europe. The vast majority speak Aryan languages, but a few speak Altaic tongues and this is indicated in the table—though it has little bearing on race. The first group (*A*) of nations is largely *Nordic*, the second (*B*) has considerable *Mediterranean* blood but little Nordic, the third (*C*) is almost wholly *Alpine*.

TABLE H

RACIAL COMPONENTS IN EUROPEAN NATIONS

Nation	Mediterranean (Brunet Dokeph)	Nordic (Blond Dokeph)	Alpine (Brakeph)	No. of Races
A. Largely Nordic				
Scandinavia	Absent	Most	Negligible	1
Finland	"	" (Altaic)	Few in south	1
Lithuania	"	"	Few	1
Poland	Few	North	South	2
Germany	Negligible	North	South	2
Britain	West	*East	Negligible	2
Ireland	West	*East	Negligible	2
France	South and S.W.	*North	East & centre	3
B. Largely Mediterranean				
Spain	Most	*Few in north	Absent	1
Italy	South	Negligible	North	2
Greece	Southeast	Absent	*Most	2
C. Largely Alpine				
Switzerland	Absent	Few in north	Most	1
Austria	"	Absent	All	1
Czecho-Slovakia	"	"	*All	1
Yugo-Slavia	"	"	"	1
Hungary	"	"	*All (Altaic)	1
Turkey	"	"	" "	1
Rumania	Few	Few	*Most	2
Bulgaria	Negligible	*East	"	2
Russia	"	Northwest	Most	2

*An asterisk indicates noteworthy changes in composition since the time of Christ.

The last column in this table is of some interest for those people who are concerned with "Purity of Race." In each of the three sections I have placed first the nations built up essentially of one race. Among Nordic peoples only Scandinavia and perhaps Finland and the little East Baltic States can claim to be essentially "pure Nordic." Britain, Germany and Poland are "half-caste," if this term may be applied to a nation! (However, the two races still live in fairly different habitats.) As for France—the leader in civilisation—she is the most mixed nation in Europe! Spain is the only fairly pure "Mediterranean" Nation. (Portugal is said to have much negro blood, but accurate data are not available.) It is in the centre of Europe that we find the largest series of "pure races." The members of the Old Austrian Empire (1700-1914), who fought among themselves like the famous Kilkenny Cats, were nevertheless all Alpine.

C. *Population Movements*

One of the most significant features of modern populations is their tendency to migrate to regions of greater attraction than their homelands. Thus about one million Chinese were passing annually into Manchuria before the recent warfare there. In earlier days before about 1820 emigration from Europe was not on a large scale. Between 1790 and 1820 the total influx to the United States was about a quarter of a million. Thereafter it was only about 15,000 a year until 1830; but then it increased rather rapidly, especially after the Irish potato famine in 1846. The unsuccessful revolutions in Central Europe about the same time also sent many of the most vigorous peoples to new lands, so that about 1846 the United States was receiving about 200,000 new settlers each year. The migration then increased still further, reaching half a million a year during the 'nineties, and almost a million a year around 1905. The following table shows the nations from which the immigrants came in the period 1871 to 1895. (Figures represent millions.)

British	Irish	German	Italian	Scandin.	Slavs	Others	Total
1.6	1.3	3.1	0.6	1.2	1.0	1.5	10.3

From 1911 to 1915 there was a vast influx of southern and central Europeans who contributed 67 per cent. of the immigrants, while those from northwest Europe were only 18 per cent. New legislation redressed the proportions, so that in 1926 about 42 per cent. were from northwest Europe and only 10 per cent. from the relatively new sources of immigrants in central and southern Europe. Thus we may say that migration into U.S.A. was chiefly from the northern countries of Europe till 1846; then from Germany and central countries during the last decades of the 19th century. In the 20th century the southern Europeans became more and more important migrants to North America, until Americans became alarmed at the influx of illiterate and ignorant settlers from the south and curbed it by law. A large migration of south Europeans to South America still continues. Since 1909 (excluding War years) Brazil has received about 100,000 a year and Argentina 200,000 a year. However, about half of the latter return to Europe when harvests are over in the south. Enough has been written to show that European populations are so large that great numbers are willing to try their fortunes elsewhere. So much for modern conditions.

It has been pointed out by Fleure[5] that even in Medieval times there was such a difference in the possibilities of various European lands that migrations on a considerable scale may have occurred. He has divided the continent into three zones, which I have endeavoured

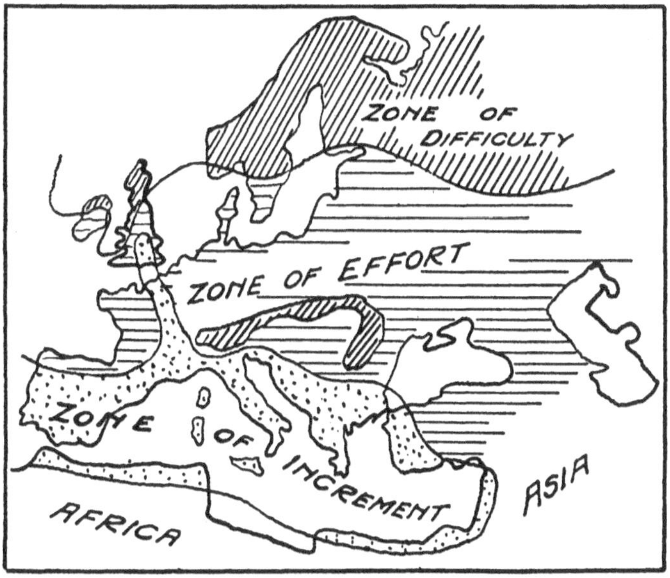

Fig. 23.—Three Zones of Environmental Control, somewhat generalised, and based on Fleure.

to chart in Fig. 23. To the Mediterranean region he gives the name "Zone of Increment," for here it was easy to get returns from human industry. The absence of forests and the attractive climate favoured the development of an early civilisation, where, if there was no great wealth, there was also no great poverty. Food, clothing and shelter are the necessities of civilised life. Less food, less clothing and less shelter are necessary on the warm coasts of the Mediterranean than in the regions to the north. Furthermore, Fleure places an extension of the Zone of Increment along the "Way of Light," as I have indicated.

[5]Fleure, "Human Geography of Western Europe," 1928.

To the region of Plains and Peneplains in Central Europe he gives the name "Zone of Effort." Life developed in small clearings in forests, or on limitless plains where it was necessary to wander. In neither case did community life or a "civic spirit" readily develop. Here the relatively easy olive-wine culture was replaced by the more arduous wheat culture, and even in the Middle Ages large forests still separated the settled regions. Still further to the north lay the "Zone of Difficulty," which contains the Coniferous Forests and the rugged elevated peneplains of the North. (The Alps also come in here.) Here population is very sparse to this day. Hunting and lumber are the main resources—though a new industry, "Tourists," is greatly helping the settlers in the highland areas of this zone.

Right through history there has been a tendency for folk to move from north to south. The north exports men, the south seems in time to engulf and possibly to enfeeble them. The Goths and Lombards were energetic Nordics; yet Ripley states no trace of them remains among the people of North Italy to-day. The Vandals in Algeria, the Goths in Spain, seem also to have vanished in spite of their relatively long control of the southern lands. In the north, however, the Franks and Saxons are still clearly distinguishable from the original inhabitants. No doubt this is in part due to their being nearer their homelands, but Italy is not much further from the Nordic area. On the other hand, in Britain the short, dark westerner (Mediterranean) flocks to the "lower levels" in the cities of the Saxon area. This seems an example of Fleure's "Export of Mountain Folk."

Huntington has pointed out that health conditions vary very strikingly in north and south Europe. In Finland the winters are inclement and deaths are then more numerous. In Italy the summers produce more deaths and the winters are agreeable. On the other hand, France and adjacent lands have on the whole pleasanter climates throughout the year. This will almost certainly lead to migration toward the central western lands from both south and north. Such a migration is of almost national importance in the United States, for in California the proportion of retired Easterners attracted by the equable climate is already large enough to attract notice.

In a study on Race and Culture ("Geographical Review," 1921) I suggested that the zone of civilisation in Europe has "marched northward," possibly as the result of a slight increase of temperature of the order of two or three degrees Fahrenheit. I admit that it is difficult to substantiate this theory, for temperature changes leave less evidence than do rainfall changes. But the following table emphasises the fact that the regions of major power in or near Europe have been placed farther and farther to the north and in general (though not always, *e.g.*, Spain) in cooler and cooler lands, during the last 2,700 years.

TABLE J

EMPIRE AND TEMPERATURE

Century	Nation	Present temperature
B.C. 7th	Assyrian	73° F.
6th	Chaldean	68°
4th	Macedonian	65°
A.D. 0	Roman	62°
7th	Frankish	54°
12th	Holy Roman Empire	50°
16th	Spanish	60°
18th	French	55°
19th	English	50°
20th	Russian (?)	40°

The writer believes that the invention of a simple cooling apparatus which is now on the market may in time revolutionize world settlement. Just as "Central Heating" has made life possible in huge communities in northern U.S.A. and Canada, so "Central Cooling" will for the first time give the white man a chance to live healthily and comfortably in large numbers in the hot, wet and productive tropics. This, of course, has little bearing on Europe, save that it is the absence of hot, wet conditions in the latter continent which has very definitely determined the vigour of its peoples.

D. Present Populations and Powers in Europe

Assuming that every European enjoyed the same standard of living and the same opportunities we could perhaps measure the importance of a nation by its population figures. Since this is not true, population only gives a first approximation to the order of importance of the nations in Europe. Yet since Europeans (like the rest of the world) are slowly being standardised, presumably *totals* of population will become increasingly of importance as time goes on. It is, therefore, necessary to study the present distribution of population with some care. At first the general controls in Europe will be discussed, and the populations of the respective *nations* will engage our attention later.

In the map Fig. 24 the present populations have been divided into six grades according to densities. These are shown in the following table:

TABLE K

POPULATION AND ENVIRONMENT

No.	Density	Examples	Character of Region
I	Under 16 per square mile	A, B	Cold tundra, coniferous forest or steppe lands
II	16-64	C, D, E, F, G	Cold farmlands, coniferous forest, warmer steppe lands, southern mountain land
III	64-128	H, J, K, L, M	Fair pasture and farmlands
IV	128-256	N, O, P, Q, R	Good farmlands, some industrial areas
V	256-512	S, T, U, V, W	Industrial belt, much good farmland
VI	Over 512	X, Y, Z	Almost purely industrial towns

It is clear from the map that the value of the land from the human point of view falls off to the north and east. There is a belt of the poorest kind (Class I) all along the north, which is primarily the result of undue cold. In the southeast the determining factor, as we have seen, is drought. This teaches us the paramount importance of the two great climatic controls of temperature and rainfall. So also the second class of sparsely settled country is found next to Class I and its low density is due to exactly the same causes. There is similar cold country in central Europe on the mountain areas, and so we see long narrow "sparse-lands" due to the Alps, Carpathians and Balkan Mountains (G). In central Spain is another drought area (F) akin to E in the southeast (Fig. 24).

The next section consists almost wholly of good farmlands—where there are few factories. Here are grown large crops of wheat, rye, maize, etc., as we shall see, but it is not the densest farmland—which is largely the region of dairying and lies in wetter country. Of this "moderately settled" country (64 to 128 per square mile) the largest areas occur in Russia (H) and Rumania (I). But much of Italy, Central France and Ireland belong to it.

There now remains the central and western part of Europe, nearly all of which has a population of more than 128 people to the square mile. Here there is usually some manufacturing mingled with the dairying and denser farming which cluster around industrial towns, if the soil is at all suitable. South Britain, most of France, north Germany, Austria and southwest Russia belong to this class.

The really dense population of Europe (over 256 to the square mile) is found in two larger regions (S, T, U) and a number of small

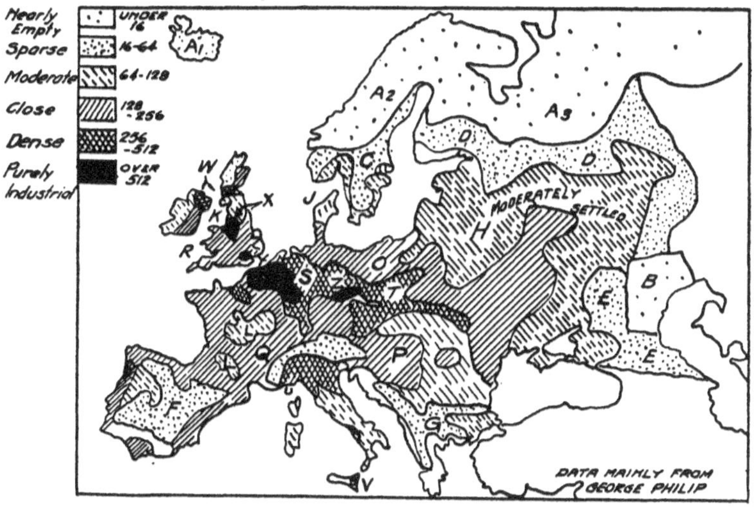

Fig. 24.—The Density of Population per square mile in Europe. Regions A, C, D are too cold, and B, E, F too dry for close settlement. S, T, W, X, Y, Z are densely populated, chiefly through Coal. The rest, H to R with U and V, have notable settlement based chiefly on agriculture.

patches (W, X, Y) scattered through western Europe. Nearly the whole of this population results from certain special geographical conditions which occurred in Europe some 200 million years ago! As is explained in another chapter (p. 65) these places all lie in the Great Coal Belt (Fig. 14) which extends from Glasgow (near W) to the Sea of Azov (near E). In three special regions (X, Y and Z) the industrial towns are clustered so close together that they seem to be growing into three colossal towns—linked by streets of houses, almost continuously. To this development of modern industrial life the world *Conurbation* has been given. One of these occurs in south Lancashire around Manchester and Liverpool. This conurbation (X) has actually been named "Lancaston," and rivals in density (over 512 per square mile) the crowded lands of China and India. Another (Y) appears to be filling up much of Belgium from Lille to Liege. A third (Z) surrounds the Saxon coalfield and includes the cluster of factory towns near Chemnitz.

One other very densely populated area remains to be noted. It is labelled U and occupies the Lombard Plain in North Italy. Here there is no coal, but many rivers flow down to it from the Alps, which have brought large amounts of rich soil into the deep Fold-basin (p. 56) and supply it with plenty of water for irrigation. The north Italians are energetic, frugal people, and the farms are fertile and support the densest farming population in Europe. Elsewhere in Europe only *factory* populations are as dense as that found in this very large area of agricultural population in northern Italy.

The richest regions in Europe as regards the amount of coal available are (1) Northern England, (2) Belgium and East Germany, and (3) Silesia. Political boundaries cut across the two latter regions, so that France, Poland and Bohemia possess outlying parts of two of these great coalfields. It does not always follow that a region with a large coal supply is a large coal producer. For instance, China and Canada produce very little coal yet, though they rank next to the United States as regards available coal supplies. Since the population of European countries seems to be doubled as a result of a good coal supply (as the writer has shown elsewhere), we may close this section with a list of the chief coal producing countries in Europe.

Coal Produced in 1928 (in million tons)

Germany, 317	Britain, 242	France, 56
Poland, 40	Czecho-Slovakia, 36	Russia, 35
Belgium, 27		

There are 26 national units of importance in Europe to-day, starting with Russia with a territory of 1,710,000 square miles and ending with Albania which consists of a mere 11,000 square miles. While we may not agree that total numbers of people determine the power and importance of a nation, yet Providence (or National Evolution) has so arranged it that these 26 countries fall readily into four rather well-marked groups. They are classified in the following table. The last column indicates approximately if the nation is mainly Industrial, Agricultural or Pastoral.

TABLE L

RELATIVE IMPORTANCE OF EUROPEAN NATIONS

		Population in millions	Area in 1000 sq. mi.	Density per sq. mi.	Type
First Rank	1. Russia	125.6	1710	74	Agr.
	2. Germany	63.1	182	349	I and A
	3. Britain	48.2	122	394	Ind.
	4. France	40.7	213	192	A and I
	5. Italy	40.6	118	345	A and I
Second Rank	6. Poland	29.2	150	194	A and I
	7. Spain	22.2	190	116	A and P
	8. Rumania	17.4	122	143	A and P
	9. Czecho-Slovakia	14.4	56	257	I and A
	10. Yugo-Slavia	12.0	100	120	A and P
Third Rank	11. Hungary	8.4	40	210	A and P
	12. Belgium	7.8	11	710	Ind.
	13. Holland	7.4	13	570	A and I
	14. Austria	6.5	32	200	A and P
	15. Greece	6.2	56	110	A and P
	16. Sweden	6.1	173	36	A and P
	17. Portugal	6.0	34	176	A and P
	18. Bulgaria	5.5	42	130	A and P
Fourth Rank	19. Switzerland	3.9	16	244	I and P
	20. Finland	3.5	125	28	A and P
	21. Denmark	3.4	16	212	Agr.
	22. Norway	2.6	125	21	A and P
	23. Lithuania	2.2	32	166	Agr.
	24. Latvia	1.8	25	72	Agr.
	25. Estonia	1.1	23	48	Agr.
	26. Albania	0.8	11	73	Pas.

There can be no doubt as to the paramount importance of the first five nations. In population and in area Russia is easily first. However, its density is rather low, which, as usual, means that the people are rather agricultural than industrial. To a lesser degree the same is true of France (the second in area) which is also rather low (for the first group) in density. Germany, Britain and Italy have rather similar areas, densities and populations. Although Italy is much less of an industrial country, her dense agriculture in the Lombard Downfold (p. 56) has largely balanced her lack of coal.

It is of interest that four out of the five Second Rank Powers have evolved owing to the Great War. They have all four grown immensely in area and population—though none of the four is a rival of a First Rank power. The greatest density is shown by Czecho-Slovakia, which is also the chief industrial power in this group. There is a well-defined drop (from 12 millions to 8.4 millions) between Yugo-Slavia and Hungary. Hence, I have placed Hungary in the Third Rank. Although their total populations are all much the same in this group, a glance at the last column shows that Holland and Belgium differ greatly from the others in density. The former is rich through her colonial possessions, whose products are manufactured largely in Holland. Belgium, of course, has the densest population in Europe—and is traversed by a chain of Coal Towns (from Lille to Liege) which have almost become one gigantic town. Sweden is almost as large as Germany, but her population is so sparse that she falls to Third Rank with Holland and Belgium. In the Fourth Rank are a number of small nations together with two large northern lands, Norway and Finland. The latter share the disabilities of Sweden, but to a greater degree, for they have much larger areas of bare peneplain than has heavily-forested Sweden.

Probably the most important feature of the social groups in Europe is the density, charted in Fig. 24. It has been shown that this distribution depends almost wholly on Environmental Control. If, therefore, the best type of history deals with *social groups* (rather than with personalities), it is difficult to understand why Environmental Control is so little considered by many historians.

CHAPTER VI

THE LANGUAGE FACTOR IN HISTORY

A. Evolution and Migration of Languages

We have seen that while Race cannot be ignored, it was much more important in the very early stages of European History than it is now. The two cultural factors, Religion and Language, however, are still of great importance, though religious differences are luckily not so bitter as they were in the later Middle Ages. Language is perhaps the best "cement" of a nation, though as in all general rules there are some striking exceptions, notably in Switzerland and Belgium. Since language is a human invention, it can readily be transferred, and hence has little relationship to the more fundamental factor of Race. To change the characters of the Race of a nation many generations of merging are required. A change of language can be arrived at in one generation. However, this distinction between Race and Language is still not generally understood by laymen. Probably this has come about, in part, because language is the first "classifier" used in many groups to-day, as, for instance, where the missionary is the pioneer.

The problem can be well illustrated by considering what has happened in Hayti. A visitor to Hayti to-day will find the land governed by negroes, who are clearly of African origin. Their language is difficult to follow, but is found to be a sort of debased French. They have, however, no racial relationship to the French, who, indeed, are hardly found anywhere in Hayti. How can this be explained? If we follow up the history of the negroes we find that they were introduced into Hayti as slaves about 1500. About 1700 the island was handed over to France, primarily because a number of French buccaneers had settled there. In 1804 the negroes rebelled and drove out or massacred all the French settlers. But in the course of a century or so they had lost their own language and learnt a new language from a *few overlords of higher culture who have since vanished.* This sort of thing has happened many times in Europe, notably in the British Isles with regard to the languages Gaelic and Welsh.

A survey of the languages of Europe shows that the vast majority belong to the Aryan group of languages. There are three or four belonging to the Altaic group,[1] one, Basque, has no allies, while Arabic, spoken in some southern ports (as in Malta), is Semitic. The evolution of language is a very controversial question, but in general perhaps it may be said that primitive languages are complicated, have many inflections and usually consist of long words, in which the various concepts concerned are "packed away" in a sort of shorthand, which is not easily interpreted. An early Aryan language is Latin in which *Habe-ba-mus* means "We were having." Here the "have" root is clear, but it is not easy to explain why the particle "*ba*" should indicate past tense, or why "*mus*" should indicate "we." The English phrase "We were having" is *analytical*, with the different ideas more clearly separated. English, of course, still uses many suffixes and prefixes.[2] The Latin language is called "Amalgamating and Suffixing." It is also marked by six "cases" to show position, possession, etc. Some other languages, such as the Wolof of West Africa, go much further and have 24 such "cases" (T. G. Tucker, "History of Language," 1908).

In other languages, such as Altaic, the roots are "glued" together. The phrase "not to be made to love oneself" is in Turkish agglutinated to *sev-in-dir-il-me-mek* (*i.e.*, love-self-compelled-not-to). This is clearly a cumbrous method of speech, compared with the concise *separable* words of the English method. The Semitic method is still further removed from our procedure. The root is usually a group of three consonants, such as Q-T-L (kill). Thus *Qutila* means "he was killed," while the vowels are varied for other forms, *e.g.*, *Yaktula*, "he kills." Such a scheme is called "Internal modification." Basque is still more involved—and like some languages of the American Indians, it even packs the direct and indirect objectives into the word; *e.g.*, *Da-kar-kio-t*, meaning "It-carry-to-him-I." Perhaps we may assume that these languages, somewhat in the order enumerated, represent more and more primitive methods of speech.

Since we are dealing here with the evolution and migration of culture it is most probable that a form of the "Zones and Strata" (or "Age and Area") hypothesis is involved—just as the writer has

[1]A discussion of the relations between various Aryan and Altaic languages will be found in Chapter XXVI.

[2]The writer believes that English is a well-advanced "species" of the Teutonic "genus." But Teutonic is an earlier genus of Aryan than some of the Eastern genera. Conceivably Persian, a similarly well-advanced species, derives from a *later* Aryan "genus" than Teutonic.

shown in his discussion of race migrations, and as Wissler has shown for cultural migrations. By this theory, marginal languages are primitive, while those nearer the centre of dispersion are developed later. It is discussed further on page 109. The zones of language shown in Fig. 25 run as follows: most marginal, Negro; followed by Basque (?), Hamitic, and Semitic; then comes the Aryan group (1) Gaelic type, (2) Brythonic type, (3) Slav type, (4) Persian type.

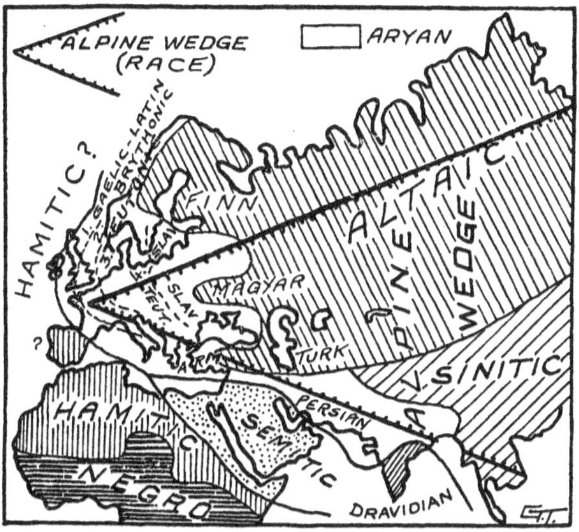

Fig. 25.—A generalised map showing early Zones of Language in and near Europe. The four Belts of Aryan Speech are numbered 1 to 4. Since the race movements have been almost wholly westward, the marginal western languages are on the whole the most primitive. Note how the Alpine Race, within the "Wedge," includes Aryan, Altaic and Sinitic languages.

The relations between Aryan and Altaic are not clear to philologists. But the writer is inclined to think that all the original Alpine folk of *central Asia* probably spoke a common primitive language. From this common tongue the Aryan, Altaic and Sinitic groups of language developed and differentiated—as their speakers moved away from the common cradleland. This is supported by the fact that the *Alpine Race* (Fig. 25) is equally important in each of the three realms of the Aryan, Altaic and Sinitic speeches.

It has been shown by Rhys and Jones that a Hamitic language akin to that of the Berbers and Pharaohs was probably spoken by the early Mediterranean folk in Britain. Seeing that Berbers and early Britons are racially akin this is a very plausible view. Hamitic, of course, antedates Semitic in North Africa. Semitic in turn is being thrust to the south by Aryan languages, while the Altaic to the north are being driven northward and eastward by Russian, an Aryan language. An analogous fact is that Elizabethan and Stuart songs and phrases are more familiar to the settlers in the rugged Appalachians of America than to the people of the English Midlands where the songs originated. Here again we have an illustration of the "Age and Area" Law; that the primitive types are preserved in the margins, *farthest* from their place of origin (Rhys and Jones, "Welsh People," 1908).

B. *The Aryan Languages*

The term Aryan should only be used to describe the group of allied *languages* spoken practically throughout Europe as well as in Armenia, Persia and northern India. Aryan emphatically does not mean "European" nor should it be given any *racial* meaning whatever.[3] The practice of using the term "Aryan" in contrast to "Jewish" (as indulged in by German officials) is plainly ludicrous. Most Jews speak Aryan languages; even Yiddish being emphatically a German dialect written with Hebrew characters. As will appear later, the Jews are racially indistinguishable from the mass of Central Europeans among whom most of them dwell. The term Jew can only be contrasted with Gentile or Christian, while Aryan is to be contrasted with Altaic, and Nordic with Alpine or Mediterranean. One cannot profitably contrast green cheese with cosmic rays!

When the links among Aryan languages are studied, it is easy to see certain groups which are naturally classed together. Others, however, like Albanian, refuse to be readily classified. The classification usually adopted agrees rather well with the marginal arrangement suggested earlier. The two major groups are the *Kentum* and *Satem* groups, so named from the words for "hundred" in Latin and Sanskrit. Furthermore, the Kentum group is again subdivided according to an interchange of the sounds K (or Q) and P. This can be illustrated by the word for "five" in Latin (quinque, the K type) and in Greek (pente, the P type). Meillet in 1908 divided the Indo-European languages in a somewhat different fashion. He adopts a western

[3]The writer pleads guilty to such a use of the word "Aryan" in his earliest publication on language. The term "Aryan-speaking people" is permissible. Some writers prefer the term "Indo-Aryan" as the general term for the language family extending from Ireland to India. (See Meiller, "Dialectes indo-europeens.")

group (Italic, Celtic, Germanic) and an eastern group (Slav, Albanian, Armenian and Persian). Greek, in his opinion, inclines sometimes towards Italic and at others to the Persian type. However, the distribution is much the same as in the other classification. In the following table the first mentioned subdivisions of the Aryan languages are set out so as to demonstrate their geographical position.

TABLE M

SHOWING DISTRIBUTION OF ARYAN LANGUAGES

Early Kentum Type		Later Satem Type	
Largely Medit. Race	Partly Nordic	Largely Alpine	Largely Alpine
Primitive K West Europe	Later P Central Europe	East Europe	S.W. Asia
Gaelic	Welsh	Lithuanian	Sanskrit
100 = Keud 5 = Keug	100 = Kant 5 = Pump	100 = Szimta 5 = Penke	100 = Satem 5 = Panch
Latin	Greek	Slav	Armenian
100 = Kent 5 = Kink (or Pink)	100 = Katon 5 = Pente	100 = Sito 5 = Pietz	100 = 5 = Pese

The geographical order given above is that indicated on the map (Fig. 25). It will appear in later paragraphs that the primitive Mediterraneans (Picts, etc.) in Britain were driven west by the Gaelic-speakers and the latter by the Brythonic speakers, and these in turn by the Teutonic speakers. This leads us to the important conclusion that Race and Language have *both* migrated outwards from Asia—though language has moved faster than race. Gaelic, for instance, is to-day spoken only by a few short, dark dokephs (Mediterraneans) in the Scottish Highlands. We know from the Gaelic Place-Names that Gaelic (or Goidelic) travelled right across Europe, but we are reasonably sure there were no short, dark Mediterraneans in central Europe. Hence it was probably carried westward by Alpines, and then transferred to the people in Scotland by a *few Alpine overlords* who have since entirely vanished. The analogy with the French language in Hayti is rather close (p. 105).

As stated earlier, we can be sure that Gaelic did not evolve among the Scottish Highlanders. It is by no means clear, however, as to who did first speak the primitive Aryan tongue. The word *Wiro* has been coined for this hypothetical tribe. Were the Wiros of Alpine or Nordic race? The study of the original roots of the language seems to indicate that the cradleland was an inland country, and Schrader in 1863 suggested Bactria (north of Persia), where he pictured a primitive pastoral occupation founded on cattle, but not nomad. The cow, horse, pig and goat were known, but each of these have rather different habitats. Some live on the plains, others in the woods, so that the junction of woods and grasslands is indicated.

The names of trees are of little value in the Aryan problem, since they are transposed very readily to similar types. For instance, in Australia native trees are called oak, mahogany, box and turpentine; but they have not the slightest connection, either botanical or ecological, with the trees originally so named. Eels, tortoises and honey bees were also familiar to the early Wiros. Such objects are less easily confused than trees, and perhaps they corroborate the choice of *warmer* sites rather than, for instance, a Baltic site for the cradleland. It may safely be concluded that any area in the vast region between the Volga and the Hindu Kush would fulfil the conditions as deduced from primitive "word-roots."

The writer feels sure that the *ecology* (*i.e.*, distribution) of languages will supply much new data as regards their movements and origins, when it is fully investigated. Indeed, the map in Fig. 25 is a pioneer attempt in that direction. It is seen in the Table that the Kentum languages are marginal, and so may be accepted as earlier than the Satem languages. But a bygone Kentum language, Tocharish, has been discovered in Bactria, whose position is difficult to explain. Probably it is an "inlier" of primitive speech protected by isolation from modification into later forms. Hittite inscriptions in Armenia are being deciphered very slowly, and they seem also to have some affinities with the Kentum languages, which have long since died out in western Asia.

What language did the original Nordics speak? Since the Teutonic-speakers are the most vociferous among the Nordics, it is generally assumed that the Nordics must always have been Aryan-speakers, even if it cannot be proved that they originally used Proto-German!

But the Finnish nation is largely *Nordic in race*, though speaking an Altaic tongue allied to Turkish and Tatar. It has not yet been possible to learn which was the original language. But the ecological evidence would indicate that the group of people furthest from the cradleland would be more primitive, and so have adopted languages more remote from their mother tongue. I can only again refer to the negroes in Hayti and Highlanders in Scotland for parallel examples. This line of evidence would lead us to believe that the blond Teutons of the Baltic learnt Aryan from the much more numerous peoples to the south, *i.e.*, the Alpines. Their close racial allies in Finland presumably did not change their Altaic speech. Alternatively, the Finns may have spoken Aryan and changed to Altaic, or conceivably the

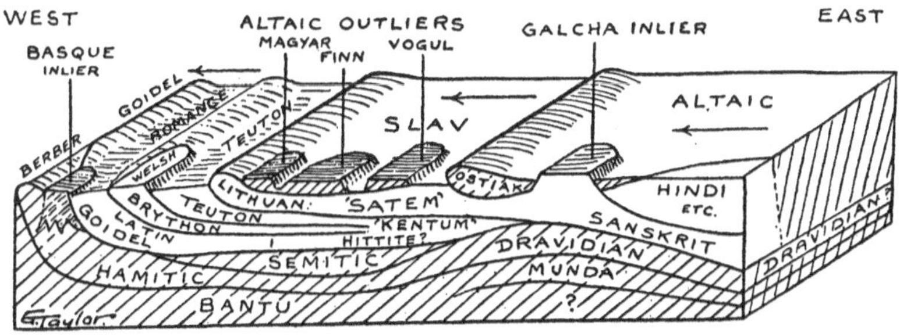

Fig. 25a.—Block Diagram (from west Europe to India) showing the "Zones and Strata Concept" applied to Migrations of Language. The earliest languages are found in the West. Notice the "Inliers" of older speech surrounded by younger speech (*e.g.*, Basque); and the "Outliers" of later speech covering older speech (*e.g.*, Magyar). The Layers are much generalised.

two tribes of Nordics concerned may have learnt entirely different languages before they reached Europe. The problem is interesting and important, but at present seems insoluble.

It is impossible to discuss the Basque problem adequately in a paragraph. The Basque-speakers live in the Pyrenees Mountains on the confines of the Old World, just where one would expect a primitive language to survive. It is perhaps more than a coincidence that the "Alpine Wedge" just reaches to their territory (Fig. 90), but there is no connection between Aryan and Basque. Some have supposed a slight connection with Finnish, others with another relic-speech in the Caucasus called Abkasian. Unfortunately, the Basque-speakers do not agree with any definite racial criteria. They are brakeph in France and dokeph in Spain. It is claimed that they usually have a pointed chin and other minor characteristics of no anthropometrical importance. Their territory once extended over a much wider area, especially in Spain, but it cannot be said that any explanation of the Basque survival meets with general acceptance.[4]

Languages evolve and migrate in a series of "Waves" and are therefore amenable to investigation by a technique which the writer has used very extensively in the last decade. The usual method of showing linguistic relationships by interlocking *circles* is much inferior to the Block-Diagram (Fig. 25a). This demonstrates that the "Zones and Strata" concept[5] applies to languages as it does to other culture-facts. The diagram shows on the surface (which represents Europe and the West of Asia) the linguistic *Zones* much as they appear in the map in Fig. 25. On the front edge of such a diagram we can, however, show the buried *Strata* of bygone languages. Moreover, it enables us to introduce the valuable concepts of "Inliers" and "Outliers" from the technique of the geologist.

The most western speech shown in the diagram is Berber, a Hamitic language spoken in Morocco. It is overlaid by waves of "Kentum" languages (Goidel or Gaelic, Latin, Brython or Welsh, etc.). These in turn have been overwhelmed by the "Satem" languages (Slav, Hindi, etc.). The relations of bygone Sanskrit to the living Lithuanian and Galcha (of the Pamirs) are suggested. Furthermore outlying and isolated areas of Altaic speech such as Magyar are represented. These are examples of later migrations covering the older speech-waves such as Slav, and are well-described as "Outliers." The opposite condition is shown by Basque, which is a very ancient language, almost overwhelmed by later languages, but projecting as a deeprooted "island of antiquity." This is what the geologist means by an "Inlier." Finally the probable relations of the Aryan Waves in Southern Asia to the primitive Dravidian and Munda languages are suggested, though this admittedly is outside the province of the present book.

---

[4]In a recent article the writer suggests that Basque is akin to some Amerind languages because it was carried into Europe by similar early Alpine migrants from Asia. (Journ. Pol. Sci., Toronto, November, 1935.)

[5]This new tool in history, ethnology and linguistics is fully described in the writer's recent paper "Geography the Correlative Science" in the Canadian Journal of Economics and Political Science, November, 1935.

C. British Place-Names as Clues to Migrations

We have learnt that languages change their habitats (and habits) relatively quickly. It is somewhat of a paradox that under certain circumstances ancient words persist longer than almost any other attribute of a people. This is the case with place-names, especially of such natural features as mountains and rivers. Newcomers establish fresh homes, villages and towns, but they cannot alter the landscape, and the first names are apt to persist. The word "Chicago"

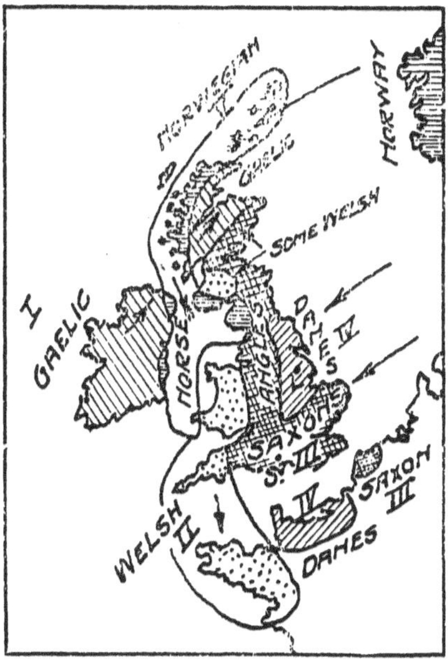

Fig. 26.—The Distribution of Place-Names in the British Isles. The figures show order of migration. *S* is Southampton, *L* is Lincoln. (After Isaac Taylor.)

will tell us that the Amerinds loved "onions," long after the last Indian has vanished from the shores of Lake Michigan. Some of the most careful studies of place-names have been carried out in Britain, and a summary of some of the British data will show how this liaison between etymology and geography can considerably advance our knowledge of early migrations.[6]

In the western part of Wales, notably in the long "finger" of

[6]"Introduction to Survey of English Place-Names," Mawer and Stenton, 1924.

Carnarvon, there still persists a general knowledge of Welsh. It has died out in eastern Wales, while not for many centuries has Welsh been spoken in Midland England. Yet if we examine the place-names, we find almost all of them in Wales are of Welsh origin, while many of the villages, hills and rivers in western England are Welsh. Indeed, we shall see shortly that river-names of Welsh origin are common throughout central and western Europe. No one doubts that this indicates the migration of Welsh-speakers from the Swiss Alps (and probably further east) to the western shores of Britain. Be it noted that it does *not* necessarily indicate the movements of the *same race* or tribe through all this distance. Language, after all, is only a form of culture.

Such words as Walcott and Walton are common in England. The former name is found in Berkshire, Leicester, Norfolk, Lancashire, Warwick and Shropshire. Of the latter there are five in Lancashire and two in Surrey. It is most probable that these names mean "Home of the Welsh." The analogous Bretby and Birkby are Scandinavian names in north England which signify "Home of the Brythons" (*i.e.*, Welsh). So also the word *combe* means "a small valley" in Welsh and is very common in the southwest of England. Cornwall means "the Welsh of Cernwy;" the latter name possibly meaning "Horn" (cf. Latin, *cornus*) and alluding to the shape of Cornwall. The common prefix "Eccles" in the north of England is from the Welsh *Eglwys* (cf. French, *église*) and means that a church was once established there. River-names vary a great deal. For instance, Avon (from *Afon*), and Dover (from Dwfr) both mean "water" in the original Welsh. So also Pen, Barrow, Brig and Mynd are Welsh words for "hill" found in parts of Western Europe as well as in England. Asser, the biographer of Alfred, gives Welsh names of such places as Dorchester (*i.e.*, Durngueir), which may indicate that Welsh was spoken so far east even in A.D. 875. No doubt the British huts and villages—often made of wickerwork plastered with mud—were largely destroyed in the Saxon invasions about A.D. 500. Moreover, it is generally assumed that the British lived in the uplands, while the Saxons preferred the valleys, so that there would be little likelihood of Welsh (*i.e.*, British) names of settlements being perpetuated. (The areas of Welsh place-names are shown in Fig. 26.)

There is a good deal of difference of opinion as to whether there was a large migration of Gaelic-speakers immediately preceding the Welsh-speakers. Some writers believe that the Goidels (Gaelic-speakers) arrived in Ireland by sea from Spain or France. A number of Celtic names belonging to the *older speech Gaelic* are common in England, but Ekwall believes they were brought over by Scandinavians who had spent some time in Ireland. The Gaelic word Ergh (a hut) is common in the north of England, and appears as Er (Winder, Docker) or erg, arg, ark (Mansergh, Arkholme, etc.). Possibly Ireby and Ireton indicate immigrants from Ireland. The views of Rhys, Jones and others are quoted elsewhere.

Turning now to Anglo-Saxon names, they are naturally extremely common in eastern England. Places ending in *-ing* are characteristic of the eastern shores, from York to Sussex, where the Saxon invaders first landed. Such names indicate the settling of a region by a *group* linked to one chief. Thus Reading probably means the group *controlled* by a chief "Reada" rather than consisting merely of his sons. Still later (and found chiefly in the centre of England) are the *-ingham* suffixes, such as Wokingham, which means the "home of Wocc's people." It is most instructive to plot on a map of England the distribution of the names ending in -ing and in -ingham as given by Isaac Taylor. Of the former a score occur in each of the four counties near London (Kent, Sussex, Middlesex, Essex), while only a few extend beyond the Southampton-Lincoln line. On the other hand, the later names in -ingham are densest in the "cuestas and vales" (p. 36) between Dorset and Norfolk. They are much less abundant in the extreme southeast. Hence they denote settlements of *second* generation settlers in England (Fig. 26).

The Danish migrations came later, about A.D. 800, and they overran Lincoln and the adjacent eastern counties. Their favourite suffix was *-by* (a farm, cf. Scottish *byre*). In one small area south of Lincoln there are forty villages all ending in *by*. Derby, Whitby, Grimsby are well-known English towns whose names derive from this period. The Danes penetrated along the rivers, but avoided the Fens to the south. After Alfred defeated them they left western England in peace, and their names rarely cross Watling Street—which runs up the centre of England. The Norwegian attacks were chiefly on the north and west coasts of Britain. Mention may be made of Thurs-o (Thor's Isle), Laxford (Salmon Fiord), Laxey (Salmon Isle), Scaw-Fell, and the names in Cumberland ending in -thwaite. All these are convincing evidence of the settlements of the Northmen of different speech from Saxons or Danes (Fig. 26).

Of particular interest are the Welsh names so common in Lowland

Scotland. Here *Pen* and *Aber* indicate hill and bay, whereas in the Highlands the words are Gaelic (as in Ireland) and are Ben and Inver. Isaac Taylor gives a dozen Scottish examples, including Arbroath, Aberdeen, Abernethy, Aberfel y.[7] In Wales, of course, the prefix is very common. For instance, Barmouth in Wales does not mean the "Bar at the Mouth" of a river, but is *Aber-mawddach*, the estuary or mouth of the Mawddach River. Lovers of the legends of King Arthur will remember that the exploits of his British (*i.e.*, Welsh) knights around A.D. 400 often take place in the Scottish Lowlands, when that region was occupied by the Welsh.

It would be strange if these many invasions on the northern side of the Channel were not paralleled on the south side; and, indeed, the north of France shows much the same nomenclature as England. In Brittany, Welsh (Breton) is still spoken to a greater degree than in Wales. Many Welsh fled there from Great Britain about A.D. 500 and founded Little Britain (Brittany), though doubtless many of their kin were there before. In Brittany we find Avons and Abers, in Normandy are Tofts and Becks (Yvetot from Ivo's Toft; Carqueboeuf from Kirkby) and on the "Saxon Shore" near Calais are numerous other "Frenchified" names. Here Sandgate becomes Sangatte, while Herosfleet is now Harfleur. The Normans of Normandy probably came from Denmark rather than from Norway—since the place-names are more allied to those of the former country (Fig. 26).

D. Place-Names on the Continent

The languages spoken in Europe by the peoples before the Aryan-speakers arrived can only be surmised. We may be reasonably sure that Basque is one such language. Since Gaelic is the most western Aryan language, it is likely to have preserved some indication of the speech which it displaced. Rhys and Jones[8] have studied those features in which it differs from the structure common to other Aryan tongues. They find a score such "idiosyncrasies," all of which agree with the early Hamitic languages of North Africa. If we turn to the place-names we find a very suggestive correlation. Many of the regional names in the west of Europe and *also in North Africa* end in the non-Aryan suffix *itan* or *etan*. Since these names have been adopted by a famous shipping line they are familiar to us: Britannia,

[7] "Words and Places," London, 1896.

[8] "The Welsh People," London 1900, in the Appendix.

Mauretania (Morocco), Lusitania (Portugal), Aquitania (Guienne), and various tribes in Spain known to the ancients as Bastetani, Oretani, Sedetani, Jacutani. All these tribes of pre-Aryan times were short; dark "Mediterraneans." We can hardly escape from the belief that *etan* indicated "place" or home. Isaac Taylor believed this was of Basque origin, but the writer would prefer some undiscovered Hamitic root as the origin of the suffix *etan*. These well-known names, however, do seem to be the earliest place-names in Western Europe.

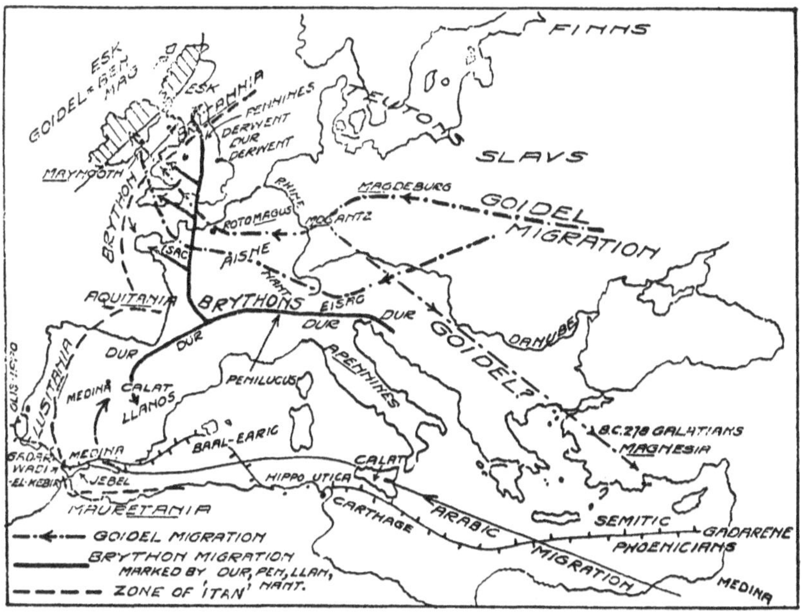

Fig. 27.—Bygone European Migrations traced by Place-Names. Hamitic, Arabic, Brythonic and Goidel migrations are illustrated. (Data from Isaac Taylor.)

Turning to the historic period it is very easy to trace the migrations of folk by their place-names. Two groups of Semitic-speakers dominated the Mediterranean Sea, though at very different periods. Before the Greek Traders spread along all the coasts about B.C. 800 the Phoenicians from Tyre and Sidon had planted their trading posts there. The following etymologies have been suggested for some of their towns. Carthage is Kartha-hada (old town), while Cordoba is perhaps Kartha-baal (Baal's town). Balearic also preserves the name of their god Baal. Utica (in Tunis) is Old Town, as also perhaps Othoca in Sardinia. Cadiz was Gadar, a root familiar to us in the Biblical phrase "Gadarene Swine." Hippo (whence came Saint Augustine) means the fort; and Lisbon is probably Olis-Hippo. The Greeks and Romans had replaced the Phoenician trade and that of their descendants the Carthaginians by about B.C. 100.

About A.D. 700 a new migration of Semitic-speakers spread along the southern Mediterranean coasts. Descendants of these Arabians or Saracens still inhabit the African region,[9] but their names are also spread all over Spain, Sardinia and Sicily. But if they had vanished

Fig. 28.—The spread of the "Bronze Swordsmen" from Hungary, B.C. 1500-1000. They spoke Goidelic in Peake's opinion. The "Iron Swordsmen" were in Kuban about 1100 B.C.

from history as completely as the users of the *-etan* suffix, we should still know a good deal about their movements. Their forts were called Kalat, and this word is preserved in Spain as Calatrava, Alcala, Calatayub (Job's Castle), and in Sicily in similar names (Fig. 27). Medina (a town) is preserved in place-names near Madrid and Cadiz. Wadi means a river, and so we have in Spain *Wadi-el-Kebir*, the great river, now written Guadalquivir. Trafalgar (Taraf-al-ghar) and Gibraltar (Jebel-al-Tarik) are famous names meaning the "Cape of the Cave" and the "Hill of Tarik" respectively. The raids of the Moors into France and the Alps are perhaps indicated by Maurienne near Grenoble and by Monte Moro and Mattmar (Moors' Meadow) near St. Bernard.

[9]Many people in or near the Morocco region still speak the older Hamitic speech (Berber) though Arabic has displaced it in large areas.

E. The Keltic Migrations and Place-Names

The two most western migrations of Aryan speech are the Gaelic (Goidelic) and Welsh (Brythonic). These have considerable resemblance to each other, though the K of Goidelic has changed to the P of Brythonic in many words (p. 109). They are hence classed together as Keltic languages. Peake has attempted to link various other culture-facts with these migrations.[10] He postulates the K speakers as men of the Bronze Age, equipped with leaf-shaped swords, who spread westwards from Hungary about B.C. 1500 (Fig. 28). These swords have been dug up throughout Western Europe except

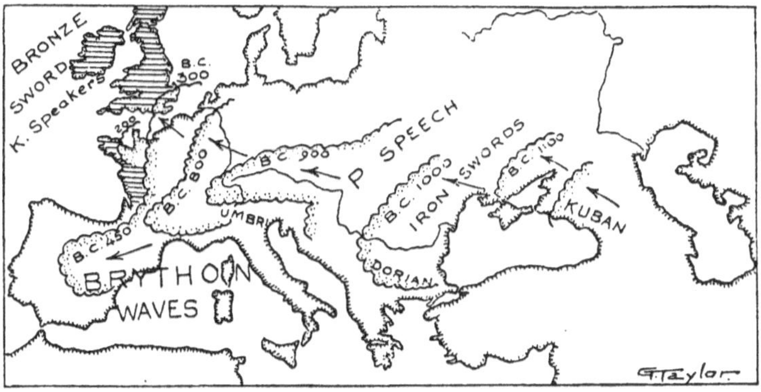

Fig. 29.—The spread of the "Iron Swordsmen" from Kuban, B.C. 1100 to 200. They spoke Brythonic (Welsh) in Peake's opinion, and they drove the Goidel-Speakers ever to the west.

in Spain; and Goidelic place-names have much the same distribution, as we shall see. Somewhere about the eleventh century B.C. the art of making iron swords was developed in the Caucasus region near the Kuban River (Fig. 29). These iron swordsmen drove the bronze swordsmen to the west in turn, displacing them everywhere east of the Seine. Peake equates them with the P speakers or Brythonics and he thinks they developed the Hallstatt Culture in Tyrol. Perhaps about 600 B.C. the Goidelics entered Britain, to be followed about B.C. 200 by the Brythons. It is only in these far western islands and in Brittany that any trace of their speech now survives. We shall now find that the place-names support these conclusions on the continent, as well as they do in Britain.

---

[10]H. Peake, "The Bronze Age and the Keltic World," London.

Following Isaac Taylor, whose general conclusions hold, though later research does not accept some of his illustrations, we can trace the Brythons with ease back to central Europe (Fig. 27). The Welsh word *Avon* (river) is found 24 times in Britain, 14 times in France, 4 times in Spain and 3 times in Italy. Probably the *Av* portion occurs in Sava and Drava in Yugo-Slavia. So also Welsh *Dwr* with the same meaning occurs (as Dour, Derwent, Adour, Tur, Torre, Dur and Thur) some 44 times in Britain, 13 in France, 7 in Spain, 10 in Italy, 5 in Germany. The Welsh *Cefn* (Keven, a ridge) is found in North England for reasons stated earlier, as Cheviot, Chevy Chase, Chevington, and on the continent in the Cevennes of France. The word *Pen* (headland) occurs in Pennine and Apennines, while it is somewhat doubtfully suggested that Penilucus on Lake Geneva is the Welsh Penyllwch, *i.e.*, "Head of the Lake." *Dun* (a fort) can be traced right across Europe, in Swiss Yverdun, Thun, and in Segodun (Belgrade). In France we have Autun, and Lugdun (Lyons), while *Novis-dun* or New Fort is now *Noyon* and Fir-dun (Man's fort) is Verdun. *Llan* (a plain) is perhaps found in the French Landes and Spanish Llanos. Milan (Medio-lan) probably means Middle of the Plain. *Nant* (a valley) is nearly as common near Mont Blanc as in Wales. *Givent* (a plain) appears as Venta, Vendee, Vannes, Veneti, Benevento, in various parts of Western Europe. It is clear from the above examples that Brythonic speakers traversed Europe from Hungary westward. Their name for themselves was *Cymry* (pronounced Kimri), and it is curious that two great tribal names in Europe were Kimbri and Kimmerian. The latter name takes us far east to the Black Sea, but it would probably be too much to assume it is the same word as the Cymry of Wales.

Turning now to the Goidelic names (Fig. 27), they are far less common and are found chiefly in the east of France and in Germany. ***esk***, a root meaning water, perhaps indicates Goidel speech, though Welsh has the analogous *Wysg* (a current). Isaac Taylor gives as other forms Esky, Axe, Ease, Ash, and Is, Uck, Oc, Ose. "Thames," he says, is *Tam-Ese*, *i.e.*, Broadwater. Altogether, he gives a total of 49 places so named in Britain. In France are Isac, Isere, Aisne, Aes, Oise, Ouessant; in Holland Yssel, Scald*is* (Scheldt); in Russia Don*as*ter (Dniester); in Hungary Tib*is*cus (Theiss); in Germany Isen, Isar, Eisach; in Italy Ates*is* (Adige), Issa, Osa, etc.

A better test word is the Goidelic *Magh* (a plain) which is not found in Brythonic. It occurs in 100 names in Ireland such as Maynooth, and is common on the continent as Magdeburg, Mogontiacum (Mainz), Noviomagus (Nimegen), Rotomagus (Rouen), Noiomagus (Nemours). In North Italy 3 similar *Magh* words appear, while the so-called Gauls, who invaded Asia Minor in B.C. 278 and founded Galatia, probably named Magnesia, Magydus, Magaba and many others. The Goidels seem to have occupied Germany in very early days, and their place-names were probably the first bestowed on many places. If this general conclusion be true, it seems unlikely that they reached Ireland (where Erse, a variety of Goidelic, is widely spoken) without first traversing Britain. In the writer's opinion, therefore, we are reasonably safe in assuming an occupation of England by Goidelic-speaking folk for several centuries between B.C. 1000 and B.C. 200. Enough data has been discussed to give the reader some idea of how long-vanished peoples have left some of their vocabulary permanently preserved in the names of places in which they sojourned.

It must be confessed, however, that there is a large gap in our knowledge of the primitive languages of Europe. Let us assume that the Mediterraneans of the west spoke some form of Hamitic and that the Alpine folk of about B.C. 1000 were a "K" group who spoke Goidelic—there still remain all the earlier Alpine waves, such as the Danubians of 2000 B.C. or the Men of Ofnet who may go back possibly to 8000 B.C. (Fig. 20a). The writer believes that some progress will be made even in this problem as the zones of primitive languages are more and more fully studied and arranged in sequence in other parts of the world. For instance, I have stated that Basque has some slight affinities with the language of the American Indians. The latter are in part early Alpine migrations from Asia, and have preserved a whole host of languages which may possibly be akin to those spoken by the first Alpine tribes to enter Europe.

F. Present European Languages

A discussion of present-day nationalities is much the same as a discussion of languages; for, as we have seen, *racial* differences are so little understood that they hardly come into the question of nationality. When a region is described as "irredenta," it usually means that the unredeemed peoples speak a language different from that of their rulers, and akin to that of a neighbouring people. But this alone is not sufficient, else Switzerland would be a cluster of Irredenta areas, while Belgium would split down the middle. Goodwill is even more important than language in preventing "Irredentism." We shall discuss the problems of such minorities in several of the chapters which follow; but it is as well to realise that in Medieval times Irredentism of the type here described was almost the rule rather than the exception. Of course the *troubles* connoted by "Irredentism" are almost wholly of later date.

We may rapidly traverse this question before closing this chapter on Language. In Britain the national limits ignore the small survivals of Welsh, Gaelic and Erse. The majority of folk in the Irish Free State speak and will continue to speak English. The attempt to revive Erse is rather comic, since it seems likely that the language of the Pharaohs is the ancestral tongue of the main short, dark dokeph peoples of Ireland! Erse is as much an introduced tongue as English. In France the Bretons are so few in numbers that only a few extremists desire autonomy. Although their Brythonic tongue differs greatly from the Romance of the French, it has no allies save a steadily disappearing group in Wales. It is different with the Walloons in Belgium and the French in Switzerland. They might conceivably be drawn to their big sister with a common culture and history across the French border. The separations, however, occurred so long ago, indeed before France became a united people, that this union is not likely to take place.

France, Italy and Spain all speak Romance languages—which only differ from each other much as Languedoc differs from Languedoil—though to a greater degree (Fig.91). But this is no basis for union, for when two dialects cease to be mutually intelligible it would appear that they might as well be of different linguistic stocks, so far as their use as a "national cement" is concerned. And, indeed, peace and goodwill are still better "cements" than a common language. Thus the folk in Roussillon in the extreme southeast of France speak Catalan, *i.e.*, the language of Barcelona, rather than that of Marseilles. However, there has been no attempt to join Spain, since the union of Roussillon with France goes back to 1659. Similar considerations affect the nationality of the Galicians in Spain, who speak Portuguese.

Turning now to the central and eastern parts of Europe where the "melting-pot," so to speak, is still bubbling (Fig. 122), we find that every country has "overlaps," which time can smooth out quite as readily as in Switzerland and Belgium if only the powers and peoples concerned are patient. Since these problems will be considered later, it will perhaps serve to embody the salient features relating Language and Nationality in a Table which also indicates many of the larger areas still isolated or unconnected with the main body. In the last column dissatisfied communities within the nation concerned are listed. Such folk as the Germans on the Volga have no reasonable hope of union with Germany, and are not considered in this Table as Irredentist.

TABLE N

SHOWING NATIONS, LANGUAGES AND MAIN IRREDENTA

Language Group	Nation and Main Language	Peaceful Minor Language Groups	Irredenta
Teutonic	English	Gaelic, Welsh, French (in Jersey, etc.)	None
	Ireland (English)	Erse in Connaught	None
	Swedish	Lapps in north	None
	Danish	Icelandic	None
	German	Poles in Mazuria	
	Austria (German)	None	
	Switzerland (German)	French in west Italian in south Romanche in s. east	None
Romance	French	Germans in Alsace Breton, Catalan in S.E. Italians in Corsica	None
	Belgium	South French, North Flemish	None
	Spain	Basque in North Catalan in N.E. Portuguese in Galicia	None
	Italian		Austrian in Trentino Slovenes near Trieste
	Rumania	Germans in centre	Magyars Bulgars in S.E.

TABLE N—*Continued*

Slav	Russian	Kalmuks in S.E. Germans in East Ukrainians in S.W.	Varied groups in Caucasus area
	Polish		Germans in Corridor White Russians in E. Ukrainians in E.
	Czecho-Slovakian	Germans in Bohemia	Magyars in S.
	Yugo-Slavia (Serbish)	Catholic Croats (?)	Magyars in N. Bulgars in S. Albanians in S.
Others	Turkey		Greeks, Bulgars
	Greece		Many Bulgars in East

CHAPTER VII

THE HISTORICAL APPROACH

A. The Time-Space Graph as an Aid to Analysis

It is a truism in scientific research that if a mass of information can be reduced to a graph not only will the data be more clearly presented and the relative importance of various sections of the study be stressed, but a number of new ideas almost invariably develop. Let us see if we can use the graph to help us along these lines. The historian is mainly concerned with *time*; and it is unfortunate that many of the best historians often consider the *space* factor of so little importance that they omit all maps whatsoever from their books. Some of them frankly dread all diagrams as too "objective"! The geographer, on the other hand, is primarily concerned with the "localisation" of his facts; and as he deals in general with the present, the time factor is of little importance. Our present study, however, deals both with geography and history, so that both these factors (time and space) should be adequately employed in our graphs. The development of what I have termed a Time-Space Graph can be easily understood by noting Figs. 30 and 31.

The diagram shown in Fig. 30 differs from the ordinary historical chart in several ways. The Time Scale is, of course, fixed, and is given by distances from the top to bottom of the chart as usual. However, since the later events are "built up" on a foundation of the earlier, it has seemed better to place the beginning of the chart at the bottom and build up a series of *historical strata* in the natural manner, *i.e.*, the lowest are the oldest. The chief difference in the graph, however, lies in the spacing of the vertical columns—representing the respective countries. An attempt has been made to assign more width to the important countries and less width to the less important. Thus Russia is the chief country in Europe and it is accordingly given three columns, which are labelled North Russia, Muscovy (Central Russia), and Ukraine. If the area of a country alone were concerned, Russia would need still wider space, but it is not expedient to show it more than three times as important as any other country. For similar reasons the tiny countries of Holland, Belgium and Switzerland have only one-third of a column each. Norway is large but relatively empty, so that it shares a column with Denmark. The countries are placed side by side as nearly as could be made to agree with their *geographical* position, having regard to the historical associations also. In the extreme right is the region of Barbary. It has been inserted here because it is closely associated with Egypt and Anatolia through much of its history—though it would geographically appear on the left of the diagram.

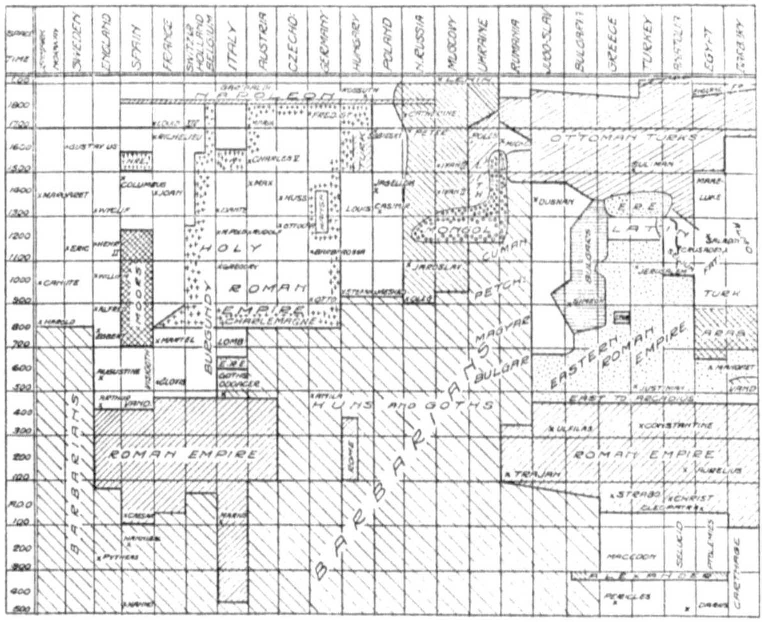

Fig. 30.—A Time-Space Diagram showing "quantitatively" the chief incidents in the history of the European Nations. The dates and countries of some of the more important leaders are indicated.

In the body of the chart are inserted the names of many of the outstanding people in the centuries and countries concerned. But the chief value in the diagram is that it enables a person with little knowledge of European history to realise at a glance what are the *leading sections of the study*. The large region labelled "Barbarians," shows that little is known of the history of many of the nations even so late as 900 A.D. The outstanding importance of the Roman Empires (both western and eastern) is seen in true perspective. As regards Europe as a whole, the graph shows that the Empire based on Constantinople (330 to 1450) is at least as important as that based on Rome, which was destroyed 1000 years earlier. The Holy Roman Empire and the Turkish Invasions are both of paramount importance. Many readers hardly realise the huge time interval between Julius Caesar and William the Conqueror, until they see it graphically expressed in some such form as this. The darkest period of the "Dark Ages" (450-1066) is indeed almost as long as all history since the

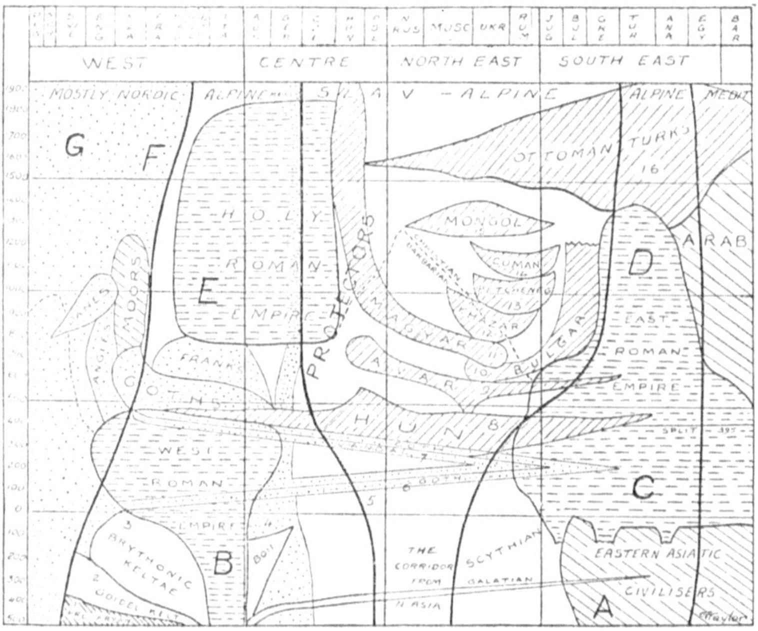

Fig. 31.—A Time-Space Diagram (based on Fig. 30) showing the main migrations across and into Europe. An attempt is made to show the Races involved. The importance of the Alpines (especially the Slav-Alpines) is indicated.

Norman Conquest—yet it is a sealed book to many, perhaps most, educated people.

The second Time-Space graph (Fig. 31) is derived from the one just described. It was obtained by tracing the salient boundaries in a generalised fashion from those given in Fig. 30. The minor details are ignored, and certain racial and cultural data have been added. The resulting graph has been found of great value by the writer and his students as a guide to the Analysis of European History. The continent is divided into four generalised regions, West, Centre, Northeast and Southeast. These agree approximately with the corresponding positions of the countries given in Fig. 30. Thus "*West*" includes the seven left-hand columns from Denmark to Italy. The time scale is divided into periods of five centuries each. Certain of the features shown on Fig. 30 are repeated here to true Time-Space scale, such as the "Western," "Eastern," and "Holy" Roman Empires. But many additional features are plotted, especially in regard to Barbarian invasions.

We may commence our study of Fig. 31 by noting that the historical units charted are classified either by letters (A to F) or by numbers 1 to 16. The former deal with more or less stable governments, the latter with barbarian or hostile migrations. Thus the first "civiliser" in Europe appeared in the southeast long before the time of Christ (bottom right corner). This is labelled A, and in general refers to the Persian, Greek and Macedonian Empires (see Fig. 33 for details). The next great civiliser was Rome far over to the west, labelled B. About the time of Christ, Rome acquired great areas in the southeast (C) which also acted as nuclei of civilisation for several centuries. Then in 395 the Empire was divided, and within a century the Western Empire fell. The Eastern Empire (D) continued, however, for a thousand years and was the civiliser for eastern Europe through all that period.

After several centuries of turmoil the Holy Roman Empire (E) (with its derivatives the Teutonic Knights, etc.) began to "carry the light." Moreover, it acted as a buffer between the turmoil of the East and the quieter regions of the West. The phrase "Protector Belt" has therefore been written across the eastern half of the "Holy Roman Empire." The next national groups are F and G, which are respectively the Romance Nations (France, etc.) and the Nordic Nations of Britain and Scandinavia. This freedom from the terrible Asiatic invasions was probably of much more value to the countries of France and Britain than their "proud Nordic inheritance." For it enabled them to progress more rapidly than could either the "Protector Countries" or the devasted lands to the east (Fig. 2).

We may now consider the units numbered 1 to 16, which illustrate (in time and space) the major migrations within Europe and invasions by non-European peoples. These invasions seem to be clustered in the West in the early centuries and in the east in the later centuries. This, of course, follows from the fact that civilisation appears in the east and gradually passed over to the west of Europe. In other words in Europe Civilisations and Invasions are in general mutually opposed, as all history teaches.[1] In the first five centuries of our graph while Greece and Macedonia were the chief centres of light in Europe, various barbarian groups lived in the west. This period in history has already been referred to in Chapter V. The zones of Pre-Aryan (*i.e.*, Hamitic-speaking) peoples appear as number 1 on the chart. The Goidelic and Brythonic folk together forming the Keltic-speakers are labelled 2 and 3. Some of the former (Galatians) invaded Asia Minor, and the latter (Boii) invaded Italy, and these are indicated in the graph. The great part of northwest Europe was occupied by Nordic Barbarians, and the northeast with Slavs (Sarmatians?) or Scythians.

In the second five centuries (0 to 500 A.D.) we see Rome controlling a large part of Europe. But there are still large groups of Nordics in the northwest and Slavs (and Scythians?) in the east who are still unsubjugated and barbarian. The main feature of this period is the wanderings of the Goths, Vandals and Huns, which presaged the break-up of the Roman Empire. These three nations migrated right across Europe—as the graph (Fig. 31) shows—during these five centuries. The Goths and Vandals started in the west and centre, the Huns appear first in the Great Eastern Corridor from Asia, along which a continuous series of invaders was to follow them.

The third period (A.D. 500 to 1000) shows at first all Europe (except near Constantinople) tormented by the passage of Nordic tribes in the west, and by the first five of the great Asiatic Hordes in the east. Perhaps the main feature of the whole graph is the emphasis it places on these Asiatic invasions "one per century, for eight centuries." (This may indicate a regular recurrence of drought years in the plains of Central Asia.) How could the peace-loving Slav peoples, who inhabited most of this area, make much progress in civilisation with this everlasting march of hostile nomads across their territories? Avar, Bulgar, Magyar, Khazar and Petcheneg all appeared along the north coasts of the Black Sea (Fig. 121). All over-ran the Ukraine, and many despoiled Rumania and Hungary during these centuries. We see the Eastern Roman Empire being gradually squeezed to death between the Bulgars and Slavs to the north and the new menace of the Saracens (Arabs) and Seljuks on

[1]Perhaps the chief exception to this rule was the Moorish invasion of Spain.

the east. Its final ruin, however, is delayed till the Ottoman Turks appear.

The fourth period (1000 to 1500) shows us that the protected Nordic and Romance countries are now becoming stabilised. Gradually the Magyars joined the Poles, Czechs and Germans to form the "Protector Belt" or line of "Marches" against later Asiatic hordes. The Mongol invasion was the most striking feature in the east, as the expulsion of the Moors was in the west. In 1453 the Ottoman Turks crushed the Eastern Roman Empire, which had endured since B.C. 148. In the last period the chief feature is clearly the advance and expulsion of the Ottoman Turks. The Slav countries are at last left in peace, and the Holy Roman Empire in turn ceases to be even a mere political figurehead at the frown of Napoleon Bonaparte.

B. The Dawn of European History

In an earlier chapter dealing with the races of Europe we have learnt something of the beginnings of European history. As early as B.C. 4000 the continent contained Mediterranean tribes in the south and west, Alpine tribes in the centre and east, and Nordic tribes in the north and northeast. So much we can deduce from Archeology. But this is not true history—that begins later, and at very different periods in various regions. History depends mainly on written records; and hence that period where a barbarian people is coming in touch with a civilised people is usually where their history begins. For this reason our most authentic records of early Europe are taken from Egyptian writings. Thus we read that Merneptah of Egypt about 1220 B.C. was attacked by tribes from Sherden, Shekelesh, Teresh and Ekwesh. These are taken to be the Egyptian forms of Sardinian, Sicilian, Tyrsenian (Etruscan) and Achean (Greeks). But long before this date civilised trading-communities had settled in Crete and Greece, which were the first parts of Europe to pass from barbarism to civilisation. Let us then discuss briefly where civilisation began.

It is a paradox of man's progress that the scientific investigation of civilisation should be carried out by the *latest heirs* of civilisation. While there is some doubt as to whether civilisation began in Egypt, Mesopotamia or Turkestan, there is none about the state of Northwestern Europe or North America some four thousand years ago. Both were occupied by tribes whose main interests have been described as getting, begetting and forgetting. Not till about the time of Charlemagne, on the one hand, or Columbus, on the other, was much time devoted to preserving a record in either of these regions.[2] To-day they maintain a strong lead in historical investigation. Yet it naturally results that in the Old World the nearer *borderland areas* have received most attention. Thus Egypt is better known than Mesopotamia and the latter better known than Turkestan. There is a widespread belief that civilisation began in Egypt, because it has been most carefully investigated as regards ancient historic times. However, many archeologists think it likely that the typical Egyptian civilisation derived from Mesopotamia, perhaps when the "Falcon People" reached Egypt about B.C. 3500. So also the writer has long predicted—from his studies of racial migrations and their isopleths—that the Mesopotamian cities will be found to have derived most of their early culture from still earlier cities in the vicinity of Te Anau and Turkestan. We need not pursue these hypotheses further; it is sufficient to know that the earliest European civilisation must have come from the Near East, *i.e.*, from Mesopotamia, Egypt, Palestine and Anatolia. It will clearly repay us to study the geography of these lands to see what are the natural routes into Europe and the environments suitable for the growth of important civilisations.

C. The Build and Climate of the Near East

A glance at the inset map in Fig. 13 shows us that the Near East can be divided into two well-defined areas. Anatolia and Armenia on the north are bounded by the two parallel arcs of the "Alpine Storm" (p. 62). Between them lies the plateau of Anatolia, about 3000 feet high in the west and rising to 6000 feet in the east. This is clearly not a very easy corridor whereby Asiatics may reach the European mainland—though a number of rivers lead down to the west, which render traffic possible. On the other hand, the Levantine coast, comprising Palestine in the south and Syria in the north, is part of the Arabian Shield, and so at first glance would appear to offer an easy line of access between the Mediterranean and Mesopotamia (Fig. 32). It is, in truth, easier to traverse than is Anatolia, but the Shield has been cracked by ancient earthquakes, and large fault-blocks (or graben) have sunk right across the east-west routes. (Indeed, the

[2]The almost undecipherable Maya inscriptions are probably exceptions to this statement.

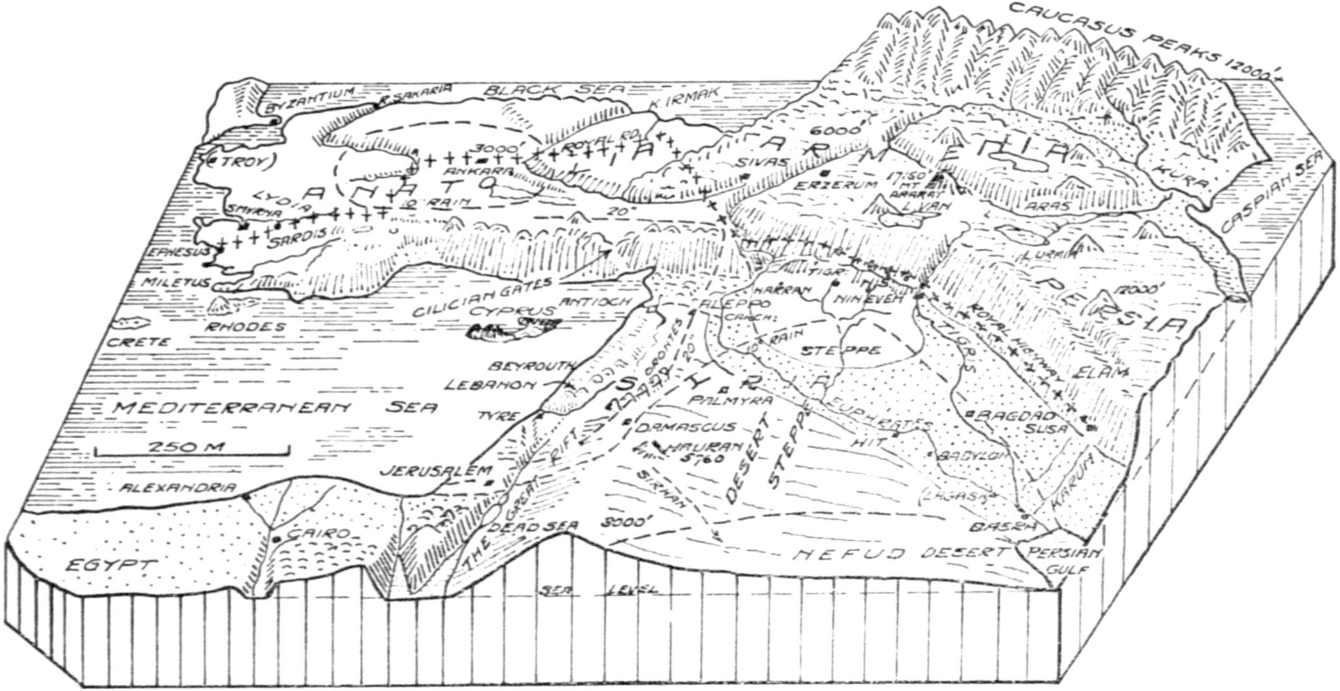

Fig. 32.—Block Diagram of the Near East. The "Great Corridor" extends from the Persian Gulf up the Tigris to Nineveh and thence below the Great Scarp to Antioch. The "Fertile Crescent" is mostly north of the 10 inch rainline.

most notable "Rift" in the world's crust starts probably near Antioch and runs southward between Lebanon and Anti-Lebanon, along the Jordan to the Red Sea and so through Africa to the mouth of the Zambesi. Thus the "Great Rift" is perhaps 3,500 miles long.) Hence the chief corridor from East to West was the "Syrian Gate," in the restricted area between the Anatolian Plateau and the Lebanon Ranges. Here also the Euphrates approaches within 100 miles of the Mediterranean and so offers a natural route southward to its mouth in the Persian Gulf. It is not surprising that Aleppo and Antioch (with Damascus to the south), some of the oldest cities in history, developed in this area.

What is Mesopotamia? As the name implies, it is a region in the "Middle of the Rivers" Euphrates and Tigris. When the Persian region was uplifted during the Alpine Storm several million years ago (p. 54) the area to the southwest was downfolded. We have seen similar examples of build in the origin of the Lombard and Hungarian downfolds. This enormous earth-hollow was naturally filled in with silt and gravel by the two large rivers. Gradually, they carried their debris further south as the head of the Persian Gulf has become silted up. Almost 200 miles of land has been added in this way since the dawn of history, when Lagash (see Fig. 32) was close to the coast. Here in Mesopotamia were offered to primitive man the three necessities of life, abundant rich soil, abundant heat, and abundant water. Other advantages were the absence of tropical jungles and of the many diseases of the hot, rainy countries. The stimulus of a varying climate was present also, for in January Bagdad has an average temperature of 48.8 degrees, although it is 92.5 in August.

There is another very good reason why migrations and trade have always kept to the route under the Great Fold Scarp (Fig. 32). No area in Mesopotamia (except close to the Upper Tigris) receives more than 10 inches of rain. Hence as soon as the rivers are left behind the country changes to desert-steppe or veritable desert. However, these regions are in the heart of the "Winter-Rain Area," hence their scanty rains come in the cool season, when evaporation is low. The winter rains result (in the north) in a considerable growth of grass which supports many nomad tribes. In ancient days before the advent of the motor and aeroplane, it was a very long and hazardous journey of 30 days from Babylon westward to Damascus. Not only nature but man placed innumerable hazards in the path of the ancient trader or migrant. It seems likely, however, that when Palmyra flourished (from B.C. 500 to A.D. 270) the climate was wetter than it is now. Breasted has named the lowland area with over 10 inches of rain the "Fertile Crescent." He includes also the actual irrigated area east of Babylon in this title. These "Plains of Shinar" only, however, contained 8000 square miles of croplands, according to Breasted. It seems certain that the Fertile Crescent was much broader and more productive up till about 200 A.D. Thereafter, if we can accept the data of the Caspian levels (p. 81), conditions changed much for the worse in this part of the world. It is customary to blame the Turk or Mongol for much which the writer feels may fairly be laid to Nature's account.

D. The Peoples of the Near East

One of the most striking facts in European history is that the great civilisations developed along the line of contact between the Alpine and Mediterranean races. This seems to the writer a sufficient comment on the foolish belief held by many to-day that a mixture of races inevitably leads to degeneration. We cannot, of course, be quite sure as to the distribution of races six thousand years ago, but probably it did not differ greatly from what we see to-day. Unfortunately, most early writers used race and language as if they meant much the same thing. They are both "classifiers" of man, but that is usually their only common characteristic. To-day in the region shown in Fig. 33 the heavy black line shows the division between high plateau and arid lowland. To-day the plateaux are largely occupied by tribes of Alpine Race and the lowlands by folk of Mediterranean Race. Probably, however, in the east Persia contained many Nordic or Mediterranean people as it does now. The languages, however, have changed a good deal. To-day Anatolia speaks Turkish (an Altaic tongue) instead of Hittite, Greek or Syrian. Armenia is still largely Aryan, and so is Persia. The Lowlands have remained dominantly Semitic—for Arabic is the chief language throughout. This is a modern equivalent of Hebrew, Phoenician, Akkadian, Aramean, and many other tongues which occur in ancient history.

At the very dawn of history about B.C. 4000, it is difficult to say whether Egypt or Sumeria (at the head of the Persian Gulf) was the leading civilised region. Egypt already had devised a calendar, Sumeria had built many small towns and soon developed cuneiform writing. Moreover, the first great ruler in Egypt was Menes (B.C. 3333); and many archeologists believe that he was descended from the "Falcon Folk" who came from Sumeria or Elam. These Sumerians, in turn, seem to have been a rather broad-head Alpine people who arrived from the north. It seems to the writer likely that their ancestral home was Turkestan. If Pumpelly is right, certain of the settlements near the Caspian (as at Te Anau) far antedated either Sumeria or Egypt. (See Huntington, "Civilisation and Climate," 1924.)

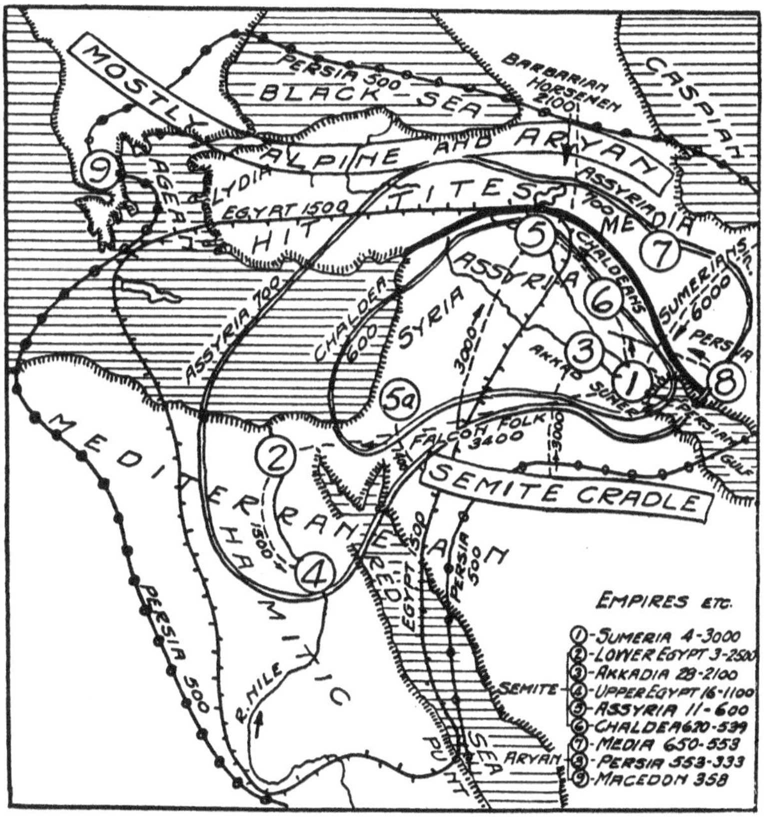

Fig. 33.—Map showing the development of the civilisations (1 to 9) in the Near East before the time of Alexander, B.C. 333. The heavy black line is the Great Scarp, separating tribes mostly Alpine (with Aryan speech) from those of Mediterranean race speaking Semitic.

Thus we may provisionally assume that Alpines (1 in Fig. 33) founded Sumeria and helped to found the "Pyramid Dynasties" in lower Egypt (2 in Fig. 33). The latter country was settled then, as now, by millions of the Mediterranean Race. Meanwhile a group of primitive nomads from North Arabia moved north of Sumeria and founded the Akkadian Empire (3 in Fig. 33) of Babylon. After a time they overcame Sumeria and adopted their method of cuneiform writing, but they kept their own Semitic language. Sargon (2750?) and Hammurabi (2100?) were their greatest rulers. Their empire disintegrated about B.C. 2100 partly owing to constant invasions from the north by barbarian horsemen. Probably these were a mixture of Alpine and Nordic tribes from the regions of South Russia and Siberia. They are known as the Gu horsemen (Fig. 21) and they destroyed the settled life of the primitive folk in Anatolia and the Balkans, just as their descendants later on smashed the Roman Empire.

Meanwhile the Pyramid Builders in Egypt (2) had given place to a sort of Feudal rule by great nobles which lasted about eight centuries (2400-1600).[3] A new dynasty then developed far up the river at Thebes, which created the first great Empire (4) throughout the Near East, about B.C. 1500. Thutmose III extended the realm to Punt in the south, to the Aegean in Europe, and to Assyria on the upper Tigris. More and more Semitic folk had been moving northward into the fertile valley of the Tigris from as far back as B.C. 3000. Others had descended from the Arabian Shield into Palestine about B.C. 1400, where they founded the Hebrew kingdoms of Saul and David about B.C. 1000. About 1100 the worshippers of Ashur (*i.e.*, the Assyrians) spread out from Nineveh (5) and in the seventh century overcame the Syrians to the southwest and the Hittites to the north (Fig. 33). The latter were of Alpine race—not far removed from the Armenians of to-day. Their language was probably an Aryan tongue belonging to the early Q type (p. 109). The fierce Assyrians (5) under Kings like Sennacherib formed the first armies armed with iron weapons. They "came down like a wolf on the fold," and about B.C. 700 conquered lower Egypt and much of Persia, as well as the fruitful "Plains of Shinar" (Fig. 32). Among other larger nations they carried off the northern Hebrews to Nineveh, as is vividly described in the Bible.

Nineveh, however, in 606 fell before the combined attacks of the Chaldeans (6), the last of the great Semitic conquerors, the Medes and Persians. Chaldeans originated somewhere near Lagash and moved north along the foothills of the Great Persian Scarp. They stand

[3]The invasion by Hyksos tribes—possibly Alpines from the northeast—occurred about B.C. 2000.

out amid ancient peoples for their knowledge of science, but were nearly as ruthless as the Assyrians to the nations they conquered. Among these were the Hebrews of Jerusalem, who now followed their northern brothers (the Israelites) into captivity. However, the Chaldean Empire neither spread so far nor endured so long as the Assyrian.

On the Persian Plateau a group of barbarian peoples called the Medes had been growing in power since B.C. 700. They spoke an Aryan tongue and were probably a mixture of Alpine and Nordic race. (The Kurds of that region to-day are of somewhat the same origin.) They had learnt to follow Zoroaster, who was one of the world's greatest religious teachers. These hardy mountaineers conquered a large Persian empire. They helped to destroy Nineveh in 606 and were powerful rivals of the Chaldeans. However, the Medes (7 in Fig. 33) only controlled Persia for a short time, for other Aryan-speaking tribes from Elam under Cyrus the Persian revolted against them in 553. In 546 Cyrus marched into Lydia and conquered Sardis. Here he came into close contact with the Greeks of Europe, who had long founded colonies in and near Lydia. He defeated Belshazzar and his Chaldean army and occupied Babylon in 539. A few years later his son Cambyses conquered Egypt.

Darius the Great (521-485) was probably the most enlightened monarch among all the empires so far enumerated. The Persian Empire now included almost the whole of the region between the Danube, the Nile and the Indus (Fig. 33). It is of special significance to us for two reasons. Firstly, it marks the first world-empire conquered by those Aryan-speakers who were later to over-run the world. Secondly, the Persians actually occupied a large part of the mainland of the Balkans and so came into armed conflict with the developing Grecian civilisations. Few episodes in history are so interesting and important as the challenge offered to the vast Persian Empire by the tiny communities of European Hellas.

E. Environmental Control in the Near East

There is little doubt that Mesopotamia is better suited to a very primitive agricultural people than Egypt. Bagdad has 9 inches of rain a year, while Cairo has only one inch. The Tigris and Euphrates flood in spring and give rise to a great variety of crops through most of the year. The Nile floods in summer, and the chief crops are grown in winter after the soil in the irrigation basins has been thoroughly soaked. Botanists point out that many of our most useful plants seem to have originated near the north of Mesopotamia, while few or none have come from the Nile Valley. Thus the lower silted delta of the rivers (near Ur and Eridu of the Sumerians) was naturally the first to be closely settled by an agricultural people. As we have seen some of these folk migrated to Egypt about B.C. 3400, where no doubt they much improved the primitive methods of agriculture along the Nile.

The key to the history of the Near East is, however, to be found to the north of Mesopotamia. The rich fields of river silt—the Plains of Shinar—do not extend much above the line joining Hit to Bagdad (Fig. 32). Both rivers have rocky valleys in their middle waters; on the Euphrates (Frat) from Hit up to Harran, on the Tigris above Bagdad. Still higher they emerge from immense gorges—where the two rivers plunge down their notches cut in the edge of the 6000 foot plateau. There is very little settlement even to-day alongside either the middle or upper sections, for there is little land suitable for agriculture. Here, however, the winter rainfall produces a good deal of grass, and it has always been the home of a pastoral people. Moreover, the country is much less rugged than the plateaux to the north, and much less dry than the region to the south. Thus one of the world's greatest Corridors skirts the scarp from Nineveh to Antioch (Fig. 32). By this route goods from India and Persia reached the Mediterranean at Antioch. It is the key point of Breasted's "Fertile Crescent," which extends from the Persian Gulf right round to Jerusalem. In the east the fertility of the Crescent is based on the gravels and silts piled into the great Downfold of Mesopotamia. In the north it depends on the winter rains, and the short streams flowing down the scarp. In the west settlement is also dependent on the Mediterranean winds, which in winter give copious rains on the windward side of the Lebanon ranges and lesser rains on Anti-Lebanon just to the east. Damascus depends on the streams which flow from the latter range only to lose themselves in the thirsty sands east of Damascus.

One of the earliest accounts of a journey through the Near East is that of Abraham. His tribe seems to have lived in Sumeria near the ancient Lagash (Fig. 32). They travelled about B.C. 2000 along the Fertile Crescent up the Tigris and then settled at Harran near the Euphrates. From this place Abraham migrated south to Hebron near Jerusalem, and later he went to Egypt, as we read in

Genesis. His later life seems to have been passed in the vicinity of the Dead Sea. We have seen how the Assyrians founded an empire at Nineveh on the Upper Tigris. We may be sure that this empire depended as much on the trade which moved along the Fertile Crescent as on local industries. The rivers here have rather swift currents, which would carry down laden rafts, but whose velocity is too strong for the boats of that day to ascend them. Hence goods were carried northward by land on donkeys—and these necessarily kept to the regions where grass could be found, *i.e.*, near the foothills. West of Nineveh developed many towns and strong posts to guard or control this valuable traffic. Nisibis was such a fortress-town (Fig. 32) later to become the last outpost of the Roman Empire on its eastern front about A.D. 363.

When Tigranes, the great enemy of republican Rome, marched down from his lands near the Black Sea he founded a city here about B.C. 75 and called it Tigranocerta. It also lay on this main Corridor between Nineveh and Aleppo. Harran is still an active settlement, and, as we have seen, goes back 4000 years, though Nineveh and Babylon have vanished. Carchemish was for a time capital of the Hittites, those warlike Alpines of Anatolia who flourished here about the 13th century B.C. The great Corridor crossed the Euphrates only 100 miles from the sea, and trade usually passed directly to Aleppo or its ancestor and so to the coast near Antioch. The road is not easy—though better than farther south, as we have seen. But the Amanus Range must be crossed either by the Bog-tche Pass, used by the Bagdad railway to-day, or by the Syrian Gate nearer Antioch. Both passes are more than a thousand feet above sea level. The rivers Orontes, Jihon and Sihon flowing into the sea near this point are cutting through fairly lately elevated territory. Hence they are marked by "juvenile" valleys (p. 59), which to the north become great canyons in the case of the two latter valleys.

Some of the most interesting people in this area were the Phoenicians. These were of Mediterranean race speaking a Semitic language, and they traded throughout the length and breadth of the Mediterranean about one thousand years before Christ. The older Aegean traders from Crete, the Cycladic islands and Greece had been dispossessed in the north by the Greeks during the period B.C. 1400 to B.C. 1200, while the Phoenicians took their place in the south. Their main centres were on the Levantine coast between Antioch and Palestine. This coast was preferable for peaceful traders either to

Antioch or Palestine. The former was always in danger from the Hittites or Assyrians who could attack it by way of the Syrian Gate. The coast of Palestine was sandy and offered no good harbours. But the coast just below the forests of Lebanon was marked by many small bays and capes and islands which gave abundant shelter for the small ships of those days.

By B.C. 1600 the Phoenicians had several trading centres such as Byblus (near the modern Beyrouth) and Arvad. The latter was on a small island without any resources—and we are told that the towns obtained their water from a spring in the sea by means of a lead cylinder and a leather pipe.[4] Tyre was also an island, while Sidon and Tripolis were similar restricted ports, in general independent of any hinterland or local resources. Some authorities believe that Europe owes her alphabet to an invention of the Phoenicians about 1500 based on Egyptian hieroglyphics. These cities were often attacked by the Egyptians (B.C. 1500) and Assyrians (B.C. 876). They founded many colonies, of which Cadiz is said to date back to B.C. 1100 or so, while Carthage was founded about B.C. 820. Tyre was besieged for thirteen years (585-573) by the Chaldean Nebuchadnezzar, and never regained its great influence. Thereafter Carthage became the centre of Phoenician trade. We shall refer to its destruction by Rome (B.C. 146) in the next chapter.

A description of Anatolia has been left to the end of this chapter. It consists essentially of a plateau 3000 feet high separated by high coastal ranges from the seas both north and south. The structure is typically that of the folds of the "Alpine Storm," as is indicated in the inset in Fig. 13. This inland plateau is protected from rain-bearing winds, so that a large area in the centre has actually less than ten inches of rain (Fig. 32). A number of salty lakes and rather barren grasslands result, which have never given rise to a flourishing civilisation. Nor has any great trading route used this plateau, although Darius about B.C. 500 built a "Royal Highway" from Susa to Sardis through it—as appears on our map (Fig. 32). It offers, however, little obstacle to nomad peoples and has been the source of their invasions from the time of the Hittites down to those of the Turks.

Only one notable gap occurs in the southern wall. This is the Cilician Gate, reaching the sea not far from Antioch, and made use

[4]See "Mediterranean Lands," by M. Newbigin, to whom I owe much in this chapter.

of by the Berlin-Bagdad Railway. For the rest the south coast of Anatolia is a frowning inhospitable region favoured by pirates in the past, and possessing no large towns to this day. On the north the Black Sea coast is not quite so unattractive. Several large rivers, like the Kizil Irmak and the Sakaria, flow down gorges to the sea. The latter is used for the railway from Scutari to Angora (Ankara), the new Turkish capital.

It is, however, on the west coast that our chief interest lies. Here the "grain" (or strike) of the folds runs east-west, so that many fairly wide valleys, such as those of the Meander and Hermos, provide space for agricultural settlement. Here also the rainfall is good, and hence the cities of Lydia and Caria were famed for their luxury. Many Greek colonies were placed on the nearest coasts of Anatolia (map in Fig. 34) about B.C. 1000; for the Aegean Sea is covered with small islands serving as "stepping stones" to Asia. Mytilene, Ephesus and Miletus were all founded by these emigrants from European Greece (Hellas). The Lydians remained at peace with these colonies until about B.C. 685 when the Lydian Gyges attacked and destroyed some of their towns. He was, however, himself defeated by the Cimmerians from the north, who may have been the same tribes as the Cimbri or Keltae. One of his successors was Croesus, who brought all western Anatolia under his rule. Ultimately he made an alliance with the rulers of Babylon and Egypt in a vain effort to curb the expansion of Cyrus the Persian. In 546 the Lydian forces were utterly crushed; and the Greeks now came into direct contact with a great Asiatic conqueror. We may defer the description of this struggle to another chapter.

CHAPTER VIII

THE RISE OF GREECE AND ROME

A. Grecian Migrations and Colonies

In the preceding chapter we have seen that there are some reasons for dating the birth of a Minoan civilisation in Crete before B.C. 3000. This gave rise to no modern nation, though it helped indirectly in the evolution of the wonderful Greek culture, which, in turn, dominated European culture for many centuries. Thus it is in Greece that we commence our study of European nationalities. As heretofore, we must have a fairly clear picture of the environment of the Greek nation before we can adequately explain its early history. The Greeks have a peculiar environment due in turn to the special morphological features of the Aegean region. It is the most "broken up" of any area in Europe; in topographic language it has met with the most "accidents." Structurally, the Aegean is crossed by the two great *mountain arcs* of Europe (Inset, Fig. 13). The main Alpine arc lies a little north, but the Dinaric arc passes from Greece proper (Hellas) across the Aegean Sea to the Greek shores of Anatolia, where it builds up the Taurus Range. This relatively simple structure, however, has been complicated by the tremendous "faulting" in very late geological time, in which two great "graben" were produced. Into these depressed areas the sea flowed and produced the Black Sea in the north and the Aegean Sea in the south. The former is deep and almost free of islands, the Aegean is dotted with innumerable islands, several of which are volcanic. Moreover, two fairly continuous festoons (representing drowned ranges) lead from Athens to Miletus in the south, and from Euboea to Troy in the north (Fig. 34). From the geological history we are prepared to hear that this Aegean region is subject to frequent earthquakes. One Aegean volcanic product—the glassy lava called obsidian—seems to have been an article of extensive commerce even in Neolithic times.

As a result of this shattering of the crust by numerous late faults and subsidences, the land consists of mountain ridges and deep valleys into which marine gulfs penetrate some considerable distance. Four such drowned ranges project like fingers from Southern Greece. A narrow fault-depression forms the Gulf of Corinth. Another helps to isolate the large island of Euboea along the east coast. Under these conditions there could clearly be no ready centralisation of primitive tribes, which were necessarily cooped up in the narrow and isolated valleys. It is easy, therefore, to explain the growth of many independent city-states in Greece. It is not so easy to explain the wonderful "Golden Age" of Greece about B.C. 400—which has probably never had a parallel in world-history.

Fig. 34.—The Birth of the Greek Nation. Barbarians (Acheans and Dorians, probably Alpine in race) about 1200 B.C. overthrew the Mykenean culture (of a Mediterranean Race). The Golden Age of the Greeks was around B.C. 444.

There is still considerable difference of opinion as to the first arrival of the northern tribes who contributed so largely to the Greek nation. Probably they filtered south (at first as small bands) from B.C. 2000 onwards. They found Greece dominated by Cretan sea-traders who had established many towns, first in the islands and finally on the mainland of Greece. Mykenae and Tiryns (see Fig. 34) were their strongest positions, and their cyclopean walls are still standing to excite the wonder of ancients and moderns alike. The most famous of their stations was Troy, founded to command the entrance to the Black Sea, perhaps as early as B.C. 2500. All these Cretan-Aegean-Mykenean peoples were almost certainly Mediterranean in race. Their religion seems to have been of a gloomy character filled with omens and sacrifices, and quite different from that about to spring up in Hellas. They were, however, wonderful artists in pottery and metal work, and the zenith of their culture in Greece—the Mykenean Age—is dated from 1500 to 1200 B.C.

The northern invaders undoubtedly spoke an Aryan tongue, whereas we may suppose that the Mykenean speech was allied to Hamitic. A survey of the movements of the tribes as recorded in Chapter V leads the writer to the conclusion that these invading Greeks were largely Alpine, akin in race to the "Danubian Folk" and "Black Earth Folk" shown in Fig. 21. Yet probably there were many Nordic members in the migration, who may have formed a sort of aristocracy of horsemen among the bulk of pastoral Alpines. Other theories have been put forward by believers in the Nordic fetish. Certainly, however, the invaders were less brunette than the aboriginal Greeks, but this character applies both to Apines and Nordics. The present composition of the Greeks shows very clearly that the nation still consists of two diverse stocks. For instance, Ripley's frequency curve[1] for "head index" demonstrates two "peaks" at 75 and 88 respectively, with very few heads at intervening positions. These head indices are characteristic of Mediterraneans and Alpines, and even after 3000 years the two stocks seem to have kept apart. If the Aryan-speakers had been mainly Nordic the frequency curve would surely have shown a peak near 79, whereas the number of individual indices in the graph is lowest at this point

It is customary to describe the earliest immigrants as Acheans and Dorians. Peake assigns the former to the Q-speakers arriving about 1200 B.C., the latter to P-speakers arriving about 1000 B.C. Other authorities would put the chief migrations earlier. We may accept the theory that climatic change in the semi-arid grasslands of Russia set in movement all the tribes living in the vicinity. They could not move north into the inclement conifer forest or barren tundra, or east into still more arid Asia. The natural movement would be southward in the hope of plundering the wonderful cities of which

[1]"Races of Europe," graph on page 116 dealing with Anatolian Greeks.

rumour had no doubt made report. The build of the Balkans makes the Morava-Vardar Valley the chief corridor of this region (Fig. 13); and this led the migrants directly to the plain of Thessaly just north-east of Hellas.

The speech of early Greece differed in various localities. Thus the Dorians were a conquering tribe who displaced the Aeolians and Ionians on their march southward from Thessaly. They settled in the "Four Fingers" of Hellas around Sparta. Ionians occupied Attica and the region round Athens. The most northerly group in Greece was the Aeolian who occupied northern Hellas. The geographical structure of the whole peninsula led inevitably to the grouping of the settlements about four main City-States. These were Thebes, Athens, Argos and Sparta.

Very early in Greek history, tribes disturbed by the various invasions of Hellas fled across the Aegean Sea and founded settlements in Anatolia. Thus the Aeolians sailed to the Isle of Lesbos and then settled the coast south from Troy (Fig. 34). The Ionians colonised the coast adjoining Lydia with Miletus as their chief town. The Dorian conquerors also moved along the festoon islands in the south of the Aegean Sea and so were led to the southwest coast of Anatolia behind the island of Rhodes. Later in Greek history Greece founded innumerable Trading Posts which were similar to those already established by the Phoenicians on the Mediterranean coasts. Near the mouth of the Vardar they planted thirty towns before 750 B.C., and then they pushed boldly into the Black Sea and founded Sinope and Trebizond about 700. Byzantium seems to have been occupied by the Greeks about 658. The Thracian coasts near the mouth of the Danube soon saw their ships and many little towns sprang up (Fig. 37). Even the Scythians of South Russia were glad to trade their skins and furs for wine and textiles from the south. Olbia and Tanais represent the Odessa and Azov of to-day, and both were well known to the Greeks by 600 B.C. Even earlier did the Greeks seek the warmer lands of Italy. Cumae near Naples and Naxos (below Mt. Etna in Sicily) were frequented by the Greek seamen before 735 B.C. Sybaris and Syracuse, both near the "toe of Italy," rivalled the original Greek cities within a short time of their foundation. By 600 B.C. the Greeks had reached Massilia, and so were the first of the civilised peoples to reach the mouth of the Rhone and the "Way of Light."

B. The Grecian Victories over the Asiatics

Only three aspects of Ancient Greek History can be presented in this brief survey. The first is their victory over the invading Asiatic Hordes under Darius and Xerxes; the second is the Golden Age of Greece, and the third is the first invasion of Asia by Europeans in the forms of the armies of Alexander.

The first attack by the great Persian King Darius (Dar-aya-vush) upon Europe was directed against the Scythians in 510. He marched through Thrace (Bulgaria), crossed the Danube and then invaded the trackless plains of Russia. Everywhere the Scythians retreated before his army—just as their descendants, the Russians, met the invasions of Napoleon. The results were the same in both cases. The environment defeated Darius and he returned quite unsuccessful to Asia.

In 500 B.C. the Ionion cities near Sardis revolted against the Persians, and were aided by Athenian ships. The Greek forces attacked and burnt Sardis, the most western provincial capital of the vast Empire of Darius. During the next five years the huge armies of Darius, helped by navies manned by Phoenicians and Cilicians, conquered all the revolting Greek cities in Asia. In 492 Mardonius sailed to punish Athens, but his fleet was wrecked near Mount Athos and his army halted by wild Thracians; so that his attack was perforce abandoned. Meanwhile Cleisthenes in Greece was reforming the government and giving it that democratic pattern which in no small degree helped Greece later to withstand the Persians.

In 490 a Persian army containing subjects from 36 nations landed north of Marathon. Here Miltiades with ten thousand men attacked them from the steep hillsides and drove 100,000 Persians back to their ships. Darius died shortly after this failure and his son Xerxes (Khsh-ayar-sha) did not resume the attack until 481. His giant army, possibly approaching a million men, was "more fitted to serve as an ethnological museum than as an efficient machine for conquest" (Oman). He also collected 1,200 naval ships from Phoenicia and Egypt. He crossed the Bosporus on a bridge of boats and reached Macedonia without difficulty, where Alexander, the local prince, joined him with all his army. Leonidas, King of Sparta, was given command of the Greek vanguard of 10,000; and his defence of the Pass of Thermopylae with three or four hundred Spartans is one of the bravest deeds in history. For three days he held back the Persians and killed 20,000 of the foe. Though the Persians won, their morale had suffered greatly. However, the army advanced on Athens, which had been evacuated by the Greeks; and Xerxes destroyed most of its buildings. The Greeks resolved to rely on their navy, which was penned up in the Strait of Salamis, just off the Athenian coast. Here again the immobility due to the immense numbers of the Persians resulted in their defeat by one-third the number of Greeks. Xerxes retired in dudgeon to Asia, leaving Mardonius to crush the Greeks with a large army of Persians. At the battle of Plataea (479) near Thebes the heavy-armoured Greeks almost annihilated the forces of Mardonius, and Europe was saved from Asiatic dominion.

Athens had been the leader in saving Greece and so naturally acquired the headship of the Delian League which was instituted to withstand further Persian attacks. Gradually the Athenians gained complete control of the Confederate Navy and levied tribute on almost all the cities of the Aegean coasts and islands. This is the period when Pericles was in control at Athens and now begins the Golden Age of Greece. Pericles came to the front about 462 and he died in 429. He used the revenues of the League to strengthen and to beautify Athens; for the wonderful buildings of the Acropolis and the "Long Walls" connecting Athens with the coast date back to his time. The greatest Greek poets and philosophers flourished in this century, as shown in the adjoining table.

TABLE O

THE GOLDEN AGE OF GREECE, B.C. 500-300

Date	Leaders	Philosophers	Dramatists	Artists	Scientists, etc.
B.C. 500	Cleisthenes Aristides	Pythagoras			
450	Pericles	Anaxagoras Socrates	Aeschylus Euripides	Phidias	Herodotus
400	Philip	Thucydides	Sophocles Aristophanes		Democritus
350	Alexander	Plato Demosthenes		Praxiteles	Xenophon Pythias Aristotle

Possibly the greatest of all was Aristotle, by some called the Father of Science, who was a Macedonian pupil of Plato in his old age. Later

Aristotle became tutor of Alexander of Macedon, and so leads us to the last glimpse we can spare at the early days of Greece.

It is hard to account for the Golden Age, for there has been nothing like it in Greece since Alexander's time. It seems likely the environment played some part in the achievement. Huntington has shown that the climate was wetter, and no doubt the country more productive then; for that great "rain-gauge," the Caspian Sea, was filled 150 feet higher in Alexander's day. Possibly soil erosion has stripped many fertile slopes since then. The healthy open-air life favoured by the Mediterranean climate no doubt helped to develop the arts and crafts of the age. The rugged character of the country was not favourable to the plans of leaders who hoped to deflect the energies of the Greeks into military conquests over wide areas. The democratic laws of Cleisthenes and the example of Aristides, Pericles and other noble Greeks played a part. Probably the chief factor was the irresistible outburst of national pride when Greece drove back the hitherto invulnerable Persians. We see something of the same kind in England after the Armada was defeated; and Russia in these next few decades may be expected for the same reason to flower as never before.

The campaigns of Alexander the Great do not greatly concern us. He had been well-trained by his father, Philip of Macedon, who was the best military leader up to that time. Philip united with Macedon all the coastal plains around the north and northwest of the Aegean Sea. He died in 336 and his son and successor Alexander marched into Asia in B.C. 334. He naturally made the destruction of the Persian naval posts his first aim. Tyre and Sidon fell in 333 and then Alexander wandered through the length and breadth of the former Persian Empire for nine years. In 330 Darius was killed near the Caspian Sea, while Alexander himself died in 323 at the early age of thirty-three. His great Empire—the first won by a European monarch—fell to pieces at once.

Two Macedonian generals founded the Seleucid Empire in Persia and the Ptolemaic Empire in Egypt. Each of these was based on a much more homogeneous environment than Alexander's patchwork of nations, which extended from the Adriatic to the Indus and the Nile; and accordingly they lasted for some centuries. In Macedon itself Antigonus and his descendants occupied the throne till 168. They espoused the cause of the Carthaginians through dread of the growing power of Rome, while other groups of Greek cities allied themselves with Rome against Macedon. In B.C. 146 the natural result followed and Greece became a province of the Roman Empire. Greek culture was soon carried far and wide through all the known world. As we shall see later its focus was no longer Hellas, but Byzantium, far away to the east at the entrance to the Black Sea. Our study of the development of European Nations now passes from the eastern to the central peninsula of the Mediterranean—to Italy and Rome.

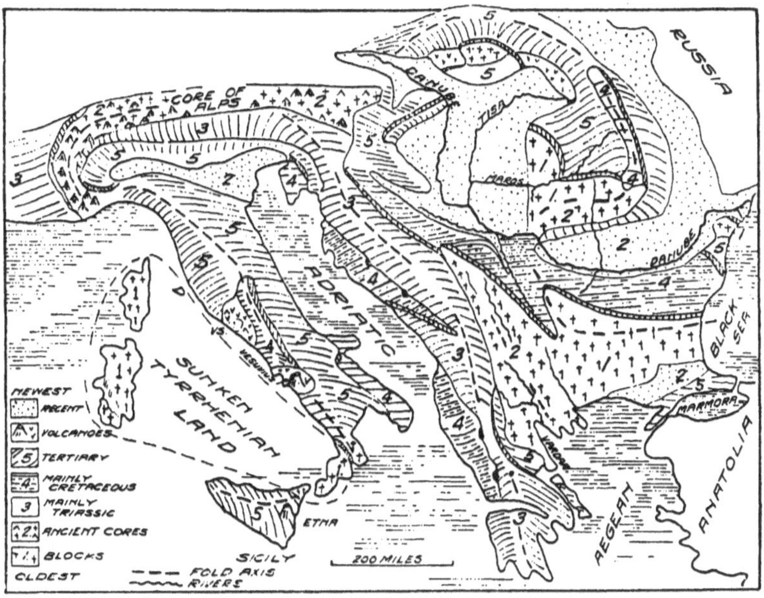

Fig. 35.—Mantle-Map of Italy and the Balkans. The Young Fold-Mountains are mostly built of Tertiary rocks (5) in Italy, of Tertiary and Cretaceous (4) in the Carpathians, of Triassic (3) in the Dinarics, and of still older rocks in the Alps.

C. *The Evolution of the Italian Peninsula*

In the chapter on the Structure of Europe the uplift of the Alps in late Tertiary times was discussed (p. 54). In another section it was pointed out that very extensive folds would naturally affect considerable depths of the earth's crust (p. 55). Obviously, less striking folds would only buckle up the youngest deposits. The southeast of Europe offers examples of all kinds of earth-folds during the "Alpine Storm." The Italian environment depends very notably on this period of great earth movement, almost all of which occurred within the last 10 million years. In Fig. 35 is shown in a simplified form the relation of Geology to Build in this part of Europe.[2] The Swiss Alps were the largest folds and involved more formations than did the other fold mountains. Thus later erosion in the Swiss Alps has exposed to view an extensive *core of complex ancient rocks* labelled "2" in Fig. 35. All sorts of granites, gneisses, marbles, schists, quartzites, etc., are involved in this core, and this complexity accounts for the bold, rugged

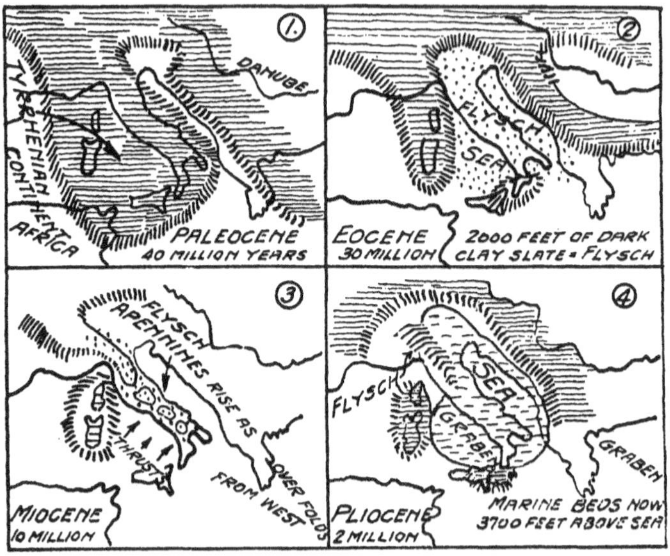

Fig. 36.—The Evolution of the Italian Peninsula. At the beginning of Tertiary times (1) an ancient land occupied the Italian area. In the Eocene the Alps began to form (2), and Peninsular Italy sank deep below the sea. In the Miocene the Flysch marine deposits became land (3) as the early Apennines. Most of this sank in the Pliocene (4), but was later uplifted. Much Faulting has occurred since Pliocene times.

and resistant character of the mountains in most of Switzerland. The scenery is indeed an expression of this type of mountain-building. Very different is the appearance of the Italian Apennines. They lack most of the ancient rocks of the Core of the Alps and their scenery in consequence lacks much of the grandeur of the Alps. The Italian backbone is built up of weak and relatively late Tertiary rocks, which readily break down under erosion—forming a slippery clay soil of little value for plant food.

[2]The best Geological Map of Europe is that by Beyschlag and Schriel, 1925, (Borntraeger) Berlin.

The four small maps in Fig. 36 are attempts to reproduce the geography of Italy at four important epochs in the evolution of the peninsula. At the dawn of Tertiary times (Fig. 36 at 1) the region was occupied by a fairly resistant block of ancient rocks, known as "Tyrrhenia." When the Alpine Storm began much of this block sank beneath the Eocene Sea as shown in Fig. 36 at 2. Dark clays covered the sea-floor forming later a soft slate called Flysch. In Miocene times this area of Flysch was elevated above the sea and formed a Dinaric Arc parallel to that across the Adriatic Sea. These early Apennines, however, did not survive, except in the North. The central and southern parts again sank beneath the sea and in Pliocene times they were covered with extensive marine deposits. The most striking final crustal movements in Europe were those which raised these marine beds to heights of 3,700 feet above the sea, thus forming the southern Apennines. These are consequently much later in origin than the northern Apennines—which possibly antedated them by ten million years. It seems likely that subsidences accompanied these up-folds, and the marked earthquake and volcanic phenomena in the area between Sicily and Anatolia show that mountain-building is by no means complete in Southeast Europe.

Such then are the main features of the Build of Italy. It is clear that a region of such late uplift cannot offer very large areas capable of close settlement—except for the large Lombard Plain in the north (Fig. 100). Here the downfold between the Alps and the Apennines is being rapidly filled with silt by the rivers engaged in eroding the steep slopes of the Alps. Its structure is considered in more detail in the chapter dealing with medieval Italy (p. 353). Rome developed on the fertile volcanic soils of the lavas from the Alban Hills. So also Vesuvius enriched with its ashes the plains around Naples, where the early Greek colonies near Cumae were planted. The "toe" of Italy is an ancient block which has survived from the submerged Tyrrhenian Land. It is akin in build to the large islands of Sardinia and Corsica. Very different is Sicily—where late Tertiary beds build up the main mass of the island, though the giant cone of Etna with its vast lava slopes has covered a large part of eastern Sicily.

On the east coast of Italy fragments of older limestone build up the "heel" and the promontory of Gargano. Structurally they belong to the Balkans, as a glance at Fig. 35 will show. The "young" east coast of the Peninsula, made in recent formations, offers few harbours for man's use. Even in ancient times the Dalmatian (Balkan) coast of the Adriatic was much frequented by ships for it has many ports. Most of the coastal plains, the larger rivers and the better ports are along the western coast of Italy. Accordingly right from the times of the Greek Colonies (about B.C. 700) the active portions of Italy have always been the south and west coasts, rather than the east. As Newbigin points out, the remarkably late evolution of the Southern Apennines has produced several peculiar features in the environment. The soft Pliocene marine shales are easily weathered and tend to slip down over the older rocks beneath. Landslides are common on these surfaces—which the Italians call "frane." The rivers become overladen with the silt which is often deposited in the complicated fault-valleys of southern Italy. Hence marshes and ill-drained plains are produced in which mosquito larvae flourish. Thus the prevalence of malaria in this part of the world can be directly ascribed to the special type of topography.

D. The Rise of the Roman Nation

In prehistoric times the peninsula was probably entirely occupied by the Mediterranean tribes, who still survive in the south and in isolated districts near Lucca and Genoa. The early history of Rome closely parallels that of Greece; for we have evidence of two major migrations of northern peoples corresponding to the Acheans and Dorians in the more eastern peninsula. The first Q-speakers may have entered about 1100 B.C. and then introduced the Latin speech. In the writer's opinion this migration was composed of Alpine tribes. Later again a second migration advanced into Italy introducing certain P dialects of which Umbrian seems to have been a representative. The place of the Mykeneans is taken in Italy by the Etruscans, whose race history is obscure. But there is some evidence for the Etruscan culture being in part derived from Mykene or Crete. Even less is known of the obscure tribe of Latins who lived south of the Tiber. They built a small fort on the Palatine Hill, and held a market for their grains and hides on the river plain below the fort. From such crude beginnings sprang the Roman Empire. Probably Rome was ruled by Etruscan Kings until about B.C. 500, but the Latins rebelled about that date and appointed their own rulers.

The early tribal history lends support to the view that Rome was established mainly by three groups. The Samnites (probably of Alpine race) from the hills combined with the Ligurians of the coast (probably Mediterraneans) and with the Etruscan overlords to form the Roman of early days (Fig. 37).

Rome had many enemies during her first years. To the north the Etruscans were kept busy warding off the attacks of the Gauls who had long held possession of the Lombard Plain. The latter were probably Alpines, speaking Aryan of the Q type, and therefore not very unlike the Latins themselves. To the south of Rome were the Greek colonies around Cumae of which mention has been made. Still farther to the south in Sicily were many strong trading posts of the Carthaginians, who had been carrying the Phoenician flag for several

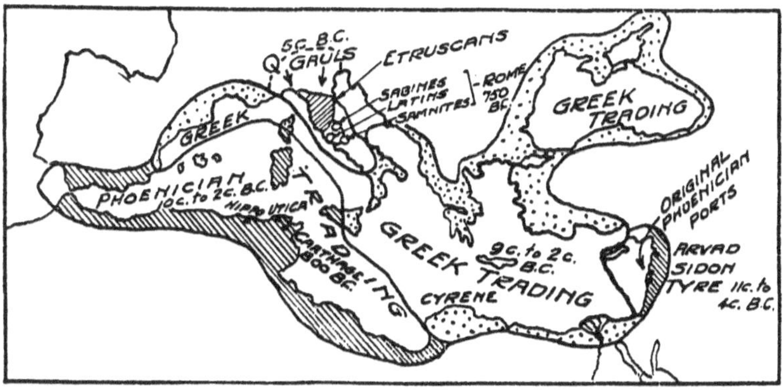

Fig. 37.—Settlements in the Mediterranean, when the Etruscans gave way to the Romans, about the sixth century B.C.

centuries into the western Mediterranean. Luckily for Rome the Greek fleet from Syracuse defeated the Etruscans about 474 B.C., and the latter never regained their former power. In 382 the Gauls ravaged the Etruscan lands and also captured Rome, but they retreated to the Lombard Plain on payment of a heavy indemnity. After many struggles with rival Italian tribes, Rome emerged mistress of Central Italy. This was about the same time as the Macedonian Empire was flourishing in Greece (p. 148). About 280, Pyrrhus of Epirus (in Greece) attempted to weld all the Greek cities in southern Italy into one strong empire. He would probably have succeeded if the Romans had not been assisted by their future deadly enemies, the Carthaginians. However, by B.C. 275 Rome was victorious against the Greek cities and controlled all the peninsula except the northern plains held by the Gauls (Fig. 38).

It is an interesting problem to try and understand why the Greeks never developed a real Empire, while the Romans with much the same origin and not very different environment founded one of the most stable Empires. Many diverse factors contributed, and it is hard to pick out the most important. It is customary to say that Rome held her Empire by her Roads. These were certainly major features which were almost lacking in Greece. But Rome is not particularly well placed to be the focus of a great Empire. Although Italy looks like a great bridge extended from Central Europe to Africa, very little "through traffic" used it for that purpose. The Alps effectively blocked most communications in the early days. The Apennines filled up a large part of the Peninsula, while the settled region in North Africa was and is a mere fringe (opposite Italy and Sicily) which soon gives place to the great Sahara. What is nearer the truth is that Rome learnt through necessity to form an efficient army and navy, and then found it more advantageous to conquer foreign "granaries" than to beat the sword into a ploughshare and settle down to grow food for her soldiers in Italy. We may examine the map (Fig. 38) to see how the Empire gradually absorbed all the coasts of the Mediterranean and their hinterlands.

It was clear that Rome and Carthage must come into conflict as soon as their territories touched in Sicily. In 264 the First Punic (*i.e.*, Phoenician) War began; and by 238 Rome had crippled Carthage and obtained control of the islands of Sicily, Sardinia and Corsica. In the Second Punic War Hannibal invaded North Italy, crossing the Alps in 218 with an army of some 40,000 Carthaginians. For a time he was victorious, and was joined by many of the recently conquered subjects of Rome. He spent 15 years in Italy vainly endeavouring to wear down the Roman resistance. At last he was recalled to Carthage and met defeat at the hands of Scipio in 203. After a lengthy period of peace the Third Punic War broke out. The city of Carthage was razed to the ground and her widespread Empire destroyed in B.C. 146. These three great wars resulted in large additions to the Roman Empire. After the First War the Gauls in northern Italy were subjugated, and their country became the province of Cis-Alpine Gaul. (See (1) in Fig. 38.) They were thus the first barbarians to be incorporated in the Empire. After the second war the Carthaginian colonies in Spain were occupied by Rome (2 in Fig. 38). These two provinces contained rich agricultural and mineral lands of great importance to the young republic. Behind Carthage there were notable

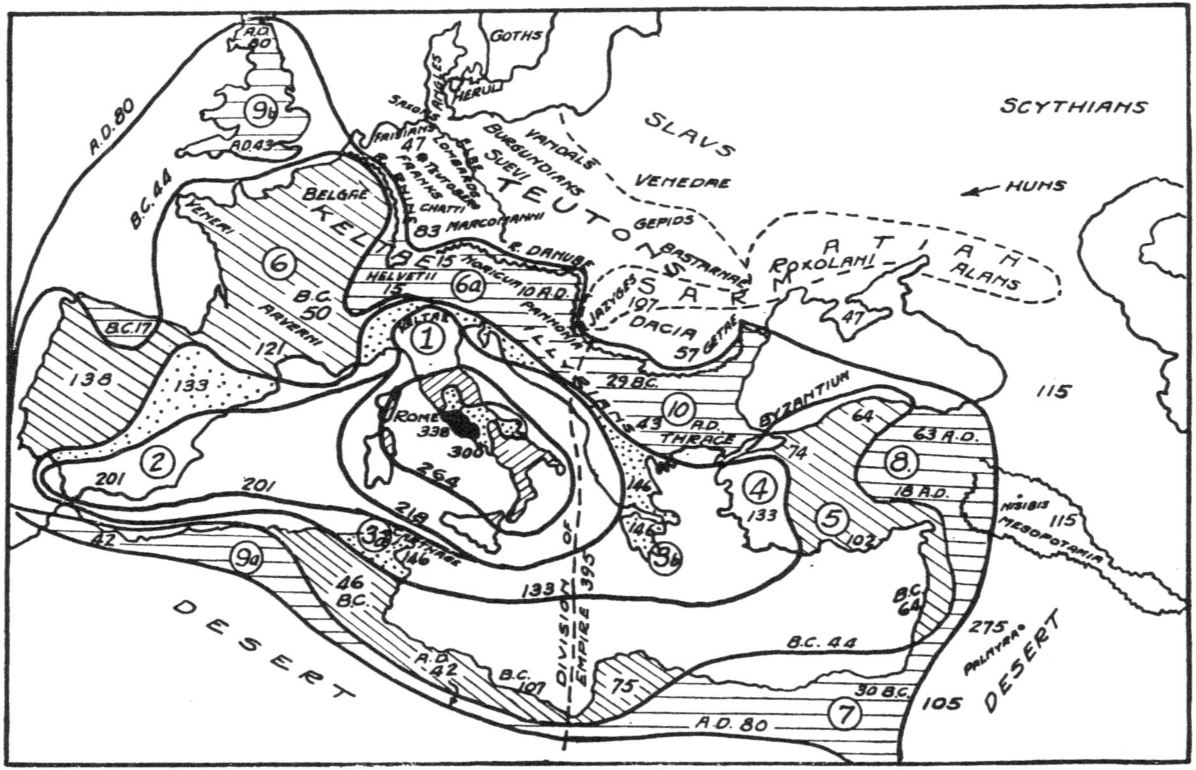

Fig. 38.—The Growth of the Roman Empire to its zenith about A.D. 100. The figures show the order of the conquests. The chief tribes involved in the later Folk-Wandering are indicated in their early habitats.

grain-growing areas which fell into the hands of Rome in B.C. 146. Hence by this time the Romans controlled most of the shores of the Western Mediterranean.[3]

Meanwhile Rome had also been expanding in the Eastern Mediterranean. Macedonia and Asia Minor joined in the Punic Wars on the side of the Carthaginians, and shared in their defeats. Epirus was acquired in 168, Macedonia in 146, while southern Greece was "protected," but became officially a province somewhat later. The Greek Coast of Anatolia (4 in Fig. 38) was acquired in 133. By this date Rome had therefore conquered the western provinces of Alexander's former widespread territories. It is stated that these Punic Wars cost the Romans 300,000 men and practically exterminated the agricultural class among the Romans. Their places were supplied by the vast hordes of slaves which Rome gained in her wars on less powerful nations.

A most important event in Roman History followed shortly after this great expansion of Rome. The restless barbarians in central Europe migrated southwards towards the rich lands of the Mediterranean at the close of the second century B.C. The main tribes concerned were the Teutons and the Cimbri. The former were probably Nordics, the latter partly Nordic and partly Alpine. The Cimbri, as we have learnt (p. 119), were probably representatives of the Brythonic Kelts, who now imitated their cousins, those Goidelic (?) Kelts who had attacked Rome under Brennus in 390. About B.C. 113 the Cimbri were ravaging the region of the Danube and they defeated a Roman army near Graz. They then moved northwest across the Rhine and passed into Gaul, where they again defeated the Romans in several large battles. In 103 they joined the tribes of the Teutons near Rouen and agreed to cooperate in an attack on Rome. The Teutons marched down the Rhone Graben, but were met by Marius and crushed in 102 at Aix in Provence. Marius then hurried east and met the Cimbri, who had entered Italy by way of the Brenner Pass. Another overwhelming victory at Vercellae (east of Turin) removed all fear of the invasion of Italy for several centuries. About this time also Phrygia and Cilicia in Anatolia were added to the Empire (Fig. 38).

Meanwhile the fleet of the Romans had been suffered to deteriorate and the pirates of the Cilician coast took advantage of Rome's neglect

[3]See Chapter XXVI for a further study of geographical control in the Mediterranean area.

of her navy. Pompey was given a strong fleet and soon crushed the pirates. In the next few years he conquered Mithridates and other powerful Anatolian Kings, and in B.C. 64 set up a "court" in Antioch, the key of the East (p. 133). Within a year or two Syria was annexed, and the Roman rule was also extended right across Anatolia to the Black Sea. Julius Caesar was given command of Gaul in B.C. 54. In this region the Germans had recently crossed the Rhine into Gaul and had defeated the Sequani (Fig. 85). The Helvetii also migrated westward from Switzerland, but were crushed by Caesar in the "Gateway to Paris" near Autun. Then he attacked the Germans in Alsace and drove them out of Gaul. The Belgae and allied tribes in the north were also conquered and sold into slavery. Similar victories followed his marches into Brittany against the Veneti. In 55 and 54 he made two punitive raids into England, but was driven back by the Britons under Cassivellaunus. Caesar made several crossings of the Rhine, but no permanent territory was held beyond that river. The last of the Gauls to surrender were the Arverni who lived in the rugged highlands of the Auvergne (Fig. 90). They were besieged in Gergovia and Alesia and finally subdued by B.C. 51. Caesar was murdered in B.C. 44 on the eve of a journey to extend the Roman Empire into Mesopotamia.

E. The Zenith of Roman Power

At the death of Caesar, Rome ruled all the region characterised by a Mediterranean Climate, save Egypt. Here the Ptolemies, successors of Alexander, still reigned as allies of Rome. Rome had also penetrated up the "Way of Light" and conquered the whole of Gaul. She here left her familiar Mediterranean environment for a region marked by the forests of the wetter northwest (p. 75). But Rome was not to expand very much further. Many of her non-Mediterranean provinces were of the nature of marches, *i.e.*, outposts against the turbulent bands of Barbarians to the north and northeast. However, some regions were relatively civilised—as in the case of Noricum (Tyrol, (6a) in Fig. 38) which traded considerably with Rome. However, in B.C. 16 they attacked the Romans and after their defeat were incorporated in the Empire. The Danube was soon reached and it was gradually realised that this river was the best limit to the extension of the Roman arms.

In the north Drusus occupied the coastlands as far as the Elbe, but the Germans rebelled against the new Roman taxes and in A.D.

9 three Roman legions were destroyed in the dense Teutoberg Forest just north of Osnabruck. Thereafter for many years no attempt was made to conquer the lands west of the Rhine—and they never became "Romanised." This battle of Teutoberg is of immense importance in our study of nationality—for it led to the main subdivisions of the Nordic and Alpine peoples of northern Europe. Those west of the Rhine mostly adopted Roman ideas and Romance languages; those east of the Rhine kept their Teutonic laws and languages. Rome was stronger than Race or environment. The northwest boundary of the rugged centre of Europe runs from Troyes in France to Magdeburg in Germany. This is approximately the racial boundary also; the Nordics living in the northern lowlands, the Alpines in the southern highlands. But the Rhine runs at right angles across these boundaries. Rome made the Rhine her frontier; and language, culture and nationality followed Rome.

Between the Rhine and the head of the Danube was a region with no well-defined natural frontier, and here the Romans built a remarkable line of defence called the *Limes Germanicus* (Fig. 70). This forested region (now Wurttemberg) between the rivers was only sparsely settled by Germans, and in A.D. 83 Domitian placed a line of blockhouses along a road from Strasburg to Ulm. Hadrian built a palisade between the two rivers; and a little later a mound and ditch was placed east of the early palisade. This ended to the south in a stone barrier, now known as the "Devil's Wall." This ditch and wall protected the Empire for nearly two centuries. Similar walls were built across northern England by Hadrian and across the Dobruja (near the Danube mouth) by Trajan.

Southern Britain had become familiar with Roman culture through intercourse with the tribes of Gaul long before A.D. 43. In that year Claudius invaded Britain with a large army and easily occupied the Lowlands. The rugged "Oldermass" (p. 35) was occupied by fierce tribes who resisted Rome for thirty years. The discussion of Roman Britain is given in some detail in a later chapter, but by A.D. 80 most of England was under Roman control (Fig. 5).

During the same period Thrace was joined to the Empire. For a time the Thracian Kings were allies of Rome, but in the reign of Vespasian (A.D. 43) their region was made a Roman Province. The Thracians spoke a language akin to Greek, and seem to have been originally mainly of Mediterranean race. They practised tattooing and worshipped Dionysus; and it is stated that some survival of the famous Dionysiac rites can still be recognised in the folk games near Adrianople (*vide* Ridgeway). Two areas of less importance in our study were also conquered in the first century after Christ. These were Cappadocia (in central Anatolia) and Egypt. Cleopatra (last of the Ptolemies) had been confirmed as Queen of Egypt by Julius Caesar in B.C. 47. Unfortunately for her, she joined forces with Antony in his struggle with Caesar's successor Octavius (Augustus). Their fleets were destroyed at Actium (off Epirus) and Cleopatra committed suicide. Egypt in B.C. 30 was thus almost the last of the Mediterranean coasts to become a province of Rome. Actually, Mauretania was the last, for Claudius annexed the Numidian Kingdom in A.D. 42 (Fig. 38).

Octavius (Augustus) was the first of the Roman Emperors; for the republic ended with the death of Caesar. During his reign the Empire was peaceful on the whole, and it was marked by the birth of the world's greatest teacher. Jesus was born near Jerusalem probably in the year B.C. 4. His teaching was to take the place of the *Pax Romana* as a civiliser when the latter was destroyed by the Barbarian invasions in the 5th century. A few words may now be given to Jewish history. In A.D. 70 the Jews revolted against Vespasian, and his son Titus besieged Jerusalem. It is said that nearly one million Jews perished when the city was destroyed. However, Jerusalem was rebuilt, but in A.D. 132 the Jews revolted again. After this second unsuccessful revolt was ruthlessly quelled, the Jews dispersed in still larger numbers, and have since had no specific country of their own. Babylonia added greatly to the number of her Jewish inhabitants, but many spread along the Mediterranean to Byzantium, Rome and Spain. Others reached southern Russia in the seventh century—where they converted the influential Khazars (p. 417). It is generally believed that the two groups of modern Jews (the Sephardim of Mediterranean race and the Ashkenazim of Alpine race) derive from Spain and South Russia respectively.

Trajan, who lived from A.D. 52 to A.D. 117, is usually regarded as the Emperor who brought the Roman Empire to its maximum of power and prosperity. The regions which he conquered, however, did not long remain part of the Empire, which was now unable to hold what it had seized. For many years the Dacians (of Rumania) had been crossing the Danube and ravaging the Roman province of Moesia. Indeed, in A.D. 89 Decebalus, their King, defeated several Roman armies. In 102 Trajan built a stone bridge and so crossed the Danube at Turnu Severin. He soon captured the Dacian capital at Sarmisagethusa, near modern Temesvar. Later he completely defeated Decebalus, and Dacia became a Roman province in 107. Lying so far from Rome, and beyond the protection of the Danube, it was peculiarly susceptible to attack. In 256 the Goths drove out the Romans, who thus suffered their first major loss of long-held territory.

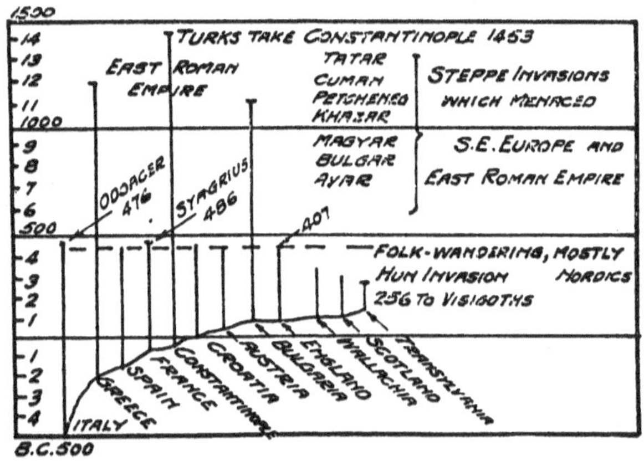

Fig. 39.—A Graph showing the Gains and Losses of Roman lands from B.C. 500 to A.D. 1453. The chronology of the Asiatic invaders is indicated.

The gradual expansion of Rome is illustrated in the graph in Fig. 39. The sloping line at the foot of this graph shows the rapid growth in territory from B.C. 200 to A.D. 100. Greece, Spain, France, Byzantium, Noricum (Austria), Bulgaria, England, Wallachia, Scotland and Transylvania (Dacia) follow in quick succession. Unfortunately, the decline of the Roman Empire was even more rapid, as the graph shows. By 410 the four last-gained provinces (Dacia, Scotland, Wallachia and England) had been abandoned. They were too far from the heart of the Empire to be protected when Rome itself was threatened. During the closing years of the fifth century occurred the "Debacle." Three European provinces, however, escaped invasion, namely, Bulgaria Byzantium and Greece, and carried on the proud Roman Tradition for nearly a thousand years as the Eastern Roman Empire (Chapter XXIV). The Barbarian Invasions and the crash of the Western Roman Empire are so important in our study of the European nations that they must be reserved for Chapter X.

CHAPTER IX

THE FOLK-WANDERING AND THE FALL OF ROME

A. The Decline and Fall of Rome

During the last two centuries of her history, the Western Roman Empire was constantly engaged in repelling the attacks of the restless barbarians on her widely extended northern and eastern flanks. Less care was necessary elsewhere, for on the west was the ocean, and on the south the empty sands of the Sahara. But to the north were the Nordics, perhaps (considering their small numbers) the most valiant warriors whom the world has seen, the Sarmatians and Scythians; and to the east the civilised Parthians and Persians, who were never conquered by Rome. During the evil times of the second century the Germans and Goths repeatedly defeated the Roman armies. In 268 the Alamanni ravaged north Italy itself. So menacing did these enemies of the Empire become that Aurelian (about A.D. 284) surrounded Rome itself with a massive wall 13 feet thick, which still in large measure survives. Many of the later emperors ruled only for a few months, being puppets in the hands of the armies. The latter were mostly recruited from outlying parts of the Empire, and controlled by non-Roman officers. We gradually see arise the curious spectacle of the Empire protected against barbarian invaders by soldiers and officers much more akin to the enemy than to the natives of Italy. Let us glance at the tribes who lived beyond the frontiers of Rome in the early centuries of our era.

Far to the northeast of civilisation were the Finn tribes, probably of Nordic race. Allied to them in race but speaking Aryan languages were the Goths, who dwelt in southern Sweden. Cousins of the latter were numerous Nordic tribes, the Vandals, Burgundians, Suevi, Lombards, Saxons, etc., who all lived near the mouths of the Oder and Elbe (Fig. 38). In the basin of the Vistula were the ancestors of the numerous Slav peoples of Europe; and south of them were the Sarmatians, a group of tribes of eastern origin, speaking a tongue akin to Persian. Further east were perhaps small groups of Scythians, who were probably not far removed in race and culture from the Sarmatians, but probably included a larger proportion of Alpine blood. The ancient legends seem to indicate that the Scythians were forced out of southwest Asia (possibly Persia) by a people called the Massagetae. These Scythians in the 7th century B.C., displaced the Cimmerians; and both of these tribes are referred to in the Bible where they are known as Ashkenaz and Gomer, respectively. The Western Scythians for the most part seem to have died out, or merged in the Sarmatians before the time of Christ. The Sarmatians gradually moved westward, and about the zenith of Roman power extended from the Alans on the Don to the Jazyges on the Theiss. In the east the Romans hoped to conquer the Road to India—but it was too great a task and they finally made peace with the Asiatic rulers and obtained Indian goods *via* Mesopotamia and Persia. The actual break-up of the Western Roman Empire was mainly due to the invasions of the western *Nordic* barbarians and to them we shall give the most attention.

After the conquests of Caesar and his immediate successors, the Keltic Wave (p. 119) was mostly incorporated in the Roman Empire. There is a great deal of confusion as to the exact race and speech of the Keltae. To the Greeks and Romans the northern barbarians were all fair and stalwart; and no doubt the group-name "Keltae" included Nordics and Alpines alike. Thus the Teutons were at times called Keltae. But the word Keltae should be used for continental tribes speaking Aryan of the P or Q types, *i.e.*, for the Goidelic and Brythonic waves before they entered Britain.[1] The Senones (of Sinigaglia) and Boii (of Bologna) were early Keltic tribes in the basin of the Po. We have referred to the invasions of Northern Italy by Senonian Keltae in B.C. 390. In B.C. 280 similar tribes invaded Macedonia, and others reached Galatia in Anatolia, where they submitted to Rome in B.C. 189. The Belgae were members of the group of P Keltae; and they spread all over the region from the Seine to the Elbe, invading Britain about B.C. 250. Thus a small amount of "Alpine" blood reached Britain in these invasions.

The Teutons lived apparently in Sweden while Rome was developing. About B.C. 600 they expanded into north Germany and displaced the Keltae in the south and west, while the Slavs were pushed to the east.

In a circle around the Baltic from south Sweden to Holland we find the cradles of a famous group of nation-breeders. Goths, Vandals,

[1]Some writers use the word "Keltae" for the P speakers (Brythonic) only. It is wrong to speak of the short dark "Mediterraneans" of Britain as "Keltae." The term "Keltic Fringe" has no *racial* significance—though it emphasises the position of a group of primitive Aryan languages.

Burgundians, Suevi, Lombards, and Angles were but petty tribes (Fig. 38), but each of them wrote its name large in history during the sixth and seventh centuries. Thus the immediate enemies of Rome were the Teuton tribes (Marcomanni, etc.) in the north, the Sarmatians (Jazyges) and Dacians in the centre and the Parthians in Persia. The long line of defence extended from Antonine's Wall (A.D. 150) in Scotland to Nisibis in northern Mesopotamia, a distance of 2,500 miles. The Arabian desert protected the Empire from serious invasion in the southeast.

War with the barbarians and civil war among the claimants for the Imperial Throne occupied the Empire almost continuously from 180 to 284. Persians were quelled in A.D. 166, and the inroads of the Nordic tribes along the Danube were blocked in 170. The last of the Sarmatian attacks occurred about 175. The Goths about this time marched south to the Black Sea, and threatened Dacia. They were held back by constant warfare until 256, when Dacia was abandoned to them. A German tribe, the Alamanni, even reached Ravenna not far from Rome, and ravaged North Italy until they were defeated by Claudius. In 273 the Romans succeeded in destroying the power of Palmyra in the Syrian desert, where Zenobia had defied the Roman armies. In 284 a skilful soldier named Diocletian was made Emperor and he and his successor Constantine halted the decay of the Empire. Diocletian divided the unwieldly Empire into four principalities under two Co-Emperors, each with a "junior" to help him. Their capitals were at Nicomedia (on the Sea of Marmora), Sirmium (near Belgrade), Milan (Italy), and Treves (in Lorraine). The son of one of these leaders was Constantine, one of the best known of the later Roman Emperors.

In 326 Constantine started building a new capital of the Roman Empire at Constantinople. Here he was nearer the battle front against the Goths and Persians. Here also he felt that the Christian religion, which he favoured, would meet with less opposition than in pagan Rome. Constantine died in 337, and civil war between his three heirs, Constantine II, Constantius, and Constans, followed for several years. The eastern rulers made peace with the Goths, who became Christians of the Arian type (p. 184). About 370 the onslaughts of the Asiatic Huns drove the Goths of Dacia to seek refuge across the Danube (Fig. 40). They were treated harshly by the Romans, and in revenge destroyed Valens and his army in 378. Theodosius, a Spanish soldier, became one of the joint regents, and in 395 he split the great Empire into two parts between his two sons, Arcadius receiving the East and Honorius the West. This separation became more and more definite in ensuing centuries.

The record of the Bordeaux Pilgrim[2] gives us a good idea of land travel through the Empire about the time Constantine was founding his new city (*i.e.*, A.D. 333). He proceeds eastward from Bordeaux *via* Toulouse and the Narbonne Gate to Arles and Orange on the Rhone. Thence he climbs the Genevra Pass in the Cottian Alps and descends to Milan, which was already one of the chief cities of Europe. Passing across the Great Plain he reaches Verona and Aquilea, soon to be

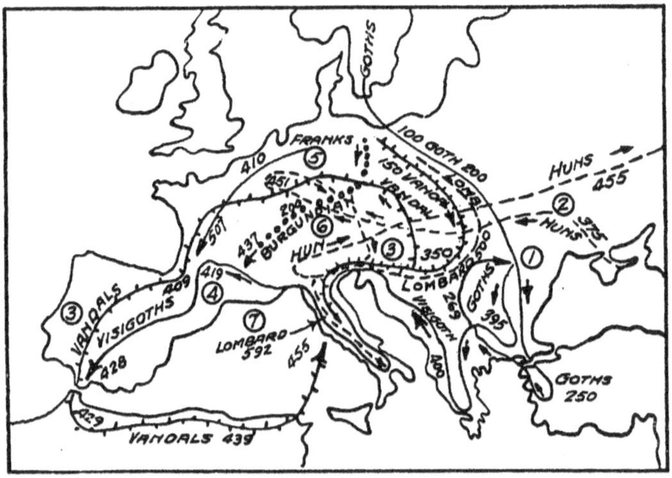

Fig. 40.—The Folk-Wandering during the Break-up of the West Roman empire. Their order (1 to 7) is Goth, Hun, Vandal, Visigoth, Frank, Burgundian and Lombard.

destroyed by the Huns. Thence he climbs the Pear Tree Pass and journeys along the Drave down the great road to Sirmium (Mitrovitza) and Singidunum (Belgrade). Here he uses the northern end of the Morava-Vardar route to Nish (p. 63), where he branches eastwards to Sardica (Sophia) and so to Byzantium. He crosses the Bosporus and passes through Nicomedia. "Annibalianus, who was once King of the Africans, is laid there," writes our chronicler. He follows the great military road through Ankara and the Cilician Gates to Antioch. Here he returns to the Phoenician coast and visits Tyre and Sidon before reaching Palestine and his main objective, Jeru-

[2] C. R. Beazley, "The Dawn of Modern Geography," Vol. I, pp. 57-66, 1897.

salem. Such a journey carried out in security under Constantine was hardly possible in later centuries until the time of the Crusades.

From 400 onwards the Emperors were practically figureheads in the power of the strongest generals. Stilicho, a Vandal (from Hungary), led the armies of the west, while Alaric, a civilised Goth, held a somewhat similar position in regard to Arcadius and the Eastern Empire. In 402 Alaric attacked Italy, but was defeated by Stilicho. The same fate befell a large army of heathen Goths in 405. But everywhere along the northern barrier the barbarians poured into the defenceless Empire. Vandals, Alans and Suevi crossed the Rhine and marched across Gaul into Spain. This date marks the real "break-up" of the Roman Empire. The Roman troops in Britain were at once recalled to defend Italy. Stilicho, the sole strong defender of Rome, was executed in 408, and in 410 Alaric again invaded Italy with his Visigoths and actually captured Rome.

A period of anarchy followed until 476, which is usually adopted as the last year of the Western Roman Empire. So complete was the divorce between Rome and Constantinople that the Emperors at the latter capital made no real effort to aid Italy until the time of Justinian. Boniface of Africa and Aetius from the Danube now took the places of Alaric and Stilicho, and fought for the control of the Roman powers. Boniface sought the aid of the Vandals, but although he defeated Aetius, his allies declined to leave Africa and founded a Vandal Empire which lasted a century. Aetius fled to the Huns (p. 168) and later succeeded in gaining control of northern Gaul. Here he resisted the attacks of Attila and his huge army of Huns, Gepids and Goths. In 451 with the aid of the German tribes he signally defeated them near Chalons (Fig. 40). This was a memorable battle, for the victory was due as much to the recently civilised German tribes as to the remnants of the Imperial armies. After ravaging Italy Attila died in 452 and his army fell to pieces. The European units (Gepids and Goths) combined against the Asiatic Huns and drove them east from the Hungarian plain.

The Roman puppet-emperors learnt nothing from the ruin of their predecessors. As Stilicho died, so perished Aetius at the hands of the Emperor. Again one of the rival parties appealed to the Vandals, who were now settled near Carthage. Their King, Genseric, attacked Rome in 455, and ravaged the Italian coast, much as Hannibal of Carthage had done 600 years before. Ricimer, an Arian German, was the next king-maker. He held back the Vandals, but much of his energy was exhausted in repelling rivals at home. He died in 472 and was followed in quick succession by the Burgundian Gundobad, and the Pannonian Orestes. In 475 the latter made his young son, Romulus, emperor. The German soldiers at Ravenna supported the Herulian Odovacar, who deposed Romulus in 476, and thereafter ruled Italy ostensibly as the representative of the Eastern Roman Empire. According to tradition, Rome was founded by a Romulus in 753 B.C. The Empire ended with another Romulus in 476 A.D.

B. The Migrations of the Teutons and Kelts

The Roman Empire, one great factor in the evolution of the Western Nations of Europe, has now been described. It remains to discuss in some detail the movements of the northern peoples which have already been briefly referred to. These Nordic migrations in some cases materially altered the racial composition of the nations just "about to be born." In all cases they strongly affected the history and culture of the European peoples, even though little trace of their sojourn can be detected in the modern nation. In the record of the Greek explorer, Pytheas (B.C. 333), he mentions the Gutones, who seem to have lived in the "amber" district in or near south Denmark, where they settled after leaving their homelands of South Sweden in which the names Gothland and Gotarike still survive. Among other tribes of the *East* Teutons were the Gepids, Rugi, Vandals and Burgundians, who spoke a common language and were dominantly Nordics. To the *West* Teutons belonged the Angles, Saxons, Lombards and Heruli; also Nordics but differing somewhat in laws and language. These Nordic tribes, however, often made war upon each other, and naturally tended to drift apart in interests as their later migrations led them to very different parts of Europe.

In Gaul proper (*i.e.*, France) the Keltae inhabited the north and centre. They were always slowly moving westward under the thrusts of the Teutons of Germany. Many of the latter had reached France before the time of Caesar and become incorporated with the Belgae, who therefore differed somewhat in culture and race from the Keltae of central France (Fig. 70). The latter were mainly Alpine in race, the former were probably mainly Nordic. The migrations in this region of northwest Europe can best be set forth in a table. Here the various districts are arranged in a series of five west-east zones.

TABLE P

HABITATS OF GERMAN AND KELTIC TRIBES BEFORE THE TIME OF CAESAR

Zone	Area	Before 500 B.C.	500-50 B.C.
A	South and West Baltic Coasts.	West Germans	Partly East Germans
B	Belgium....................	Keltae	Keltae and Germans
	Holland and N.W. Germany..	Keltae	Germans
	Oder Basin.................	Dacians?	Germans
C	Seine Basin.................	Keltae (Belgae)	Belgae
	Bavaria and Bohemia........	Helvetii (Keltae)	Boii (Keltae) Germans (at end)
D	Saone Basin.................	Ligurians ?	Keltae
	Upper Danube..............	Empty ?	Keltae
	Middle Danube.............	Illyrian and Sarmatian	Boii Keltae (at end)
E	Basin of Po (Italy)..........	Ligurians	Keltae

A consideration of the above table shows that the tribes were all moving southward or southwestward. The impetus came from the restless Germans. The tribes in the southern zones were "squeezed" between the Germans and the expanding Roman Empire, and soon vanish from history. Among these disappearing tribes were the Ligurians of the Po, and the Sarmatians of the Danube. Very little is known of the Dacian tribes of the Oder—for they had retreated to the Transylvanian region before Rome came into close contact with them. Many of the data upon which the above table is based consist of place-names and personal names which often give a clue to the language spoken by the tribes concerned, but do not, of course, help us as to their race.

We have seen that Caesar closed the Rhine frontier to the German tribes—except such as he recruited for his regiments. Later the Romans were confronted by a rival in the person of Marbad, King of the Suevi. He tried without much success to organise the German tribes between Mainz and Bohemia. Somewhat later Arminius, who had been trained in a Roman army, led a revolt against Varus and defeated him in the Teutoberg Forest (p 158). Except for punitive raids, however, the tribes of most of Germany were not greatly menaced by the Roman Empire, which preferred to seek new territory from less valiant warriors.

C. *The Huns, Visigoths and Ostrogoths*

The Goths after their arrival in north Germany about A.D. 100 offer one of the rare examples of a large migration for a time *towards* the east. This is against the almost universal current of migration in ancient and medieval times—which sets from east to west. Early in the 3rd century they were living in the Ukraine (South Russia), where they soon split into a eastern group, the Ostrogoths, and a western, the Visigoths (Fig. 40). In 251 they crushed the Romans at Abritta (in Bulgaria), where the Emperor Decius was killed. For many years they roamed through the Balkans, destroying many famous monuments, as at Athens and Ephesus in Asia Minor. But Claudius defeated them at Nish in Servia in 269; and Aurelian made peace with them, abandoning Dacia to their settlements. About this time the famous bishop Ulfilas converted them to Christianity. It is of importance that he favoured the simpler *Arian* creed and not the Athanasian creed adopted by Rome. This led to much bloodshed later. The Ostrogoths increased greatly in power, so much so that about 350 their King, Ermanric, ruled from the Baltic to the Black Sea.

The first of the great Asiatic invasions of medieval times was now advancing on Europe. We know little of the origin of the Huns. They belonged to the Altaic-speakers, and were originally members of the Mongolian division of the broad-headed Alpine Race. They attacked the Chinese before the time of Christ, and the Great Wall was built about B.C. 200 mainly in the hope of restraining their invasions. Repulsed in the east they moved westward in great bands of nomad cavalry. About A.D. 84 they reached the sea of Aral (Turkestan) and then attacked and conquered many Finnish tribes who lived near the Ural Mountains. About A.D. 372 they crossed the Volga, and this irresistible pressure in the east was transmitted through the Slav and Nordic tribes of Eastern Europe to the weakened defences of the Roman Empire. The Ostrogoths were overwhelmed and incorporated in the Hun migrations. The Visigoths had time to remove themselves across the Danube and seek protection from the unwilling Romans. The ferocity and mobility of the Huns struck terror into the Europeans. The flat noses, small eyes and beardless faces of some of their forces seemed to mark them off as different from ordinary people and supported the belief that they were familiar with magic. When their leader, Attila, had conquered the region between the Volga and the Danube he ravaged Persia and Syria. In 446 the Byzantine Empire paid Attila large sums to prevent his attacking them. In 451 he invaded France with half a million men, but was defeated near Chalons (p. 320). Many of the Asiatic hordes returned to Asia after Attila's death in 453, but others remained near their old headquarters in Hungary and helped later Asiatic hordes to attack the western civilisation until the 10th century.

The history of the two branches of the Goths differed considerably after the defeat of the Huns. The western branch of Visigoths under Alaric and Ataulf (p. 321), retired to Gaul and in 419 occupied Aquitaine (Fig. 40). Gradually their armies conquered most of Spain, whence they drove out the Vandals by 428. The Gothic Kings were ultimately defeated at Poitiers in 507 by a later Nordic invasion—that of the Franks under Clovis. In Spain, however, their power lasted until 711, when Roderick was overwhelmed by Moslem invaders from Morocco.

The Ostrogoths after the battle of Chalons retreated to the Hungarian Plain, but they soon moved into Italy to drive out Odoacer (or Odovacar) at the instigation of the Byzantine rulers. About 493 Theodoric, King of the Ostrogoths, founded a Gothic empire in Italy, nearly a century after the Visigoths had occupied southern Gaul. Theodoric gave Italy a better government than did any of the rulers in the Roman Empire, but on his death Justinian (the Eastern Emperor) sent his general, Belisarius, to bring back Italy within the bounds of the Empire. After various campaigns against Totila and other Gothic leaders the Byzantines were successful in 552, and the Ostrogoths left Italy. Little is known as to their subsequent history.

D. *The Vandal and Frankish Migrations*

Among the many Nordic tribes on the southern coasts of the Baltic were the Vandals, who about A.D. 150 were chiefly settled between the Oder and the Vistula. They took part in some of the migrations of their kinsmen, the Goths. During the 4th century the Romans allowed many of them to settle west of the Danube in Pannonia. In 405 with other Germans under Radagaisus they made the first determined attack on the Roman Empire and besieged Florence. After Stilicho defeated them (p. 342) they invaded Gaul which they ravaged for several years, until in 409 they passed through the weakly defended passes of the Pyrenees and divided Spain among their allies (Fig. 40). Thus the Suevi took the northwest, the Alans[3] the southwest, and the Vandals themselves the south. They were not left unmolested long—for the Roman authorities soon put in practice one of their favourite plans. They persuaded the Visigoths to leave Italy

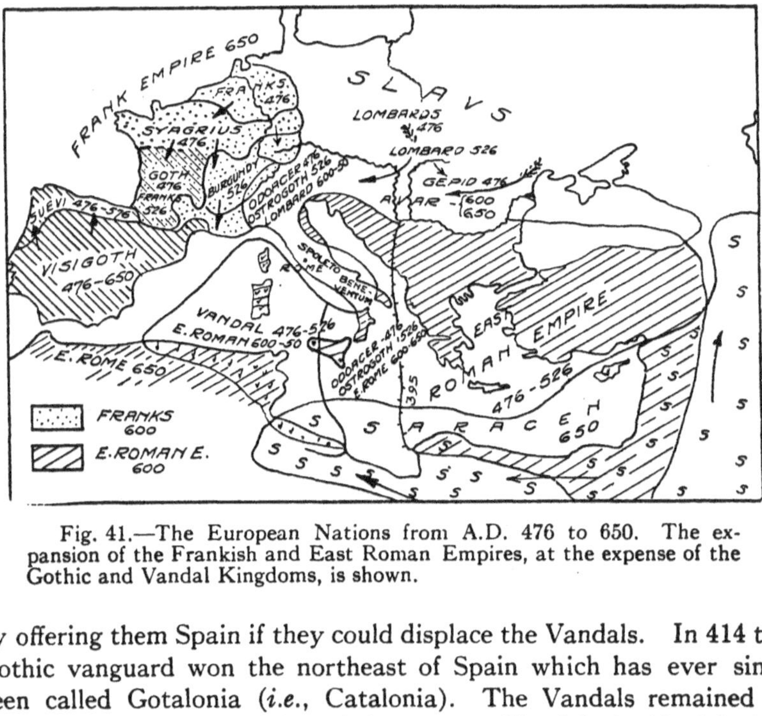

Fig. 41.—The European Nations from A.D. 476 to 650. The expansion of the Frankish and East Roman Empires, at the expense of the Gothic and Vandal Kingdoms, is shown.

by offering them Spain if they could displace the Vandals. In 414 the Gothic vanguard won the northeast of Spain which has ever since been called Gotalonia (*i.e.*, Catalonia). The Vandals remained in security in the south, giving their name to Vandalusia (Andalusia). Their fleets ravaged the Balearic Isles, and as we have seen they migrated into Africa at the invitation of Boniface in 429. In 439 they took Carthage and imitating the ancient Phoenicians their fleets soon conquered Sardinia and Sicily. In 455 they sacked Rome carrying off her chief treasures to Africa (Fig. 40). They became

[3]The Alans were Sarmatians who were disrupted by the Huns. The western group joined the Vandals, while the eastern group ultimately became the present-day Ossetes of the Caucasus.

Arian Christians about this time. Under Justinian the Byzantine power increased so greatly that for a time it seemed that the old Roman Empire of Augustus might be restored. Belisarius conquered the Vandals in 534 and their shortlived empire came to an end (Fig. 41). They were thus the first of the Nordic tribes to attack the Empire, and the first to lose their independent conquests. Their early collapse was no doubt partly due to the fact that they had wandered farthest from the Nordic homeland, and occupied a hot, arid environment which differed most definitely of all the Nordic conquests from that of the coastlands of the Baltic.

In the Atlas Mountains especially at their western end there are many somewhat blonde folk among the Kaybles. So also blondes occur on the Riff coast among the Berbers. One possible explanation is that these fair people are descendants of the Nordic Vandals, though if so it is curious that they are least abundant in the east where the Vandals settled longest. Dolmens and similar stone monuments are very abundant in this part of North Africa as they are in Spain and France, and it is possible that the blonde people go back far beyond the Vandal period to an early Neolithic migration which may have included the extinct Guanches of the Canary Islands. The suggestion that they are a local development due to a special environment does not appeal to the writer, and Ripley leaves the matter an open question.

The noblest destiny of all the barbarian peoples was that reserved for the Franks, who were, indeed, the founders of the Holy Roman Empire, which in a sense replaced the Western Roman Empire. They first appear in history as outlying members of the western Teutons who settled along the Rhine about B.C. 250. The best known tribes were the Batavians in the north and the Chatti (*i.e.*, Hesse) in the south. The former belonged to the *Salian* or seashore Franks, the latter to the *Ripuarian* Franks. In 410 Stilicho recalled the Roman troops from Britain and the Rhine to defend Italy against Alaric, thus leaving the Franks free to expand into Gaul. They occupied the basin of the Moselle, and when Aetius conquered the Burgundians many Franks also settled along the west bank of the Rhine near Worms. The Salian Franks about A.D. 400 moved into the valley of the Sambre. After the defeat of Attila more Franks poured into Roman Gaul and when their Merovingian leaders first came into notice they were encroaching on the basin of the Seine. Meanwhile, a last fragment of the Roman Empire persisted in the north of France (Fig. 41). Here Aegidius and his son Syagrius at Paris ruled in the name of Rome for a whole decade after Odovacar had deposed the last Emperor, Romulus, in Italy (p. 166).

In 486 Chlodwig, the Merovingian ruler, defeated Syagrius, and the last relic of Western Rome vanished. In 496 he became a Christian of the Roman (*i.e.*, Athanasian) type. This aligned him against the Visigoths, who were Arians, and who controlled the southern half of France until 507. He occupied much of his reign in successful struggles with the Alamanni on the upper Rhine and the Burgundians near Geneva. In 531 the Franks defeated the Thuringians of central Germany, partly with the help of the Saxons who lived on the lower Weser. By 600 they controlled all central Germany and their boundaries extended practically to the lands of the Slavonic peoples near Brunswick. The further expansions of the Franks belong to the history of France and are described in Chapter XVII.

E. *The Burgundians and Lombards*

Both these Western Teutons lived in the plains of north Germany before the Folk-Wandering. The Burgundians came from the lower Oder and the Lombards from the lower Elbe. Their later paths crossed, so that the Burgundians slowly moved southwest to the Rhone Valley and the Lombards southeast into Hungary and Italy (Fig. 40). About the year A.D. 200 the Burgundians occupied southern Germany near Worms, thrusting the Chatti (Franks) to the northwest. Here Aetius attacked them in 437 and after their defeat settled many of them near Lake Geneva. They gradually spread into the Valley of the Rhone and founded a kingdom which for a time rivalled that of the Franks. Indeed, Gundobad was for a time Patrician (*i.e.*, Governor of Italy. The Burgundians were for the most part Arians, and this gave Chlodwig, the Frank King, an excuse for interfering on behalf of orthodox Burgundian Christians. Although Gundobad was defeated in 500 by the Franks, his kingdom survived until 534. In the 9th century the Burgundians again became independent and gradually increased in power till the time of Charles the Bold in the 15th century. But these later events belong to French History (Chapters XV and XVII).

The last of the Nordic invaders to carve out an empire on the continent were the Lombards.[4] Their early history was much like that of their kindred already described. Their early name of Lango-

[4]See Chapter XXV for an account of the Varangians.

bard refers to the long battle-axe (hal-berd) with which they were armed. They succoured the son of Arminius in A.D. 47 when he fled to their lands on the lower Elbe. When the great Teutonic movements began during the second century the Lombards moved southeast up the Oder Valley, and the Slavs poured westward to occupy their abandoned lands. In 380 the Lombards settled north of the Carpathians and in the 5th century they migrated still farther south into Hungary. Here they became vassals of the Heruli (from Denmark) and received the Arian form of Christianity. However, in 495 they drove the Heruli from Hungary and became independent.

During the 5th century the Tatar Avars attacked the Lombards and their kinsmen, the Gepids. This did not prevent the Lombards from uniting with the Avars in 568 to overthrow the Gepids. The Avars thereupon were allowed to occupy the Hungarian Plain—while the Lombards under Alboin invaded Italy (Fig. 104). Indeed, many of their troops had already helped Narses—the general of Justinian—to overthrow the Ostrogothic empire in Italy a decade earlier. They fixed upon Pavia as their capital, and in 592 they were only prevented from sacking Rome by the bribes of the Roman bishop Gregory. The representative of Byzantium now governed from Ravenna, but he had no forces capable of withstanding the fierce Lombards.

Now began a period of anarchy in Italy which lasted until the time of Charlemagne (Fig. 41). The Lombards had had no experience of imperial power, and for many years gave themselves up to raiding and pillage. The large towns like Ravenna and Naples protected themselves by strong walls. The Byzantine forces retreated to the seaports and the Pope was isolated in the territory immediately around Rome. Many of the Lombard leaders created separate dukedoms and gave little or no allegiance to the King of the Lombards at Pavia. The Dukes of Benevento and of Spoleto were perhaps the most important of these independent leaders (Fig. 41). Struggles between the Popes, the Lombards and the Byzantine Empire continued throughout the 7th century. Duke Grimoald of Benevento became King about 650 and succeeded for a time in uniting the southern and northern Lombard dukedoms. But about 682 Arianism practically disappeared, and the Lombards (especially in the south) made peace with the Pope. The independence and licence of the northern Lombard peoples, however, prevented them from combining with Rome and Benevento in the south. This regional cleavage persisted through later Italian history.

About this time Venice, which had been founded by refugees from Aquileia in 452, rose into some prominence and was generally anti-Lombard and on the side of the Pope. The latter, moreover, had invoked the aid of the French ruler, Pippin, against his enemies the Lombards. In 754 and 756 Pippin and his Frankish soldiers defeated the Lombards and forced them to return Ravenna and other districts they had taken from the Romans. Pippin was concerned to help the Pope, and had no interest in the shadowy claims of Byzantium. His action in conferring Italian lands upon the Pope was a vital factor in the history of the Roman church. In 774 Charles, son of Pippin, invaded Italy and captured Pavia. The Lombard Empire thereupon came to an end. However, the southern regions around Benevento were permitted to retain a Lombard Duke under the suzerainty of Charles the Great.

F. Changes Consequent on the Fall of the Empire

It is obvious from the preceding pages that there was no sharp break in the history of Europe at 476. During the century since the Visigoths defeated Valens in 378 Rome had experienced several major defeats at the hands of the German barbarians. The Vandals broke the Roman barriers in 405, as did the Visigoths in 412, and the Franks a decade or two later. By this time Rome had no army which was not composed in large part of the blood kin of her attackers. One main factor in her fall was clearly her dependence on mercenaries—a practice which went back to Caesar and even earlier. The crass stupidity of many of her later emperors also contributed to her downfall. These facts are obvious enough, but it is by no means clear why the Goths and their allies left their homelands so long before any Asiatic invasion menaced their lands. Possibly, as Huntington suggests for the later Vikings, Scandinavia and the Baltic experienced inclement seasons for many decades. Crops perhaps would not ripen, pastures were poor and the Goths may therefore have determined to attack the rich settlements of the south, of which they undoubtedly had some knowledge. But it is curious that they made so little stay in North Germany—which the Slav tribes gladly entered. Presumably the latter were more easily satisfied than the Goths.

The German Barbarians were in conflict with the Keltae (who probably spoke Brythonic) long before Caesar came in contact with them. He made considerable use of the natural antagonism between the two groups, so that the Keltae soon became Romanised and acted as buffers against such Teutonic tribes as the Alamanni and Marcomanni. It was not these latter (who fought Rome during her strength) but the more northern and hitherto less aggesssive Teutons—the Vandals, Franks, Saxons and Burgundians who later shared the Empire between them. The Goths, Vandals and Lombards moved to the southeast from the Baltic—as if to keep a considerable zone of western German tribes between them and the formidable Romans.

Thus it was along the lower Danube, *i.e.*, at the furthest point of the Empire from the Gothic cradlelands, that these conquering Nordics first invaded the Empire. However, they had become fairly peacefully settled in Dacia when Constantine moved his headquarters to Byzantium (A.D. 330), so that he could keep a watchful eye on these potential enemies. The Asiatic invasion by the Huns only indirectly determined the Fall of the Empire. The final decisive thrust was made at the heart of the Empire by the Vandals and other Germans striking from Hungary at Rome about 405. They moved down into Italy by the low gate between the Alps and the Dinaric Mountains of Yugo-Slavia. No doubt the Pear Tree Pass (p. 63) was one of their main routes. The withdrawal of the legions to protect Italy naturally left Britain open to the Picts and Scots, and Gaul at the mercy of the Franks and allied Germans.

The most obvious result of the ruin of the Empire was the decay of civilisation in the West of Europe. Commerce was at a low ebb, though during the fourth century both sea and land had been thronged with merchants exchanging Mediterranean and Eastern products for the goods of the north. Rome almost fell into ruins and was for a short time deserted. Bridges were never repaired, and the elaborate network of imperial roads lapsed into decay. Forests grew thick, and sheltered not only wild beasts but wilder men. Only one institution illuminated the Dark Ages and that a relatively new one. The Christian religion survived the barbarian onslaughts—and, indeed, the Goths were at this time actually Christians, though the Vandals and Franks were not. Hence the monasteries behind their strong walls kept alive some love of literature and humanity (Chapter X). Three centuries were to elapse before some of the advantages of the invasions became apparent. The robust virtues of the northern peoples by this time were in a sense wedded to the culture and crafts of the Romans to produce the vigorous Frankish civilisation associated with Charlemagne.

As regards racial changes one cannot do better than quote Ripley in his discussion of the subject in his classic work "The Races of Europe" (p. 254): "Are there any vestiges in the population of northern Italy of that vast army of Teutonic invaders which all through the historic period and probably since a very early time has poured over the Alps and out into the rich valley of the Po? Where are those gigantic, tawny-haired, 'fiercely blue-eyed' barbarians, described by the ancient writers, who came from the far country north of the mountains? Even of late there have been many of them—Cimbri, Goths, Ostrogoths, Visigoths, Saxons, Lombards. Historians are inclined to overrate their numerical importance as an element in the present population. On the other hand, many anthropologists, Virchow, for example, have asserted that these barbarian invaders have completely disappeared from sight in the present population. Truth lies intermediate between the two."

"It is, of course, probable that ancient writers exaggerated the numbers in the immigrant hordes. Modern scholars estimate their numbers to be relatively small. Thus Zampa holds the invasion of the Lombards to have been the most considerable numerically, although their forces did not probably exceed sixty thousand, followed perhaps by twenty thousand Saxons. Eighty thousand immigrants in the most thickly settled area in ancient Europe surely would not have diluted the population very greatly. We can not expect too much evidence in this direction consequently, although there certainly is some. The relative purity of the Piedmont Alpine type compared with that of Veneto is probably to be ascribed to its greater inaccessibility to these Teutons. Wherever any of the historic passes debouch upon the plain of the Po there we find some disturbance of the normal relations of physical traits one to another; as, for example, at Como, near Verona, and at the mouth of the Brenner in Veneto. The clearest indubitable case of Teutonic intermixture is in the population of Lombardy about Milan. Here, it will be observed on our maps, is a distinct increase of stature; the people are at the same time relatively blond. The extreme broad-headedness of Piedmont and Veneto is moderated. Everything points to an appreciable Teutonic blend. This is as it should be. Every invading host would naturally gravitate toward Milan. It is the focus of all roads over the mountains. Ratzel has contrasted the influence exerted by the trend of the valleys on the different slopes of the Alps. Whereas in France they all diverge, spraying the invaders upon the quiescent population; in Italy all streams seem to concentrate upon Lombardy. The ethnic consequences are apparent there, perhaps for this reason."

"With the exception of Lombardy, the blood of the Teutonic invaders in Italy seems to have been diluted to extinction."

In another chapter he writes as follows (p. 30): "The Roman colonists in Gaul and Brittany have disappeared, to leave no trace. The Vandals in Africa have left no sign—neither hide nor hair, in a literal sense. Aquitaine was held by the English for three centuries, but no anthropological evidence of it remains to-day. The Tatar rule in Russia and the Saracen conquest of Spain were alike unproductive of physical results, so far as we can discover. Both alike constituted what Bryce aptly terms merely a "top dressing" of population. The Burgundian kingdom was changed merely in respect of its rulers; and spots in Italy like Benevento, ruled by the Lombards for five hundred years, are, in respect to physical characteristics, to-day precisely like all the region round about them."

"The truth is that migrations or conquests to be physically effective must be domestic and not military. Wheeler rightly observes, speaking of the Eastern question, that 'much that has been called migration was movement not of peoples, but of power.' Guizot's eighth lecture upon the History of Civilization in France contains some wholesome advice upon this point. Colonization or infiltration, as the case may be, to be physically effective must take place by wholesale, and it must include men, women, and children. The Roman conquests seldom proceeded thus, in sharp contrast to the people of the East, who migrated in hordes, colonizing incidentally on the way. The British Isles, anthropologically, were not affected by the Roman invasion, nor until the Teutons came by thousands. There is nothing surprising in this. In anthropology, as in jurisprudence, possession is nine points of the law. Everything is on the side, physically speaking, of the native. He has been acclimated, developing peculiarities proper to his surroundings. He is free from the costly work of transporting helpless women and children. The immense majority of his fellows are like him in habits, tastes, and circumstances. The invader, if he remains at all, dilutes his blood by half as soon as he marries and settles, with the prospect that it will be quartered in the next generation. He cannot exterminate the vanquished as savages do, even if he would. Nay more, it is not to his advantage to do so, for servile labour is too valuable to sacrifice in that way. Self-interest triumphs over race hatred."

CHAPTER X

THE EXPANSION AND DIFFERENTIATION OF CHRISTIANITY IN EUROPE

A. The First Three Centuries

One of the most potent factors in the development of the various nationalities of Europe has been Religion. When the Roman Empire dropped the torch of Civilisation it was picked up by the young Christian Church. Yet in spite of its noble purpose it is at least arguable that Christianity in its present imperfect form has produced more disunion than union among the peoples of Europe, though tribal, national and economic factors complicated the issues then as now. History indicates rather clearly that the so-called Christians through all the centuries since the time of Christ have fought more vigorously in support of rather controversial and relatively unimportant aspects of Christianity than they have ever done to advance the simple, self-evident truths of the Great Teacher. Probably the most disastrous wars in all our period (excluding the Great War) were the Religious Wars in central Europe during the 16th and 17th centuries, at the outset fought ostensibly to decide which was the true fashion to worship the Prince of Peace. For instance, Robinson (1917) states that Great Britain and Ireland are perhaps the only countries in which the profession of Christianity was not at one time or another spread by persecution and death. Even to-day when a general tolerance in regard to religion is advocated by all people of reasonable intelligence, the cleavages due in large part to bygone religious differences still tend to hold the nations apart. In an earlier chapter (p. 22) it was indicated that Religion played a part only second to Language—and far more important than Race—in determining national distinctions and antagonisms. It is clear, then, that we must examine in some detail the spread of the Christian Religion, more especially as it replaced the *Pax Romana* as the great civiliser of central and northern Europe. It is important to remember that Christ taught precisely in the century when the Roman Republic was changing into the Roman Empire.

Man has always felt the need of a religion. The savage endows all visible objects with powers to help or thwart him. He makes propitiatory offerings to placate these powers. This type of religion (*Animism*) still had its votaries about the time of Christ in the most barbarous regions of Europe. Even in the more civilised countries it had its place in the minds of the people, though it was subordinate to a belief that certain superior beings exercised control over all the interests, duties and occupations of man. At first these beings were "*spirits*" indefinite in form and shape; later they were considered to be *anthropomorphic*, *i.e.*, shaped like man. Thus arose the *Polytheistic Religions* of which many obtained in the Roman Empire. These developed in turn into *Monotheism*—in which, however, many subordinate agencies have usually been recognised. Monotheistic religions, acknowledging one supreme authority, stress the importance of the ethical aspects of their teaching. Some are more or less *national* communions linked with a special people—such as Brahmanism, Judaism, Confucianism, etc. Others are *universalistic* and appeal to all men alike. The chief are Buddhism and Christianity. Islam is partly national and partly universalistic; and so is perhaps a transition religion between the two last classes.[1]

Christianity naturally derived much of its philosophy from earlier religions. Christ was a Jew, and much of his teaching was that already practised by the most devout and enlightened Jews. A few authorities believe that Christianity was to some extent a popularisation of Essenism—which in turn owed much to the sacred lore of Persia (the Avesta) based on the teachings of Zoroaster. The latter is thought to have lived about B.C. 1000, perhaps in the mountains of Armenia. Buddha was born near Benares about B.C. 568, and his teachings had spread into Judea to some extent before the time of Christ. One of the important rivals of Christianity was another cult which had also spread from the east a century or so earlier. Mithras was one of the deities in the Persian Religion, subordinate to Ormuzd, and was associated with Light and Victory. His worship spread far, especially among the Roman soldiers, and was carried by them after the end of the first century A.D. all over the Empire just about the time when Christianity was becoming more widely known in Rome. Mithraism was supported by some of the Emperors—for it taught the "Divine Right of Kings," but it reached its zenith about 250 and died out in the 4th century.

One fundamental difference between the Christian and Jewish

[1]This is a summary of the views of J. E. Carpenter, "Encyc. Brit.", 1910, see article on "Religion."

religions was, of course, the fact that the Gentiles were equally welcomed with the Jews in the Christian fold. Soon the cult spread to Phoenicia and Antioch, and Paul became the first great missionary, travelling through Anatolia and Greece. Later he was taken as a prisoner to Rome. The destruction of Jerusalem drove the Christian teachers to distant lands. Later tradition has it that Andrew traversed southern Scythia and reached the Caspian Sea. Thomas and Bartholomew may have taught in India. John laboured in Ephesus and Patmos. Mark went to Alexandria and Peter to Asia Minor and

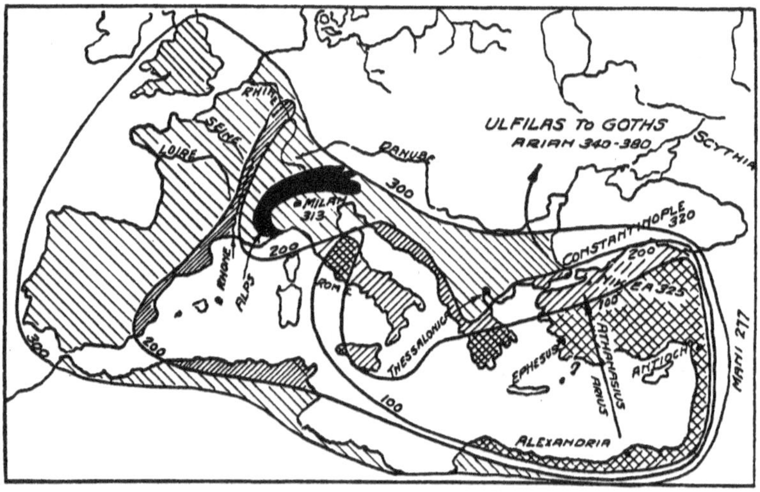

Fig. 42.—The Spread of Christianity during the first three centuries, in which it reached to the boundaries of the Roman Empire.

Rome, where he, like Paul, suffered martyrdom. The distribution of notable Christian communities within the Roman Empire at the end of the first century is shown by the isopleth 100 in Fig. 42.

During the second century the Christian Faith reached Carthage, and about 200 A.D. a meeting attended by 70 African bishops was held there. We have seen (p. 64) how readily merchants and Roman armies made use of the Rhone Valley, so now Christianity soon traversed the Way of Light and reached Vienne and Lyons in A.D. 177. Italy, Greece and the whole of Asia Minor now contained numerous settlements of Christians. From time to time there were periods of persecution when it was customary to imprison and otherwise punish Christians because they refused to conform to the state-cult of emperor-worship. However, the Emperors Commodus (180), Severus (211), and Philip (244) treated the Christians with more humanity. In Alexandria, Origen and other teachers drew upon Greek philosophy to help them expound Christianity. In 250, however, the Emperor Decius hoped to strengthen the decaying Empire by compelling all his subjects to conform to the old orthodox cult, which in turn was partly based on Greek philosophy. Many Christians suffered death, but the Faith still flourished and had spread through most of the Empire long before Diocletian's persecutions in 303. Both he and Galerius filled the prisons with Christians, ostensibly for breaking the ordinary civil laws. During the internecine struggles between Maxentius and Constantine about 312 it is related that Constantine became convinced that he could win only if he bore the symbol of the Cross on his banners. In 313 the Edict of Milan gave religious freedom to the entire Roman Empire.

It might be supposed that all would go well with the Christian Church after it became the orthodox religion strongly supported by the all-powerful Emperor. Unfortunately, struggles between sects now arose to trouble civilisation, and these were inevitably linked to struggles between political parties. In 325 a great Council of Bishops met at Nikea in Asia Minor. They devoted much of their session to an endeavour to decide between the views of Arius and Athanasius, both teachers in Alexandria. The Arians believed that Christ was subordinate to the Father, the Athanasians that he was "of the same substance" as the Father. The latter party gained the victory, but the Arians were not convinced; and as they converted large numbers of pagans in central Europe, who in turn migrated to France, Spain and Algeria, the schism became greater as time passed. Furthermore, the cleavage between the Church of Rome and that at the capital of the Empire (which Constantinople became in 330) grew more and more pronounced. A Council at Sardica (Bulgaria) settled nothing, but it is important to note that the Athanasians were supported by the Bishop of Rome and acknowledged his authority over their communion. On the other hand, the Emperor Valens supported the Arians at Constantinople. Now one party was in the ascendant, now the other. Pelagius opposed Augustine, Cyril defied Nestorius; so that though the theological problems were often changing the Church could never attain a large measure of agreement.

Large bodies of Christians broke away from the Mother Church, especially in Asia Minor, and founded the Nestorian Church. The break-up in the East, as Professor Scott points out, was partly due to underlying and persisting local differences submerged in the palmy days of Roman power. Rome and Constantinople next quarrelled in the matter of Image-Worship. The East opposed the practice and punished folk who attributed supernatural powers to images. The Popes at Rome supported the Image-Worshippers, who also won the Empress Irene to their side in 787. An almost complete break occurred in 867 when the Patriarch Photius in Constantinople accused the Pope in Rome of various unorthodox innovations. He summoned a Council which "deposed" Pope Nicholas. The latter retorted by excommunicating Photius and his followers. Though there were later attempts to heal the break they had very little success; and after 1054, when Pope Leo excommunicated his opponents, the Roman Catholic Church moved along a different and much more progressive path than did the conservative Greek Orthodox Church.

It is a curious coincidence that major events in religious history have occurred at intervals of about 500 years. Thus if we adopt B.C. 1000 as about the period of Zoroaster, then B.C. 500 was the date of the rise of Buddhism. Five hundred years later came Christ; while Mahomet (strictly Muhammad) was born in A.D. 571. The great schism was 1054, as noted above. The Protestant break occurred in 1517. Surely by 2000 we shall have made some progress toward the inevitable World State, and be less hampered than now by the horizontal cleavages of Industrialism and the vertical cleavages of Nationalism.

B. *Christianity among the Keltae and Teutons*

The various early subdivisions of the Church having now been briefly described, it remains to show how the new converts to Christianity were affected by these differences and schisms. It seems possible that converts among the Galatians (*i.e.*, Keltae) of Asia Minor carried Christianity to the people of Southern Gaul. Pothinus from Smyrna became the first bishop of Lyons about A.D. 150. In 244 seven missionaries landed at Narbonne and converted the people about Limoges. Paris was reached about 205 A.D. It was perhaps while Constantine was in Gaul that he conceived he beheld the figure of the Cross in the sky with the legend "By this, Conquer." In 336 Athanasius was exiled to Treves (Lorraine), and there led many people to adopt the monastic life. About 370 the great monastery of St.

Martin was established near Tours. Martin did much to extend the monastic system throughout France before his death (ca. 400).

Britain was able to send three bishops to the Council at Arles in 314, though the names of the earliest missionaries in Britain are not known. Probably the bulk of the Britons were not converted, but retained their Druidical practices. After the withdrawal of Rome the Christian Brythons withdrew to the west. One of their earliest monasteries was at Llancarvan, in Glamorgan, founded by St. Cadoc, who later

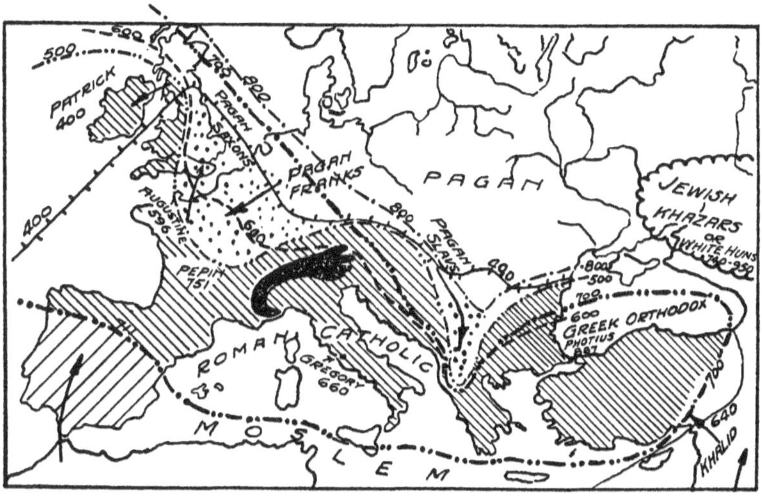

Fig. 43.—The Spread of Christianity from 400 to 800, showing the separation of the Greek Catholic and Roman Catholic Churches. Note the dangerous attacks by Moslems and Pagans about 700.

died among the heathen Saxons. St. David founded a similar institution (in the southwest corner of Wales) which had close affiliations with the Menapii on the adjacent coast of Ireland (p. 40). He died about 547. Ireland had never been subjected to Rome, and there was no well-established civilisation there in which the Church might take its accustomed place. However, in 431 Palladius was sent to the Scots, who were still in their original home in Ireland. Saint Patrick, who followed Palladius, was born "near the Irish Sea" about 389. He was carried off in a foray to Ireland, and later made his way to the monastery of Lerins near Toulon. About 432 he returned to Ireland as a missionary and subsequently converted many of the pagan Irish. He established many small monasteries and schools and died about 461 (Fig. 43).

Meanwhile, St. Ninian had laboured in southern Scotland probably before A.D. 400; but St. Columba was the chief missionary to Scotland. The latter was born in 521 and began his mission to the Scottish Highlands in 563. He built his first church on Iona, a little island off Dalriada, the new settlement of the Scots in Scotland (p. 49). In 565 he set out to Inverness to convert the Picts, and soon won a position of influence among them. At the time of his death (597) his church of Iona was the chief among all in Scotland and Ireland. England was still controlled by pagan Angles and Saxons.

Ethelbert, King of Kent, married a Christian princess from Paris and this led to Gregory (Bishop of Rome) sending forty monks to Britain in 596. They were under the leadership of Saint Austin (or Augustine, who is not to be confused with Augustine of Hippo), and he became first Archbishop of Canterbury. The new religion rapidly spread through the various small Saxon kingdoms, and soon came in contact with the converts who held by the Iona form of the religion. However, an Assembly at Whitby in 664 settled most of the disputes in favour of the southern church.

On the continent the Teutons (Germans) were much less advanced in culture than the Keltae, who, as we have seen, soon became Romanised. They were less inclined to agriculture and rarely lived in towns. They roamed from place to place, living by hunting and by pasturing their flocks and herds. During the many battles between Rome and the Barbarians it seems likely that numbers of Christians were taken prisoners by the pagans. It is difficult to give the exact date when the German tribes along the Roman frontier became Christian. In the fifth century there were Bishoprics at Treves, Cologne, Metz, and Toul among the Belgae and adjacent Teutons. Further east were the Bishoprics of Coire (Switzerland), Lorch on the middle Danube, and Pettau and Laibach near the River Drava.

One of the most famous of the early missionaries was Ulphilas, who was mainly responsible for the conversion of the Goths. He was born, perhaps in Dacia, about 311, and went to Constantinople as one of a convoy of hostages. He modified the Greek alphabet to suit the Gothic speech and translated the Bible into Gothic. He thus prepared the first written account in any Teutonic language which has survived to our day. In 341 Ulphilas attended the Council at Antioch just while the Arian creed was in the ascendant. It was this form of Christianity which he taught to the Goths (Fig. 42). Many of his followers settled south of the Danube in Moesia (Bulgaria). When the Huns overwhelmed the Ostrogoths, the Visigoths under Fritigern fled to Roman Moesia, accepted Christianity, and laid down their arms. As narrated elsewhere they were badly treated by the Roman leaders and inflicted a crushing defeat on Valens, the Roman Emperor, near Adrianople. Ulphilas died in 381, just while the Council at Constantinople was to some extent succeeding in reconciling Arian and Athanasian difficulties. Somewhat later the Visigoths under Alaric levied tribute on Rome in 410, but he respected the most sacred shrines.

Some of the Goths retired to the province of Noricum (Austria) where Christianity had been largely destroyed by barbarian invaders. These tribes were visited by the Apostle Severin about the middle of the fifth century. By this time the citizens of outlying Roman towns had shut themselves within their walls for protection against roving hordes of Barbarians. Severin had lived as a hermit near Vienna, and he now visited Barbarians and Romans alike. He was often able to persuade the conquerors to adopt peaceable methods in taking over the towns abandoned by the central Powers at Rome. His death occurred about A.D. 482. Although Severin was orthodox (*i.e.*, Athanasian), most of these invaders found the Arian faith more in accord with their old mythology, and it was several centuries before they joined the orthodox church.

About the middle of the fifth century Salvian declares that the Saxons, Franks, Gepidae and Alans were heathen, while the Goths, Heruli, Rugians and Vandals "were no better than Arian Heretics." Of all these tribes, the Franks became the most famous (p. 171). Until he was 30, Clovis, their leader, worshipped the gods of the barbarians. He married Clotilda, a princess of the orthodox (Athanasian) faith; and in 496 he and 3,000 of his warriors were baptised by St. Remigius at Rheims. This success for the Church of Rome was especially important—for the rivals of the Franks in the west, the Visigoths, were strongly Arian. Thus the victories of Clovis and his immediate descendants were acclaimed by the Bishop of Rome as victories for the orthodox Roman Church. In Italy at this time the Arian Ostrogoths were opposed to the Bishop of Rome, and it was not until the time of Justinian that the Goths with their Arian creed abandoned Italy. Meanwhile, Clovis drove the Arians out of most of Gaul after his victory over the Visigoths at Poitiers in 507 (Fig. 43).

Many teachers carried Christianity eastward from France during the seventh century. Lupus, Aloysius and Amandus worked among the heathen of the Frisian Lowlands. Columban founded a monastery in the Vosges Mountains, and one of his followers was St. Gall, who died in Switzerland, near Lake Constance, the scene of his labours, in 640. By the year 700 the Faith had spread to the Franks, Frisians, Alamanni and Bavarians. The Irish apostles founded scores of monasteries in Switzerland and adjacent countries, and had an honourable place in this period of enlightenment (Fig. 43).

Somewhere about 718 Winfrid, an English missionary, travelled through Bavaria and Thuringia and founded many monasteries. He took the name of Boniface and boldly preached to the wild Saxons, who still held strongly to their ancient tree worship. His assistants were enrolled from the English Church rather than from the older Irish Church, which was not nearly so closely associated with Rome. In the 8th century Christianity was more nearly extinguished in Europe than at any other time. The Moors (or Saracens) were ardent followers of Mohammad and carried their arms north almost to Paris in 732. A great battle in the region of Poitiers rolled back the invader into Spain. About the same time followers of Mani (of Persia) appeared in western Europe and their beliefs spread considerably, especially in south Europe. The Pagan Slavs were also migrating in large numbers down the Morava-Vardar corridor, and added one more factor which tended to separate the Greek and Roman Churches (Fig. 43).

Charlemagne, King of the Christian Franks, was perhaps the first medieval ruler to advance against an enemy with what amounted to a policy of killing or converting him. He might perhaps be termed the First Crusader. His savage pagan enemies were the Saxons, who lived in north Germany, and whose brothers in England had long become Christian. The wars between Frank and Saxon extended from 770 to 804. The Saxons then accepted Christianity, but it is significant that seven centuries later this part of Europe was the centre of that opposition to the Roman Church which produced the Protestant Communion.

Willibrord, another Englishman, converted the Frisians by 739; while centres of missionary enterprise developed at Paderborn (Westphalia), Fulda (Hesse) and St. Gall. The remaining Teuton tribes were slow to become Christians. The Pomeranians joined the Church about 1128, while the Old Prussians (a Lithuanian people) waited till the 13th century (Fig. 44).

In the Scandinavian regions early in the 9th century the missionary Ansgar was favoured by Harold of Denmark. He was, however, driven from that country and proceeded to Sweden in 831. Before his death in 865 much of south Sweden was converted. In Norway, King Olaf about 1019 was the first to organise the Church on a permanent basis. From Norway Christianity spread rapidly to Iceland and even to Greenland, of which a bishop was appointed about 1100. It must be understood that in these outlying northern lands it was several centuries after the introduction of Christianity before pagan customs generally died out.

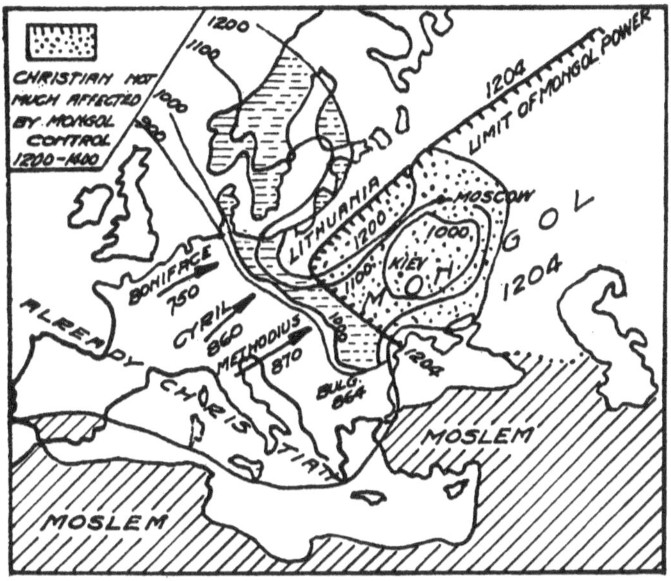

Fig. 44.—The Spread of Christianity from 800 to 1200, now reaching the Slavs and Scandinavians.

C. *Christianity among the Slavs*

The early history of the Slavs is unknown. They seem to have occupied lands abandoned by the Teutons in their advances on the Roman Empire (p. 168). Often the Slavs were vassals to the fiercer Goths or Huns, but when left alone their main interests were agricultural rather than warlike. Gradually they built towns, such as Lubeck, Kiev and Novgorod; but here, as in Scandinavia, in outlying regions their idols and temples were venerated till fairly late in medieval times. Perun, Triglav and Svantovit were many-headed gods each worshipped in parts of the vast territory of the Slavs. The first of the Slav peoples to become Christians were naturally those most closely in touch with the Empire. It is very important in our present study to realise that Slavs bordered on outlying portions of both the Roman and Greek portions of Christendom, though the latter were closest.

The Bulgars were a composite nation who seem to have consisted of a Finnish aristocracy (who came from the Volga in the 7th century) mixed with a Slav lower-class which migrated into Bulgaria in the 6th and 7th centuries. As is so often the case, the popular tongue conquered—so that the Bulgars are now classed as Slavs. For several centuries they were a constant menace to the Byzantine powers. There are many legends as to the methods by which the Bulgarians were converted. One report is that early in the 9th century a Bulgarian princess was captured by the Byzantines. She lived many years in Constantinople and became a Christian. Ultimately she returned to Bulgaria and Bogoris, her brother, was persuaded to become a Christian, mainly to avert pestilence and famine. However, Spinka (1933) states that the conversion of Bogoris (Boris) was forced on him by the Emperor Michael in 864. The latter invaded Bulgaria while its army was out of the country aiding Ludwig the German; and Boris had to submit to Michael's demands. For a time, indeed, Bulgaria seemed disposed to join the Western Church, mainly for political reasons, just as Moravia about the same time was negotiating with Byzantium. But in both cases the *adjacent* Church ultimately won.

The second Slav region to be converted lay far to the west. Moravia at this time was a powerful kingdom, where Charlemagne by fire and sword had endeavoured in 801 to promote Christianity. In 863 Radislav appealed for help to the Byzantines, who sent to them Methodius and his more famous brother, Cyril. The latter had already visited the Khazars of South Russia as a missionary. They composed a Slavonic alphabet and translated the Scriptures for the Slavs—just as Ulphilas had helped the Goths five centuries before. The Pope recalled the brothers to Rome, and they persuaded him to support their missionary work; which thereafter linked Moravia to Rome rather than to Constantinople. In 907, however, Moravia was invaded by the pagan Magyars who were the scourge of civilised kingdoms for several generations. Otto the Great defeated them about 950, but they showed no inclination to become Christians until 997. Then King Waik was baptised under the name of Stephen—and the organisation of the Magyar kingdom into dioceses under the Archbishop of Gran (near Buda-Pest) soon followed (Fig. 44).

Bohemia was occupied by the Czechs about the 4th century and in 844 some of its chiefs were baptised. About 871 Borziwoi, the pagan ruler of Bohemia, visited the Christian ruler of Moravia. Here he met Methodius, who persuaded him (by political arguments apparently) to accept the new faith. Later struggles between Wenceslas, a Christian, and his brother Boleslav resulted in the success of the latter, and the Christians were almost all driven from Bohemia. The Emperor Otto, however, compelled Boleslav to desist from persecution, but the paths of the bishops of Bohemia remained thorny for a century. Indeed, in 1060 Methodius was pronounced a heretic in spite of his excellent work in civilising the Bohemians.

The great pillar of the Eastern Church—and, indeed, in a sense the successor of the Byzantine Empire—was the enormous Slav Territory now known as the Soviet Republic. In 865 and in 907 Russian pirates crossed the Black Sea and attacked Constantinople. However, as happened so often, the new faith was introduced by way of a princess to the notice of the pagan ruler. Olga became Christian in 955 and though her son Sviatoslav remained a rough Russian pagan, his son Vladimir became interested in the various southern religions. Christians, Moslems and Jews attempted his conversion, and in 987 Vladimir (quite in the modern spirit) sent out an embassy of enquiry to see how the religions worked in their own lands. We are told that the envoys were not favourably impressed by German, Roman, Jewish or Moslem cities, but the superb spectacle of Saint Sophia in Constantinople excited their greatest admiration. Vladimir threw the local god Perun into the Dnieper at Kiev, and Christianity spread with great rapidity through his dominions. Moscow became Christian about 1100, but the region to the north held by the Lithuanians, and that to the south held by the Petcheneg Turks, lagged considerably behind the Slav area.

The history of the other great Slav nation, that of Poland, has been very different; mainly because the Polish contacts have been with Germany and the west rather than with Constantinople. About 966 Mesko (Mieceslav) married a Bohemian Christian, and the King and many of his nobles are said to have become Christians. A few years later a bishop was stationed at Posen under the guidance of the Archbishop of Magdeburg. About this time the Polish ruler Boleslav attacked the pagan Prussians and ransomed the bones of the martyr

Adalbert. These were placed in a shrine at Gnesen, and Adalbert has been greatly venerated by the Poles as their patron saint.

The last period in the conversion of the pagan Europeans is concerned with the work of the Military Knights on the Baltic coasts (Fig. 45). During the siege of Acre in Palestine in 1190 a few Baltic merchants gave great help to the sick and dying. This kindly action gave rise to a body of Teutonic Knights who fought valiantly in Pales-

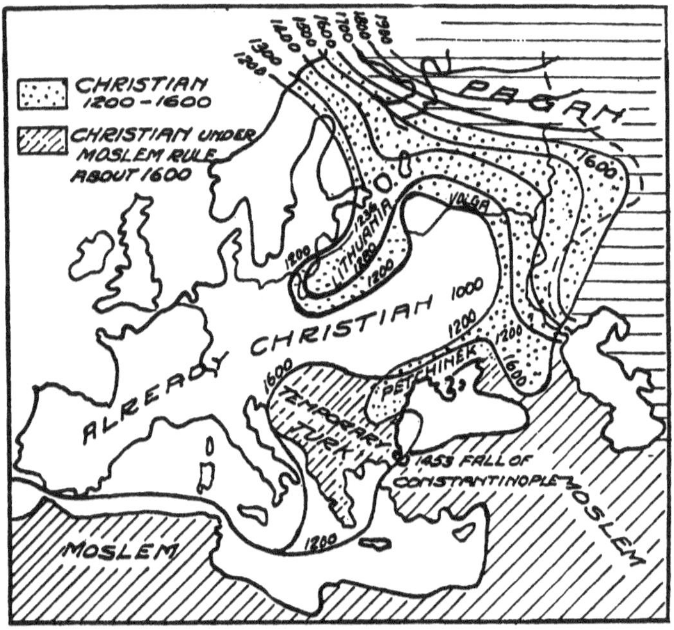

Fig. 45.—The Spread of Christianity from 1200 to 1900 extending to the Lithuanians and the North Russians, etc.

tine during the later Crusades. With the failure of the Eastern Crusades they were persuaded to turn their attention to the heathen elsewhere. In 1211 they helped the King of Hungary against the Cuman Turks, and built a castle at Brashov in Transylvania. In 1228 they migrated to Kulm in Prussia, and began to carry eastward the German culture by land, while the Hansa ships were doing the same on the sea. Farther to the east the Knights of the Sword were doing similar work in Livonia, but the two orders were merged in 1237.

About this time the Knights acknowledged no overlord but the Pope, and they gained control of almost all the coasts from Stettin to

Leningrad. During the 14th century their Court at Marienburg was one of the most brilliant in Europe, and the Knights of the order carried out very vigorously their duties of conquering the heathen Lithuanians. However, when the latter nation became Christian at the order of Vladislav Jagellon in 1386, the main purpose of the Order was accomplished, and thereafter it was of political rather than religious significance. At the close of the 14th century there were many lurking-places of paganism among the Lithuanians, and other primitive tribes of northern Europe. The Great Mongol invasion of 1204 naturally halted the advance of the Church, but the Mongols were not bigoted rulers, making less attempt to alter the religion of their subjects than did most of the western conquerors. When they were expelled from most of Russia about 1500 practically the whole of Europe was Christian.

D. *The Rise of the Protestant Churches*

The previous pages have shown us that the early Christian Church weathered the Arian controversy, though its leading exponents, the Goths and Vandals, were as bitterly opposed to the orthodox Catholics as to the heathen. Developing in part from this controversy, but chiefly resulting from environmental factors, we have seen the Greek (Byzantine) Church separating entirely from that owing allegiance to the Pope at Rome. The boundary between the two latter branches of the Church is shown in Fig. 46. The Greek Church controls most of the Slav peoples, including the Russian nation. Its tenets are followed in most of the Balkan Peninsula and have been adopted by the Rumanians, although the latter claim that their culture is derived mainly from Rome. The boundary shown in Fig. 46 is for the year 1648. Since that date the Greek Church has gained territory in Poland and West Rumania (formerly Hungary); but elsewhere the boundary has remained much the same. However, a new schism split the Catholic Church during the 16th century which affected most directly those lands and nations which had been the last to join the Roman Church. Other factors, however, such as race and language, and especially politics, contributed to decide whether a nation should join the protesting churches of Germany and adjacent lands.

Long before the teachings of Zwingli and Luther in 1516 and 1517 aroused determined opposition to Roman Catholicism there had been outspoken criticism and often open defiance of Rome. It is interesting to mark on the map the main "Foci of Revolt" in the Middle Ages (Fig. 46). As we saw earlier (p. 179), Mithraism vanished before the advance of Christianity; but other creeds, also in part based on the ancient beliefs of the Aryan tribes, persisted in out of the way regions. The main tenet of the Aryan "religion" was the conflict between the Power of Good (Ormuzd) and Power of Evil (Ahriman). This concept is found in the works of Zoroaster, and other early teachers. About A.D. 220 Mani in Persia preached Manicheism, a modification of Zoroaster's creed, which gained many adherents in Anatolia and Persia. Furthermore, the belief in celibacy, priesthood and saints, etc., while familiar enough to Mediterranean people, held

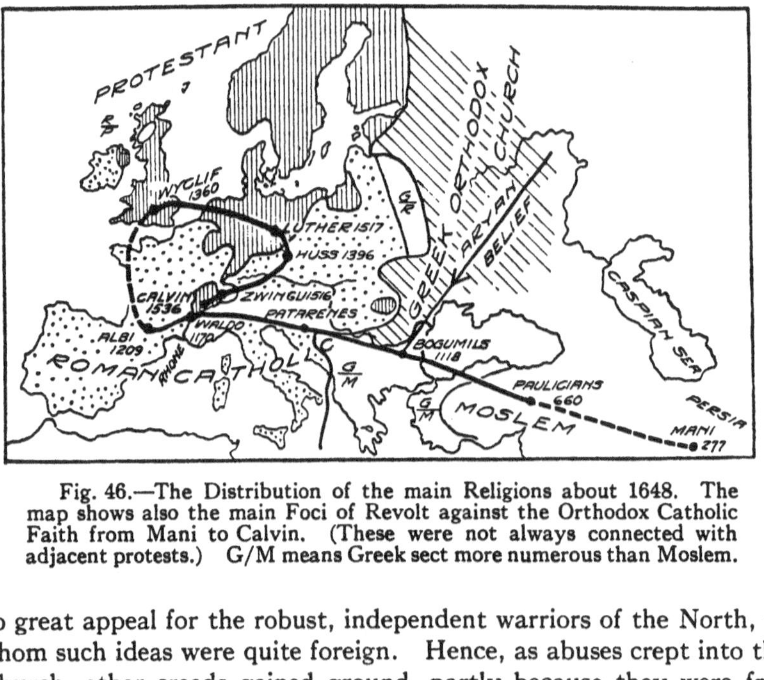

Fig. 46.—The Distribution of the main Religions about 1648. The map shows also the main Foci of Revolt against the Orthodox Catholic Faith from Mani to Calvin. (These were not always connected with adjacent protests.) G/M means Greek sect more numerous than Moslem.

no great appeal for the robust, independent warriors of the North, to whom such ideas were quite foreign. Hence, as abuses crept into the Church, other creeds gained ground, partly because they were free from errors practised by many Roman Catholics, partly because they incorporated something of the old pagan teaching. About the 5th century the Paulicians, a sect which rejected the Orders of the Church, spread through Asia Minor. Many migrated to Thrace in 752 and later to Albania. They were an offshoot of the Manicheans, and were persecuted by the Byzantine rulers; for instance, Theodora about 850 killed 100,000 Paulicians. The later adherents were called Bogumils, and also followed the teachings of Paul of Samosata. They denied the divine birth of Christ and the validity of sacraments and ceremonies. They refused to fast, and rejected monachism. So far they differ little from modern Unitarians, but they added many Manichean superstitions, such as the conflict between Michael and Satan—in their creed both sons of God. Bogumil himself was a priest who lived in Bulgaria about A.D. 960 and made many converts to this faith. The Cathars, Patarenes, Waldenses, Anabaptists and Dukhobors have all much in common with these Paulician sects (Fig. 46).

Early in the eleventh century a congregation later known as Albigenses developed in the vicinity of Toulouse in France. They seem to have been vegetarians and vowed to chastity, while their other rules agreed largely with those of the Paulicians. They were protected by various nobles in the south, notably by Raymond of Toulouse. The Catholic authorities persecuted them relentlessly, and in 1209 a crusade was preached against them which united the northern French nobles against their Provencal protectors. The Catholic party gained the victory by 1229; and not only was Toulouse absorbed by the King of France, but the unique character of the Provencal culture was destroyed. The sect was never again able to offer any opposition to the Roman Church.

A branch of the Bogumils called Patarenes in the eleventh and twelfth centuries settled in Bosnia and also migrated to Piedmont. They strongly objected to the abuses of the Church and their views were in part adopted by the followers of Waldo of Lyons. In 1170 Waldo took a vow of poverty and his followers preached from an unorthodox Provencal edition of the Bible (Fig. 46). After the Albigenses were destroyed, their allies, the Waldenses, withdrew into the rugged Alps of Piedmont and Vaud. Hence they are also known as Vaudois. They were often attacked by Catholic armies, especially in 1487 and 1650. They helped the House of Savoy against France in 1696, and gradually they were left in peace. They still maintain separate congregations mainly in the mountain valleys southwest of Turin near Pinerolo.

The protesting group best known to English readers is, of course, that which followed Wyclif in England. John Wyclif had a distinguished career as a teacher at Oxford, and was employed by the King on a mission to Bruges. It was not till 1376 that he declared that an unrighteous clergy could be deprived of their property by civil power. The University, the Court and the people protected him against the attacks of the Roman Church. He continued to publish numerous articles advocating Church Reform, and died in peace in 1384. It is not certain how much Wyclif contributed to the English translation of the Bible, which appeared just before his death. There does not, however, appear to be any marked connection between Wyclif and the earlier southern reformers. His followers, the Lollards, were never of much importance in England, but his writings were taken to Bohemia by friends of the Queen of England (a Bohemian princess), and in large part determined the revolt of John Huss.

John Huss (in the words of J. S. Black) was "the chief intermediary in handing on from Wyclif to Luther the torch which kindled the Reformation; and was one of the bravest of the martyrs who have died in the cause of honesty and freedom, of progress and of growth toward the light." About 1400 he began lecturing at Prague (the capital of Bohemia) on the writings of Wyclif. A few years later he wrote a pamphlet against forged miracles. In 1410 he was excommunicated by the Pope in spite of Royal and popular petitions in his favour. He was summoned to answer his critics at a council at Constance. Here he was convicted of heresy, and in spite of a solemn covenant that he should be allowed to return to Prague, he was burnt at the stake in 1415. His adherents were called Hussites and under John Ziska fought valiantly against the orthodox powers. For many years the Protestants in Bohemia—where Calvin's ideas spread widely—seemed to be victorious, but they were crushed in 1620 at the battle of White Mountain.

The preceding accounts have shown how the main "Foci of Revolt" against the orthodox church spread through Anatolia, Bulgaria, Bosnia, Italy, South France and England.[2] This rather definite movement to the northwest was reversed when the chief opposition centered in the person of John Huss in Bohemia (Fig. 46). Two men, Zwingli and Luther, were mainly instrumental in bringing about the final and successful fight against the abuses of the early Roman Church. Zwingli was ordained as a priest in Glarus, Switzerland, in 1506. He was a fine Greek scholar and was much attracted by the writings of Erasmus. In formulating his opinions his principal reliance was on the Bible, and he strongly opposed much of the theological dogma which had crept into the Roman Church. He, like Luther, objected to the sale of indulgences, and was so strongly in favour of

[2]This slow movement of "New thought" from the East perhaps naturally follows very closely the great land-route used by the Bordeaux Pilgrim (p. 164). Ideas and trade often use the same corridors.

marriage for the priesthood that he married in 1524. He was never much in sympathy with his co-worker Luther; and they met only to agree to differ in 1529. Zwingli was killed in a battle between the Reformed and Catholic cantons in 1531.

Martin Luther entered the University of Erfurt (in Saxony) in 1501. Here he heard many of the new humanist views (p. 198), and possibly imbibed something of Hussite socialism. But in 1505 he became a monk, and seven years later professor of theology at Wittenberg. In 1517 he protested against the sale of Indulgences and in accord with custom nailed his protest to the door of the Church at Wittenberg. The opposition of the Church to a publication which cut down their revenues was natural and prompt. But the Court protected Luther and the matter was shelved for a time. In 1519 he appeared to justify his views against the arguments of Eck. It was clear that he sympathised with the opinions of Huss, and he was excommunicated. Luther replied by burning the Papal decree. At the Diet of Worms in 1521 he boldly announced his views in defiance of Emperor and Pope, and was only saved from punishment by his flight to a secret refuge. In 1524 broke out the Peasant's War which was naturally mainly motivated by social grievances. But they hoped for support from Luther, and it is a clear indication of his unwillingness to break altogether from authority that he came out strongly on the side of the upper classes. He married a former nun in 1525, and in 1526 was very busy revising the services of the church to suit the Reformers. The League of Schmalkald arranged among the Protestant Princes united many powerful regions, such as Saxony and Brandenburg, to defend the Reformed Religion. Luther was not directly concerned with any such military preparations, and he died in 1546 without participating (as did Zwingli) in the religious wars.

It is one of the paradoxes of history that Henry VIII of England should have been an important factor in the Reformation. His attitude to the Church was not in the least based on his religious convictions, for he was essentially a Catholic all his life, but on selfish and political reasons. He broke from Rome because the Pope would not permit his divorce from a Catholic Princess. In 1534 Henry was declared the only supreme head of the Church of England. Later he seized the property of the monasteries and gave their sites to his favourite followers. In Holland the Revolt against Catholic Spain broke out in 1567; and the Armada directed by Spain against England in 1588 was part of the same great struggle of Catholicism versus Reform.

In France the progress of reform is associated largely with the name of Calvin. He was born in Picardy (N.E. France) and trained first for the church and later for law. In 1534 he definitely adopted the tenets of the Reformed church, some seventeen years after Luther's bold action at Wittenberg. In 1536 he published an outline of his religious belief, which found so little favour with the authorities in France that he found it well to reside in Geneva, where he lived all the rest of his life. He preached and practised a stern morality, which was adopted by the more extreme Puritans, not only in Switzerland and France, but equally in Scotland, England and New England (in U.S.A.). The Huguenots of France were mainly followers of Calvin, so-called because they first met at the Gate of King Hugo[3] in Tours. The Protestant cause was soon espoused by many powerful nobles, and was as usual complicated by political rivalry. The religious wars broke out in France in 1562, a few years earlier than in Holland, and lasted until 1629. Religious persecution followed, culminating in the Revocation of the Edict of Nantes in 1685. As a result nearly half a million of the most valuable French citizens left France. The cause of Protestantism never recovered from this blow and France has remained essentially Catholic ever since.

It now remains to study some of the other geographical aspects of the Reformation. At the beginning of the religious wars about 1560 the revolt against Rome was strongest in Scotland, England, Holland, the Scandinavian countries, Germany and Switzerland. All except England were countries which had received little or no culture directly from Rome (Fig. 38); and England as a whole revolted mainly to serve the private ends of Henry VIII. We may also note that the area of revolt agreed very closely with the boundaries of the Nordic Race. Only in southern Germany and Switzerland were the Alpine peoples in the majority. But since the Nordic north of France was mainly Catholic, while Central France (Alpine) was the stronghold of the Huguenots, it seems fairly clear that race played little part in this very notable cleavage in the Romish Church.

There were in 1560 strong Protestant minorities in France, Bohemia, Poland and Hungary, and it was in these countries that the issue was so doubtful that at the close of the Thirty Years' War the religious pattern might have been very different. The Calvinists, the successors of the Hussites in Bohemia, revolted against Hapsburg rule in 1618. But already dissension (the curse of religion) prevented their

[3]An alternative etymology is "Eidgenossen," *i.e.*, Oath-Companion (of Geneva).

co-reformers, the German Lutherans, from coming to the help of the Bohemian Protestants. The Emperor smashed the Hussites at White Mountain in 1620 and Bohemia was lost to the Reformed Faith. The King of Denmark led the Protestant armies in the second phase of the war, but Wallenstein, the Catholic general, was too strong for the Danes, who soon withdrew their support from the German armies. However, a much more powerful general came forward from the north to aid the Lutherans. This was Gustavus Adolphus, King of Sweden, who rallied the Protestants from 1630 to 1632, until he was killed at the Battle of Lutzen. In the fourth phase of the war political rivalry between France and the Hapsburg Emperor brought the Catholic cardinal, Richelieu, to the aid of his religious opponents. The Treaty of Westphalia in 1648 closed the struggle, and the Protestant nations emerged with liberty to worship as their governments might determine.

In Poland the Protestant church made small progress and was strenuously opposed by the Jesuit Order, so that Poland remained safely within the Catholic fold. In Hungary the Reformation aroused much more attention. By 1557 Calvinists and Unitarians had become quite numerous in Transylvania. So many of the Magyars joined the new religion that the country was split into two political parties with the German Catholics on one side supporting the Emperor, and the Magyar Protestants defying him. Owing to the struggles in Germany the Magyars were not very strongly attacked, and later the task of driving out the Turk occupied Austria to the extent of her powers. As a result there are still to-day in Hungary about 27 per cent. Protestants as opposed to 64 per cent. Catholics. In conclusion we may say that the religious boundaries have not changed materially in Europe since 1648, as a glance at Fig. 46 will show.

References to Works on the History of Christianity

1. Case, McNeill (*et al*). *Bibliographical Guide to the History of Christianity.* Chicago, 1931.
2. Gougand, L. *Christianity in Celtic Lands.* London, 1932.
3. Heussi and Molert. *Atlas of Christianity.*
4. Maclear, G. F. *Conversion of the West (Slavs and Celts).* London, 1879.
5. Merivale, C. *The Continental Teutons.* London, n.d.
6. Robinson, C. H. *Conversion of Europe.* London, 1917.
7. Spinka, M. *Christianity in the Balkans.* Chicago, 1933.

CHAPTER XI

THE RENAISSANCE IN TIME AND SPACE

A. The Development and Spread of the Universities

It is hardly possible to define in a few words what is meant by the Renaissance. It involved a period of time marked by a rapid evolution of culture—but the time and the character of the culture differed considerably from place to place. It is this latter aspect of the problem—the migration and modification of the Renaissance as it progressed—which makes the geographical treatment of considerable interest. Throughout much of its development it is closely linked with Humanism and with the Reformation. It would be foolish, however, to look upon it as a factor in national differentiation of the same importance as language or religion. It was rather a phase in human progress which was largely—though not wholly—confined to the Western half of Europe; at any rate in the earlier, more significant stages of the Renaissance. Yet its roots are to be found in the Southeast—and a later phase of its flowering is to be perceived in the Northeast. In the present brief survey of the Renaissance, our attention will be confined mainly to two phases, first, the growth of a more liberal and modern frame of mind, which is labelled "Humanism," and, secondly, the progress in the Arts will be illustrated by the development of architecture. Both subjects illustrate very clearly that shift of the area of notable social evolution from the Southeast to the West which marked the close of Medieval Times.

It has already been stated that it is difficult to date the beginning or end of the Renaissance. Still there are considerable differences in its stages of development and it will be found helpful to consider these for a moment before discussing them separately. The following four periods seem to be fairly well-defined, at any rate in the areas including Italy, France, England and Germany. They are illustrated in the four maps (Figs. 48 to 51).

TABLE Q

PHASES OF THE RENAISSANCE

A.	1100 to 1350—Rise of Humanism.
B.	1350 to 1450—Definite revolt against orthodoxy. Birth of Modern Literature.
C.	1450 to 1550—The Reformation and the Spread of Printing. (Literature in abeyance.)
D.	1550 to 1650—General Development of Science and "Modern" Literature.

Since the centres of the New Learning were to a considerable extent associated with the Universities, we may commence our study by noting when and where the early Universities started.

A very good illustration of the effect of environment on history is offered in Fig. 47, which shows the gradual development of University after University in the west of Europe. The outstanding feature is

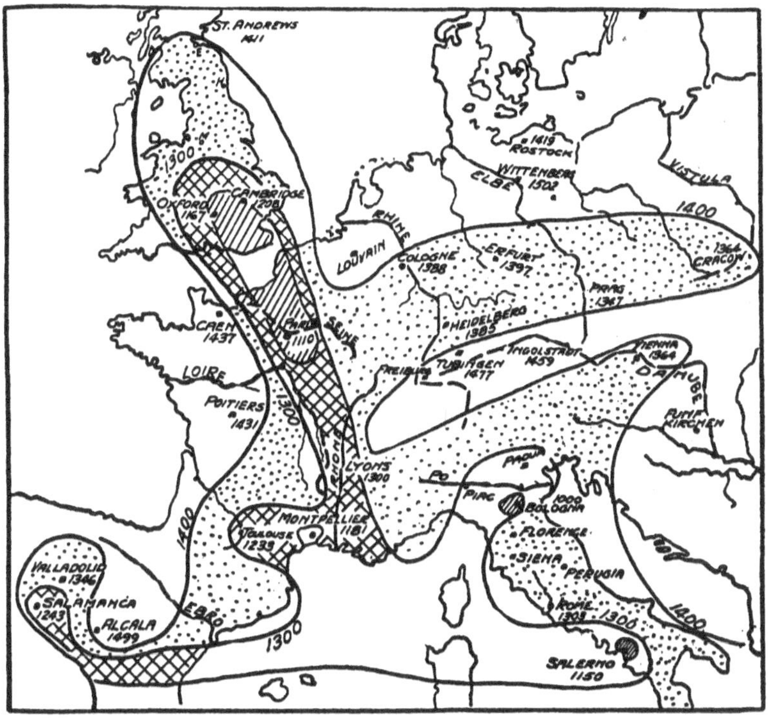

Fig. 47.—The Spread of the Universities from Italy to the north up the "Way of Light." Bologna, Salerno, Paris and Oxford were the earliest.

the importance of the "Way of Light," which functions here just as it did in regard to civilisation and Christianity. The earliest Universities had a very "indefinite" origin. Usually they were associated with some monastic teacher who acquired a wide reputation for his learning. After the Folk-Wandering, as we have seen, Learning was mainly confined to the safe refuges of the large monasteries, and Boetius (d. 524), Bede (d. 733) and Alcuin (d. 804) were famous teachers in the Dark Ages. It was not till long after the time of Charlemagne that western Europe had leisure to set about collecting good teachers in various places to form the nuclei of Universities. Since almost all the teaching before the rise of Humanism was confined to expounding works dealing largely with religion which were written in Latin, it was natural that the region near Rome should be the early centre of such interests. When a widespread desire for other disciplines, such as Law or Medicine, arose, the Italian centres again had some distinct advantage in language and situation, even though Latin was practically a universal tongue among the educated.

To Salerno—to-day a little town of no importance near Naples—is usually given the honour of producing the first European University. Probably its situation where three cultures met, *i.e.*, Latin, Greek and Arabic,[1] played no small part in determining the start of a School of *Medicine* about the end of the eleventh century. Possibly of earlier date was the *Law* School at Bologna, but in both cases it is impossible to say where the ancient school of the type common in the Roman Empire (and later) merged into the forerunner of a modern University. At Paris the study of *Logic* was so popular that a special school was opened about 1100. To this school came Abelard, and his lectures attracted scholars from all over western Europe from 1100 to 1140. Oxford University seems to have started about 1167 as the result of the recall of many English theological students to their benefices by King Henry. University College was founded to accommodate students in 1249. In 1209 some students migrated to Cambridge from Oxford, and several of the monastic Orders erected buildings here shortly afterwards.

Montpellier University in the south of France was founded before Cambridge, about 1181, and was at first chiefly concerned with medicine. Salamanca in Spain dates back to 1243, and Coimbra in Portugal to 1309. By the end of the thirteenth century the old Romanised lands of Italy, Spain, France and England had all instituted centres of University Teaching as shown in Fig. 47. Students from outlying countries flocked to these institutions, especially to Paris—where they tended to congregate in special communities known as "nations." It was only a matter of time when these communities were strong enough to found centres in their own countries. Thus Charles of Bohemia

[1]Thus the physician Avicenna (Ibn-i-sina) lived in Bokhara and Persia (980-1037), and his writings were of much influence among the Moors in Spain. Later, another Arab, Averroes (Ibn-Rushd), lived in Seville (1126-98) and taught Aristotle to Christian scholars in the 12th century.

founded the University of Prague in 1347, and for a time it rivalled Paris itself. In Poland, Cracow University was founded in 1364. Several German Universities started about this time, such as Vienna (1364), Erfurt (1379), Heidelberg (1385) and Cologne (1388).

In the next century Scotland obtained her first University at St. Andrews in 1411; and in Brabant Louvain was founded in 1426. In the north Upsala and Copenhagen date from 1477 and 1479. Wittenberg and Koenigsberg (Prussia) were much later, in 1502 and 1544 respectively. Wilna (Poland) was about the same date, but Russia (of the Soviets) had to wait till 1755 (Moscow). It is interesting to notice that though Germany lagged behind France in the spread of Universities—just as it had done in Christianity—the "lag" had now been reduced to a short period. Indeed, Germany had about as many Universities as France by the fifteenth century, though none of them was as famous as that of Paris. There are now more Universities in Germany (21) than in France (17), which is natural, since the population of the former is much larger. England has eleven Universities, Scotland four, Ireland three, and Wales one.

B. *The Development and Spread of Humanism*

During the Dark Ages the emphasis in education had been laid mainly on theological questions. The authority of the Roman Church in the west was so great that its pronouncements were in general accepted without question. Interest in purely human affairs was stifled; and teaching based on the "unorthodox" literature of the pre-Christian period practically died out. About 1340 we see a general interest aroused in "unorthodox" subjects, which was in some degree due to the discovery of the great value of many forgotten Latin and Greek texts of Classical times. To this new "unorthodox" field of learning the name *Humanism* has been given. At first it was opposed by the Schoolmen at many of the Universities. But naturally, humanism soon became "orthodox;" and, indeed, the meaning of the word has in later years altered somewhat. For our present purpose it represents a field of thought rather closely indicated by the more common word "Humanities."

Long before 1340 there were cultured folk who realised that ideas and theories must be examined and criticised from every angle before they are to be accepted as "orthodox." We have already seen that this was the case with the early "protestants" against the Catholic dogmas (p. 191). But in other fields, the names of Abelard,

Frederic of Sicily, and Roger Bacon stand out—each a critic in a different country. Dante also can fairly be classed as a humanist rather than a critic in the period before 1340. The main centres where the seeds of humanism were planted are indicated in Fig. 48. Abelard (1079-1142) has already been referred to. He was not concerned with the revival of the classic ideas, but his lectures at Paris and elsewhere prepared the way by questioning the philosophy behind orthodox dogmas. In many cases his philosophic inquiries strengthened the accepted dogmas, but his critical attitude gained him the enmity of many churchmen, so that he was condemned for heresy in 1121.

Fig. 48.—Some Fore-Runners of the Renaissance from 1100 to 1340. Note their association with the "Way of Light."

Frederic of Sicily (1194-1250) was a "modern" who lived four or five centuries too soon. In his career we can see quite clearly the effect of environment on thought and indirectly on government and nation. By descent he was a German of the royal family of Hohenstaufen (p. 269). His mother was a Princess of Sicily and his first wife came from Spain, his second wife from Jerusalem and his third from England. In 1215 he was crowned King of Germany, and five years later he became Emperor of the Holy Roman Empire. In 1229 he also became King of Jerusalem. When we remember that his favourite home (Sicily) was the meeting place of Moorish, Greek and Italian cultures, and that Frederic supported many new ideas with all the weight of his authority, we can realise his importance in the intellectual life of the 13th century. He was often referred to by contemporaries as *Stupor Mundi*, the "Admiration of the World." He was the first to encourage the writing of poetry in Italian instead of Latin. He founded a University at Naples in 1224, and codified the laws of Sicily in a manner far ahead of his time. It is, however, chiefly owing to his remarkable criticisms of the Pope and the Romish Church (which he published in 1239 and 1245 somewhat in the form of the modern "Open Letter") that he stands out as an innovator.

Roger Bacon (1214-1294) was a contemporary of Frederic, but he lived mainly in Paris or Oxford. His claim to our notice rests on his publication of many writings dealing with almost all the branches of science known at that time. Philosophy, theology, grammar, mathematics, the fundamentals of physics and especially the value of experimentation in science were all discussed by him. He vigorously condemned contemporary methods of teaching, both by the Church and by laymen, and in consequence spent many years in prison. It seems clear that his example did not lead to any great increase in the investigation of science; since for a successor of equal note we have to wait until Copernicus, nearly three centuries later. The fourth name in our list is Dante (1265-1321). It is worthy of note that all these four "pioneers of intellect" flourished on or near the "Way of Light" which we may perhaps extend to include the region between Southern Italy and England (Fig. 48). Dante is with some reason considered as the first great writer of the Renaissance. There is no doubt as to his greatness, and he wrote mainly in Italian. His "Divine Comedy," written about 1300, is in a sense a picture of medieval times and manners. It breathed no spirit of revolt against authority, but exhorted its readers to a better way of life, illustrated by the characters of well-known people of the times. Dante was hopelessly involved in the internecine warfare of the Guelphs and Ghibellines (p. 269). He was banished from Florence, but found refuge with the Ghibelline rulers of Verona. He hoped for the Union of all Italians in one great nation; and though this event did not take place until 1870, it is often stated that the national admiration of Dante's verse helped to link all the Italian parties, since it was one of the few subjects on which all were agreed.

During the next century (1350-1450) occurred the revolts of Wyclif and Huss, which we have already discussed (p. 193). It was marked also by the discovery of the rich treasures of Greek literature, and by the production of famous literature in the national languages (as opposed to Latin). As before, Italy showed the way, but the chief figure is perhaps Petrarch (1304-1374)—who wrote mainly at Avignon, actually in the "Way of Light" (Fig. 49). He also was affected by the rival factions in Italy, but, luckier than Dante, he became one of the most honoured of Italians during his lifetime. His chief claim to our notice is that his works often succeed in uniting the deep religious feeling of the medieval Church with the elegance and erudition of the Classic writers. His best known writings are his Italian love poems; but he also wrote some of the finest patriotic odes, which played their part in the development of the Italian Nation.

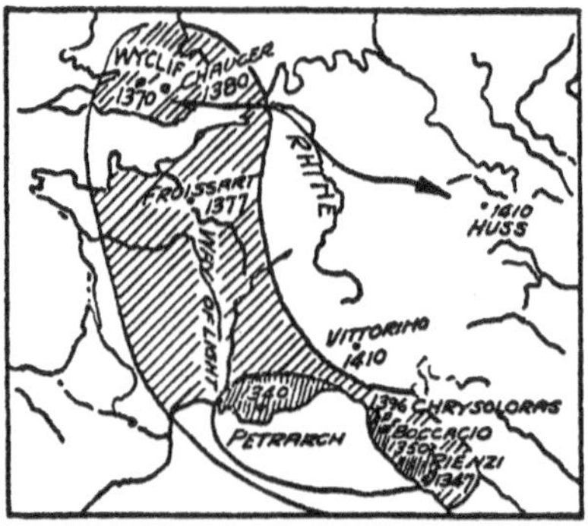

Fig. 49.—The First Period of the Renaissance from 1350 to 1450. The Revolt against Orthodoxy.

Worthy of some note in this brief survey of the spread of new ideas is the attempt of Rienzi in 1347 to form a Republic in Rome on the pattern of the splendid institution described in the Latin writers and poets. For a time Rienzi was successful, and he was supported by the Pope and also by the poet Petrarch. But he used his power without intelligence and his brief attempt to improve the evil governments in Italy was quite abortive. A contemporary of Rienzi and Petrarch was Boccaccio (1313-1375), who is sometimes described as the first of the modern novelists. He was one of the earliest Italians to learn Greek and to study the immense field of Greek literature. Naturally he was proud of his share in the appointment of the first Greek teacher at the University of Florence. In 1395 Chrysoloras, a Greek from Constantinople, was appointed to the chair. He was a notable example of the migration of learned men who spread over Western Europe as the result of the Turkish attacks on Constantinople in the 15th century. Mention should here be made of two great names in northern letters. Froissart wrote his Chronicles of French History in his native tongue, and Chaucer in England proved that English was at least as good a vehicle as Latin in the realm of poetry. Chaucer was essentially of the modern spirit. He showed that the homely, friendly English life around him in 1380 was just as worthy of description as the deeds of heroes hallowed by antiquity (Fig. 49).

C. Later Developments of the Renaissance

The Second Period of the Renaissance (1450-1550) saw the invention and spread of Printing, which may well occupy our at-

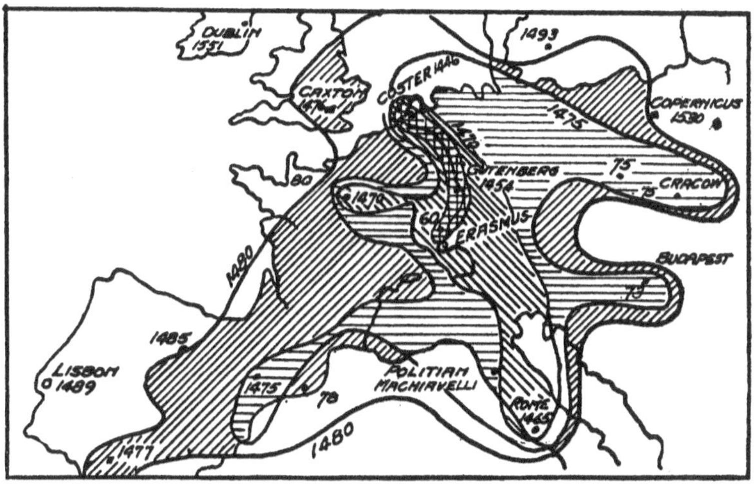

Fig. 50.—The Second Period of the Renaissance, from 1450 to 1550. The Spread of Printing is shown by means of Isopleths for 1460, 1470, 1475 and 1480. (Data mainly from Hessels.) Note the importance of the "Rhine Way."

tention for a short time. Two towns are rivals for the credit of being the birthplace of printing from movable type. Printed pages, derived from single wood-blocks, were fairly common long before 1440; but the first use of movable metal types seems to have been made between 1440 and 1446 in Haarlem, Holland. The inventor was an oil-merchant, Janszoon Coster, and legend says that an unfaithful servant stole his type and carried it to Mainz. Gutenberg's work is well known and started at Mainz, as far as we know, about 1454. Thence it spread all over Europe as shown by the isopleths in Fig. 50. Rome was reached in 1465, Buda-Pest in 1473, Cracow in 1475 and Westminster (London) in 1476. The map is quite interesting for it shows that the

"Way of Light," *via* the Rhone, is now being succeeded by the "Rhine Way"—as regards the spread of new ideas and processes. Perhaps we saw the first indications of this shift in the spread of the later ideas leading to the Reformation (Fig. 46). They also moved along the route from Oxford to Prague—which more nearly approaches the "Rhine Way" than the "Way of Light." By 1480 all the important capitals of Europe had printing presses—and in 1488 even Constantinople, which had only recently been sacked by the Turks, had its press.

Four famous men of this period concern us in our survey of the Renaissance, in addition to those particularly concerned with the Reformation (pp. 193-7). The first is *Politian* (1454-1494), an Italian humanist and poet who flourished at the court of Lorenzo de Medici. He had an unrivalled knowledge of the Greek and Latin authors, and wrote tasteful verse both in Latin and Italian. Like Erasmus, he lived at the zenith of the appreciation of classical scholarship—whch was soon to have serious rivals as greater interest in national literature and science developed.

Erasmus (1466-1536) has been described as the apostle of common sense in an age of passion and unreason. It is no wonder that Catholics termed him Lutheran, while Protestants thought him afraid to attack the Pope. His chief interest was in translating and expounding classical and early theological manuscripts. He was long associated with Froben's Press in Basel, for which he edited many important works.[2] He recoiled from ignorant priest and bigoted Lutheran alike, and preferred an independent life of scholarship to anything that Church or Prince could offer him. His chief claim to our notice was his edition of the New Testament in Greek (1517?) published side by side with a careful Latin translation. This book clearly showed that the common Latin Bible (the Vulgate) was often a careless and incorrect translation. Its publication can be compared with that of Darwin's "Origin of Species" in its indirect but powerful attack on the ignorance and superstition of many of the orthodox party.

A contemporary of Erasmus was *Machiavelli* (1469-1527), who was concerned most of his life with the government of Florence. His chief work is the "Principe" (1513), which is said to be based in part on the ambitious and unscrupulous life of Cesare Borgia. In this work "he founded the science of politics for the modern world, by concentrating thought upon its fundamental principles" (Symonds). The teachings

[2]At Basel also taught Paracelsus (1490-1541). He despised the medicine of Galen and Avicenna, and used for diseases a mixture of mineral drugs and mysticism.

of the humanists were interpreted in the light of the difficulties of contemporary Italy. Machiavelli frankly believed that a strong, almost a despotic, ruler, was the best for a country's prosperity. It has been well stated that his book forecasted the absolute monarchs who characterised the next few centuries.

Our fourth contributor to the New Learning in the sixteenth century is *Copernicus* (1473-1543). Not until 1500 do we find that the pioneers in progress have moved east of the Elbe. Luther (1517) is one example (p. 195), and the Polish astronomer is another to show that the nations far removed from the Roman heritage were beginning to make major contributions to civilisation. It is but fair to mention that Copernicus was trained in Italy under Novara. In 1543 he published his epochal book on Astronomy, which proved that the earth rotated on its axis and revolved around the sun. Before his time it was generally assumed that the earth was stationary. Copernicus, however, did not realise that the planetary orbits were ellipses—but his discoveries removed the earth from its "central" position in the universe. Modern data, of course, show us that the earth and man himself play inconceivably insignificant roles in the Universe—but this somewhat humiliating but salutary belief was unknown in Medieval Times!

It will be noted by reference to the preceding maps (Figs. 49 and 50) that the field of intellectual progress had left Spain in a "backwash" little touched by new opinions. This no doubt helps to explain the fact that in Spain originated the most effective opposition to the Reformation and to the other heterodox aspects of the Renaissance. Lack of space does not permit us to dwell on this aspect of our subject—but the reactionary Emperor, Charles V, drew much of his power from Spain; and Ignatius Loyola, the devoted supporter of Papal Power, was a Spanish soldier before he founded the Society of Jesus in 1543. (In the seventeenth century this Order was the most powerful among the many Catholic institutions—and its able and learned missionaries have spread the Catholic Faith all over the world.)

In the last century of the Renaissance (1550-1650) a remarkable division of the intelligentsia of Western Europe into two rather well-defined areas becomes apparent. This division is indicated in Fig. 51 which shows that a line drawn from Oxford to Florence separates the main "Field of Letters" from the main "Field of Science." By 1550 printing had supplied thousands of books—where formerly only a few manuscripts were available. New ideas could now be disseminated all over Western Europe in a few months—whereas formerly eager learners were compelled to journey to the teacher's home to benefit from his studies. Naturally there was a great "flowering" both in Letters and in Science which in indicated in Fig. 51.

It is not necessary now to consider in detail the contribution of these famous pioneers. Let us consider the field of Science first. *Vesalius* was born in Belgium—but carried on most of his teaching in Italy. He is sometimes described as the Father of Human Anatomy.

Fig. 51.—The Third Period of the Renaissance from 1550 to 1650. It is marked by the spread of Science and Literature, which are separated by a rather well-marked line of division.

One of his famous successors was Harvey, the Englishman who discovered the circulation of the blood in 1620. In the fields of Astronomy *Brahe* (a Dane) built a well-equipped observatory on Hven Island in the Sound, and there for the first time carried out the accurate and continuous observations so necessary in his science. His data were used by Kepler at Prague to prove in 1609 that the planets moved in ellipses. In Italy *Galileo* about the same time was inventing thermometers, clocks and telescopes, and discovering fundamental principles in astronomy, physics and meteorology. Since these in many cases contraverted the ideas of Aristotle they aroused much opposition. In 1616 *Galileo* was censured by the Church for his astronomical teaching, which seemed contrary to Holy Writ. In 1632 he published a more elaborate treatise, for which he was haled before the Inquisition in June, 1633, and under threat of punishment recanted his views on the earth's motion. It seems likely that in this sort of treatment we have the explanation of the separation of the two fields of Science and Letters (Fig. 51).[3] On the whole, the western area was Catholic. Here it was obviously safer to work in the time-honoured field of Letters. To the east of our line the nations were mainly Protestant, and progress in Science was tolerated and indeed honoured by many rulers.

This brief account of the rise of Modern Science may fitly be closed by a reference to the work of *Descartes* (1596-1650) which dealt with almost all contemporary departments of physics and philosophy. In his youth he was a soldier, who seemed to have adopted that profession (as Dampier did that of a buccaneer) because it enabled him to travel and study mankind. He was born in the "Gate of Poitiers," but after 1629 he dwelt in Holland, where he wrote most of his books (Fig. 51). It is of interest that he never published certain essays which might be construed as unorthodox, for he had no mind to suffer the fate of Galileo. It would be impossible in our limited space even to list the topics studied by Descartes, but his attitude to education is illuminating. It has been said that contempt of erudition and aesthetics is characteristic of the Cartesian school. Descartes objected strongly to the emphasis laid on the study of Latin and Greek; so that we may fairly assume that the period when his philosophy became the vogue marked the decline of enthusiasm for the classics—an enthusiasm which so materially helped to bring about the Renaissance.

The western half of our survey (Fig. 51)—dealing with Letters—does not concern us perhaps so greatly. It is true that in their own fields *Corneille* in France, *Shakespeare* in England and *Camoens* in Portugal have never been equalled. But they represented the flowering of a literature already in full growth at an earlier period. Glancing back over our recent discussion we find that the earlier stages of the Renaissance were on the whole more peaceful and were characterised by the poetry of Dante, Petrarch and Chaucer. The middle period was marked by bitter political and religious struggles, an environment in which literature does not in general flourish. Much of the last

[3]Some readers may feel that this chart savours somewhat of "picked data." The author plotted the names impartially from several authoritative accounts of the Renaissance. He sees no reason to doubt that a real regional distinction obtained in late 16th and early 17th centuries, for the reasons stated.

period, 1550-1650, was more peaceful, or at any rate, more stabilised; though truly it ended with the worst struggle of all, the Thirty Years' War. At any rate, Poetry and Drama reached its zenith in western Europe in this last period.

However, three men whose writing greatly affected Public Opinion must receive some attention. They are Rabelais, Cervantes and Comenius. The "Gate of Poitiers" saw the birth of Rabelais (1490-1533) as well as of Descartes. *Rabelais* left the cloisters to become a physician in Lyons in 1532. His unorthodox "Pantagruel" annoyed the Church and led to his taking refuge in Metz in 1546, but he soon returned to France and published several further volumes of his fantastic story of Pantagruel and Panurge. He described with his satirical pen the foibles and follies of his age, but it is impossible to state clearly what was his outlook on religion and the world. Tolerance and good humour were not characteristic of Medieval Times, but Rabelais certainly excelled in both. Montaigne (1533-1592) also derived from the West of France, spending most of his life near Bordeaux. His influence on literature was profound, since he practically created that whimsical type of essay of a kind which is more familiar to Englishmen from the pen of Lamb. It cannot be said that he greatly influenced public opinion, but almost all later writers, including Shakespeare, benefited from his essays.

Of more importance were the writings of *Cervantes* (1547-1616), who was born at Alcala near Madrid. In 1571 he took part in the famous battle of Lepanto (p. 370), where he received three wounds. He was captured by corsairs in 1585 and spent five years as a slave in Algiers. Later he was imprisoned for debt, and probably wrote part of "Don Quixote" in jail about 1600. The second volume appeared in 1615. As everyone knows, it is one of the first and finest novels extant —but it is a novel with a purpose, that of ridiculing the current romances dealing with the Age of Chivalry. As in the case of Montaigne, his influence was exercised on literature rather than on the course of world thought.

Our last example was of a very different type; and is the only one on the wrong (*i.e.*, east) side of the division (Fig. 51). *Comenius* was born in Moravia, but wrote most of his books at Lissa (north of Breslau). He is of interest to us because his work illustrates the birth of new ideas of worldwide importance among the east-central European nations. He was the chief pioneer of modern education, and his books thereon (1631) were read all over Europe. He believed that languages should be taught by conversation and that children should learn from pictures. Geography, Handicrafts, and Science were all necessary features of education in his opinion. These principles are only nowadays being generally accepted; and, indeed, many conservative schools, especially in England, still follow Vittorino (who at the dawn of the Renaissance emphasised classics and athletics) rather than Comenius, the best teacher at the close of the Renaissance.

D. *The Development and Distribution of Architecture*

We have now followed the spread of Renaissance ideas through the chief countries of Europe. A brief survey of the way in which more material products of the period, such as the great churches, illustrate human progress in the later Middle Ages, will be found of interest. The flowering of Renaissance literature occurred about 1600, but no period in Architecture, since the Golden Age of Pericles, has produced such marvellous monuments in stone as those which were erected about the year 1300. In this century appeared the fanes of Rheims, Exeter, Cologne and Milan, and though many noble buildings appeared later, it would not be easy to prove that they were finer samples of architecture. Rome was not built in a day, and the gradual evolution of the Gothic Architecture—precisely during that period of mental evolution known as the Renaissance—is well worth a few words at the close of this section devoted to General Factors in the Evolution of the Nations.

If we compare a little more closely the spread of Renaissance ideas with the spread of architecture, we find that Romanesque Architecture was common during the eleventh century, well before the literary Renaissance began. This developed into Early Gothic Architecture about 1100, and the finest period of the Gothic had been reached about the time of Dante. The Third Period of the literary Renaissance (1550-1650) was therefore contemporaneous with the waning in favour of late Gothic Architecture and its replacement by the Renaissance style. This comparison in time appears at the left of Fig. 52.

The Christian Churches at the Fall of the Roman Empire were usually in the form of *Basilicas*. Such a church may be simply described as a large rectangular hall, much resembling in shape the nave of a Gothic church. Often an aisle ran along each side of the nave—and in the larger structures an apse (*i.e.*, a semi-circular projection) was added at the altar end of the nave. The whole structure was very massive and the nave was usually supplied with many small windows with rounded arches. A simple church of this type survives which was built at Banos in North Spain as early as 660. Meanwhile, two quite different styles were developing around the Mediterranean Sea. The Byzantine type was characteristic of Constantinople after it was founded by Constantine in 330. In large examples, such as Saint Sophia (550), we find a square hall carrying a huge dome—with four accessory apses or chapels which make it resemble a Greek Cross in

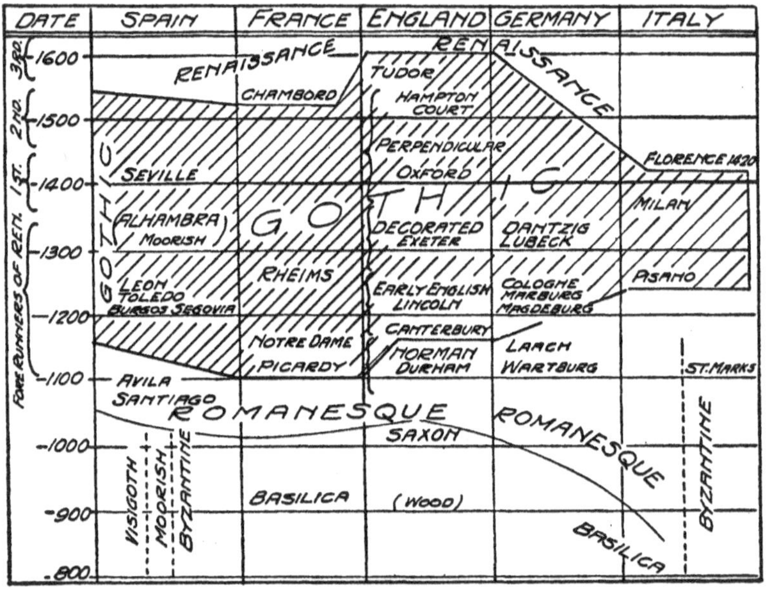

Fig. 52.—A Table showing the Evolution of Architecture in various European lands. Note the correlation with the times of the Literary Renaissance in the left-hand column.

plan. The interior is everywhere enriched by mosaics, and the well-known Cathedral of St. Marks at Venice (1100) is of the same general type (Fig. 53). Our map[4] shows that this Greek or Byzantine style spread to Kiev by 1020 and to Novgorod by 1050. There are a number of early Byzantine Churches in the south of France, such as Angoulême (1130).

The Crusades occupied the twelfth century and resulted in a veritable early renaissance in the development of architecture, while the

[4]These maps are based mainly on the dates given for notable examples of the types discussed in the "Encyclopedia Britannica."

Moorish invasions of Spain also gave rise to great modifications of Spanish buildings. The Basilica changed gradually into the Romanesque (or Lombard) style, which is well illustrated by St. Ambrogio at Milan. In this type the facade (front) is generally characterised by many heavy small round arches over the windows and doors. The exterior is more enriched by ornamental detail than is the Basilica; and towers are at times added to dignify the facade. The map (Fig. 54) shows that this type originated in central Italy and spread quickly up the Rhine to Worms and Laach by about A.D. 800. In the north-west of Europe it is often called the Norman style. Fine examples were built in London, Durham and Tewkesbury before 1120. About the same time the Romanesque style spread into Spain.

Fig. 53.—Some examples of the Spread of Basilicas, and of Byzantine and Moorish Architecture.

Fig. 54.—The Spread of the Romanesque (or allied Norman) Architecture from Rome to the north and west.

Gothic architecture developed in direct response to the environment (Fig. 55). In the darker northern lands the heavy Norman naves with their small windows were shrouded in gloom. The walls were therefore pierced by very large windows, covered usually with pointed arches—which were probably copied from buildings in Syria. The use of buttresses to support the weight of the heavy roof became necessary now that the walls were so much less solid. Strong roof vaulting also developed in the form of arches supported on clustered pillars. In this fashion the Norman style changed into Early Gothic in the region near Paris about 1137. The best known Gothic Cathedral, that of Notre Dame in Paris, was built about 1166. At an early date this type spread into Spain, for it was used in Burgos about 1146.

During the 13th century (as already noted) magnificent early Gothic cathedrals sprang up all over England and France. Very few large churches of the Gothic style are found in Italy. Milan Cathedral is the most striking exception, and this was built about 1386. But it persisted longer in England perhaps than elsewhere; passing from "Early English" to "Decorated" and so to "Perpendicular" as noted in Fig. 52. The last subdivision is called "Tudor" and flourished during the middle of the 16th century (in Elizabeth's time), but it was confined to England. It is stated that the word "Gothic" was given to this noble type of northern architecture by Rafael, who was not favourably impressed by the specimens with which he was acquainted.

Fig. 55.—The Spread of Gothic Architecture from the vicinity of Paris to Spain and England.

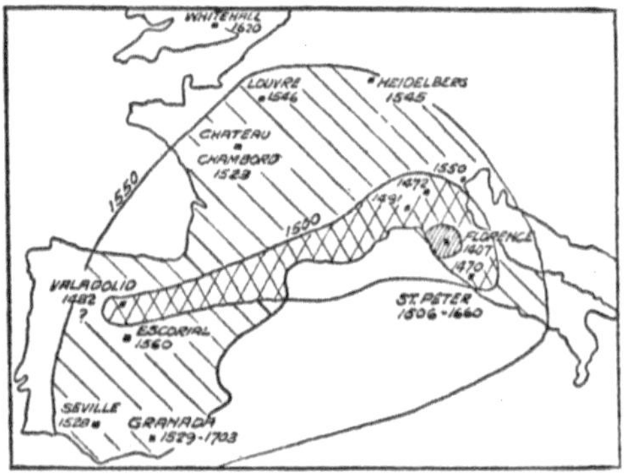

Fig. 56.—The Spread of Renaissance Architecture from the north of Italy into Spain, France, England, etc.

In Florence about 1420 Brunellesco and others were experimenting with designs which went back to classical times. This brought about a Renaissance type of architecture—strictly comparable in origin with the Renaissance in Literature (Fig. 56). In the new style the dome was used as in Byzantine churches. Pillars now replaced clustered columns; the ceiling of the nave was made flat instead of being vaulted, and the external walls were built of columns with pediments and niches like those of the temples of ancient Rome. St. Peters in Rome, built between 1506 and 1660, is the finest example of its kind. The fashion spread, about a century later, to France, where a château at Chambord (of 1523) is perhaps the first large building. Many fine buildings of the same type in Spain also date from the 16th century. In England Inigo Jones built several Renaissance structures, such as

Whitehall in London, early in the 17th century. We may therefore conclude that in the north of Europe the ushering in of the Renaissance buildings was contemporaneous with the close of the Renaissance in Literature (Fig. 52). If we compare Fig. 55 with Fig. 56 we can see that the environment seems to have determined the Gothic and Renaissance territories. The latter is more prevalent in the sunny Mediterranean lands where indeed the classic style of architecture had originally developed some twenty centuries or more before.

PART III

THE PROTESTANT NATIONS OF NORTHWEST EUROPE

CHAPTER XII

FURTHER GEOGRAPHICAL FACTORS IN ENGLAND'S HISTORY

A. Introduction. Build and Growth of London

In Parts I and II of this book we have discussed the broad physical and cultural settings of the Nations of Europe. In Parts III, IV and V we take up the descriptions of the various nations in some detail. A few words may be given in explanation of the grouping of the nations employed. Obviously there are many possible categories, depending on which particular factor is most stressed. Since this book endeavours to link the two fields of geography and history, it seems best to adopt a grouping which will appeal to students of both disciplines. For this reason I have adopted three cultural divisions which also agree broadly with physical regions. We have noticed that the growth of slight language differences is enough to divide nations, so that they have nowadays little community of thought. One may instance Holland and Germany. But a common religion indicates a common background of culture which steps across linguistic boundaries. Thus France and Italy, though not linked by languages, are closer than are France and Germany, for their background, based on the Catholic religion, is the same. Following out this idea we find that the three most convenient divisions of the national groups are those given in the Table below. (The numbers of their chapters are added.)

TABLE R

RATIONAL GROUPS OF EUROPEAN NATIONS

	Group I	Group II	Group III
Location	Northwest	Southwest and Centre	East
Chief Religion	Protestant	Roman Catholic	Greek Orthodox
	Britain XII Scandinavia XIII Germany XIV Netherlands XV Switzerland XVI	France XVII Spain XVIII Italy XIX Austria XX Bohemia and Hungary XXI Poland XXII	Balkans XXIV Rumania XXIII Russia XXV East Baltic States (not Greek Church) XXVI Conclusion XXVII

In Chapter II the build of the British Isles and some of the more obvious relations between structure and history were discussed. In the present chapter a similar relationship in regard to the city of London is first considered. Next, the way in which environment in part determined the subdivision of England into Counties is considered. The distribution of the rival parties in the Wars of the Roses and again in the Puritan Revolution has an interesting relation to the structure of England, and so may well be referred to. The gradual change in environment which led to the Industrial Revolution is discussed in the last few pages of this chapter.

The growth of a great city naturally depends much more closely on environment than does the development of a Nation.[1] Since London is not only the capital of the English Nation, but also the heart of the British Empire and the largest city of the world, we may well devote a few pages to the relationship between Environment and Man which has led to the development of a settlement of over seven million people. In an earlier section the site of London was briefly discussed (p. 37). It occupies the centre of the youngest beds of the Tertiary Basin (Fig. 3)—where the sea has scoured out a broad estuary in the soft rocks—almost opposite the Rhine mouth and the Corridor of Brabant (p. 290). We may now turn to its local topography.

Research in the London area shows that the early Gaulish tribes who settled in the vicinity preferred the open woodlands and pastures of St. Albans (Verulam) and Colchester to the swampy flats of the Thames. The North Downs—open ridges of chalk—led the traveller from Rutupia (near Dover) towards Brentford and Westminster, where he would cross the Thames and so reach St. Albans, some 20 miles to the northwest (Fig. 3). Possibly London was not in existence when Caesar made his raids in B.C. 55 and 54. The name London is supposed to mean the "Dark Pool" (or perhaps "Fort on the Pool") and is of Welsh origin. It refers probably to the wide sheet of water which at high tide covered the Thames Flats near the site of London. These have all been reclaimed, but even to-day some of the wide marshes of the River Lea east of the city are drowned at high tide (Fig. 57). A dense forest covered the heavy clays between the Thames and St. Albans—there being only a narrow belt of sandy

[1]Excellent studies of the geographical factors controlling the evolution of Glasgow and Edinburgh have been published by the Scottish Geographical Society. The writer has just published a similar shorter account of the relation of topography and growth for Toronto ("Canadian Journal of Economics and Political Science," Nov., 1936).

heath close to the river—which was of little use for primitive agriculture. The name Blackheath reminds us of a similar heath on the south of the river. The flats of the river near London were occupied by marshes two or three miles wide with, however, small, gravelly islands scattered through them. One such islet (Saxon "ey") was later known to the Saxons as Thorn-ey; and among other similar islets were Chels-ey, Batters-ey, Bermonds-ey, etc. It seems likely that the

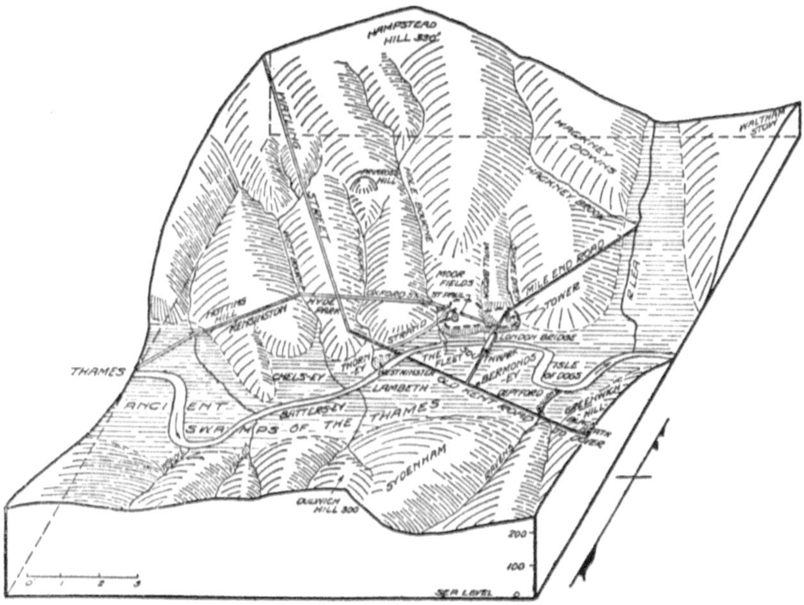

Fig. 57.—A Block Diagram of London, showing the control of Settlement by Topography. Note that Watling Street originally crossed the Thames by a Ford near Thorn-ey (Westminster). The Port developed near the Fleet Stream. The heavy broken line around the "Twin Hills" shows the site of the early London Wall.

lowest ford on the Thames was at Thorney (now Westminster), and a rough road soon led north from the ford towards St. Albans. Two small streams, West Burn and Ty Burn, here cut through the gravels to reach the river marshes. They are shown on our Block Diagram (Fig. 57); with the old road, later Watling Street and to-day the Edgeware Road, running between. After crossing the ford the road to the south (now the Old Kent Road) climbed Blackheath and so reached the Kentish ports near Dover.

Probably some time before the invasion of Claudius (A.D. 43) London became of some importance as a fishing port. The river swings somewhat to the north below Westminster and flows under fairly steep gravelly hills. These were trenched deeply by three streams which were kept flowing strongly by the springs in the gravels. These three small streams, Hole Brook, Wall Brook and Shore Ditch, have carved the gravels into a pair of somewhat isolated hills, between which lay snug little coves suitable for the small boats of that day. (The Tower now dominates one, St. Paul's the other.) Thus Wallbrook Valley was about 130 feet wide just where "Poultry" (Road) crosses it to-day. Hence, the port of London developed on the north bank about two miles below Thorney Ford, where a bank of stiff clay stands about 30 feet above the river.

The Romans during the first century A.D. built a fort on the eastern of the Twin Hills, and about A.D. 400 constructed a wall which included both hills, and ran just south of east-west gullies which nearly connected Wall Brook and Shore Ditch. It seems to have been built in a hurry—perhaps to resist invasion, for it is formed of all sorts of heterogeneous material. The Wall is indicated by the heavy line (like a dumb-bell) in the Block Diagram (Fig. 57). We do not know when the first bridge was built across the river. It was naturally constructed where firm high land came close to the river, and so London rather than Westminster Ford was chosen as its site. A timber bridge is known to have existed before 1000 (perhaps in Roman times), and Southwark had developed at its southern end. When the Bridge was built Watling Street was diverted from Westminster toward London. It entered one of the old gates, near the later New Gate, and then crossed the river, within the protection of the City Walls, to Southwark.

During early Saxon times, London was almost deserted, for the Saxons were not town-dwellers. After the Danish invasions it may have actually been abandoned, so much so that some historians claim that Alfred refounded London. He repaired the walls and added New Gate, Aldersgate and Bishopsgate, and it became a prosperous city under his care.

In 1066 William the Conqueror marched from the Battle of Hastings to Dover and then along the main road to Southwark. Here the Saxons opposed him strongly; so he left London and proceeded up the Thames nearly to Oxford, laying waste all the lands he passed through. Here he crossed the Thames and moved to Berkhampstead near St.

Albans, where the London citizens offered him the crown. To maintain his precarious conquest he covered England with strong castles. The Tower of London was begun by him, and was naturally erected close to the Bridge at the east end of the Walls of London. Later Norman Kings added to it greatly so that it now covers 13 acres. The White Tower (1078) is a fine specimen of Early English Architecture (p. 214).

In the twelfth century the tiny area of London City could be divided into four east-west belts. Along the river, in narrow streets reclaimed from the marsh, were the fisherfolk and watermen. Behind this area were the homes of the merchants on the higher ground. The third zone contained the markets of West-chepe and East-chepe. The latter is still the name of the street leading east to the Tower. The north belt was industrial and at that time contained some small orchards and isolated dwellings. The Wall on the north ran along the line of the present Underground—whose stations, Aldersgate, Moorgate, and Aldgate, perpetuate the old gates through the Wall. About 1400 London filled the area within the Walls and also extended solidly to Temple Bar about half a mile further to the west.

Thorney was probably deserted for a century or two after the Bridge was built. Its isolation rendered it suitable for a monastic retreat, perhaps before the Danish invasion. Here Edward the Confessor built the Abbey of Westminster on the gravel ridge where Ty-Burn entered the Thames. The Norman Kings seem to have strengthened the Tower on the east and Westminster on the west in part to overawe the truculent citizens of London. There was great rivalry between Westminster and the City through Medieval Times, but the foreign trade preferred London, which soon forged ahead. Henry III rebuilt the Abbey, and by degrees Palaces and mansions grew up around Westminster, but for many centuries fields occupied most of the two miles between the two towns. The road along the edge of the higher ground was called the Strand—as it is to-day. In the time of Elizabeth there were almost continuous houses along the Strand and many fine mansions stood between the Strand and the river.

The writer possesses a small county-atlas of Britain dated 1748, which clearly shows that the close-set houses of London at that time only extended west to Ty-Burn (Fig. 57), north to Tottenham Court and Islington, and east to Wapping. Southwark had hardly grown since Saxon days, though a fringe of houses extended along the southern bank of the Thames from Lambeth to Bermondsey. London now covers about 16 times the area of the city of 1748, and close-set houses extend far beyond the limits of the block diagram in all directions.

It has been well said that London City has made and unmade more Kings perhaps than any other of the European capitals. The decisive factor in almost every struggle for the crown was the attitude of the capital. Examples could easily be illustrated in connection with Edmund Ironsides, William I, Henry I, Stephen, Edward II, Richard II, Henry VI, Henry VII, Charles I, Charles II, and James II. Some of these events will be discussed later.

B. *The Origin of the Counties of England*

The salient features of the Roman and Saxon occupation have been described in Chapter II. The occupation of the east of England by the Danes occurred in two main migrations. The first extended from 787 to 870 and resulted in the conquest of most of England north of the Thames (Fig. 26). Only Wessex and the northern part of Northumbria remained free of the Danes.[2] The Britons of Wales made common cause with the Norse pirates, and Wessex nearly succumbed to attacks from east and west. King Alfred of Wessex, however, rallied all the Saxon forces and defeated the Danes at Eddington (near Stonehenge). As a result the Danes in 878 withdrew to the region known as Danelaw—*i.e.*, northeast of Watling Street (Fig. 5). This division of England was not a "natural" one, for it ran right across the "grain" of the country. Thus the *Danelaw* included the rather rugged Pennine ranges as far north as Cumberland (which was still Welsh) and Durham, which was still held by the Angles. "*Wessex*" was England south of the Severn and Thames, and *Saxon Mercia* was essentially the March formed by the Severn Basin. We cannot omit some reference to various other claims of Alfred to our admiration. He translated the history of Orosius, the chronicles of Bede, and the work of Boethius into the Saxon language. He founded the British Navy, which soon controlled the Channel. Especially in regard to administrative districts of England is the pioneer work of Alfred and his immediate successors worthy of our attention.

The counties of England have a local importance which is totally disproportionate to their size. In other lands, such as Australia and

[2]There was a second invasion by Danes, beginning in 991, which resulted in a Danish conquest of the whole kingdom from 1013 to 1042.

U.S.A., counties are of little importance save as administrative units. There is naturally no common factor in the culture of the folk inhabiting each of them, but this is not the case in England. In many cases the counties express an original natural grouping of families or

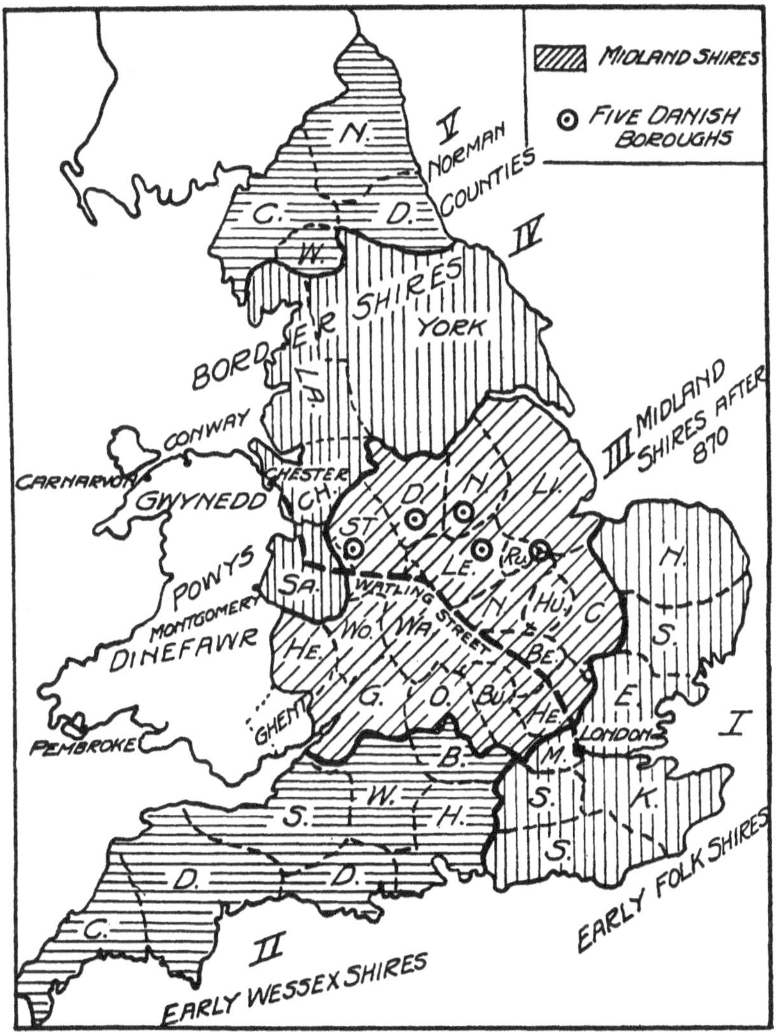

Fig. 58.—The Evolution of the Counties of England, which are arranged in five groups in order of their subdivision. The main Welsh districts about 1086 are also shown. (Ghent should be Gwent.)

clans about a common centre; and until the rapid mixing of modern days, something of this early historical grouping clung to the counties. Devon, Somerset, Norfolk, Essex and Kent, each has its local history, its famous sons, and even its real differences in dialect.

A survey of England shows us that the counties belong to five different classes (Fig. 58). Naturally the oldest counties are found in the southeast, and the latest in the northwest, for that represents the trend of progress throughout England's story. The first class of shires represent the early well-marked settlements of groups of Nordic invaders and occur in the southeast. Thus the Angles inhabited the region between the Wash and the Stour, and this was early divided into two "shires" (or "shares"), those of the "North Folk" (Norfolk) and the "South Folk" (Suffolk). *Kent* was originally the "open ground" (Chenth, *i.e.*, the North Downs) between the great forest of Andred (which occupied the hollow of the Weald, p. 37) and the marshes of the Thames. It was settled largely by Jutes from Denmark. *Sussex* was the similar open ground of the South Downs (Fig. 58) occupied by the South Saxons. *Essex* had a similar story based on the settlements of the East Saxons. Between the two were *Middlesex* (around London) and *Surrey*, the "South Rege" or region to the southwest of London.

The second group of shires developed in the south of England. Possibly in point of time these areas in Wessex were the earliest shires (called counties by the Normans) carved out of the larger kingdoms. Their names do not refer to clans of Nordic invaders, but usually to local sites. Thus *Hampshire* is said to mean the "Home-share;" and lies around the estuary of the Itchen, where the West Saxons established themselves about A.D. 500. Half a century later, a shire (or share) was carved out around Wilton (to the northwest of Hampshire) and this became *Wiltshire. Somerset* and *Dorset* refer to the lands of settlers (settan) around the "Sea-Meres" and Dor-chester, respectively. *Devon-shire* is the share of the old British tribe of Dumnonii, and the name *Cornwall* (p. 114) reminds us that it was part of Wales long after the rest of South England belonged to the Saxons. Since 1337 Cornwall has always been held by the King's eldest son, who is Duke of Cornwall. *Berkshire* is said to be the "forest-share" (containing many *berroc*, *i.e.*, box trees) south of the Thames.

The third group of shires occupies central England (Fig. 58). They are all of much the same area, and each surrounds an important town after which the county is named. In general the boundary is about 15 or 20 miles away from the town—at such a distance that it could easily be reached in a day's journey over the poor roads of early England. These 15 shires can be divided into two sections according as they lie west or east of Watling Street (Fig. 58). All belong to the old border kingdom of Mercia, but those east of Watling Street were long under Danish control. The Danes established "Five Boroughs" or strongposts at Derby, Leicester, Lincoln, Stamford and Nottingham, but about 917 these were conquered by Ethelfled, the Saxon Princess of Mercia. Each became a shire (except Stamford) of the same type as the Saxon shires which grew up around Gloucester, Hereford, Worcester, Warwick, Stafford, Buckingham, etc. Other former Danish districts of the same general type were Cambridge, Bedford, Hertford, Huntingdon and Northampton, all of which lie between London and the Five Boroughs. The tiny division of Rutland did not become a shire until the time of Henry I.

The fourth group of shires consists chiefly of the rugged country in the northwest, which long resisted complete conquest by invaders—whether Roman, Saxon, or Dane. *Shropshire* has an interesting history. The word means the "share of the scrub" region, and is named from "Scrobs-burg," a town founded by King Offa in 780, which has since become Shrewsbury. *Cheshire* (around Chester) was much larger than it is now when, in 1086, William compiled his invaluable Domesday Book. It then included the Welsh county of Flint. It was a *County Palatine*, which means that its ruler acted somewhat independently of the King, so as to control more quickly the difficult situations bound to arise in borderlands. Right up till 1541 Cheshire had a special government, but no representation in Parliament. *Lancashire* in 1086 was not a separate county, but was partly administered by Cheshire and partly by Yorkshire. Not till 1177 were its boundaries laid down, when it was given to the King's second son. Throughout medieval history it remained a conservative, poorly populated region, and as we shall see it usually supported the longer-established party in various struggles, and resented innovations.

Yorkshire has an area of nearly four million acres, and is five times the size of the average county. It is divided into three Ridings (originally Tredings, *i.e.*, Thirdings). The eastern third was occupied by Angles in the Derwent Valley, the northern third by Angles along the Ouse; the western third was long held by the dark Britons—who are still to be found in the Forest of Elmet near Leeds.[3] York (*Eboracum* of the Romans) became the chief centre early in Danish times.

The fifth group consists of those four northern counties—defined after the Norman Conquest—which therefore in a sense were never "shires." *Northumberland*, as the name implies, was the Anglian conquest north of the Humber. By the Danes it was split in two at the Tees, so that the south part became Yorkshire—while the north still retained the old name. This northern part was again divided into Durham and a still smaller Northumberland between the Tyne and Berwick. *Durham* was a special area controlled by the Bishop of Durham, to whom William I gave palatine powers akin to those of the Earls of Lancaster, Northumberland, etc. *Cumberland*, as the name implies, was for centuries inhabited by Welsh folk (Cymry) and formed part of the Welsh region of Strathclyde (p. 48). It was often invaded by Norse pirates from the west and by Danes from Yorkshire. At the time of the Domesday Book it was still unsubjugated. In 1092 William Rufus transferred a number of settlers from the New Forest in Hampshire to Cumberland, and many castles were built here for protection against Scottish forays. Henry I separated the wilder "western moors" of Yorkshire and added them to the rugged area of Cumberland around Appleby. Thus originated *Westmoreland*, the last of the counties of England.

A few words may be given to the events which led to the union of England and Wales. Domesday Book (1086) omits Wales from its survey. During Norman times the borders were held by powerful barons who built many strong castles, as at Shrewsbury and Montgomery, to maintain their authority along the March from Chester to the British Channel. In the 11th and 12th centuries the Welsh had lost most of their civilisation and lived as savage clans in the mountains, much as did the early Highlanders in Scotland. About 1100 the Earl of Clare led a force of Flemings and English to Pembroke and there he founded "Little England." Henry II tried again and again to conquer Gwynedd without success (Fig. 58). King John had better fortune and in 1211 Prince Llewelyn of North Wales was imprisoned. However, the Welsh rebelled again, and from 1246 to 1277 a second Llewelyn was master of Wales.

In 1277 Edward I conquered North Wales and built the castles of Conway and Carnarvon to maintain his authority. In 1284 he

[3]E. H. Carrier, "Historical Geography of England," who gives a good account of the evolution of the shires.

created the counties of Flint, Carnarvon, Merioneth, Cardigan and Caermarthen. The Welsh Barons still held the rugged central country, while "Lords Marcher" (akin to the Mark-graves of Germany) held the border. Of these Mortimer (Montgomery), Bohun (Brecknock) and Clare (Glamorgan) were the best known. During the wars with France Glendower raised Wales against the English from 1400 to 1410. Henry VII placed Wales under English law, and about 1536 Dinefawr (southeast Wales) was divided into six counties. Of these the most eastern, Monmouth (formerly Gwent), was attached to England.

C. The Civil Wars of the 15th and 17th Centuries

The so-called "Wars of the Roses" (1450-1471) followed directly from the power acquired by certain northern nobles in England during later Plantagenet times. Clearly when the royal power dwindled the powerful barons (who were almost independent in their strong castles along the Welsh and Scottish borders) felt that they might gain much by supporting rival claimants. Moreover, certain of these border rulers were of royal blood themselves, and owned large estates in many parts of Britain besides their holdings on the Border. The feebleness of Henry VI and his lack of an heir encouraged the Duke of Somerset, a descendant of Edward III through John of Gaunt (Duke of Lancaster), to aspire to the throne. His rival was Richard, Duke of York, who was allied to the border family of Mortimer, as well as descended from Edward III. He was supported by the Earls of Salisbury and Warwick. Open warfare broke out at St. Albans, where Somerset was killed in 1455. However, civil war still went on between the factions of Lancaster, led by Queen Margaret, and of York, who rallied the southeast to his side. Richard of York was slain at Wakefield (1460), but his son, Edward, Earl of March, carried on the contest and was helped by the Earl of Warwick, "the King Maker." Margaret won several minor battles, but was crushed at Towton, near York, in 1461, the bloodiest battle in England since Hastings (1066).

Ten years later Warwick turned upon King Edward IV and endeavoured to restore the mad Lancastrian King Henry VI. In 1471 Edward won at Barnet and again at Tewkesbury, and the hopes of Lancaster vanished. Edward's son (Edward V) was murdered by his uncle, Richard III, who, however, only reigned two years. In 1485 his enemies supported Henry Tudor, whose father was of Welsh descent, while his mother was one of the Somerset family. He was therefore descended from Edward III through the Duke of Lancaster. At Bosworth (near Leicester) Richard was killed, and Henry VII, first of the Tudor Kings, came to the throne. It would be incorrect to state that this civil war involved the nation as a whole; mostly it was a struggle for supremacy among the leading earls and their immediate followers for the acquisition of supreme power. But in general we may say that the northwest, the conservative Older-Mass, supported the Lancastrian party, while the centre and southeast (the Younger-Mass) were generally favourable to the revolts of the Yorkist party. At the end of the contest Henry Tudor landed in Pembroke and rallied many Welshmen to his side, and once more Lancaster gained the upper hand.

It is more important in our study of national development to notice that the Wars of the Roses made it relatively easy for the Tudor monarchs to acquire absolute authority. The most powerful barons had largely been killed or ruined in the Wars, and the Church was soon compelled to give most of its powers and wealth into the hands of Henry VIII. The middle class was growing slowly in power, but was not able to challenge the absolute rule of the Kings for nearly 200 years. During this interval, largely owing to the wise rule of Elizabeth, the trade and prosperity of England were rising to a height never before attained. The distinction between southeast England and northwest England, which has been emphasised throughout this brief study, was becoming more marked. The forests were replaced by farms, and were also cut down for use in the various crafts and industries which flourished in the southeast. The north and west had relatively few dense forests or farms, but this region was in general suited best for grazing. The age of coal was to alter this distribution entirely (p. 234), but not for another two centuries after the great Civil War, which must now engage our attention.

There are two outstanding features in the Civil War which are of importance in this particular study of national development. The political and religious reasons for the war belong purely to the field of history, and though we cannot ignore them, they are necessarily of less interest to us than are the effect of the war upon the constitution, and the way in which the strongholds of the two opposed parties were distributed throughout the kingdom. Henry VIII had been almost absolute in his power, while Elizabeth identified herself so closely with the nation's welfare that she might perhaps be described as the best type of absolute monarch. James I, with his Scottish upbringing, was not only unfamiliar with his English environment, but had little personal charm and entirely lacked the astuteness of Elizabeth. His son, Charles I, was unable to realise the importance of the growth of a powerful middle class based largely on the flourishing trade with the continent and with the new lands in America. It was his misfortune that he firmly believed that the English Church should draw closer to Rome, whereas a large proportion of English people were steadily moving in the direction of nonconformity, even with the English Church. The attitude of the obsequious Anglican Church under Laud contrasted unfavourably with the independent creed of Calvin and his followers, the Scottish Presbyterians. As usual, there were financial aspects in the struggle. Charles needed money to carry out his ideas, and found it impossible to collect it in the teeth of the opposition of Parliament led by Hampden and Pym.

The actual struggle was begun by the Scotch, who were resolute in their defence of the teachings of Calvin. They raised an army under Leslie, a general trained in Sweden, and occupied Newcastle in 1640. Trouble broke out in each corner of the realm. It is true that in 1641 Charles visited Scotland and for a time seemed to quell the disaffection there. But the Irish rising in 1641 had followed the impeachment and execution of Strafford by the Parliament. The Catholics in Ireland in certain areas massacred the hated Protestant "settlers" wherever they could lay hands on them. Meanwhile in England a party of moderate men more afraid of bigoted nonconformity than of the approach to Rome gathered behind Clarendon, Newcastle and Falkland to aid the King.

In May, 1642, Charles marched to Hull in an unsuccessful attempt to seize the arsenal there (Fig. 59). He then raised his standard at York, while the Parliamentary headquarters remained in London. In the autumn the two forces faced each other across the Trent Valley. Once again we see the age-old division of England—based on its fundamental structure—appearing in the struggles of two parties of Englishmen. Almost all the country northwest of the Exeter-York line consists of the Older-Mass, including the folded transition area of Carboniferous rocks called the Pennine Range (Fig. 3). This region was almost solidly in favour of the King and represented the conservative people in England (Fig. 59). The only notable exception was that the elder University of Oxford strongly supported the King. On the other hand, all the southeast of England (plus the Cheshire Gate) stood by the Parliament. It included the Younger-Mass (Fig. 3) and was at that date the chief manufacturing part of England. Pembroke, as we have already noted, was always isolated from the rest of Wales and was known as "Little England" (p. 228). It also stood by Parliament. The alignment was in some respects like that of the parties in the Great American Civil War, where the manufacturing and popu-

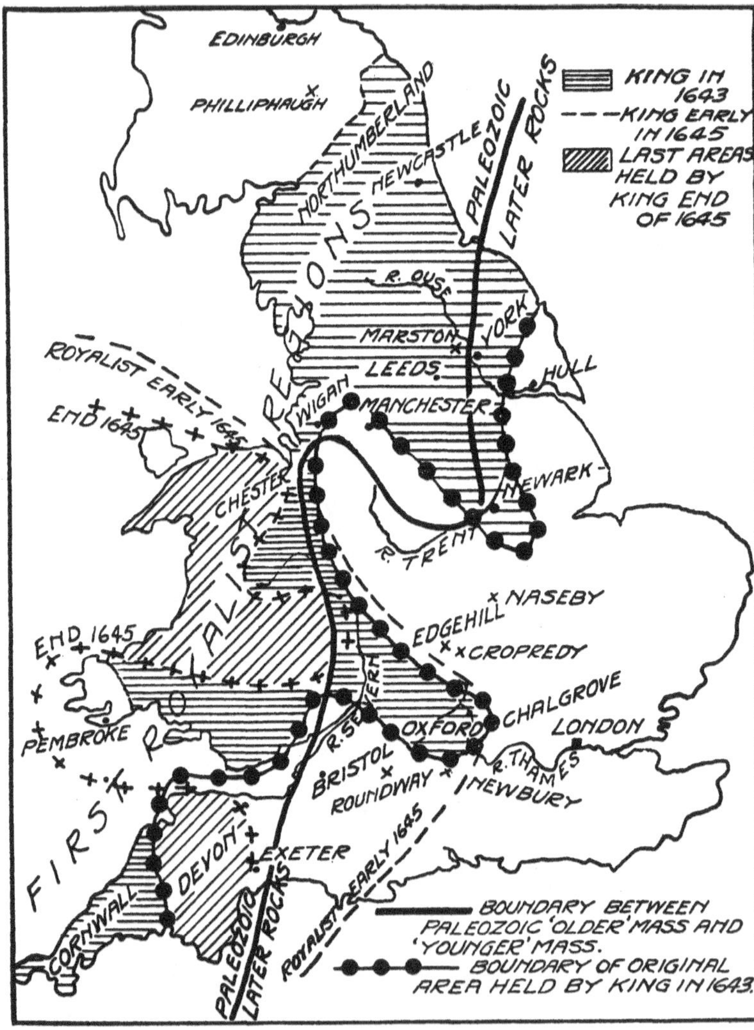

Fig. 59.—The Distribution of the Parties in the Civil War (1642-1646). The strength of the King was in the western Older-Mass (with Oxford), while that of the Parliament lay in the eastern Younger-Mass.

lous north was opposed by the less populous but more conservative and more military south.

During October, 1642, Essex, the cautious leader of the untrained Parliamentary army, manœuvred near the Stratford Avon. Charles marched from the north towards Oxford, met Essex and fought the indecisive battle of Edgehill. Prince Rupert thereupon menaced London and captured Brentford just to the west of the city, while Charles fortified Oxford. In 1643 the Cornishmen rallied to the aid of the King, for Cornwall was still almost feudal in its loyalty to its local chiefs. It had been isolated from the rest of England not only by its position but because the old Welsh culture still persisted in part. Hence the Cornish were conservative to a degree and almost unaffected by the progressive ideas of Parliament and London. Grenville and Hopton won all the engagements against the Puritan forces, and at Roundway (1643) Waller's army was crushed. Hampden was killed at Chalgrove, and Bristol, the second city in England, surrendered to the King.

However, though the Royalists won in the field, they lost greatly by negotiations. Parliament agreed to bring the English Church into close relations with Scottish Presbyterianism, which brought the Scottish army to their aid. Charles, on the other hand, made terms with the Catholic rebels in Ireland, who agreed to invade Scotland. By this arrangement an English Royalist army under Ormond was set free to help Charles. Meanwhile, Cromwell, the greatest of the Puritans, was coming to the front in East Anglia. He smashed Prince Rupert and his Welsh supporters at Marston Moor (1644) and so saved the North for the Parliament. But the Irish and Scottish Catholics defeated the Scottish Presbyterians at Tippermuir (near Perth), while Waller was again defeated at Cropredy (near Oxford) and Essex badly beaten in Cornwall. However, Charles was foiled in his attack on London at Newbury and again retreated to Oxford. The indecision of the Puritan generals—who were in fact loath to crush the King!—brought Cromwell rapidly to the front.

In 1645 dissensions among the Puritans and the victories of Montrose in Scotland over the Presbyterians again encouraged the King. He marched north from Oxford and captured Chester, and turned thence to the east. But Cromwell and Fairfax met him at Naseby, within a few miles of Edgehill where the war started (Fig. 59). The Royalist army was finally routed here, and Charles fled to Wales. There was no further serious resistance to the armies of Parliament in the south, and Montrose was crushingly defeated in September, 1645, at Philiphaugh. This really ended the war; though in 1648 an army of Scottish sympathisers invaded England. They were completely routed by Cromwell near Wigan in Lancashire, and in January, 1649, King Charles was executed by order of the Parliament.

D. *Change in Population Distribution due to the Industrial Revolution*

We may well conclude our discussion of the effect of environment in England by a brief reference to the Industrial Revolution. The course of social evolution described in the following paragraphs has developed in much the same way in every part of Europe where a coal supply has been found. The present writer does not understand how

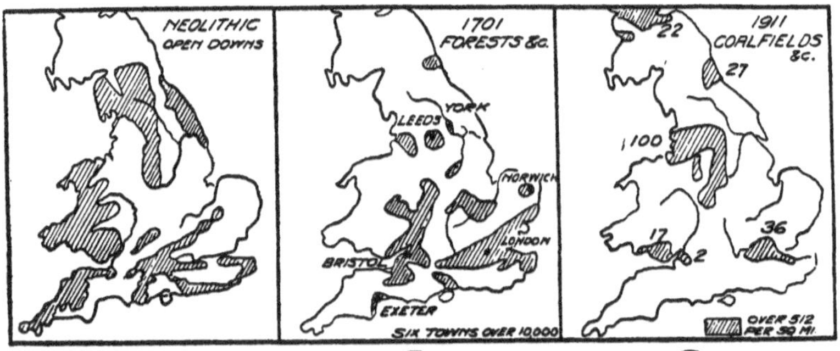

A. B. C.

Fig. 60.—Variation in the Distribution of Population with varying Environment. *A* shows that the Neolithic Folk lived in the open Uplands (after Fleure). *B* shows Agriculture in the Vales. Ruled areas over 128 per sq. mile (after Muir). *C* shows modern population based on Coal. Figures represent number of towns over 10,000 (after Muir).

any student of human development can ignore the control exercised by the *build* of a country in face of the evidence in Britain. We have seen that Neolithic man with his feeble tools could not cope with the forests of oak, beech, etc., which covered the lowlands of England (p. 39). Hence he kept to precisely those areas in which to-day the population is most sparse. A reference to Fig. 60A shows that the Neolithic folk lived on the "Older-Mass" and on the grassy downs which clothed the limestone cuestas. To-day these uplands are called the Cotswold, Marlborough Downs, North and South Downs, York Moors, etc.

Gradually, however, the fertile vales were in large part cleared of forests and their place taken by farms or pastures. This led to a much denser population in precisely those areas left empty by the Neolithic folk. Hence Fig. 60B is nearly the complement of Fig. 60A. London on the Thames Estuary and Bristol on the Severn Estuary were the two largest towns situated at the outlet of the two largest populous areas. Only four other towns had over 10,000 inhabitants. They were Exeter, Norwich, York and Leeds.

In medieval times iron was worked where charcoal was plentiful, *i.e.*, near the forests. Hence Birmingham arose near the Forest of Dean and sent goods down the Severn. Kent produced much iron with the timber of the Anderida (or Andred) Forest of the Weald, while the steep streams of Sheffield supplied water-wheels, the first form of machine power. It was not till 1720 that pit-coal replaced charcoal for smelting in any large quantity. Wool was, of course, the chief export of England in medieval times, and cloth-making was carried on in the many small towns adjacent to the sheep pastures. Thus Winchester, London, Worcester, Oxford, Norwich and Northampton all produced cloth. Gradually regions specialised, such as Norfolk for worsted and Colchester for baize, blankets at Hereford, serge in Devon, frieze at Worcester. The rapid streams of the limestone cuestas in the west gave adequate water-power for many cloth mills, so that Gloucester and Stroud soon excelled in the best cloth. Later Bradford in Yorkshire began to replace Norwich as a centre for woollen manufacture, mainly because there was no water-power in the flat lands of East Anglia. After 1840 the use of power looms, based on a coal supply, gave Yorkshire a great advantage over either Norwich or Gloucester, and Leeds is now the chief centre of the woollen industry. Somewhere about 1650 some Flemish spinners of cotton emigrated to Manchester. It was the use of the steam engine about 1790, again based on a coal supply, which caused the rapid growth of the Lancashire cotton trade. (See Carrier, *loc. cit.*).

Thus the use of coal determined a third change in the distribution of population. Clusters of towns called conurbations have arisen, of which the group linking Leeds to Liverpool is one of the largest in the world. As we saw in Chapter II the coal supplies (5 in Fig. 3) are found above (*i.e.*, to the east) of the oldest rocks of England, and below (*i.e.*, to the west) of the "Younger-Mass." This is clearly demonstrated in Fig. 3. Liverpool, Manchester, Sheffield and Birmingham, which were hardly known as towns in 1701, are now rivalled only by London. The Severn Valley, so populous in 1701, is now relatively empty compared with the Cardiff and Swansea coalfield (Fig. 60). Though

London has grown enormously, Kent and Essex are by no means relatively as important as they were in 1701. Norwich, Exeter and York are towns with a history rather than the vigorous centres of two centuries ago. Bristol and Leeds—which are close to coalfields—have, however, held their own.

There seems little doubt unfortunately that the change from the way of life of the medieval worker to the factory and slum life of the Industrial Revolution was for the worse, as regards the bulk of the people concerned. But conurbations and millionaires need not necessarily result from the exploitation of Nature's resources. Luckily cheap motor cars, long-distance electric power and enlightened socialism are some of the many factors which must greatly improve the industrial environment in the future.

E. Environment and Nation in Ireland

Ireland consists of much the same rocks as are found in the north of England, but they have not been warped into a long ridge or anticline (as in the Pennine Ridge, Fig. 3). Ireland, indeed, is shaped somewhat like a tray with a broken rim (Fig. 61). The central part is a flat lowland, built up largely of lower Carboniferous rocks like those of the Pennines, but unfortunately the rich coalfields are almost absent. At the sides (north and south) are elevated areas of rocks akin to the "Older-Mass" (p. 35). Granites and gneisses are common in the northern uplands, while granites and Paleozoic rocks build up the southern uplands. There are no significant areas of rugged mountains save in Wicklow (E.), Kerry (S.W.) and Connemara (W.). The importance of Dublin in the low gap of softer rocks between the rugged rocks of the Older-Mass is obvious in Fig. 61. The coastline reminds one of that of England, for the two great bays in the northwest and southwest are where the waves of the sea have cut away portions of the softer central plain. The constant rains mentioned in the chapter on climate (p. 73) have led to the development of many swamps in the lowlands and moors in the uplands. The country is crossed by many rivers, of which the largest, the Shannon, is almost a chain of lakes.

The effects of this rather peculiar environment are marked. Lying on the extreme edge of Europe, it was originally occupied by the most primitive tribes—consisting essentially of the Mediterranean Race. We do not know when these aborigines began to speak Gaelic (Erse), but it may well have been not much before the time of Christ, if we follow Rhys and Jones (Fig. 25). Brythonic (Welsh) never gained a hold in Ireland in the way it did for a time in Scotland. Nor did the

Roman Empire approach Ireland. Its isolation and the poverty of its resources both combined to keep it racially much more homogeneous than either England or Scotland. We may say that essentially it is all Mediterranean. But the numerous invasions and plantations, though not changing the population in the east as did the Saxon migrations in Britain, have affected eastern and especially northeast Ireland. In Ulster and around Dublin the population is distinctly less dark, and somewhat more brakeph. The history of Ireland easily accounts for this.

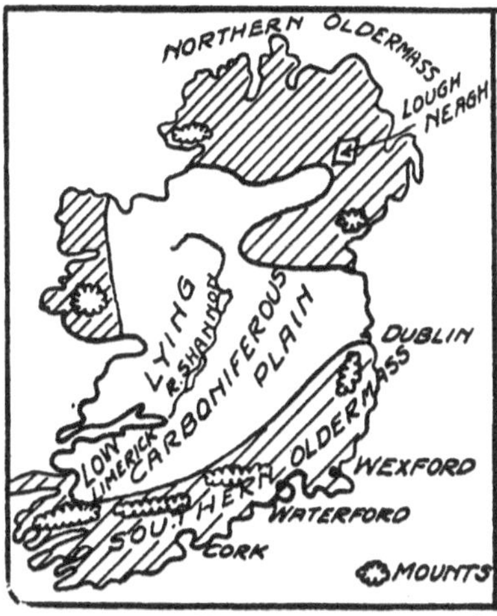

Fig. 61.—The Build of Ireland showing the Older-Mass surrounding the central Plain.

Fig. 62.—Some of the main historical divisions in Ireland.

It is important to remember the differences between the early history of Ireland and of England, for it largely explains the never-ceasing antagonism between the two regions. The Christian Church of Ireland developed by St. Patrick had a long and honourable career before the Saxons became Christians in England. The feudal system so completely established by the Normans in England was contrary to the system employed in Ireland, where there were many independent chiefs (or kings) who usually ignored the claims of the titular overlord, O'Neill of Ulster. Furthermore, the Northmen do not appear to have coveted Irish lands, for they only occupied a few coastal towns from about 800 to 1000. Their chief sites were Dublin, Wexford, Waterford, Cork and Limerick. A native prince, Brian Boroimhe, attacked the invaders and completely defeated them at Clontarf (near Dublin) in 1014.

Towards the end of the 12th century there were many quarrels between the old Irish Church and the later Roman Church. At this time there was an English Pope (Nicholas Breakspeare) on the throne of Saint Peter, and he gave Henry II a commission to subdue the unruly Irish. Many filibustering Norman knights under Strongbow aided the King in his conquest, so that by 1172 he was declared Lord of Ireland. By the middle of the 13th century the "Englishry" controlled the eastern littoral and the central lowland, except the area near the bogs and lakes along the middle Shannon. However, the isolation of Ireland soon resulted in the Norman overlords merging with the Irish and throwing off almost all allegiance to England. So much was this the case that about 1500 English rule did not extend much beyond the "English Pale," *i.e.*, some 50 miles north and 25 miles west of Dublin (Fig. 62).

Henry VIII attempted to end the reign of misrule and tribal rivalry which existed throughout Ireland. The Geraldines of Kildare were the most powerful family at this time, and rose in rebellion against the King and especially against his drastic changes in the Romish Church (p. 195). Mary Tudor carried on the struggle, although she was a Roman Catholic. She formed plantations of English settlers in Queens County (Maryborough) and Kings County (Phillipston).

Later the Desmonds of Kerry were overthrown, but the O'Neills of Ulster flamed in a great rebellion which taxed Elizabeth's resources. Early in the reign of James I the O'Neills again rebelled, but about 1607 they were crushed, their lands confiscated and about one million acres in Ulster were handed over to Scottish and English settlers. Thus eleven centuries after their original migration, the Scots returned to Ireland. But now they had lost their Irish tongue, and above all had adopted the dour Calvinistic type of the reformed religion. There has been little friendship ever since between the Ulstermen of the North of Ireland, and the Irish Catholics of the rest of the country. In the Revolution Ireland naturally supported Charles with his Catholic sympathies. As a result Cromwell and his generals in 1649 ruthlessly put down the Catholics in Ireland, and many further plantations were assigned to Cromwell's supporters throughout Ireland. These bitter struggles were largely ended when in 1921 and 1922 Ireland was divided into two entirely distinct governments. The Irish Free State comprises the Catholic region and has the same constitutional status in the Empire as Canada. Ulster also has a local Parliament, but is much more closely linked to England.

CHAPTER XIII

SCANDINAVIA AND ICELAND

A. The Build of Scandinavia

Both structurally and racially the Scandinavian countries follow logically after Britain. Reference to Fig. 10 will show that Norway is built up of the same rocks as Scotland (Caledonia). Indeed, the "Caledonian Folds" cover more ground in Norway than in Scotland itself. The people of Scandinavia are of exactly the same stocks as those Nordic invaders who form the bulk of the population of Eastern England, Lowland Scotland and the Western Isles of Scotland (p. 43).

Three closely associated nations are grouped under the name of Scandinavia. These are Sweden, Norway and Denmark. Their lands consist essentially of two large peninsulas, which are separated by a belt of straits. The "roots" of the two peninsulas merge rather indefinitely into the mainland of Europe, so that here the national boundaries are not marked by any clear natural feature. Elsewhere the boundary of Scandinavia is formed by the Atlantic Ocean on the west, and the Baltic Sea on the east. The main peninsula is 1,100 miles long and about 250 miles wide. Denmark is only 200 miles long, and the southern end of this peninsula forms part of Germany.

The whole area can readily be divided into three structural divisions. All three are found in Sweden, two in Norway and only one in Denmark. All down the east of the larger peninsula is a broad belt of lowlands and uplands rarely rising over 1,000 feet (Fig. 63). This consists of very ancient rocks forming part of the primitive continent of the North Atlantic ("Arctis") of which the Scottish Highlands and Iceland also originally formed a part (Fig. 10). Overlying this to the west is a belt of slightly newer rocks (called the *Glint*) which is largely of Cambrian age or thereabouts. It is thus very similar in age to the rocks of South Scotland, Cumberland and Wales. It contains the highest part of the country, which consists essentially of a 5,000 feet plateau rising in the south to peaks over 8,000 feet. Glittertind, 8,400 feet, lies 150 miles northeast of Bergen, and rivals Galdhopig as the highest point in northern Europe.

Still further to the west—forming an outer fringe to the plateau along the Atlantic Coast—is a peculiar area of ancient rocks like those in the east, but thrust bodily over the younger Glint from the west in the far distant past. This constitutes what may be termed the "Fiord

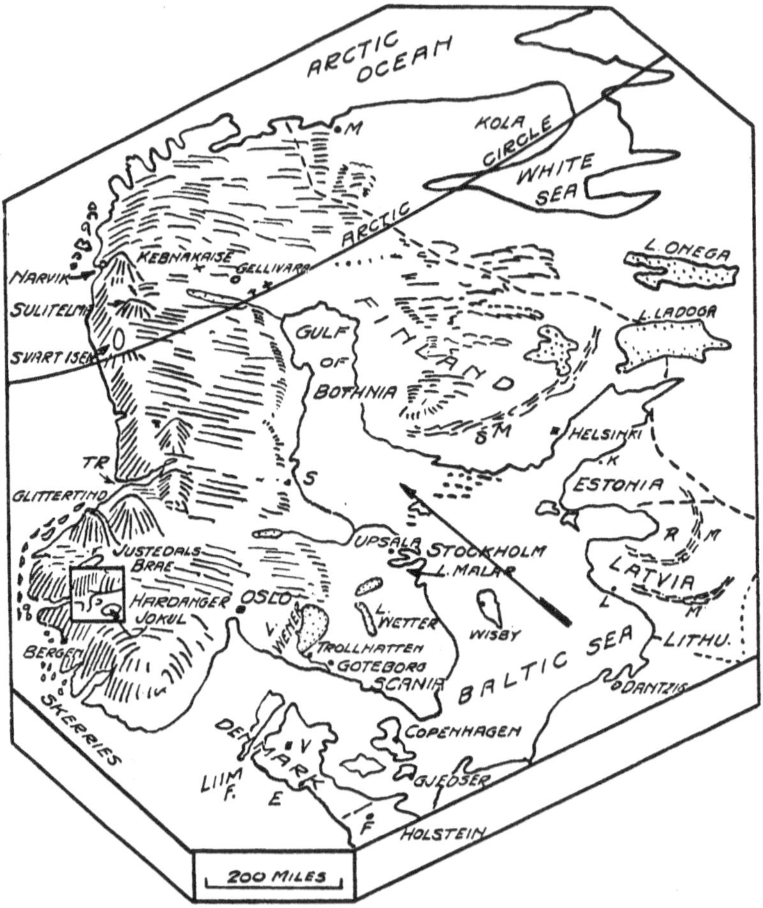

Fig. 63.—A Block Diagram of Scandinavia and the East Baltic States. The small square includes the area which is shown enlarged in Fig. 64. M indicates major Moraines.

and Skerry" Belt. It is one of the most rugged coasts in the world, as if a long narrow plateau had been cracked by a thousand quakes, gouged out by a thousand glaciers, and then in part drowned by the sea. The outer fragments of this shattered plateau consist of countless small rocky islands, islets (skerries) and crags, which form what the Norwegians call the "Skerry-Gaard." Deep channels separate the skerries, and penetrate far up the gloomy canyons of the fiords. This extremely picturesque coast is one of the most favoured tourist resorts in the world (Fig. 63).

Quite unlike the preceding belts in origin are the lowlands of Denmark, and the southern tip of Sweden is of the same type. It consists mainly of recent deposits, largely moraine from the huge ice cap of the Great Ice Age. But on both sides of the Sound there are harder limestones, which form picturesque white cliffs some 400 feet high at Moen, south of Copenhagen.

The Drowned Belt. In late glacial times, perhaps 10,000 years ago, the *Yoldia Sea* extended right across the land, from the North Sea by way of the Gulf of Finland to the White Sea and the Arctic Ocean. Marine deposits, often 600 feet above the present sea (and containing seashells called *Yoldia*), were then distributed over the lowland parts of Sweden and Finland. The large lakes of Sweden, Vener, Vetter and Malar, as well as the lakes of Finland and Russia (Ladoga, Onega) are relics of this sea (Fig. 63). Since those times, however, the land has been rising. In the north of the Gulf of Bothnia this elevation is about 40 inches in a century. Near Stockholm the rise amounts to 20 inches in the same time. Thus the sands and clays in the present bays—which slope very gradually out to sea—will be the fields of the future. In those areas already abandoned by the sea the speckled alder (*Alnus*) is the first growth, to be soon followed by pine forests. These marine deposits have formed a large part of the good soils of southern Sweden.

Erosion of the Plateau. The Plateaux of Norway extend from latitude 70° in the far north almost the whole way down the coast to the Skagerrak Strait. They originally formed one lowlying shield (or peneplain) which was raised fairly recently high above sea level. The general level in the southern part reaches 4,000 or 5,000 feet, but in the northern part it is distinctly lower. In the Great Ice Ages, beginning some 800,000 years ago and extending perhaps to 20,000 years ago, almost the whole of this region was at times covered with an Ice Cap. Only the highest knobs projected as "nunataks" (or rock islands) above the level of the ice. The glaciers deepened the valleys already cut by the rivers. They excavated the deep fiords, especially along the weak fault-planes in the rocks which had apparently been developed by crustal pressure. As a result these deep valleys were cut right across the uplifted peneplain and now divide the plateau into four or five distinct sections.

Divisions of the Plateau. In the north is a long flattened ridge sometimes known as Kiolen (*i.e.*, the keel of a boat). This rises to 7,030 feet in Mount Kebnekaise, the highest in Sweden. It is very narrow in places, and the railway to Narvik (bringing iron ore from Gellivara in Sweden) crosses to the sea where the plateau is lowest. Somewhat to the south is Mount Sulitelma (6,150 feet), which is a little north of the largest of the icefields which still persist on the Norwegian Plateau. This icefield is called Svart-is (*i.e.*, Black Ice) and is some 30 miles long and covers 230 square miles. A few nunataks, of which Snetind (5,246 feet) is the chief, project through the ice. Many small glaciers flow slowly from the icefield down to the fiords to the west (Fig. 63).

The middle section of the plateau is the lowest, rivers from each side having cut back into the highlands until their headwaters are close to each other. Near Trondhjem the valleys are lower, and railways run east to Sundsvall; or south by the Glommen River to Oslo, with a rise of only 2,000 feet. The other railway across Norway (through Finse) rises to 5,000 feet (Fig. 64).

South of Trondhjem the plateau rises again to the Dovrefeld with Snehatten (7,570 feet). Another deep cross-valley separates this mountain mass from the Giant Mountains (Jotun-fjeld). There is an interesting lake in this valley, called Lesjeskagens Vand (2,000 feet), from which one river flows north to the ocean, while another flows south to Oslo. Jotun-fjeld was not well known till 1820, though it contains the highest mountains in northern Europe. Only a few, however, are higher than 8,000 feet, and the snowline here lies about 5,600 feet. (In the Swiss Alps the snowline is about 8,800 feet.) The two highest points, Galdhopig (*pron.* "Gallerpig") and Glittertind (8,400 feet) are only seven miles apart. They are surrounded by many small icefields, and rise about 4,000 feet above the surrounding upland valleys. To the south the elevations decrease again, few of the mountain domes in Hardanger Fjeld reaching 5,000 feet.

The writer has travelled in all the seven continents, but has visited few places as interesting as Finse (4,010 feet). Here the Bergen railway is built at the level of permanent snow (Fig. 64). From the large hotel at the station we look across to a view which must closely resemble what all northern Europe looked like in the Great Ice Age. A small lake (Finse Vand) is all that separates the hotel from the ice-cap or Jokul (*pron.* "Yerkel") which extends right across the view to the south for a distance of nearly 10 miles. The ice-cap is circular—like a flat dome rising from the high plateau to a height of 6,500 feet. My first visit was made through a thick mist. One rose only about 1,000 feet in two miles, walking over alternating boggy or rocky uplands. Piles of stones called Varrer marked the faint rough track. Here one crossed black bog with scattered grass and some heath and flowers; there were large expanses of rock, scraped bare by a vanished ice-cap.

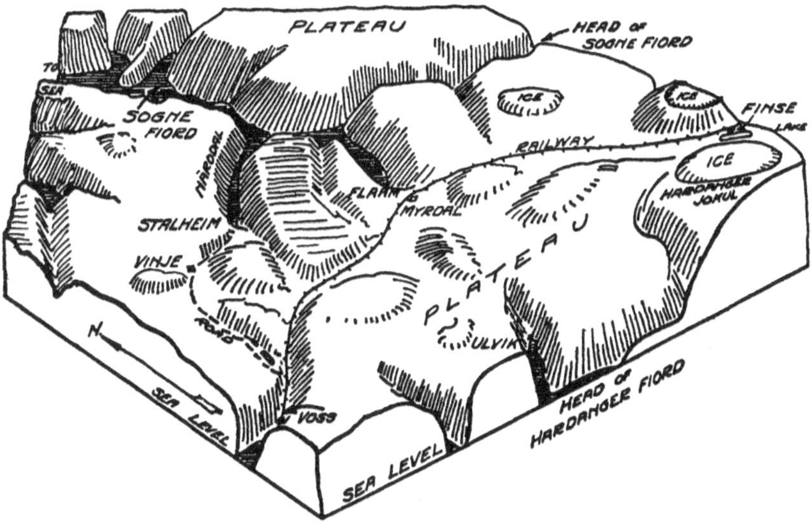

Fig. 64.—A Block Diagram showing the Fell and Fiord scenery in south-west Norway. Note the small stagnant Ice-Caps. The plateau (Fell) is about 4,000 feet above the sea.

Again were numerous tarns and pools with grassy ridges between. Near the ice-cap was only a tumbled waste of rock debris (ground moraine). Crossing various shallow streams issuing from the gentle slopes of the Jokul, I readily made my way a considerable distance up the ice. It is indeed quite easy (except for mists) to cross the Jokul in several directions. The ice was rather spongy in texture, and was grooved by numerous water channels. Here and there were heaps of silt and gravel, while to the east the black knob of the Kong Peak formed a veritable rock island or nunatak. All around for miles was the same desolate landscape of the plateau, diversified with rocky knobs and shallow tarns—but without trees or bushes and with only scattered patches of heath or grass. Truly the plateau of Norway is a region incapable of supporting man or beast. Perhaps its chief value is as a catchment for water—which in its descent can be harnessed to produce electricity in the future and so replace coal in Norway.

The Surface of Denmark. This small country has only an area of about 16,000 square miles, or a little more than that of the state of Maryland. It consists of three large islands, Zealand, Funen and Lolland, with the larger peninsula of Jutland. It is one of the flattest countries in the world, for the highest point (in East Jutland) is only 550 feet above sea level. Moen, south of Copenhagen, is, however, nearly as high. The country shows many broad meadows and rolling hills, and the soil is largely "boulder clay" resulting from the debris left on the retreat of the Great Ice Cap to the north some 20,000 years ago.

The northern part of Jutland is boggy heathland, which is still only sparsely populated. Along the west coast is a belt of sand dunes some 200 miles long which projects north as The Skaw and here carries a great lighthouse. Formerly the forests were largely of oak, but these have been replaced largely by beeches, which are now the commoner. In recent years pine trees have been largely planted. The coast is penetrated by many narrow gulfs called fiords, though they have little in common with those of Norway. Liim Fiord extends right across the north of Jutland, but its shores are flat and it is not often visited by tourists.

B. *Relations of the Three Kingdoms*

In earlier chapters in this book an account has been given of the Folk-Wanderings mainly as they concerned the tribes of the German Plain. Thus the Goths, after wandering over almost the whole of southern Europe, made their most permanent settlement in Spain, while the Franks occupied France, the Anglo-Saxons England. In the present section we are concerned with similar migrations on the part of the Scandinavian peoples. Some reference has already been made to these movements in connection with the history of Britain, where the Danes and Norwegians settled. But many other countries were affected by their fierce Viking forays and were forced to yield territories to them.

Little is known of the history of these lands before A.D. 800. Tribes almost wholly of Nordic origin (but probably speaking Finnish) settled north of the Baltic in Neolithic times. They lived a hazardous life, one may imagine, depending on hunting, fishing and the scanty crops made possible by the short summer. The land was almost wholly divided between barren rocky slopes and plateaux and forest-covered lowlands. Everywhere man had to struggle valiantly for a bare living, so that Scandinavia as a whole belongs to those Regions of Difficulty already discussed in Chapter V. A second important factor in their history was the distance of these lands both from the Mediterranean and from Asia. Hence Scandinavia both by its inherent poverty and by its isolation has been preserved from the attacks of the invaders who so greatly handicapped much of southern and eastern Europe. Never throughout their history have these three lands been seriously invaded, though there have, of course, been isolated naval and land battles on their margins. A much more significant fact in their history has been that *export of surplus warriors* which developed into the Viking and Norse Raids of the eighth to the thirteenth centuries.

Norway has an area of 125,000 square miles and Sweden is much larger, with 173,000 square miles; but these figures are somewhat misleading. The sparsely-settled regions—with less than 5 folk per square kilometre (about 12 to the square mile)—are shown in Fig. 65. Hardly any other part of Europe except the conifer forests of Russia is so empty as these blank lands in Scandinavia. If we ignore them, then Norway's really useful lands (*i.e.*, equivalent to those listed for other lands in our Table, p. 103) drop to about 36,000 square miles, and Sweden's to 75,000 square miles. So small are Norway's agricultural resources that her population is much less than that of tiny Denmark —though the latter is also agricultural and not industrialized to the extent of Holland or other populous countries.

The geographical isolation of Denmark from the other two Scandinavian Kingdoms is obvious, for the Sound rolls between them. It is true that the railway trains cross on ferries from Sweden to Elsinore and from Germany (Warnemunde) to Gjedser in the south. But there is enough difference in environment to keep Denmark and Sweden apart. Fig. 63 shows us that Norway and Sweden are only linked by a narrow isthmus of good land near Oslo. The rest of their common boundary is mostly traced on the tundra of the plateau described on an earlier page. This sets a more definite limit to a country even than a belt of sea of the same width. However, in spite of these obstacles of sea and plateau, the history of the three Scandinavian countries—as well as of Iceland and Finland—is closely interwoven.

Since Norway and Sweden are, as we have seen, divided by a broad barrier, it is not surprising to find that they have had different outlooks through much of historical times. Norway's trade has always been on the Atlantic, leading her to the Faroes, to Iceland, Greenland and Britain. Sweden—even after she gained Scania (the southern province)—has been shut in by Denmark and she faces Germany and

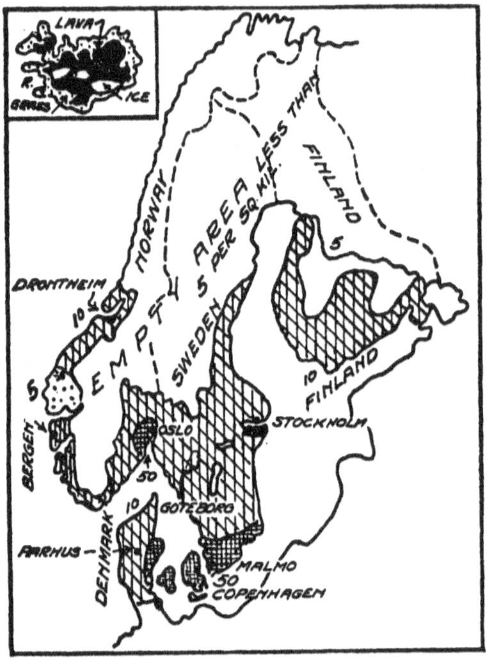

Fig. 65.—The Density of Population (per square kilometre) in Scandinavia. The Plateau (Fell) is almost empty. Inset is a map of Iceland showing the small area not covered by ice or recent lavas. Drontheim is the same as Trondhjem.

Russia, with whom, accordingly, she has had most contact. The Norwegians were driven to the sea—where they were even more independent of each other than in their isolated farms, each on a "talus fan" at the foot of a steep mountain valley (Fig. 64). There are three important centres of settlement in Norway to-day: Oslo (250,000) in the south in what is by far the most populous district; Bergen (97,000), the nearest large port to Britain, and Trondhjem (Nidaros), where the central plateau sinks to a lower level. Both the latter towns are now connected to Oslo, the capital, by rail. The only other two towns of over 25,000 are Stavanger and Drammen. In Sweden, Stockholm has 474,000, Goteborg 236,000, and Malmo 119,000. There are eleven other towns each with over 25,000 inhabitants. In Denmark, Copenhagen has about 600,000, while Aarhus and four other towns have from 76,000 to 25,000 inhabitants. The relations of the three kingdoms are shown in a summarised form in the following Table:

TABLE S

MAIN HISTORICAL EVENTS IN THE THREE SCANDINAVIAN KINGDOMS

	Norway	Denmark	Sweden
A.D. 400		Jutes to England	Goths to Spain
500	Finns in S. Norway		520—Hygelac attacks Franks
600	Legendary history	Legendary history	Legendary history
700		Saint Willibrod to Denmark. First Danes to England	
800	Harald's enemies to Iceland	826—Ansgar and Christianity	830—Ansgar to L. Malar. Rurik in Russia
900	To Greenland and America Rolf to Normandy		Varangians to Byzantium, Olaf (993-1021) and Christianity
1000	Olaf 1016-30 and Christianity. To Hebrides, Sicily and England	Second Danes to England. Canute's Empire (1014-35)	
1100	Conquer S. Italy		1134—Sverker unites Swedes. 1159—Finland invaded
1200	Hebrides lost	Valdemar wins N. Germany. Iceland to Denmark	1255—Stockholm founded
1300	Crown to Sweden. Margaret rules Norway	Wars with Hansa. Margaret rules Denmark	1397—Margaret rules all Scandinavia

TABLE S—*Continued*

1400	Danish Aristocracy in Power		
1500		Nobles take Church Lands	1520—Bloodbath, and Gustavus Vasa rebels
1600		Denmark loses in Religious Wars. 1660—Denmark loses Scania	Charles IX Protestant. Gustavus Adolphus
1700		1801—Nelson defeats Denmark	1721—Lose outlying lands
1800	1814—Sweden gains Norway	1814—Denmark gives up Norway. 1865—Holstein to Germany	1809—Lose Finland 1818—Bernadotte king
1900	1905 — Peaceful separation.	1918 — Iceland independent.	

C. *Migrations of Norsemen from Regions of Difficulty*

We may sum up in a few words the early history of the Scandinavian lands. In the Stone Ages Denmark was settled by primitive folk who lived on the shores and have left us huge kitchen-middens containing the refuse of their feasts and many of their tools and weapons. In later Neolithic times the tribes moved inland to the sandy soils and heaths, just as in Britain. At first they avoided the clays which were covered with forests. However, as metal tools were introduced they occupied the clay areas also and cut down many of the trees. About 1043 Copenhagen was first occupied, and during the 12th and 13th centuries many small towns arose at the heads of the various gulfs or "viks."

In Norway somewhat similar conditions obtained. The shores of the gloomy fiords were first settled, while the inland regions of the Glint (Fig. 63) were gradually occupied as population increased. In the Iron Age the settlers moved up the valleys into the more sunny areas and began to cut down the forests. The 11th century saw the end of the Age of Vikings (*i.e.*, "the sons (Ing) of the Fiords (Vik)"). By this time, according to Ahlmann, the whole of the attractive portions of Norway was fairly occupied.

In Sweden the early settlements were in the southern promontory called Scania. The Iron Age seems to have begun here about the time of Christ, and silver, glass and Roman coins also occur among the antiquities of this time. The Swedes made use of Runic inscriptions, which were Gothic words written in a modification of the Greek alphabet. During the later Iron period the language of the Runes seems to have changed from the early general Teutonic type to a special Scandinavian type. The folk in Scania were mentioned by Tacitus (A.D. 100) under the name of Suiones, while Ptolemy (A.D. 150) tells us of the Goths. The northern parts of Sweden after A.D. 900 were inhabited by Lapps of entirely distinct race—and akin to the round-headed Samoyedes of Asia. Beowulf, written about A.D. 700, mentions that Scania was controlled by the Danes; while somewhat north were the Gotar, who also owned Oland Is. Still further north beyond thick forests were the Svear (Swedes), living around Lake Malar, who seemed to have recently dominated the Gotar. Probably there were other Nordic tribes, but of Finnish culture, still further to the north, for some of the place-names indicate such a settlement.

There seems little doubt that nations living in such a region of difficulty as Scandinavia would be very quickly and strikingly affected by relatively slight changes of climate. An increase of cold or of rain, but little above the average, would ruin their scanty crops and drive the folk out on piratical raids. (Just the opposite sort of a change has a similar effect in the borderlands of the deserts, where increase of heat and drought leads to bandit forays.) Huntington has given much thought to this problem and has drawn up a graph of climatic change for the Christian era, which is of great importance in our present study. His data are based largely on the "growth-rings" of the giant trees of California. From these variations in growth it is possible to deduce the changes in climate in that region. There is little doubt that the whole of the northern Temperate zone is linked—as regards weather—by the constant procession of cyclonic storms. These move across northern U.S.A. and a few days later reach northern Europe, and so pass on to the east in their endless circle around the North Pole. Hence the data for California probably represent fairly closely contemporary conditions in Scandinavia. (See "Character of Races," 1924.)

In Fig. 66 are given these changes in the rainfall in the northern hemisphere. The wet years were those which drove the Northmen to seek sustenance for their families elsewhere. The dry years on the whole gave them sufficient crops on their tiny farms, and they stayed at home. During the first four centuries Scandinavia was very wet and crops must have been scanty. It seems likely that the only important agricultural region in those early times was Scania. Here the Goths and their allies lived, and from Scania we may believe they poured out over northern Europe causing the Goth migrations which were so numerous in those centuries (p. 168). The fifth, sixth and seventh centuries were dry. We may deduce that population increased greatly in the more attractive areas of Scandinavia during this interval; and that migrations were correspondingly reduced. Then bad years of heavy rains and ruined crops began in the eighth century and continued until about 1100 (Fig. 66). These were just the times of the great Viking raids of history, which are listed in the preceding table. England was invaded in 787, France in 789, Ireland in 795, Frisia in

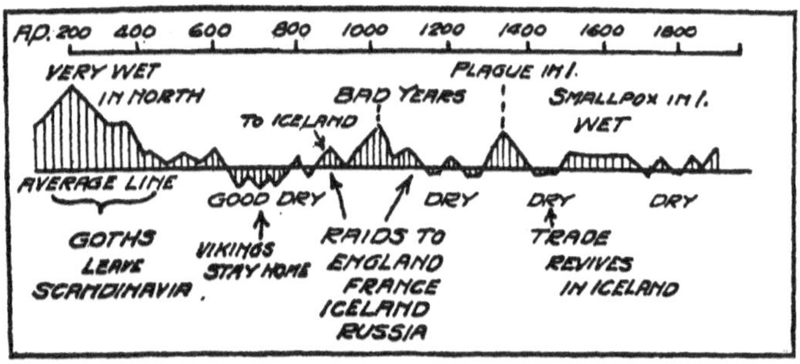

Fig. 66.—Rain Variation in the Northern Hemisphere (California) compared with Scandinavian History (after Huntington).

800, France again in 841, and Russia in 859. As conditions became more desperate in Scandinavia the raids gave place to actual migrations (Fig. 67). Thus Danes came over in large numbers to England between 866 and 930, leading to the historic struggles with Alfred of Wessex, which have already been referred to.

None of these Scandinavian raids resulted in the permanent conquest of an outlying land except in the case of Iceland. In all other cases the Northmen ultimately merged with the more abundant native peoples; no doubt adding strength to the resultant mixed nation, but not really altering the race or culture very obviously.[1] In addition to the British Isles, they penetrated continental Europe by way of the Elbe to Hamburg, by way of the Scheldt to the middle districts of the Rhine and the Somme; by way of the Seine up to Paris and up to Burgundy; the lower region of the Loire, the Garonne, Northern

[1]The following paragraphs are based on the valuable account in the official "Handbook of Norway and Sweden," Intelligence Dept., No. 1214.

Spain, the northwest African coast, the mouth of the Rhone, and even the Italian seaboard were visited by them. Such was the European range of the early rovers who set out over the North Sea, though it is not always possible to distinguish between those from Denmark and from Norway.

Concurrently there was a movement from Sweden across the Baltic and far beyond. Though the statement is disputed, there appears to be no doubt that in the ninth century the relations between Sweden and the Trans-Baltic lands became greatly strengthened, that Swedes (known as Varangians) pushed up the great rivers of Russia (Fig. 131) and that they established themselves in territory in the district of Lake Ladoga and Novgorod (862). They then found their way to the Black and Caspian seas and to the region of the Caucasus. The foundation of the Russian Empire is dated from the settlement of the Novgorod territory by Rurik and the Rus, *i.e.*, the sea-warriors of the east coast of Sweden.

The Viking Age is usually taken as terminating about the time when King Charles the Simple (of France) ceded to the Northman Rollo (or Rolf) in 911 the territory of the lower Seine. This had been a favourite goal of the raiders, and it is still called Normandy, *i.e.*, the home of the Northmen. The overseas voyages did not cease abruptly, and they continued long after this date. Thus the expeditions across the northern seas to Iceland and Greenland belong mainly to the tenth century (Fig. 67). The famous voyage which reached the North American coast at "Vinland" (Nova Scotia probably) took place in the year 1000. (These events are discussed more fully in a later section.)

The effects of these voyages upon the social and political history of Scandinavia were profound. In them may be found the germ of the modern eminence of Norway in the mercantile marine of the world, or again of the distinguished part played by Scandinavians in exploration, especially in the polar regions (Nansen, Nordenskiold, Amundsen, Stefansson, etc.). But at and after the immediate period with which we have been dealing, the Northmen both taught and learned from those with whom they came in contact in their wanderings. When they settled they sometimes showed (as in Normandy and Russia) a capacity for government and for peaceful organization which was less ardent among the previous inhabitants of the settled countries. In certain methods of warfare and, above all, in shipbuilding and navigation they were the superiors of other Europeans.

The activities of the Vikings, however, sapped a good deal of strength from both Norway and Sweden. Though they brought wealth into Scandinavia from overseas, they did not leave their taste for warfare there. The period of some three centuries following the close of the Viking Age leaves a general impression of civil dissension and weakness on the one hand, and yet on the other of striving in various directions towards nationality and ordered government.

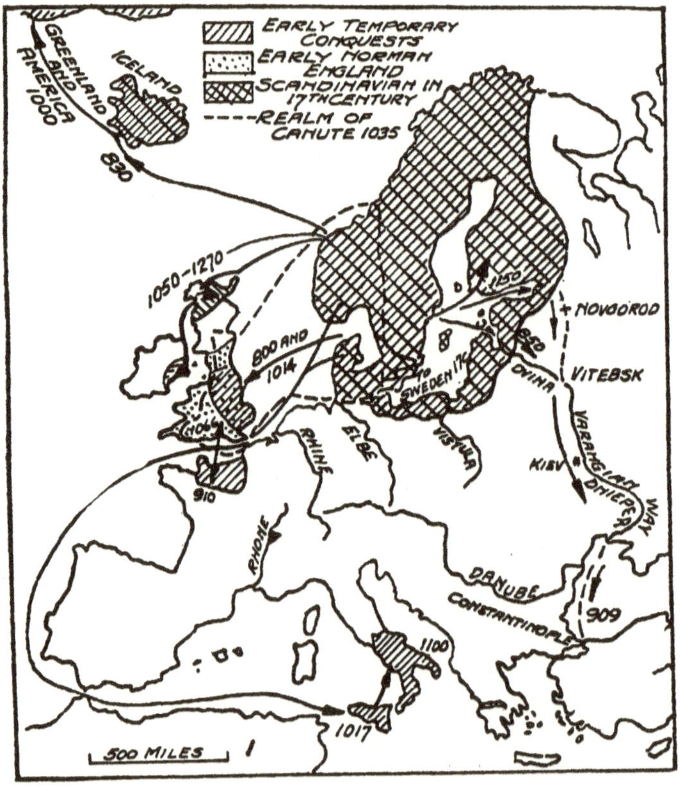

Fig. 67.—Conquests of the Scandinavians from 800 onwards. Their greatest power occurred in the 17th century.

D. *Medieval and Modern History of Scandinavia*

We have seen that despite the small area of Denmark, it is almost as well endowed in resources as Sweden and is better than Norway. Hence the importance of the three nations in European history does not at all agree with their areas on the map. In fact, Denmark was the leading state through the Middle Ages, though Sweden came to the front in the 16th century and is still the predominant Scandinavian nation. In this brief survey we can only touch upon the various territorial changes in the three kingdoms. At the outset Norway with strong inducement to take to the sea came most into contact with other nations. Denmark, however, naturally had closer relations with the civilisation of Charlemagne and his German successors than did either Sweden or Norway. Accordingly King Canute (Knut) of Denmark was the first great figure in their history. He fought many battles with Edmund Ironsides, King of Wessex, and on the death of the latter in 1016, Canute was elected King of all England, while in 1018 he succeeded his brother as King of Denmark. In 1028 the Kings of Norway and Sweden attacked him, but he defeated them and conquered a large part of Norway, as well as parts of Prussia (Fig. 67). He was one of the first early monarchs to link himself closely with the powers of the Church for the benefit of his subjects. He is placed second only to Alfred among the early Kings of England for his wisdom and initiative.

The three countries were separated politically until the middle of the 14th century. In 1134 King Sverker united the Swedes (Svear) with the Goths and in 1150 Eric led a crusade against the pagan Finns of Finland, which from that time until 1809 was closely connected to Sweden. In Norway between 1030 and 1217 there was a long period of civil dissension between various claimants to the throne, but this was of little importance outside Norway. From 1050 onwards Norwegian settlements had been maintained in the Scottish Hebrides, but they were wisely abandoned by Magnus about 1270. In Denmark there was also a period of conflict between various rulers, but about 1157 Valdemar the Great obtained the throne. He subjugated the piratical Wends—a Slav tribe who lived near the Elbe—and converted them to Christianity. His successor Valdemar the Victorious conquered the South Baltic coasts and helped the German "Knights of the Sword" to conquer the Esthonians. In addition to Denmark he ruled Scania, Hamburg and Lubeck. About this time the Hansa Towns were rivals for power in the Baltic area (Fig. 74).

Various marriages among the royal houses resulted in Margaret of Denmark becoming regent of all three kingdoms in 1397, though this "Union of Kalmar" permitted each country to keep its own laws and lands. But the Danish nobles attempted to keep all the authority in their own hands. Revolts and civil wars followed until the massacre of many Swedes in Stockholm in 1520 ("The Bloodbath") led to the independence of Sweden under Gustavus Vasa. For centuries after 1397 Norway remained practically a backward province of the Danish kingdom.

The Reforms of Luther were welcomed by the people of Scandinavia and in 1536 the Catholic Church in Denmark was abolished and Protestants appointed to the Bishoprics. Christian IV entered the Thirty Years' War but was defeated by the Catholic forces. Sweden also expanded in Esthonia in 1578 where the Teutonic Knights had been attacked by Russia. The famous King, Gustavus Adolphus, fought several wars with Denmark, Poland and Russia before he entered the Thirty Years' War. As a result, in 1631 he conquered the Catholic army near Leipsig. On his death in 1632, his strong government was continued by his chancellor, Oxenstierna. At the Peace in 1648 Sweden obtained Upper Pomerania, Wismar and Bremen, thus controlling the mouths of the Oder, Elbe and Weser (Fig. 139). In 1658 Charles X restored the southern province of Scania to Sweden. However, the ambitions of Charles XII led to a coalition against him of many European Powers, so that by 1719 Sweden had lost all her Trans-Baltic Empire.

The last notable phase of Scandinavian History is linked to the Napoleonic Wars. Gustavus III as an absolute monarch naturally opposed the French Revolution, but he was assassinated in 1792. His son, Gustavus IV, at first took up a neutral attitude, but later he joined the losing side, so that he lost not only Wismar, Pomerania and Finland, but also his throne in 1809. In 1810 the new King, Charles XIII, chose as his heir the Napoleonic general, Bernadotte—surely the most curious succession in the history of Europe. In 1813 Bernadotte led a northern army against Napoleon, his former Emperor, and his ally, Denmark. After Napoleon's defeat in Russia in 1814 Norway was taken from Denmark and ceded to Sweden as payment for Bernadotte's help. In 1818 this son of a French lawyer became King as Charles John, and his family still occupies the throne.

In 1905 the friction between Sweden and Norway led to a referendum in which the Norwegians voted to dissolve the Union (by 368,211 votes to 184) and a Danish prince was chosen as their King. It is said that Sweden now feels that her "back is again bared to attack as it was before the union." Something of the same friction developed in Denmark in regard to the German province of Holstein. In 1848 the German-speaking population rose against the Danes and a German army came to their help. However, the Danes defeated the Prussians, who withdrew. In 1863 the Danish King died and Holstein again became a source of trouble. The Danes were crushed by Prussia and Austria and ultimately not only Holstein, but a good deal of Danish Schleswig was annexed to Prussia. This boundary was rectified, however, at the close of the Great War.

E. Environment and National Development in Iceland

The history of Iceland shows a remarkable relation between environment and nation which has been the subject of research by Ellsworth Huntington.[2] This large island comprises about 40,000 square miles, of which only about 6,000 square miles is usable. It lies in latitude 65°N. just south of the Arctic Circle and would be entirely covered with ice were it not for the southwest winds and warm Gulf Stream (described in Chapter IV). As it is, perhaps one-tenth is covered with ice, and most of the rest consists of barren fields of lava poured out by the many volcanoes. (See inset, Fig. 65.) Its environment is clearly even more unfavourable than that of Norway, yet it has a larger Roll of Fame (proportionate to its population) than almost any country in the world. For instance, it has produced three times as many authors (proportionately) as France or Germany.

As we have seen it was settled by Norwegians in the 9th century. By 930 A.D. there were perhaps 30,000 people in Iceland and there has been little change in their composition or numbers till modern times. Even in 1920 when there were 95,000 people, only 200 were not Scandinavian. The temperature does not rise above 50° in summer, but is so uniform that it only falls to 30° in winter. Iceland's resources are grass-pastures and fishing; and the crops other than grass are negligible. In Lord Bryce's opinion Iceland's record of literature stands next after the Golden Age of Greece. Moreover, for four centuries it was the only independent republic in the world. It is one of the three outstanding countries as regards the small proportion of illiterates. It is of note that printing was introduced in 1531, more than a century before it reached Norway.

Huntington believes that Inheritance, together with the physical and social environment, explain Iceland's supremacy in world progress. He shows that the hazardous life of the fishermen has weeded out all but the most capable and cautious for many generations. This is indi-

[2]"The Character of Races," New York, 1924.

cated in Fig. 68, where the death rates for various nations are compared. Thus the young male Icelanders of 20 years have the enormous death rate of 175 as compared to the low rate (90) of the men in an inland country like Switzerland. (In this graph the male death rate is given as a percentage of the female death rate which is used as a criterion in each country.) "Thus the original cause of Iceland's greatness was the repeated selection to which its people were subjected. This produced a peculiarly homogeneous sober-minded and competent group of people. Within a century or two there ensued a wonderful outburst of genius, which produced the famous sagas." Huntington concludes that a selected inheritance, when isolated, protected and

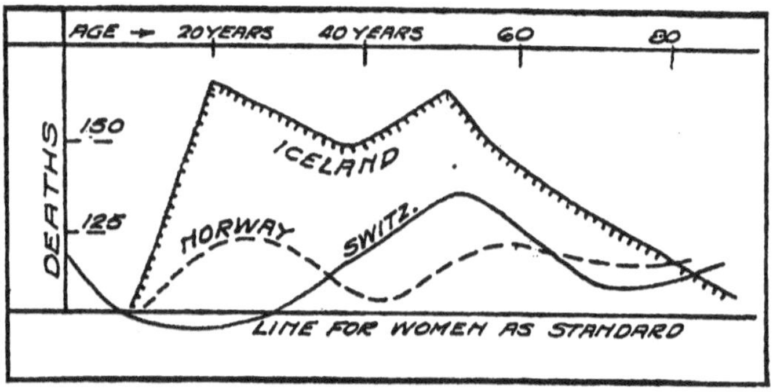

Fig. 68.—Death Rates in three countries, showing the risky occupations of Men in Iceland (after Huntington).

kept up to the mark by further selection, seems to be able to persist indefinitely.

Huntington also gives an interesting comparison between the changes in climate (Fig. 66) and the history of Iceland, which is worth a brief study. When the Norse nobles and their retainers migrated to Iceland from 880 to 920, Norway was cold and rainy. After 930 A.D. Iceland had a drier interval in which its peculiarly democratic Parliament, the Althing, was instituted. The Althing, however, left enforcement of the Law practically to the influence of public opinion About A.D. 1000 the climate was colder and stormier than in the preceding and succeeding centuries. This was a period of great poverty and discontent which made the people ready to adopt Christianity. The Golden Age of Iceland now began, and though we may believe that the new religion played a part, so also no doubt did the improved climate. In the 12th century large trees grew in Iceland which could be used for building houses, though to-day there are few birches or willows over ten feet high. In that happy period fruit trees were raised and grain cultivated, and there was little pack ice even on the north coast.

Again evil conditions began to affect Iceland, for the Black Death in 1348 seems to have been determined in part by the unusually wet conditions about that year (Fig. 66). For a time the rains were so abundant that crops failed year after year in Norway (and also in England during the 14th century). The trade of Iceland declined so that by 1440 there was almost no commerce with Norway and little with England. It is of interest to note that nothing is known of Iceland's own colony of Greenland after 1410 when the last ship from Greenland returned to Norway. When John Davis reached Greenland in 1585 only Eskimoes remained, and there were only meagre traditions of a former population of 10,000 Europeans.

In 1264 after much civil strife Iceland came under the control of Norway. In 1280 the rule of Iceland was transferred to the Kings of Denmark. They governed the island somewhat despotically until 1874, when a separate constitution was granted with two divisions of the local Althing. In 1918 Iceland was recognised as a separate kingdom and Denmark is concerned only in its foreign affairs.

CHAPTER XIV

GERMANY, THE HEIR OF THE HOLY ROMAN EMPIRE

A. The Build of Germany and its Historical Implications

In the last chapter we saw that the Build of Scandinavia in part resembled that of the northern half of Britain, which is natural since in past geological times they were certainly connected. The same relationship also holds good between Southern Britain and Germany. Occasionally the trawlers in the North Sea bring up fossil land animals which show that in very late Tertiary times there was land connecting Northern Germany and England. The little island of Heligoland is the last fragment of this lost area. Germany extends from the North Sea southward to the northern crests of the Alps, which just reach into Bavaria. It includes, therefore, many of the relic-stumps of the old Armorican Ranges which were described in Chapter III. It is rich in coal, since the great Carboniferous Trough (p. 65) runs along the edge of these relic-stumps.

The Mantle-map shown in Fig. 69 will make the Build of Germany readily intelligible.[1] In the middle east are the low plateaux of the Ardennes and the Taunus, through which the Rhine has cut its way as the original plateau slowly rose across its path. These plateaux are two of the relic-stumps mentioned above. They consist of very ancient rocks and are accordingly labelled 1 in the Mantle-map. Of much the same ancient formations are the two ridges which flank the Great Rhine Graben further south, which are labelled Vosges and Black Forest. These old rocks perhaps originally formed an arch of which the centre collapsed many hundred feet. Down the trough (Graben) thus formed the Rhine found its way, and has covered this valley some 20 miles wide with deep deposits of gravel, shingle and soil. Much of Bohemia in the east of our diagram also consists of similar ancient rocks surrounded by up-thrust blocks (Horsts) which build a wall around Bohemia, as indicated in Fig. 123. Lastly, in the very centre of the map there is an isolated horst of the same relic type, which is called the Harz (the old name was "Hercynia"). Sometimes Hercynian is used to define all these relics of the Permian mountains

[1]Based on the "Geological Map," of Europe by Beyschlag and Schriel, 1925, Berlin.

(instead of Armorican). These four regions correspond to the Older-Mass, which we studied in some detail in England.

Unfortunately the remaining formations are somewhat more complicated in their structure than those in England. Overlying the ancient relics are a series of strata belonging to the Triassic and Jurassic ages. They form "Mantles" as I have previously termed them,

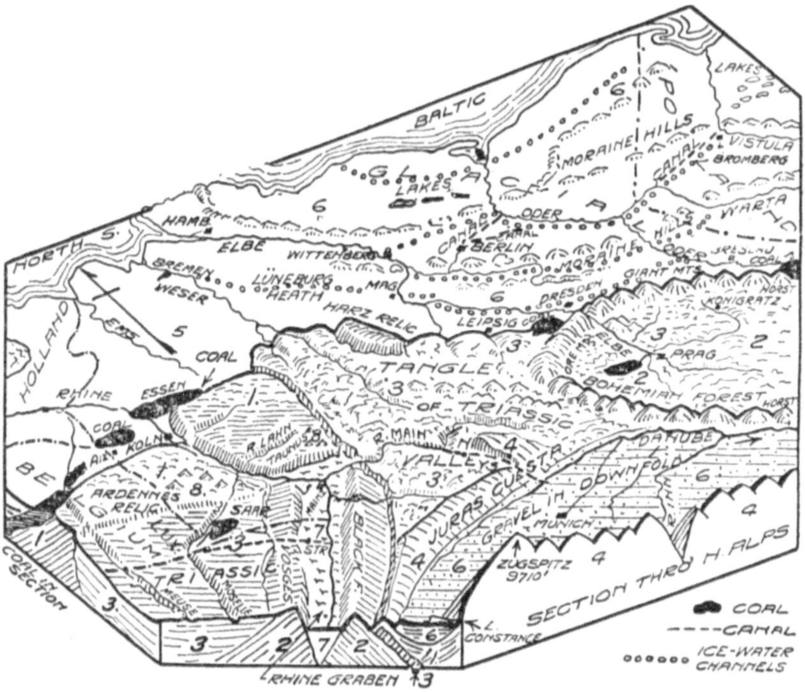

Fig. 69.—Mantle-Map of Germany. 1 and 2 represent Relic-Blocks, Coalfields (2a) fringe these on the north, 3 are Triassic, 4 is the Jurassic Cuesta, 5, 6 and 7 are shallow alluvial deposits. Small extinct volcanoes occur at 8. A geological section is given on the front edge.

more or less horizontally laid around the relic-blocks. No doubt in places large portions of this Younger-Mass have been eroded, especially where they were elevated by late geological movements. Thus most of Lorraine, traversed by the Meuse and the Moselle, and now in France, consists of these Triassic rocks (3 in Fig. 69). So also much of the region between the three relics of the Harz, Black Forest and Bohemia is part of the Triassic Mantle. This last area is cut into innumerable "mature" valleys (p. 59) by the waters of the Main, Weser, Saale, etc., and the whole country is sometimes described as the "Tangle of Triassic Valleys." This disunion certainly helped to maintain the many isolated feudal units which characterised Germany in medieval times.

The third Triassic area is found in the north of Bohemia around Prague and the Upper Elbe. Lying smoothly on the Triassic strata and just north of the Danube is a series of sloping Jurassic beds (4) which form a rather prominent scarp or Cuesta of the type which we have already studied in England. This Cuesta dips to the south under the soft Tertiary (and later) beds of the Danube Downfold. The Downfold is capped by much glacial gravel, etc. (6) carried down by the great glaciers of the Alps some 40,000 years ago. The Alps as stated just touch the south of Bavaria. Here peaks of 9,000 feet occur, and they are built up of Jurassic and later rocks, which have been buckled up in the Alpine Storm of some ten million years ago. We may picture the Jurassic Cuesta as passing south under the Danube and forming part of the sharply folded front edge of the Alps themselves.

The last major division of the Mantle-map (Fig. 69) is situated in the north, where it constitutes a notable area of the Great European Plain. As Fig. 69 shows, there is a marked line of division between the soft beds of glacial gravels, silt and river deposits of the Northern Plain (5 and 6), and the much more ancient rocks (1, 2 and 3) of the southern uplands and mountains. This topographic boundary has greatly affected the history of Germany. Tongues of plain penetrate the elevated lands near Cologne, Munster, Leipsig and Breslau. The Plain may be conveniently divided by the Elbe into the western and eastern divisions. Both portions slope gently to the sea, but everywhere they are covered with the irregular pattern of fairly recent moraine material. In the Ice Ages the Scandinavian Ice-Sheet reached south so far that its margin climbed up the rugged uplands. The marginal moraines are found near Cologne, on the slopes of the Harz, on the Thuringian Hills, far south of Leipsig, above Dresden, and near the source of the Oder.

In general, however, the moraine hills are much more abundant and higher to the east of the Elbe. The western portion is characterised by broad expanses of ill-drained moors and bogs of which the Bourtange Moor on the Dutch border is the largest. In some cases the Ice-Sheet halted for many decades in one position. All along its front there developed a broad marginal channel, which carried the ice-waters northwest to the sea. One such halt has produced an ice-water valley (now largely dry) leading from the upper Oder near Breslau to the Elbe near Leipsig and thence south of the Luneburg Heath to the mouth of the Weser. A second similar "dry" valley runs from the Vistula to the middle Warta (Fig. 69) and thence along the middle Spree (south of Berlin) to the Elbe below Magdeburg. A third leaves the Vistula valley at Bromberg, then follows the Netze and Oder to the old Berlin canal and joins the Elbe near Wittenberg. A similar channel runs through Pomerania about 25 miles south of the coastline.

The youngest moraines are naturally found furthest north, since they were deposited as the Ice Cap retreated to Sweden. These moraines have suffered much less erosion than those at the southern margin of the sheet already mentioned. Hence there is an almost continuous wall of moraine extending from near Dantzig westward towards Berlin, and then again northwestward to Denmark. This is indicated in Fig. 69. The Ice-Sheet also left ground-moraine scattered haphazard over the surface in its retreat, and this young land surface has often not yet developed a regular drainage scheme. Great numbers of small lakes are therefore typical of the landscape especially just near Berlin and north of that city half-way to the coast. The strategy of the two famous battles of Tannenburg (in 1410 and 1914) depended entirely on the topography of these glacial lakes.

Many interesting relations which have a considerable bearing on the national development can now be pointed out. Germany consists of a flat northern portion and of a southern rugged portion. Clearly the former will grow most of the crops and sustain the largest populations. The densest industrial regions will lie near the junction of the plains and the uplands, just where the coal deposits of the Saar, Chemnitz, Essen and Silesia cap the ancient rocks (p. 52). The poorer portions of the uplands will be on the relic-blocks and rugged Alps, while the "Tangled Triassic Valleys" with their newer easily eroded rocks will give much scattered agricultural land, found mainly on the floors and on the lower slopes of the "mature" valleys. Metals are always more abundant in the more ancient rocks, and are especially rich on the flanks of the Erz (Ore) Mountains southwest of Dresden and in the Harz relic-block.

Communications would seem to be easily established in the plains and difficult in the rugged south. As a matter of fact, conditions have varied a good deal throughout man's history in this respect. The plains were largely covered with forest, which primitive man dreaded, whereas the rugged south had two well-marked relatively level areas which have always been rather easy to traverse. These are the Rhine Graben and the Danube Downfold, which were occupied long before the dawn of history. So also the sparsely wooded Cuestas and relic-blocks offered fairly satisfactory sites to primitive man. Rivers are easier to negotiate than forests and marshes, and here again the Rhine route is of supreme importance, while the Weser, Elbe and Oder early enabled small boats to link the uplands with the sea. It may be mentioned here, though we shall return to the topic later, that east-west communication was obviously not so easy, since no rivers except the Danube flow in that direction. Finally the Bavarian Alps offered a formidable bulwark to early travel, especially as they only form part of a northern ridge standing before a series of even higher ridges. For this reason the Oder valley which led up to the Moravian Gate was of great importance in medieval times. All of these barriers separated the products of the north from the totally different goods produced in the Mediterranean Basin.

B. Early Tribes of Germany and their Ethnology

Chapter IX deals with the relations between the Teutons and the Keltae, who were in closer touch with the more progressive nations of the south. The German Cradle seems to have been in Hanover, but long before 600 B.C. the Germans had spread into the lands between the river Main and Denmark, *i.e.*, as far as Thuringia and Hesse (Fig. 70). To the east the Slavs had filled in the country abandoned by the migrating Teutons (p. 167); to the west between the Weser and the Rhine the eastern Keltic tribes lived till about 300 B.C. They were then driven still further west by the advancing Germans. By 150 B.C. the Germans also encroached on the (modern) Belgian region, which, however, they occupied more or less in harmony with the Keltae. Thus there developed a somewhat mixed nationality in this particular area.

Keltae inhabited Bavaria and Bohemia during the last four centuries before Christ, but about the time of Caesar they moved south of the Danube into the region of Munich and the Tyrol. The district around the Black Forest was left practically empty. Germans moved up the Rhine Graben and also into Franconia about the same time. Since, however, the head-indices are high in this area it seems clear that there were larger numbers of Alpines still living here. Thus at the greatest extent of the Roman Empire the Germans touched the present border from the North Sea down to Strasbourg. The Roman palisade wall, built across the re-entrant between the Rhine and the Danube, ran from Wiesbaden to Wertheim and then Ingolstadt. It is shown in Fig. 70. Thus the territory of the Germans was perhaps only about 200 miles wide, for they had the Slavs to the east, and to the south the deserted lands of Wurttemburg.

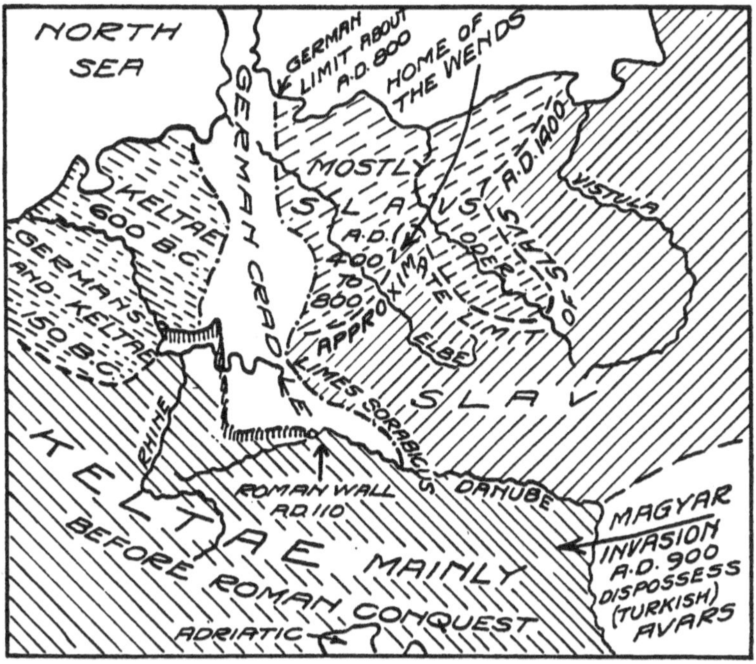

Fig. 70.—The early spread of the German tribes. They drove the Keltae to the southwest from Thuringia; and later drove out the Slavs to the east. The Roman Wall and Frankish Wall (Limes) are shown. Note the present home of the Wends. (Data mainly from Ripley.)

For a time Rome tried to overcome the Germans, east of the Rhine. But as we have seen Arminius (p. 158) led the tribes against the legions of Varus and crushed them in the Teutoberg Forest in A.D. 9. We cannot here concern ourselves further with the early history of the German tribes, but the expansion of the Franks is discussed in chapter XVII. Some reference to the final racial map is necessary. As Ripley remarks (*loc. cit.*, p. 215), "First and always as to the physical geography of the country, everything ethnically depends upon that."

The Germans living north of the Aix-Breslau line in Fig. 71 are all narrow-heads, those to the south are almost wholly broad-heads. Much the same follows if we study the Blonde-Brunette grouping. There are only 5 per cent. of brunette types north of the Aix-Breslau topographic boundary. South of Nuremburg the proportion of brunette rises to 20 per cent. or more. In the intermediate regions of Hesse and Franconia the proportion is about 16 per cent.

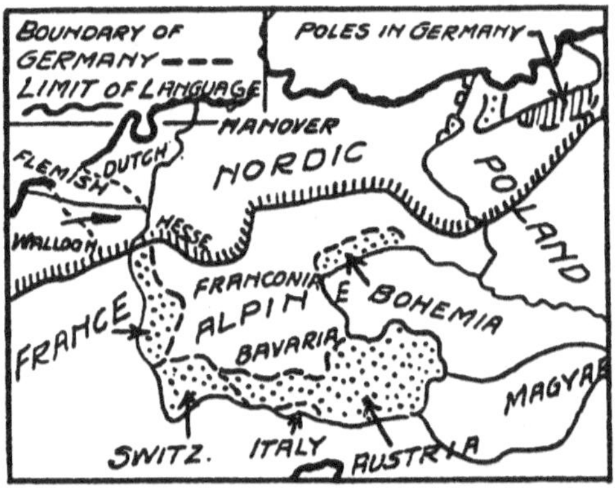

Fig. 71.—The limits of the German Nation compared with those of the German Language. Dotted areas are where German is spoken outside Germany. The approximate boundary between the Nordic and Alpine races is also shown. (After Ripley.)

The southern half of Germany is overwhelmingly Alpine in race and therefore entirely distinct from the people of the plains of Hanover and Prussia. Ranke of Munich objected to this racial cleavage, and tried to account for the broad-heads in the south as being a peculiar effect of the elevation! As Ripley suggests, Ranke endeavoured to show that the Bavarians "were still ethnically Teutonic, but merely fallen from dolicho-cephalic grace." Of course we have here only another example of a language spreading far beyond its original speakers. It is a moot point whether the Nordic Prussians learnt German from the Alpine Bavarians or *vice versa*. They are racially so different that one of the two certainly spoke some entirely different language originally. Proto-Finnish seems to be the most probable alternative; and the present writer thinks that the evidence (which is very slight) is in favour of the ancestors of the modern Prussians (who are essentially Nordic) having learnt German from some primitive Alpine folk.

There is an interesting tongue of Nordic folk projecting south into Thuringia and the valley of the Main. They seem to have avoided the infertile plateaux of ancient rock on each side (whether in the Taunus or in the horsts of Bohemia) and kept to the better soils of the "Tangled Valleys" (Fig. 69). This region between Leipsig and the Main Valley is Thuringia, sometimes called the Heart of Germany. It is certainly not the region of purest Nordic blood, but if the highest

Fig. 72.—The Marks (Marches) or Protecting Areas between the civilised Holy Roman Empire and the Slavs, Magyars, etc., to the East, about A.D. 1000. Mountains over 3,000 feet on the margin are shown in black. The chief Passes are (1) the Moravian Gate, (2) the Vienna Gate, (3) the Pusterthal, (4) the Pear Tree Pass.

type of culture develops where two good stocks mix, then Thuringia is certainly a region where a moiety of Nordic blood has been mellowed by a reasonably large mixture with Alpine blood. A somewhat similar mixed Nordic group penetrated up the river Neckar to Hohenstaufen and Hohenzollern and even to Hapsburg in Switzerland (Fig. 84). These famous cradles of German dynasties, though surrounded by solid blocks of Alpine peoples, have therefore some claims to be considered partly Nordic.

One outstanding feature of the evolution of the German people in early medieval times is the institution of the Mark (Fig. 72). This was the margin of civilised territory which bordered on the lands of the pagans such as the Slavs, or was subject to the attacks of Asiatic nomads like the Magyars, Mongols, etc. The natural response to such a condition of unrest was for the ruler of the threatened civilised lands to appoint strong military governors to hold back the barbarians. In the Roman Empire with its strong central government the whole border was patrolled, but the successors of Charlemagne were unable to adopt this procedure. They handed over an area of land, often from 10,000 to 20,000 square miles, to a suitable adherent, who was called a Mark-Graf, Margrave or Marquis. These Margraves owed allegiance to their feudal lords, but inevitably many of them became powerful enough to carve out independent kingdoms and indeed many of the 1,500 separate sections of Germany developed in this way. Moreover, it was precisely the most energetic men in the kingdom to whom the career in the Marks made the most appeal. As we shall see the leading families from whom the Emperors were chosen developed in these Marks. In Fig. 72 the dozen Marks protecting the east flank of the Holy Roman Empire about A.D. 1000 are charted. In the north is the Mark against the Danes, which in time became Denmark. The North-Mark and the East-Mark gave rise to the Saxon Emperors and later to Brandenburg, the cradle of Prussia. Bohemia became an independent kingdom but also produced a line of Emperors from 1347 to 1438. It was the Bavarian East-Mark which gave rise to Austria (Ost-Reich). Finally in the southeast were the Marks of Styria, Carinthia and Carniola, early incorporated in the Hapsburg lands.

These Marks defending the Empire were naturally based on strong mountain positions where possible, as in the case of Moravia which protected the Moravian Gate at the head of the Oder. Especially was this the case in the southeast, where they faced the Magyars, Croats and Turks. Vienna itself was where the Danube cuts across the Alpine Folds. The three more southern Marks were rendered secure by their position on the eastern slopes of the impregnable Alps, where they served to protect the passes of the Pustertal and of the Pear Tree.

C. *The Development of the Holy Roman Empire*

The Holy Roman Empire developed from the conquests of Charlemagne, the Great Emperor of the Franks. The wall he built against the Slavs stretching from Lubeck to the Danube is shown in Fig. 70.

The history of Charlemagne is given in the chapter on France (p. 321), and we may now discuss the events which followed the partition of his empire at Verdun in 843. At this time the Franks of North France had already given up the old Frankish Speech, which, of course, was much the same as German, and now spoke the French language. At this date a man who spoke no Latin (or a Romance derivative) was called Thiudisk, and this word in time became Deutsch (Dutch), *i.e.*, German. By 879 the German ruler had conquered a number of the old Belgae tribes, who spoke Romance languages, but there was no idea at that time that the boundaries of languages or nations should agree with those of states.

For a time around 882 Charles the Fat governed almost as large an Empire as Charlemagne, but his rule was unpopular, and the Eastern Franks elected for their own ruler the nephew of Charles. This principle of electing the Chief had long been observed by the Teuton tribes, but it was given up in favour of hereditary heirs in most nations of Europe. However, its survival in Germany led to the great weakness of German government, the lack of assured continuity in the Imperial succession. (Of course it was not the election which was at fault, but the total unwillingness of rivals to abide by the election. Hereditary monarchs represent an earlier phase in human progress than rulers *elected* by democratic votes.)

The descendants of Charlemagne controlled the East Franks, and some of them soon took the title of Holy Roman Emperor. They were appointed by the Holy Father at Rome, who hoped that they would follow in the path of Charlemagne and extend the realm of Christendom at the expense of the pagans. In reality most of the early rulers of the Holy Roman Empire were generally occupied in fighting the Pope, so that there was little that was Holy or Roman about their rule. Moreover, they had little imperial power, unless their personalities were strong and their private possessions (before election) were unusually large. One might without much exaggeration say that at first the so-called Holy Roman Empire was an "Unholy Collection of Rival German Electorates," and that later it became a "Private Perquisite for the Promotion of Hapsburg Policies."

Charlemagne fixed on Aix as his chief residence. This town lies near the great Rhine Route just where it debouches on the Great Plain (at Cologne) and crosses the East-West Route from Bruges to Cracow (Fig. 75). As time went on the regions from which the Emperor was chosen varied in rather a peculiar way, moving south as may be seen by reference to Fig. 73. The second group of Emperors were *Saxons*, and descended from Henry the Fowler (A.D. 919). These peoples had been the chief enemies of Charlemagne and his Franks a century before, but had now become civilised, and their courage and initiative led to their leader being chosen King of the eastern lands of Charlemagne. Henry fought against the heathen Slavs, Magyars and Danes and added considerably to the German realm. His son was Otto the Great (930), who crushed the Magyars in 955 at the battle of Lech near Augsburg. He established the East-Mark (Ost-Reich) against the Magyars and this Mark developed into Austria.

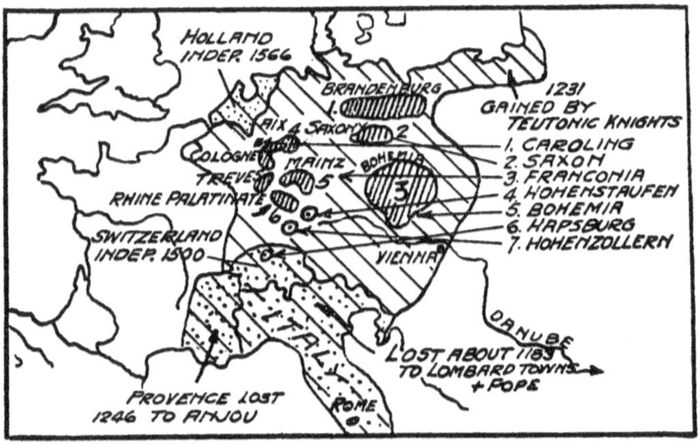

Fig. 73.—The Holy Roman Empire, showing the seven "Dynasties" which controlled it. The territories of the seven Electors are also shown by rulings. The dotted areas gradually broke away from the Empire.

Meanwhile Italy was in a state of anarchy, for the Pope controlled small areas near Rome; the Normans occupied Sicily; the Byzantines South Italy; and a so-called King of Italy tried to govern Lombardy. Otto the Great marched across the Alps just as did Charlemagne and defeated the enemies of the Pope. In 962 he also was crowned Emperor of the Holy Roman Empire, and added Italy as far south as the Papal States to his German Territories.

In 1024 the imperial crown passed from the Saxon House to that of *Franconia* further to the south, and Conrad became Emperor. In 1032 he added the kingdom of Burgundy to the Empire, so that for a time each Emperor was solemnly crowned at four places, Aix, Arles (Burgundy), Milan (Lombardy), and Rome. Otto had decreed that the appointment of the Pope was to be approved by the Emperor; and his successor Henry III (1039) actually appointed four Popes. But in 1059 the Catholic authorities decided that only the Cardinals should be allowed to choose the Pope, so that he could not become a mere puppet in the hands of a powerful Emperor.

Further trouble arose when Henry IV in 1075 refused to obey Pope Gregory VII and refrain from appointing important dignitaries to the Church. For this Henry was excommunicated, and pardoned only when he made public submission to the Pope. The Crusades filled all men's minds during the period 1095 to 1224, and this inevitably led to an increase in papal authority. In 1150 the *Hohenstaufen* family, whose ancestral castle surmounted a hill in Wurttemburg, acquired the imperial crown (Fig. 73). Their first Emperor was Frederic Barbarossa, a strong character who hoped to develop as powerful an organization as the former Roman Empire. He came into conflict not only with the Pope, but with a new factor in Europe, the Independent Towns. The cities of Lombardy (Milan, Verona, Mantua, etc.) formed a league to withstand Frederic's claims for revenue from the rich burghers. The Church in Germany and the Nobles stood by the Emperor, constituting the original *Ghibelline* Party; the towns were helped by the Pope, and often linked with the Emperor's *Guelf* opponents in Germany. At Legnano near Milan in 1176 the Emperor was defeated, and the power of the Italian cities greatly enhanced. Indeed the Holy Roman Empire never regained its authority in North Italy after these reverses.

Pope Innocent III roundly declared that "God had set the Prince of Apostles over Kings and Kingdoms," and justified his statement by appointing Frederic II, grandson of Barbarossa, Emperor in 1214. Frederic, although friendly with the Pope at first, opposed his designs for most of his reign. He preferred his lands in southern Italy (derived from his Norman mother) to the German territories, and he delegated most of the powers in the north to ambitious German nobles. In 1231 at Worms Frederic recognised the territorial powers of the secular princes, and so established the Empire "as a loosely connected congeries of ruling princes under an imperial head" (Heyck). In 1246 a further loss occurred when Provence went to the House of Anjou. The interests of Frederic in Science and Literature are discussed in an earlier chapter (p. 202). After his death in 1250 his descendants vainly endeavoured to maintain his Hohenstaufen power in south Italy, and it had already passed away in Germany. A long period of anarchy ensued in Germany, the chief secular prince being Ottakar of Bohemia. In 1269 he also ruled over Austria, Styria and Carinthia (Fig. 113). He co-operated with the Prussian Knights, and the new town of Konigsberg was named in his honour.

In 1273 a Suabian count, Rudolf of Hapsburg, was elected Emperor, on condition that he resigned his claim to any Italian territories. His ancestral castle was near the junction of the Aar and the Rhine, but he also owned much territory in Suabia (West Bavaria). In 1278 he met Ottakar in battle at Marchfeld, where he completely crushed the Bohemian King. *Hapsburg* soon acquired control of Ottakar's Austrian lands and thus balanced the loss of Italy. Thus the Popes won their long struggle, and "Emperor of the Holy Roman Empire," as far as concerned lands south of the Alps, was but a barren title. For nearly a century the German nobles chose as Emperor men of small power from the Hapsburg, Bavarian and Luxemburg Houses. In 1347 *Bohemia* secured the Imperial Crown in the person of Charles IV (originally from Luxemburg). In 1356 he issued the Golden Bull, which secured the election of the Emperor by seven Electors (Fig. 73). Three of these were the Archbishops of Mainz, Cologne and Treves, four were the secular princes of Bohemia, Brandenburg, Saxony and the Rhine Palatinate. Of equal importance was the decree that these electoral lands should not be subdivided, and that primogeniture should obtain in the secular lands.

In 1414 Sigismund, last of the Bohemian Emperors, sought to use his imperial position to bring about peace in western Europe; an object worthy of that "Unity for the Good of Western Europe" which was dimly felt to be one of the reasons for the continued existence of the Empire. He failed completely, and within a few generations the religious divisions due to the Reformation destroyed any idea of unity for the benefit of all the Empire. It was inevitable that the monarchs who became Protestants should cast off all allegiance to the Emperor, who at that time was Charles V, a strong Catholic partisan.

In 1437 Albert of *Hapsburg* became Emperor, and thereafter with one short interval the honour became essentially merely an extra title of the Austrian House of Hapsburg (Fig. 108). Its chief function probably was to offer a centre of resistance to the ever-growing ambitions of the Kings of France. In 1477 Maximilian of Hapsburg married Mary of Burgundy, through whom he acquired large territories in the Netherlands, but shortly afterwards (about 1500) Switzerland was lost to the Empire, as is described in the chapter on that country. Early in the struggles between Protestant citizens and Catholic Emperor Holland broke away from the Empire, being practically independent after 1566. Finally in the Wars of Religion the attempts of the Hapsburgs to crush the Protestant princes of Germany was foiled by the Catholic armies of Louis XIII (p. 197). Thus the Balance of Power proved a stronger factor even than Religion in determining the development of the nations of Central Europe. The later history of the Hapsburgs finds a place in the chapter on Austria, and we may now pass on rapidly to the closing scenes of the Empire. Napoleon's conquests in a sense paralleled those of Charlemagne; and it grew clear that he meant to acquire the title as well as many of the lands of the Holy Roman Empire. Foreseeing this, in 1804 Francis of Hapsburg had declared himself Emperor of Austria, and in 1806 he resigned the barren title of Emperor of the Holy Roman Empire. With the passing of the title no vestige was left of this ancient, interesting but long effete institution.

The decline of the power of Austria was accompanied by the growth of Prussia, which finally united Germany into a powerful empire; in some respects more like that of Charlemagne's than almost any of the varied phases of the Holy Roman Empire since his date. This is discussed in a later section (p. 275).

D. *The Hansa Towns and Medieval Trade*

Our survey of the history of Medieval Germany shows us that the Holy Roman Empire was controlled to a large degree by nobles whose estates lay chiefly in the "Tangled Valleys" of the mountain region. At first the Emperor tried to bestride the Alps and administer two entirely distinct environments. As we have seen, Frederic II decided to concentrate on the southern lands, and gave up the northern German lands; and finally his family lost both. While the later Emperors were quarrelling in the "Tangled Valleys" a new institution was developing in the German Plain, founded on Commerce and on the industry of the Middle Classes. For a time the commercial *League of the Hanseatic Towns* was the most powerful force from Belgium to Finland; and though it finally disappeared, it forecasted much more closely than did most medieval institutions the type of union for mutual benefit which became increasingly noticeable in later history.

It has been pointed out by continental writers that in early medieval times the major trade routes seemed to ignore Germany. There was a long-established route from the Baltic to Constantinople, *via*

Riga, the Duna, Vitebsk and the Dnieper. We have heard a good deal about the Rhone Corridor in the west. The Mediterranean trade in the south was very important, but the Baltic trade developed much later, mainly owing to the energy of the folk in the German Plain. It has been noted that the growth of London depended in part on its position opposite to the Rhine estuary. About A.D. 1000 there were German traders in London, and by 1157 there was a Guild of Cologne merchants established in the same city.

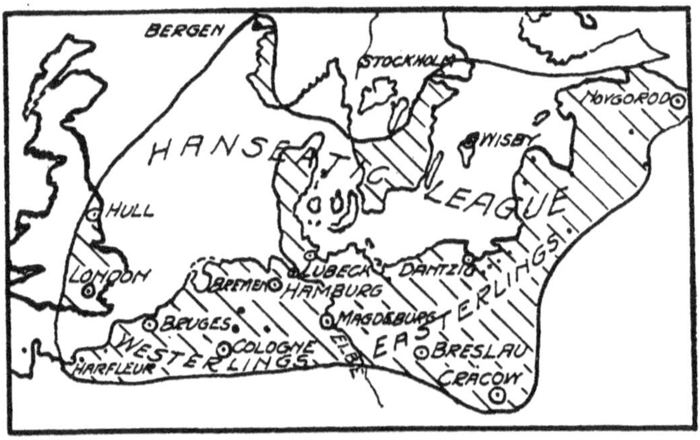

Fig. 74.—The Hanseatic League of Merchants in the 13th and 14th centuries. It offers an interesting example of almost purely Nordic control.

About this time the routes across the Alps were linking Germany to Italy (Fig. 75), but these only indirectly concerned the Hanseatic League. In the North German Plain the merchants east of the Elbe were called Easterlings, while those west of the river were known as Westerlings. (The English term "sterling" used in coinage is said to refer to the use of Easterling money as a standard about 1180.) About 1266 Hamburg and Lubeck founded Guilds for mutual protection, and Cologne joined the community about 1282. By this time the German Hansa ships were linking every port in the Narrow Seas. In the 12th century they had reached Russia, and Wisby on the island of Gothland was one of their strongest posts. In the 13th century the far-distant trading centre of Novgorod (south of Leningrad) was almost completely in the hands of the Hansa League, at this time dominated by Lubeck (Fig. 74).

The prosperity of Hansa roused the jealousy of Denmark, especially as the herrings of the Sound and the cod of Bergen (in Norway) were two of the chief trade commodities of the Hansa League. In 1227 Waldemar the Dane was defeated by the German merchants, and as we have seen no strong ruler arose in north Germany who might have curbed the growth of their authority. The traders were still for the most part within the Empire, and never united for any other purpose than for the promotion of trade. In addition to the towns mentioned, Bremen, Brunswick, Magdeburg, Dantzig, Breslau and Cracow were all important centres of leading Hansa merchants, while nearly a hundred smaller towns had their Guilds for promotion of Hansa business. These all exchanged the metals and other products of Germany for the fish, fur and wool of other lands.

Towards the end of the 14th century Denmark again attacked the League and sacked Wisby. The Teutonic Knights (p. 190) also came into conflict with them in East Prussia and Estonia, but the forces of the Knights were crushed at Tannenburg by the Poles in 1410. During the 15th century many factors contributed to the decline of Hansa. For some unknown reason the herrings abandoned the Baltic area. Strong Emperors of Germany curtailed the privileges of Hansa. Some of their western ports like Bruges were silting up, so that their trade died. England grew in maritime power and built up her own trading concerns with Germany. Russia and Poland also increased greatly in importance, while the Wars of Religion ruined the prosperity of north Germany. Perhaps the chief reason of all was that the centre of interest in trade definitely shifted from the Baltic and North Seas towards the Atlantic and America after 1500. North Germany had no important share in this trade for several centuries.

It is difficult for us to-day to understand the set-back to trade and communications which followed the Fall of the Roman Empire. Without a strong central authority no communities will maintain the through roads on which land commerce depends. "For six centuries (600 to 1200) by far the greater part of the people of England, Germany and Northern France were living in the country on great estates belonging to feudal lords" (J. H. Robinson). Very few towns survived from Roman times, and without towns one can have only the simplest forms of trade. The Crusades helped to stimulate the exchange of goods, and the more peaceful periods which gradually arrived in northern Europe promoted the growth of a citizen class, which slowly wrested town-rights from their over-lords. By and by folk produced more goods than they could consume, precisely in order to trade them for attractive articles from other countries. The woollen goods and the raw products of the north were needed in the south, and the character of this trade has already been referred to (p. 61).

In Fig. 75 the control exercised by the topography upon medieval trade is emphasised. We can divide the elevated areas of central Europe into two kinds. In the high rugged mountains such as the

Fig. 75.—The Topographic Control of Medieval Trade-Routes in Central Europe. Six main belts are indicated. 1, the Alluvial Plain of the Po; 2, the High Alps; 3, the Alluvial, etc., of the Fore-deep or Downfold; 4, the "Tangled Valleys" formed of Triassic and older rocks; 5, the Alluvial Coastal Plain; 6, the Dune and Polder Coast. Note the towns strung along the route on the southern edge of the Coastal Plain. A somewhat similar route follows the Downfold from Dijon to Vienna.

Alps, Pyrenees and Carpathians, trade was confined to narrow rough tracks suitable only for pack animals. The Brenner Pass, however, linking Germany and Venice is so low that much traffic crossed it, but no carriage road was made till 1772. The St. Gotthard Pass and Genevra Pass (shown in Fig. 75) were much less easy to negotiate.

Once north of the Alps several routes offered themselves. Along the Downfold (p. 298) it was relatively easy to journey east to Vienna and the Hungarian Plain or west to Dijon and Geneva. The route from Geneva to Lyons was rather rugged, for the gorges of the

Rhone fill the gap between the Juras and the Alps. North of the Downfold the trader could readily pass the Moravian Gate to Cracow, or go due north from Munich over somewhat rugged country to Nuremberg. These "Tangled Valleys" are indicated in Fig. 75 by lighter shading than the high mountains, though they still offered considerable difficulty to travel except near the streams. Another much used route led from Basel into the Rhine Graben and so down the Rhine to Cologne. Perhaps this was the most important of all. Nuremberg was for a time the chief trading station in Germany, though it is rather difficult to point out any great advantage that it had save its central position. One main route led down the river Main to Mainz. Another went north to Leipsig where the Great Plain was reached.

In the Great Plain an important route connected the cities which bordered the uplands. It led westward from Leipsig to Magdeburg, Hanover and Cologne, and thence through Belgium (p. 290) to Ghent and Bruges. Eastward from Cologne went the road to Dresden, Breslau and Cracow. At first communications over the Plain itself were very bad. Forests and swamps, lakes and rivers abounded, and no one cared to maintain good roads and bridges. However, the Hanseatic League helped to remedy these conditions until strong rulers arose in Germany itself. Unfortunately the Thirty Years' War with its terrible devastation ruined the prosperity of much of this part of Germany for a century or more. It is clear that Germany owes much to the Hohenzollern princes, whose energy and far-sightedness made North Germany the most progressive country in Europe.

E. The Development of Prussia

The sluggish rivers of north Germany, often burdened with silt and frost-bound in winter, flow through infertile plains in the north. So sandy are they in Mecklenburg that it was called the "Sand-box" of Germany. The piles of tumbled glacial debris and the higher moraine hills, which were drained into innumerable small lakes, meant an uphill job for the ruler of the region, even after the forests were cleared from large areas. In the east the Slav population was nearly all driven into Poland by German advances during the 5th to the 8th centuries. But south of Berlin, in the basins of the Spree and the Neisse, there still persists a block of Slav people, the Wends, who even to-day extend from Kottbus to Gorlitz (Fig. 70). By 1400 the Slav boundary was not very different from what it is now. It was in this not altogether favourable environment that the Hohenzollerns founded the strongest monarchy in Europe.

It is well to realise that the Prussians of to-day have little in common with the "Old Prussians." The latter in early medieval times were a subdivision of the Lithuanians, speaking a language not at all like German, but something like Sanskrit. They were convinced pagans, and resisted conversion later than almost any tribes in Europe (Fig. 139). As a result they were practically annihilated by the ungentle methods used by the Teutonic Knights. All that is left of them is the name of their land (now East Prussia) and a few manuscripts in their long-dead language. However, they do not seem to have differed materially in race from the Nordic folk who live to-day both in Lithuania and North Germany.

In the southwest corner of Germany astride the divide between the Danube and the Rhine is the castle of Hohenzollern. It is between and quite close to Hapsburg and Hohenstaufen, and produced a family as energetic and capable as either (Fig. 73). The Hohenzollerns levied tribute on travellers who passed nearby, as was the custom of all such powerful barons. The younger sons obtained posts at various courts, so that in 1170 Conrad of Hohenzollern moved northward to the Imperial City of Nuremberg, where he had been appointed Steward by Barbarossa. This office remained in the family, and in 1248 Frederic of Hohenzollern became Bygrave of the city. A century later the Nuremberg Hohenzollern had risen to the position of an Elector of the Emperor, and in 1346 was given control of the Mark of Brandenburg in the far northern Plain (Fig. 76). In 1486 we hear of the Court of Brandenburg being held in Berlin. It was, however, not till 1609 that Brandenburg began that remarkable growth which developed first into the powerful kingdom of Prussia, and later into the German Empire.

The first lands joined to Brandenburg were the result of rather vague marriage claims, and lay far to the west on the lower Rhine. These were a group of Territories of which Cleves (2 in Fig. 76) is perhaps the best known. They lay too far off for the Brandenburg ruler to derive much benefit from them for many years. In 1618 the much larger region of East Prussia (which had been assigned to another Hohenzollern in 1526) was added to the Brandenburg realm. This was just as far to the east as Cleves (or Kleve) was to the west, but such scattered domains were quite usual in disunited Germany. It was Frederic William, who became Elector of Brandenburg in 1640, who raised his country by energy and initiative above the hundreds of other small German principalities. At the close of the disastrous Thirty Years' War he benefited by the addition of Pomerania (4 in Fig. 76) which gave him an outlet to the coast. He also constructed a canal along one of the ice-water channels described earlier, so as to link the Spree (and Elbe) with the Oder, and so with the coast. His son in 1701 took his title from those of his lands which lay outside the Holy Roman Empire, and so became King of Prussia. The important port of Stettin was gained from Sweden in 1720.

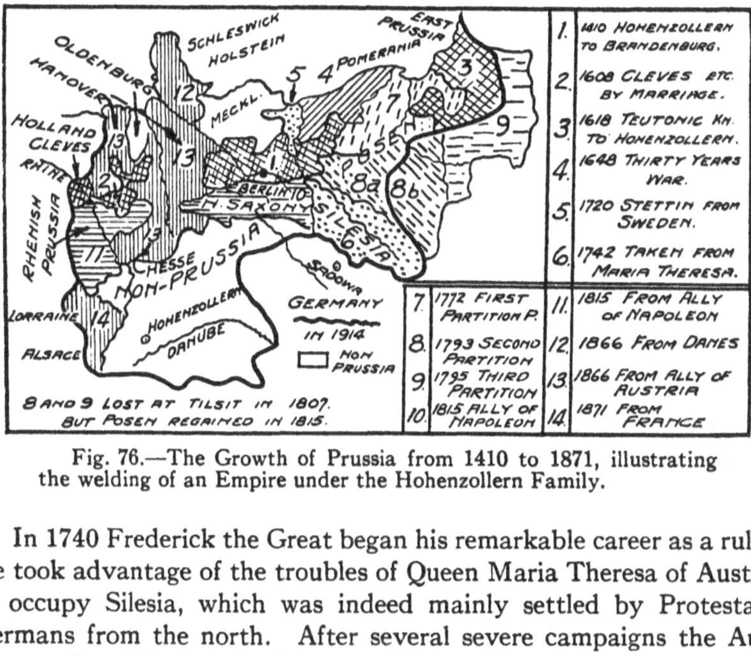

Fig. 76.—The Growth of Prussia from 1410 to 1871, illustrating the welding of an Empire under the Hohenzollern Family.

In 1740 Frederick the Great began his remarkable career as a ruler. He took advantage of the troubles of Queen Maria Theresa of Austria to occupy Silesia, which was indeed mainly settled by Protestant Germans from the north. After several severe campaigns the Austrians relinquished their claims to Silesia. Brandenberg now resembled a star-fish reaching out arms north, south and west, but still far removed from Cleves and East Prussia. We are told by some writers that the Great Elector in the 17th century already realised that north Germany could not thrive upon the land traffic from north to south, that trade in the Baltic was dying, and that it behoved Brandenburg to gain control of the Elbe and of ports on the open North Sea. It may be so; but it was not until 1866 that this long-range planning bore fruit. The very general practice of most European monarchs of grabbing land in every direction explains the growth of Prussia equally well. However these early Prussian rulers showed remarkable wisdom in welcoming Dutch engineers, French Huguenot refugees, and even Jewish merchants, who certainly helped to teach the Prussians habits of industry and scientific application which they have never forgotten.

Frederic the Great was adept at balancing one power against the other. He encouraged Catherine of Russia to subdivide Poland rather than Turkey (which would have aroused Austria), and to share the Polish spoil with Prussia and Austria. At last in 1772 Brandenburg and East Prussia were united, and thereafter the whole region became geographically Prussia (Fig. 76). This unwarranted "grab" includes, of course, the ill-omened Polish Corridor, which has always been settled by Poles, who thus link their inland country with the sea along the lower Vistula. On the other hand, no Poles ever lived in most of East Prussia, for which Germans have a sentimental regard, not in the writer's opinion very well founded. We must remember that the Teutonic Knights were in no sense politically allied to Brandenburg (*i.e.*, Prussia) before 1618, though, of course, most of the Knights were German and so also were the bulk of the settlers by this date. As regards the Polish Corridor, one may digress to point out that the Canadian Pacific Railway from Montreal to Halifax crosses a foreign country by the "Corridor of Maine," but neither Canada nor the United States seems to think it a problem likely to lead to war!

The first partition had resulted very happily for the victors, and as Poland continued her futile internal squabbles, the process of partition was repeated in 1793 and 1795 (Fig. 117). As a result the whole re-entrant of the Vistula (so unsightly in Prussian eyes) was occupied at the latter date. However, Prussia was conquered by Napoleon at Jena in 1806, and was compelled to return most of her recent gains to the Poles—who were strong supporters of Napoleon. Posen was, however, given back to Prussia after Napoleon was finally defeated at Waterloo in 1915. Prussia also willingly accepted large tracts in north Saxony and on the middle Rhine in place of central Poland (8b and 9 in Fig. 76). She thus obtained such famous cities as Aix and Cologne with thousands of German nationals in place of the deeply resentful Poles.

Prussia now controlled the Lower Vistula, the whole of the Oder, the Middle Elbe and the Middle Rhine (Fig. 76). It was obvious that Hanover was needed to fill the gap between the Rhinelands and the cradleland of Brandenberg. The first half of the 19th century witnessed an intense if peaceful struggle between Austrian Hapsburgs and Protestant Hohenzollerns for priority among the many German Principalities. One of Napoleon's greatest achievements was to reduce the number of independent German Princelings, Grand-dukes, minor Dukes, Counts, Imperial Cities, etc., from 1,500 to 50; but of all these Prussia was now immensely more powerful than any of the others. Trouble on the Danish border resulted in war between the Danes and Prussia plus Austria. As usual, though the victors obliterated the "Holstein Irredenta" by joining it to Prussia, they created another by annexing much purely Danish country. (Exactly the same plan has been adopted by Italy in the Trentino.)

The rivalry between the two great powers burst into flame in 1866; but in a short campaign the Prussians under Moltke crushed the Austrians and their Hanoverian and Bavarian allies at Sadowa (or Koniggratz). Most of Hanover and Hesse was handed over to the victors, and now the whole German Plain (except Oldenburg and Mecklenberg) became Prussian; while in Saxony, Hesse and the Rhine provinces the Hohenzollerns now extended well into the mountain lands also. Four years later Napoleon III, fearing the growth of Prussia, declared war. All the south German states rallied to the side of their compatriot, and France was decisively beaten. In January, 1871, William of Prussia was solemnly proclaimed German Emperor, and an actual united German Nation was created for the first time in history. In the Great War Germany was forced to return Alsace and Lorraine, Posen and the Corridor (with other smaller areas) to the French and Poles respectively. Germany adopted the Republican form of government after the collapse of the Hohenzollerns in 1918.

We may finish this lengthy chapter by contrasting the present boundaries of the German Republic with those of the German language (Fig. 71). As we have seen earlier, language is the chief cement of nationality, and it certainly seems to have been adopted as the criterion of nationality in many of the so-called ethnic maps, which have blossomed as the result of the Great War. However, the North Swiss who speak German have made no move to join their linguistic brothers; while the Alsatians (who also speak German) have gladly left the German Republic. On the other hand, Austria is wholly German-speaking, and to-day Europe is agitated by the possibility of a close union (Anschluss) with Germany. The Germans have spread up the elevated Horsts which bound Bohemia on the north, so that many generations of Germans have dwelt there, fairly peaceably under Austrian or Czech rule. On the Polish side the boundary seems to non-partisans to have been drawn as reasonably as most boundaries.

It is the "technique" of the administration of the Corridor which seems to the writer calculated to arouse animosity, rather than the actual boundary itself. In the south of East Prussia is a group of Polish-speakers who have actually voted to unite with Germany rather than to link with the people of their own culture to the south. Surely even this brief survey shows that given time and good-will all of these causes of friction can be adjusted peaceably, while nothing is more certain than that the victor in modern wars loses almost as heavily as the conquered.

CHAPTER XV

HOLLAND AND BELGIUM

A. The Build of the Lowlands

These two countries are so close to Britain, both in location and culture, that we are apt to forget how small they are compared with, say, Germany or France. The two together are not much bigger than one of the main provinces of either of the latter countries. Furthermore their national existence has been short, and much of their early history and indeed of their build, has already been discussed. Our description must therefore be relatively short. Reference to the Mantle-Map in Fig. 89 will show that the Lowlands form part of the Younger-Mass which lies between the Older-Mass of the ancient Ardennes Relic-Block and the similar Older-Mass of Western Britain. The Younger-Mass is in the form of a shallow Basin of which the central portion has been drowned by the North Sea. Many of the rocks of Belgium and Eastern Holland (7 and 8 in Fig. 89) agree closely with those of the London Basin, but the coastlands of the Lowlands exhibit much larger areas of the latest strata (9 in Fig. 89) than are found near London. However, it is not far from the truth to say that in the Lowlands between Luxemburg and Amsterdam, we see a sequence of rocks much like that discussed earlier as extending from London to Snowdon (p. 35). But in this case the older rocks are in the southeast and the newer rocks lie as "mantles" flung over the other as we move to the northwest.

The highest point in the two countries is in the Ardennes Relic-Block, and is only 2,330 feet above the sea, just east of Liege (Fig. 77). Let us suppose that we start here and travel northwest. In a dozen miles we descend from the ancient plateau to the Coal Basin of Liege. This occupies something like a long narrow graben, in which flow the Sambre and part of the Meuse. Still further to the north are the Tertiary beds of Brabant, while the intermediate Triassic and allied rocks are practically absent. Proceeding northward we reach a zone of clay soils of much the same age as those of Brabant, but giving rise to the famous mud of Flanders. This is at a still lower level, so that it is little above the sea.

Most of Holland, as all the world knows, consists of the Recent

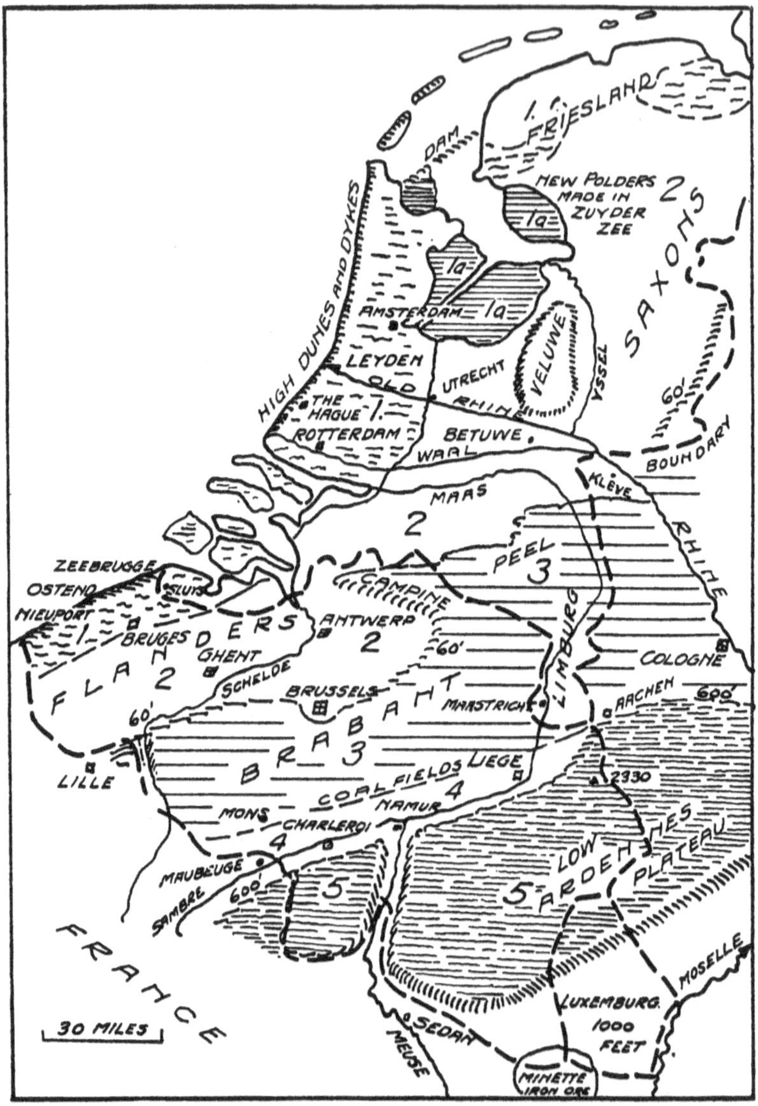

Fig. 77.—A Mantle-Map of the Netherlands showing the five major Belts. 1, the Dune and Polder Margin; 2, the Lowlands of Flanders and central Holland; 3, the sandy Uplands of Brabant, etc.; 4, the Great Coal-Belt of Belgium; 5, the ancient Peneplain of the Ardennes.

deltas of the Rhine, Meuse and Scheldt, which are built up of thick deposits of river-silt and mud washed down from a large part of Eastern France, Western Germany and Switzerland. Probably this loading of the Earth's crust has caused it to sag, so that the deltas have been subsiding through most of historic time. No problem of the relation of Man to his Environment is more interesting than this struggle of the Dutch to combat the never-ceasing attack of the sea on their sinking territory. In Fig. 78 the shoreline as it would be if it were not for man's endless effort is clearly marked. A strip some thirty miles wide along most of Holland would vanish beneath the sea if Nature were allowed to take her own course.

From this peculiar environment we can see that Holland consists almost wholly of late Tertiary rocks which have not yet been hardened to form rock. There are therefore practically no quarries for stone, or mines for coals or metals, except in the southeast corner where the long "tongue" of Limburg extends towards the Ardennes (Fig. 77). Here, however, deep-lying deposits of coal were found lying under the Tertiary beds and they are being extensively worked.

There are six fairly definite regions in Holland which can be readily picked out on the map in Fig. 77. In the north is Friesland, a region of bogs and marshes intermingled with glacial debris, of which little rises much above the sea. Out to sea is a belt of elongated islands which represent the former high coastal dunes. Since the slow subsidence of Friesland these ancient dunes stand far from land, and artificial dykes along the present coast take the place of Nature's defence of dunes. In this region dwell the descendants of the Frisians, a Teutonic tribe closely akin to those which settled in England about A.D. 500. Indeed the Frisian dialect of to-day is nearer English than any other tongue on the continent.

To the southeast of Friesland lies the old home of the Saxon tribes. The country is higher than Friesland, but mostly consists of irregular heaps of moraine (deposited by the Scandinavian Ice-Sheet at its maximum extent), as well as many bogs such as the Bourtange Moor which extends into Germany. The higher sandy land bordering the marshes is called *Geest*. Just below Kleve (Cleves) on the Rhine the river divides into the Yssel entering the Zuyder Zee, the old Rhine and the Waal. The river flats are formed of rich soil, called by the early settlers the Betu-we (*i.e.*, Better Meadow). To the north a large area of sandy Geest at a higher level is the Velu-we (*i.e.*, Evil meadow). These richer southern lands are occupied by the descendants of the

Franks, who constitute the bulk of the Dutch people. West of Utrecht we reach a third type of country, the *Polders*. These are low-lying marshes, so widespread that 38 per cent. of Holland is below sea level. They have been drained by pumping and are now devoted to intensive farming. The flourishing districts of North and South Holland and Zeeland consist almost wholly of these Polders. Along the present coast are the natural dunes and artificial dykes (300 feet wide and 20 feet high) which prevent the North Sea from flooding Holland. Massive gates at Katwyk and elsewhere permit the waters of the rivers to flow out at low tide, but they are shut at high tide to hold back the salt water. In the southeast of Holland is the Peel, a region of sandy moor and fen which is still relatively unoccupied.

In Belgium the western districts are much like those described in Holland, save that the Polder country is much smaller and the Tertiary beds in Brabant are wider and more fertile. This belt of richer land in Brabant between the wet soils of Flanders and the uplands of the Ardennes (which are sterile like most uplifted peneplains) has been a great Corridor for traffic right through later historic times. For similar reasons, lying between the Catholic French and the Protestant Germans it has witnessed more warlike invasions than any other part of the continent (p. 292). Hence it is often referred to as the "Cockpit of Europe" (Fig. 78).

The rich coalfields along the edge of the Ardennes extend westward to Lille in France and eastward with interruptions to Aix and the Ruhr coalfield. Consequently South Belgium is the most industrialised region in Europe. A string of busy but unpicturesque coal-towns includes Lens and Douai in France, Mons, Charleroi, Namur, Seraing and Liege in Belgium (Fig. 77). An entirely different landscape appears in the Ardennes. They are formed of ancient slates, etc., which were folded to form mighty mountains in Permian times (p. 53). To-day only the worn-down stumps are left, but we know that the plateau has been uplifted lately, because the rivers have cut juvenile gorges deep into the plateau (p. 60). Copses of beech, oak and pine of poor growth, together with much heather, cover the ancient rocks. There has never been much settlement here, and indeed the inhabitants were heathen until the seventh century.

B. Early Settlers in the Lowlands

The chief historical interest in this area results from its buffer position between the Romance and Teutonic cultures. But to under-

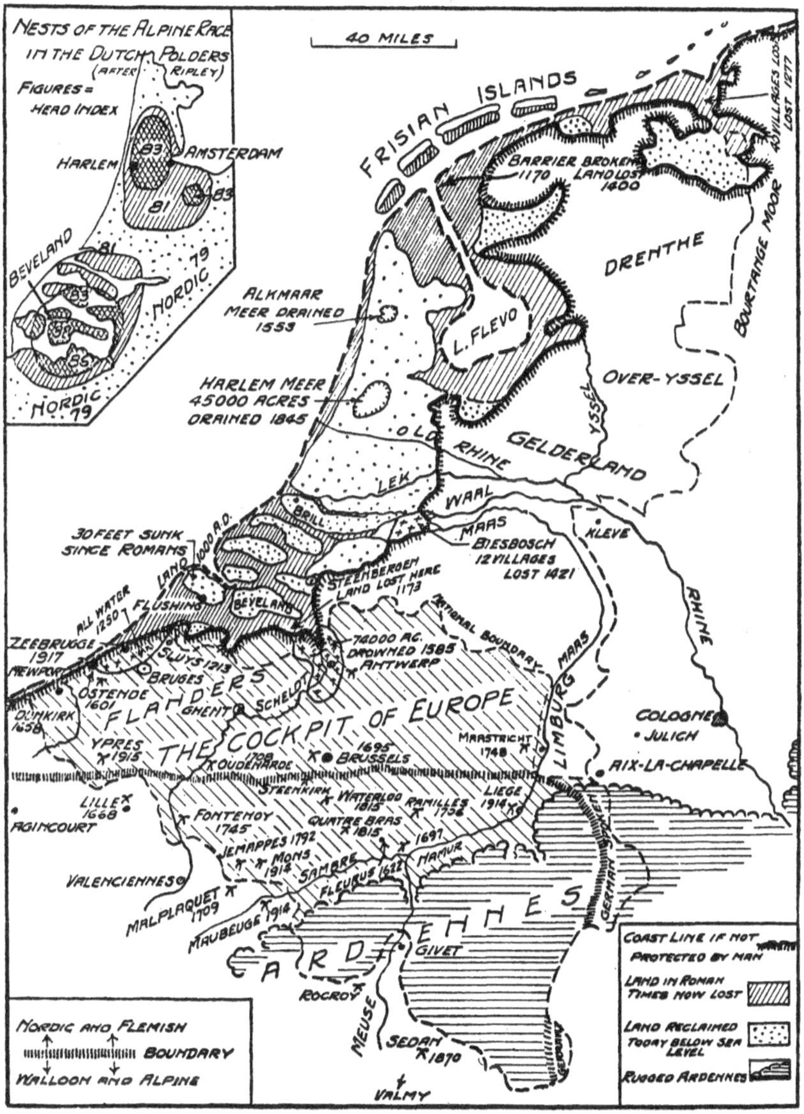

Fig. 78.—A map to illustrate the changes in the shoreline of Holland, and also the "Cockpit of Europe" in Belgium. The chief language and ethnic groups are also indicated.

stand this antagonism we must study the migrations of the tribes who lived in the Rhine Delta when Caesar brought the *Pax Romana* to this part of Europe. We cannot do better than follow Ripley in his discussion of the ethnology. The more we study the two countries the more we become puzzled as to the essential "ingredients" of nationality. The main racial boundary in Northern Europe cuts Belgium in half, and agrees fairly closely with the language boundary. South of this line from Lille to Liege (Fig. 78) live the Walloons, who speak French, and are all brakephs, *i.e.*, Alpines. To the north are the Flemish, who speak a Teutonic dialect akin to Dutch, and who are mainly Nordic. Thus as regards race and language the division between Holland and Belgium might well have run from Lille to Liege. It is, of course, religious differences which on the whole most sharply divide the two small countries: for Flanders is as Catholic as the region of the Walloons, whereas Holland is Protestant. The causes which have produced this cleavage will be made clear when we discuss the history of these two interesting nations.

In Holland the people are dominantly dokeph. The descendants of the Frisians and the Saxons as well as the Franks living in the Betuwe (*i.e.*, Batavians) all have head-indices about 79 or 80. They are Nordic exactly like the bulk of the Flemish. But there is a rather sharp change apparent as the polders of Zeeland and Harlem are reached (inset in Fig. 78). Here the cephalic index rises to 88, showing that we are dealing with some of the broadest (*i.e.*, Alpine) heads in Europe. Probably these short stocky brakephs form an "inlier" akin to the Walloons of the Ardennes, who were almost submerged by the later Teutonic waves. It is stated that Keltae (who were partly Alpine) reached the Rhine Delta by B.C. 400; and probably these are their descendants protected from later hordes by the swampy environment. In Denmark and also in Wales there are similar small "nests" of Alpines, whom Fleure prefers to link up with the Merchant-wanderers from distant Anatolia. As regards Holland the former "Walloon hypothesis" seems more probable to the writer.

The lands which we are discussing came into direct contact with Rome during the various campaigns of Caesar about B.C. 57. He ruthlessly attacked any Teuton tribes who crossed the Rhine, but protected the Belgae, who were a somewhat mixed Gallic-Teutonic group, but probably on the whole mostly Keltae. Already Belgium was acting as a Buffer State; but it was a very large "Belgium," extending from the Rhine to the Seine and south to Rheims and Bonn

(Fig. 79). Many of these Belgic tribes, *i.e.*, the Atrebates, Menapii and Catevellauni were also found in Britain (Fig. 4a). Rome advanced to the Elbe in the first decade of our era, but almost immediately relinquished her conquests beyond the Rhine as the result of her defeats by the warlike Germans. Hence Belgium received a Roman culture which was relatively absent in Holland; and this division undoubtedly helped to determine the later differences in religion.

In the later flourishing years of Rome, Belgic Gaul was split into Lower Germany (which included most of Brabant and the Ardennes)

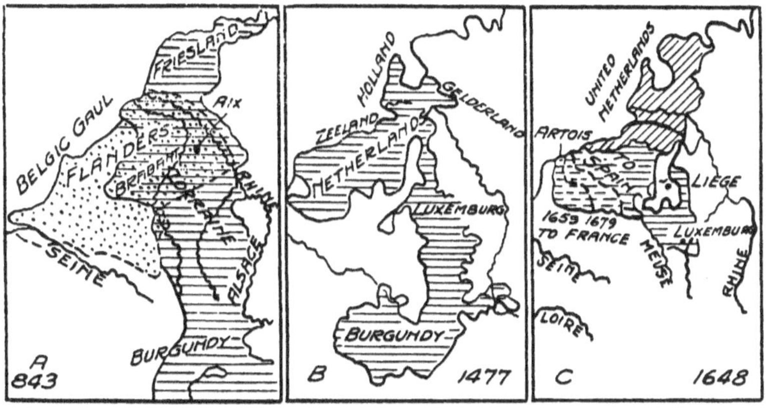

Fig. 79.—The Evolution of the Dutch and Belgian Nations.

A. The region of the Belgae (B.C. 58) is shown by dots, and Lothar's Kingdom (A.D. 843) by rulings.

B. The large aggregate of territories controlled by Charles the Bold.

C. The division into the Protestant and Catholic Netherlands after the war with Spain. Liege belonged to the Bishopric of Liege.

and Belgica, which was almost wholly in modern France. This division indeed determined the northwest boundary of Lothar's "Middle Kingdom." During the Roman control the Franks were confined to the east banks of the Rhine. After 400 they burst across the breaking barrier, and flooded all northern France. Modern Belgium was indeed precisely the area of the Salian Franks. There is no need to give the story of the rise of the Frankish Empire (see Chapter XVII), and we pass on to the Partition of Verdun when the history of these two Buffer States really begins.

After Charlemagne's death his Empire was soon split into three parts. Lothar chose the Middle Kingdom, no doubt because it contained Aix and Rome, the two great cities of the Frankish Empire. He controlled the "Way of Light" and the Lower Rhine as well as the richest region of all, that of North Italy (Fig. 92). There was, however, no other unity about his kingdom, for it lay precisely along that "shatter-belt" of culture where the Roman civilisation impinged on the still barbarian Saxons and other Germans. The later partition of Mersen gave Belgium to Charles of France and Holland to Ludwig of Germany. Most of the rest of Lotharingia also went to the German ruler, together with the whole of North Italy. However, this resemblance to the modern pattern in the Lowlands soon vanished under the political conditions of the early Middle Ages. We have discussed elsewhere the development of practically independent provinces in the troubled period of the Feudal Ages, and these Delta lands afford typical examples of the rise of such powerful vassals. Since the environment played such a large part in the struggle for liberty, we may here pause to discuss the remarkable changes in these delta lands in historical times. We shall see that the earlier independence of Holland arose from the greater ease with which she could be defended as compared with Belgium.

C. *The Changing Environment of the Delta Lands*

Where numerous large rivers enter the sea there are bound to be great changes in the coastline, owing to the development of deltaic islands, etc. When in addition the coastline is rather rapidly sinking the changes are still more marked. For instance a Roman Temple on Walcheren Island has sunk 30 feet since it was erected (Fig. 78). The various distributaries of the Rhine have varied greatly also, often as the result of man's interference. Thus the Romans in the first century deflected the Rhine near Nymegen, and cut a canal further west which ultimately became the Lek (Reclus). Yet other factors in the evolution of the coastlands were the extraordinary storms of the 11th, 12th and 13th centuries. In Roman times there was no Zuider Zee (South Sea). A large lake (Flevo) occupied the lowest area some 50 miles from the North Sea. In the 11th and 12th centuries the low barrier north of Lake Flevo was inundated over and over again, so that many of the inhabitants abandoned the district. The outlying dunes became islands, and these later were in part drowned or washed away, so that now only the long narrow string of the Frisian Islands is left. These structurally represent the continuous on-shore dunes which still exist along the coast of the Province of Holland. The worst storms occurred in the 13th century, and they removed most of the land which lay north of Lake Flevo (Fig. 78). Similar destruction occurred to the east at the Lauwer Zee and Dollart though to a much less degree.

Just as striking have been the changes at the mouths of the Rhine, Maas and Scheldt. In 1421 a terrible spring tide drowned 12 villages and formed the Biesbosch ("Field of Rushes") at the mouth of the Maas. Before 1173 the Scheldt ran north from Antwerp and joined the Maas near Steenbergen. In that year storms washed away the western banks, and converted Beveland into an island surrounded by wide sea-marshes. Many of the Saxons from this district migrated far away to Transylvania. The old channel past Bergen has since become completely silted up.

There were also changes in the opposite direction. Thus in the early 14th century there was a wide bay south of Antwerp, though the city is now far from the open sea. In the 11th century man had begun to embank the Scheldt, and much was reclaimed in the next few centuries. When Parma besieged Antwerp in 1585 nearly 75,000 acres were inundated to repel his forces, and some of this area has not yet been reclaimed from the sea (Fig. 78).

Further to the west the coast of Flanders has seen some remarkable changes. In 1000 A.D. the coast ran straight from Ostend to Brill. Yet in two centuries the coast had encroached much farther inland than to-day. In 1213 was fought the naval battle of Sluys, where to-day are meadows three miles from the sea. Bruges, as we shall see, was the chief port in Europe in the 14th century, but gradually its outlet silted up, first at Damme and later at Sluys. So also the "Old Port" for Ypres has given place to Newport 20 miles to the west. The outlets in the dunes kept for drainage were naturally the sites of towns, and Ostend grew up at the *eastern* end of the most direct gap leading to Brussels. Zee-Brugge is the seaport of Bruges, with which it is connected by an expensive modern canal. Antwerp has prospered so greatly because its tidal river (the Scheldt) is so much less likely to be silted up than the tiny streams of the early ports. Human factors have played a great part here also. The trade of the Northern Plain moved more and more into Protestant hands, and this undoubtedly was something of a factor in the decay of the old Catholic towns of Bruges and Ghent. No doubt also Hansa played a part in the decay of the West Flemish Towns.

It is several centuries since the Dutch began the giant reclamation schemes, which have added so much to their restricted agricultural areas. Alkmaar Meer (Lake) was drained as far back as 1553. About 1850 the 45,000 acres of Harlem Meer were converted into Polders in a few years with the aid of three steam pumps. The Zuider Zee is now dammed by a giant barrier across its mouth some 20 miles long. This work was begun in 1920, and in a few years from now four great Polders (1a in Fig. 77) comprising 800 square miles of good farmland will be added to Holland. A central Gulf over 16 feet deep will finally lead the waters of the Yssel, etc., to the sea.

D. *The Evolution of the Dutch and Belgian Nations*

In 953 Otto the Great (p. 272) created two duchies in the Lowlands, which were known as Upper and Lower Lorraine (*i.e.*, Lotharingia). While not strictly Marches these districts were far from the Emperor's eye, and as the headquarters of the ruling dynasties shifted ever to the south, the Delta lands were left much to themselves. These large areas of swamp and woodland had little importance in the early days of European history before trade revived. It is somewhat of a paradox that better times resulted in part from distant wars. Many of the leading Crusaders came from the Lowlands, and among them mention may be made of Godfrey of Bouillon, Count Robert of Flanders, and that Baldwin, also of Flanders, who became King at Constantinople in 1203. It was natural that the trade and new ideas resulting from the stimulus of the Crusades should especially invigorate the home lands of the leaders. Nowhere can we learn more about the evolution of the commercial cities than in Flanders. Here grew up the earliest example of an industrial population in Europe. The wool of England was brought to Flanders, and there thousands of Flemish weavers turned it into cloth. It was inevitable that these sturdy workmen should win privileges in return for the wealth which they created. The various rulers of Flanders, Brabant and neighbouring provinces granted the growing towns a considerable measure of free government in return for the taxes which they levied.

Until the end of the 14th century there was not much union among the numerous vassal fiefs of France or the Empire in the Lowlands. However, in 1384 Philip of Burgundy (Fig. 79) received Flanders and Artois as dowry from his wife. Within fifty years his successors had added, either by inheritance, purchase or "grab," Brabant, Namur, Holland, Zeeland, Hainault, Friesland and Luxemburg. Philip the Good gave the powerful bishoprics of Liege and Utrecht to his bastards, and later his son Charles the Bold acquired Gelderland. With this extensive territory Charles was much more powerful than his sovereign Louis XI of France, and no doubt he hoped to become King of a restored Lotharingia (Fig. 79). He established, with the large revenues from the Flemish trade, something like the first Standing Army in Europe. But his ambitious schemes were thwarted by rebellions in Switzerland. He was defeated at Grandson and Morat, and killed in 1477 at the battle of Nancy. His daughter Mary soon married Maximilian of Austria, and so Burgundy and the Lowlands were added to the Hapsburg Dominions. Their grandson was Charles V, who was born at Ghent in 1500.

The birth of the Dutch nation resulted from the economic and religious problems of the Reformation. Charles was well-disposed to his Lowland subjects, but he needed enormous resources to carry on his numerous wars. These he obtained by levies on the rich Lowlanders. In 1555 his son Philip became ruler, and he had no particular liking for the Dutch, and, moreover, was a bigoted Catholic, so that he introduced the Inquisition and supported it by Spanish armies. The leading Dutch nobles Egmont, Horn and Orange objected vigorously, and Philip retaliated by executing the two former. In 1569 William of Orange subsidised the "Sea-Beggars," who captured Brill, Flushing and Schiedam in 1572; and next year destroyed most of Alva's fleet. Harlem, Alkmaar and Leyden were besieged, but by cutting the dykes the environment was used to help the desperate Dutchmen, and the Spanish were forced to withdraw. In 1578 German Calvinists under Casimir came to the aid of Orange, but they made themselves so unpopular in Belgium, that the latter preferred the rule of Spanish Catholics rather than alliance with bigoted Calvinists. In 1579 at the Union of Utrecht the northern Protestant Provinces pledged allegiance to William of Orange, a splendid soldier and one of the shrewdest diplomatists in history. Although there was bitter fighting with Spain during the next seventy years, the Dutch maintained their independence, which was acknowledged at the Peace of Westphalia in 1648 (Fig. 79).

The Catholic Provinces shared the fortunes of the Spanish Hapsburgs until Charles II of Spain died without sons in 1700. His realm included "twenty-two crowns," and he left the Spanish Netherlands by will to Philip Bourbon, the grandson of Louis XIV of France. France and Austria quarrelled as to the succession in Spain; and after a dozen years of war Austria (helped by England) gained the victory.

Belgium was handed over to the Austrian Hapsburgs at the Peace of Utrecht in 1713.

The French Revolution aroused hopes of independence among many subject nations, most of whom were doomed to disappointment. But in Belgium a revolution in 1795 against Austria was successful, and the Great Powers decided to form a stronger buffer state by uniting it to Holland. In 1830 a second French revolution against the reactionary rule of Charles X in France emboldened the Belgians to revolt against Dutch control, which had not been used sympathetically. As the result of a Conference at London the Belgians were allowed to secede and they chose Leopold of Saxe-Coburg as their King.

The significance of the Belgian Corridor in the later wars of Europe may well be stressed at this point (Fig. 78). There was almost continuous fighting from 1568 to 1579, when the Walloons decided to adhere to the Catholic King. Farnese captured many Flemish cities in 1585. In 1600 Newport and in 1604 Ostend were blockaded by the Spanish. In fact fighting continued almost continuously till 1648. In the War of the Spanish Succession Marlborough won Ramillies (1706), Oudenarde and Malplaquet and drove the French out of Belgium. The whole country was conquered by the French (Fontenoy, 1745) in the War of the Austrian Succession, but was restored at Aix in 1748. An abortive revolt broke out against Austria in 1789. In 1792 the French Revolutionary Army invaded Belgium and won Jemappes, and the battle of Fleurus in 1794 ended the Hapsburg rule. Napoleon marched to attack Blucher and Wellington in 1815, and won Ligny but lost his final battle at Waterloo. In the Great War the huge armies of the Germans could only manoeuvre in open country, and so they invaded Belgium instead of striking directly at Paris across the broken cuestas of Metz and Verdun. During August, 1914, the Allies fought retreating battles at Liege, Dinant, Mons and Cambrai. Belgium was entirely occupied by Germany, except for the corner near Ypres, where fighting was continuous for four years. In November, 1918, the British at last advanced again and occupied Mons on the day before the Armistice.

In conclusion it may be noted that the two nations, Belgium and Holland, have nothing in the way of natural boundaries on the east and south (Fig. 77). Their limits are not drawn as the result of racial or linguistic factors. Economically they are somewhat different, for Belgium depends on her rich coal resources, while Holland has a much more important mercantile marine and an important colonial empire. Both, of course, have strong agricultural interests. Were it not for encouraging examples like Switzerland it would seem unlikely that Belgium would hold together permanently. Yet the Great War certainly did nothing to draw the Flemish and North Germans together, whereas there is the strong bond of a common religion uniting the otherwise diverse Flemings and Walloons. Both seem to be loyal to the present King of the Belgians.

At the southeast corner of Belgium is an interesting principality of less than one thousand square miles. In the tenth century it constituted the domain of the Count of Lutzelburg, which has since become the Grand Duchy of Luxemburg (Fig. 77). In the Middle Ages it was part of Burgundy, and as we have seen it gave rise to the reigning family in Bohemia early in the 14th century. In the 16th, 17th and 18th centuries it was bandied back and forth between France and the Empire. After Waterloo Luxemburg was linked with the King of Holland, though it was also a member of the German Confederation. After the King's daughter succeeded him in Holland the Duchy passed to a male heir (Nassau). Its chief interest to us is that it is a survival of one of the original small European States, Alpine in Race, German in speech, Catholic in religion, which has remained isolated in spite of the apparent advantages of union with one of the large nations which border it.

CHAPTER XVI

SWITZERLAND—A COMPLEX NATIONALITY

A. The Dominant Features of the Alpine Folds

Switzerland offers the strongest contrast in environment to the Netherlands, for here we find a nation founded upon the rugged ranges of the Alps. It is even smaller than the Lowlands, and even more mixed in its national "ingredients" as regards language, about equally mixed in religion, though rather less so in race. It is dominantly Protestant, but has a large minority of Catholics, as is natural since it lies on the borders of the Catholic Bloc. For this reason as well as others based on geographical grounds it may well be discussed here at the end of the Protestant countries of Western Europe.

Although it is to many people, including the writer, perhaps the most interesting country in the world, it is barely 200 miles long in an east-west direction, and only 100 miles from north to south. But the variety of landscape and the striking illustrations of human progress packed into this small area are remarkable. Naturally it offers the best examples of topographic control in Europe, since it contains the boldest mountains flanked by a broad Downfold. Since, moreover, the Alps lie close to the "shatter-belt" where Keltic and Teutonic tribes struggled with Rome, where France fought Italy and both fought Austria and Germany, the region has been in the limelight of history almost all the time from the Foundation of Rome to the present day. As usual we shall commence our study by discovering what is the build of the country and how it has affected the broader historical features.

In Chapter III some account was given of the overfolding of the earth's crust in Late Tertiary times which produced the Alps. It is important to remember that Switzerland only includes about one-quarter of the whole Alpine area. About an equal amount is found in Austria and in Italy, while Eastern France also includes much rugged Highland country including all the Pelvoux Massif.

In Switzerland there are six notable divisions of the Alpine Folds, which are shown in a generalised form in Fig. 80. Furthest to the northwest are the simply folded rocks of the Jura Mountains, which lie as an "advance guard" of the Alps proper. The Swiss-French boundary runs along their summits for a considerable dis tance. Nearer

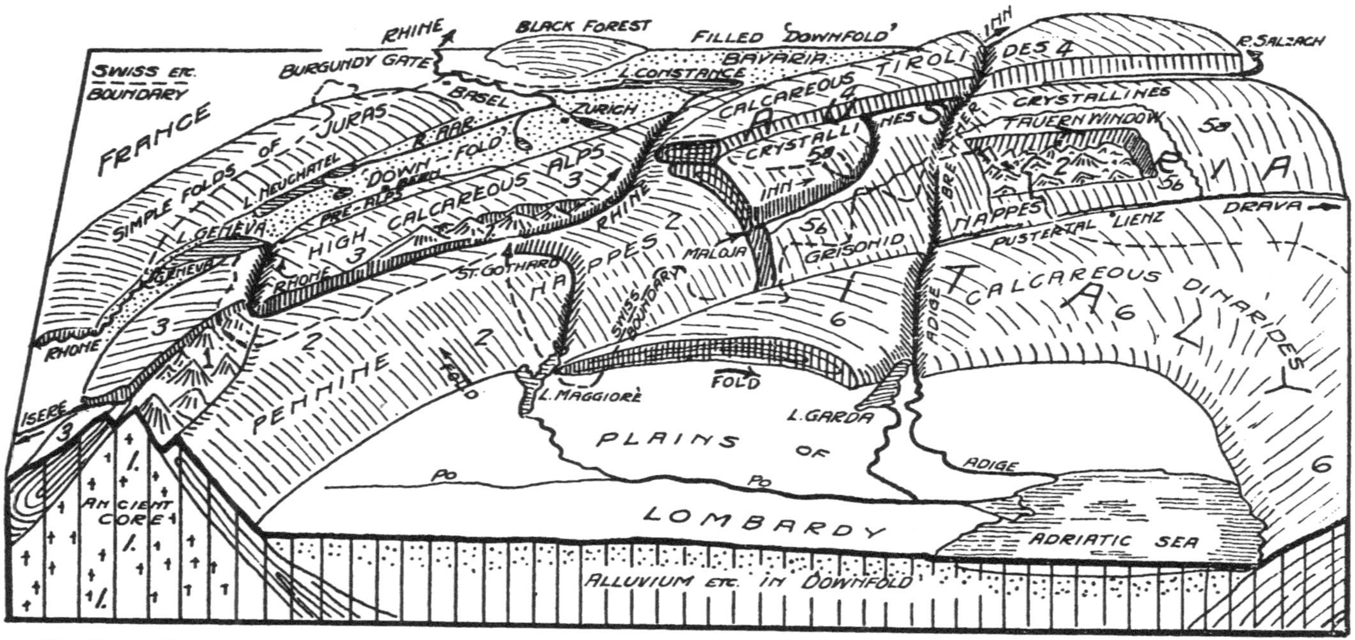

Fig. 80.—A Mantle-Map of the Alps (much generalised). The Core of ancient Rocks (1) appears only in the centre of the western Folds. The Pennine Nappes (Overfolds) cover the south flanks (2) in the west; but appear only in the "Window" in the Tyrol in the east. The High Calcareous Alps (3) are formed of Folds overlying the Pennines (2). In the Tyrol still later Overfolds (4 and 5) appear. These all fall over to the north, but the Dinarides are Back-folds and fall over to the south. The main longitudinal and cross valleys are indicated. A simplified geological section appears on the front edge of the diagram. (See inset, Fig. 106.)

the centre of the disturbance (which was in the Mediterranean, or perhaps in North Africa) is the well-defined Downfold. (Some writers give this the entirely inappropriate name of the "Swiss Plateau".) It is covered with early Tertiary beds (*Molasse*). These are fairly level in the northwest, but have been steeply folded by the Alpine pressure near the mountains. Then comes the main mass of the Alps, which (as described in Chapter III) are overfolds (*Nappes*) flung over each other like breakers towards the north. Since this profound crustal movement occurred there has been plenty of time (perhaps nine million years) for much erosion. The whole upper series of Tyrolid Nappes seems to have been stripped off the Western area, and also the tops of the Pennine Nappes; so that here the cores of ancient rock are exposed.

The Alps proper can therefore be divided into three belts (Fig. 80). At (1) we see the oldest Core, consisting of granite and gneiss, and forming the jagged peaks of Pelvoux, Belledonne, Mont Blanc, the Oberland and Saint Gotthard. Lying on the southeast side of the Core are the complicated folds of younger rock constituting the Pennine Nappes (2). Flung right over the crest and lying as extremely complicated overfolds along the northwest side of the Core are the Nappes of the High Calcareous Alps (3). Lastly in the Engadine region of Switzerland in the southeast is a part of the Tyrolid Nappes (5b), a younger series of overfolds, which as stated are found only in the eastern half of the Alps. Our diagram (Fig. 80) shows also the main features in the build of Austria, which may here be briefly referred to. The Tyrolid Nappes have been eroded in places as at the "Tauern Window," so that the Pennine Nappes (2) can be seen exposed below the Tyrolid Nappes. The Calcareous Tyrolides are still younger Nappes flung over far to the north and building the Bavarian Frontal Range (Fig. 106). Lastly the reverse folds (6) of some of the Dinaric arcs (flung to the south) build up the Dolomites, Venetian and Julian Alps. (Cf. Fig. 12 at H, p. 59; and Fig. 106.)

It is obvious at a glance that the national boundaries have no relation to this series of east-west crustal folds. A close inspection of the main river valleys will, however, show that there is a relation between the rivers and the fold-lines. In general, the river valleys are either longitudinal (parallel to the folds) or transverse (cutting across them). Thus the river Aar is clearly longitudinal for the most part, flowing along the lowest portion of the wide Downfold (Fig. 80). The Rhone is transverse just above Lake Geneva, and then longitudinal along the fold valley between the Pennine Nappes and the

High Calcareous Nappes. The River Isere rises near Mont Blanc and then flows south in a longitudinal valley for some fifty miles. So also the Upper Rhine near its source is longitudinal but soon cuts *across* the folds to reach Lake Constance. The Middle Inn and Upper Salzach occupy parts of the great longitudinal valley between the Calcareous Tyrolides and the Crystallines. To the southeast the valley called the Pustertal links the Drava to the Adige, though no river flows along the wide empty valley west of Lienz (Fig. 80).

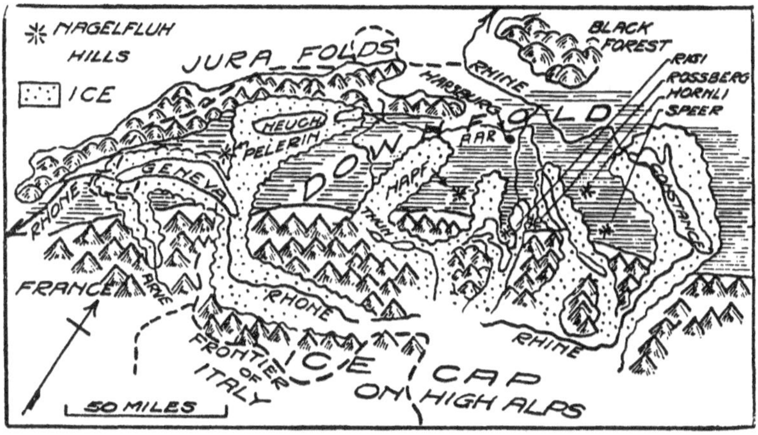

Fig. 81.—A Diagram of the Swiss Downfold showing the numerous Twin-Glaciers (dotted) of the Ice Age; and also the hills (of Nagelfluh) which deviated them. (Data from Newbigin.)

B. Topography due to Glaciers and Rivers

Most of these river valleys were filled with glacier ice during the maximum periods of the four Ice Ages. These great glaciers spilled down into the Lombard Plain, and gouged out great hollows just where they crossed the uplands and reached the plains. Piles of debris sometimes a thousand feet high, were left in the form of moraines, while the great hollows thus cut out contain the lakes Maggiore, Como, Garda, etc. On the northern side of the main Alpine Ranges the ice poured down into the Downfold (Fig. 81). In many cases it split into two tongues, each of which carved out a broad valley. Thus the Rhone Glacier sent one tongue west down to Geneva, which cut out the hollow occupied by Lake Geneva. The other tongue moved north across Moudon, and climbed some way up the Jura Barrier. When it melted its valley remained dry, since the drainage of its upper basin flowed into the large lake. Thus a very convenient route for road and railway has been produced. So also the Aar Glacier divided above Interlaken, and sent one tongue over the Brunig Pass to Lucern, and another *via* Lake Thun towards Bern. A railway uses the dry Brunig Valley. The Reuss Glacier split above Lake Lucern and produced a similar pair of outlet valleys *via* Lucern and Schwyz. The main route runs *via* the Schwyz valley to Germany. The Rhine Glacier sent its main ice supply to Lake Constance, but an overflow to the northwest cut out the Wallen See and made another main route (from the Tyrol) to Zurich.[1]

In Middle Tertiary times (long before the Ice Ages) immense areas of sands were deposited north of the Alps, and these became coarser nearer the shoreline of the Tertiary Sea. The finer sandstones thus formed are called *Molasse*, and the coarser conglomerates and fan-gravels formed from the beaches are called *Nagel-fluh*. The latter build the prominent hills seen in the Downfold, which are indicated in Fig. 81. Hornli, Speer, Napf, Rigi, etc., stand up a thousand feet above the plain, and are composed of these Tertiary conglomerates. They directed the course of the ice-lobes, and so determined the main lines of communication across the Downfold.

It is clear that the main bulk of the people will be confined to the Downfold, for there is nothing but pasture land, or at best small garden plots in the Alpine valleys. However, we may well devote a short time to a special study of these Alpine valleys, since they form so characteristic a feature of much of Switzerland (Fig. 82). Before the Ice Ages, say one million years ago, there were the same high mountains, but the deep juvenile valleys were of the V type shown at A in Fig. 82. As the climate grew colder and colder, we may picture the cold layers of the atmosphere slowly descending. By and by freezing conditions became normal much lower on the mountains, and small glaciers formed in the upland valleys (as at B in Fig. 82). These cut out arm-chair valleys or Cirques by a process called thaw-and-freeze erosion (or *nivation*).

At a still later stage the cold layers descended far lower, until great ice-caps covered most of the highlands, and huge glaciers flowed down all the valleys into the plains (as in Fig. 81). This condition is seen in section in Fig. 82 at C. The glaciers rasped away the original V-shaped valleys (as cut by the rivers) and converted them into flat-

[1]Based on the detailed discussion by M. Newbigin, "Southern Europe," 1932.

bottomed troughs. On the waning of the Ice Age (*i.e.*, as the cold layers of air ascended again to normal heights) all the glaciers vanished, except a few little "hanging glaciers," and the topography to-day is nearly as shown in Fig. 82 at D. Later river erosion (since the ice vanished) has cut a gorge in the flat floor, and this sort of a valley is very common in Switzerland.

In summer the cattle feed high up in the meadows called Albs (whence Alps). These are the relics of the early valley-slopes shown at A. Waterfalls tumble down the steep cliffs cut by the giant glaciers and give light and power to the Swiss of to-day. The floors of the glacier-cut troughs are where the Swiss passes his winter. He could not have lived in the mountains (in the pre-Trough stage) under the conditions shown at A. Lastly the post-glacial stream is rapidly cut-

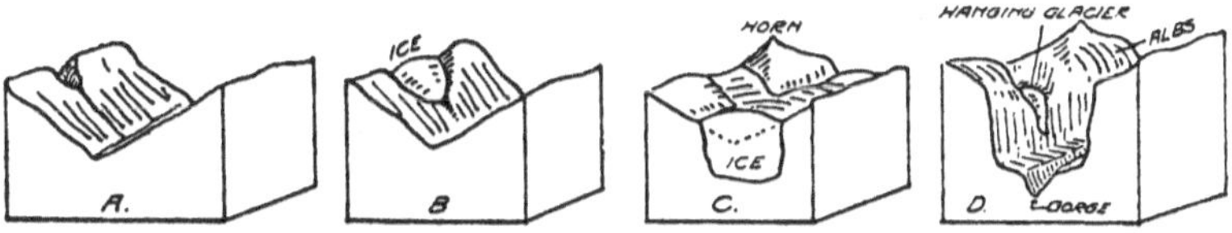

Fig. 82.—Characteristic Features in Swiss Scenery. A. Non-glaciated Juvenile river-valley. B. Cirque Glacier cutting high-level valley. C. Valley-Glaciers eroding valleys with U section. D. Post-glacial gorge cut by river.

ting a notch in the small strip of arable land in the trough-floor, and so gradually making communications as difficult as before the great flat troughs were eroded by the ice. The "Via Mala" at Thusis is such a post-glacial notch, and the Devil's Bridge crosses another near Andermatt (Fig. 83). The deep wide valleys at Lauterbrunnen and Chamounix are typical glacier-cut Troughs.

The main ridges and valleys of the Alps have thus been briefly described, and some reference to the structure of the Passes may now be given. Most of the Folds clearly run west-east in the main body of the Alps (Fig. 80). Thus the grain (or "strike") of the intermingled harder and softer layers will also run east-west. Minor valleys will naturally tend to be cut in the softer rocks, and this feature is illustrated in Fig. 83 dealing with the topography of the St. Gotthard Pass. Here three hard beds of rock are separated by two softer layers. The latter give rise to minor longitudinal valleys, which have been occupied by the Upper Ticino River (on the south); and by the Upper Rhine, Upper Reuss and Upper Rhone (on the north). The Aar almost reaches the longitudinal "strike" valley at the Grimsel Pass. In the generalised block diagram it is clear that the Furka and Nufenen Passes are also longitudinal, whereas the St. Gotthard and Grimsel Passes are transverse. In general the latter would be narrower and less accessible, if glacier ice had not formerly used these Passes, and so eroded them much more widely than did the original streams.

If reference be made to Fig. 101, dealing with the gaps in the barrier of the Alps west and north of the Lombard Plain, it will be seen that

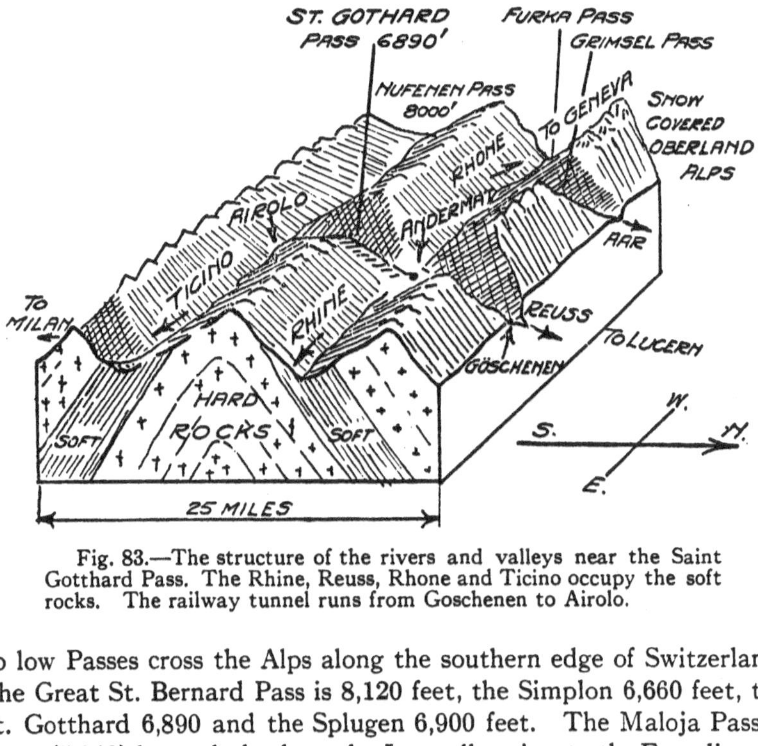

Fig. 83.—The structure of the rivers and valleys near the Saint Gotthard Pass. The Rhine, Reuss, Rhone and Ticino occupy the soft rocks. The railway tunnel runs from Goschenen to Airolo.

no low Passes cross the Alps along the southern edge of Switzerland. The Great St. Bernard Pass is 8,120 feet, the Simplon 6,660 feet, the St. Gotthard 6,890 and the Splugen 6,900 feet. The Maloja Pass is lower (5,940) but only leads to the Inn valley, *i.e.*, to the Engadine, an isolated corner of Switzerland. To leave the Engadine on the north we must climb the Albula or Septimer Passes each 7,500 feet (Fig. 88).

Of all the Swiss Passes the St. Gotthard is much the most important, and the reason for this depends on the arrangement of the river valleys. Not only does the Reuss lead up to the main divide (direct from the historic and industrial centre of Switzerland) to meet the head of the Ticino on the south side, but the great longitudinal valleys of the

Rhone and Rhine meet here also, or rather at Andermatt a few miles north (Fig. 88). This latter little town is clearly the strategic centre of Switzerland, and is a strong military post (Fig. 83). Moreover, the Aar valley also turns the flank of the high snowfields of the Bernese Oberland, and leads again to the St. Gotthard Pass by the feasible Grimsel route. No wonder that the St. Gotthard Hospice was built before 1330, for the shelter of folk using the mule path constructed late in the 13th century. The carriage-road was not made till 1830, while the railway tunnel nine miles long was opened in 1882. Of course the Brenner Pass from Italy to Austria, and the Genevra Pass from Italy to France were much more used in early days, but they do not directly concern Switzerland.

C. Development of the Swiss Region before A.D. 1000

As usual the information which we possess about the people in the Alps before the Roman Conquest is very incomplete, as it is generally based on tribal legends or on hearsay reported to southern writers. But place-names and archeological data are more accurate, while the best data of all can be gathered from the present-day ethnology. As we have seen elsewhere, governments, religions and languages may often vary, while the people resist change of race to a remarkable degree. However, in a mountain country we generally find groups of the earlier communities driven to seek refuge in the higher valleys, while the later more powerful tribes tend to occupy the better lowlands. Such an interpretation is clearly indicated by the present ethnic map of Switzerland (Fig. 84).

The outstanding racial feature of the people of Switzerland is that they have the broadest heads of any part of Europe except Bosnia and Albania. Ripley's map shows that all the districts have average head indices above 85. Switzerland is therefore dominantly Alpine, though no doubt many Nordics have moved up the valleys of the Rhine basin and mixed with the Alpines between Basel and Zurich. In Fig. 84, I have generalised Ripley's data somewhat, producing four zones, whose isopleths indicate pressure from the north. We shall find that this is corroborated by history. The mountain-refuges most remote from the Rhine Graben are the southern Juras and the Alps of the Upper Rhine and Engadine. Precisely here are found very dark Alpines, who are conveniently called *Rhetians* (from the old name of the Engadine region). Somewhat nearer the point of pressure, which we may place near Basel, is a zone including the broad Rhone valley and the Western Downfold (Fig. 80). It also includes the Rhine valley just above Lake Constance. These folk are also rather dark, but are intermediate between those of the southern zone and the northern Swiss. Almost all the rest of Switzerland is inhabited by much fairer folk; not, however, true blondes, who are not found much south of the Great Plain of Germany. These fairer people it is customary to assign to the *Helvetians* of the times of Caesar, but it would be more accurate to call them a mixed Helvetian-Alamannian group, as we shall see.

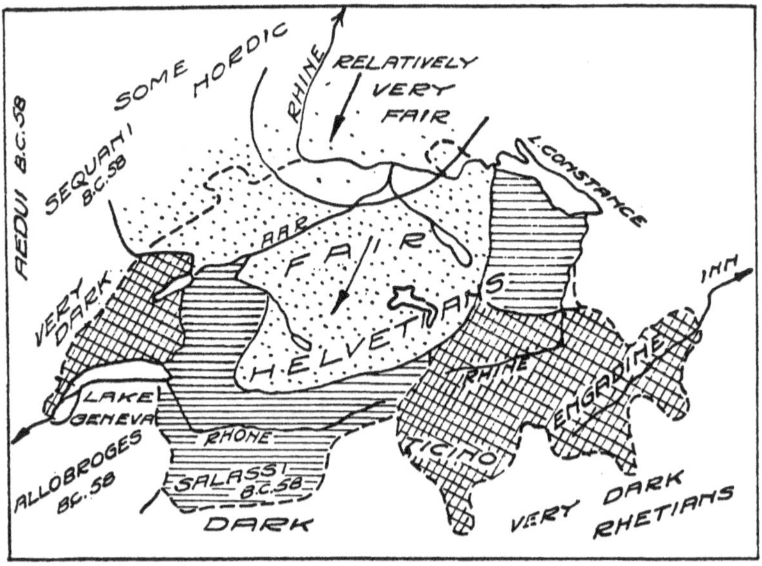

Fig. 84.—Race in Switzerland. The fair Helvetians have driven the dark Rhetians southward into the mountains. Both are essentially Alpine.

Even in the later Ice Ages Paleolithic Man had settled in Switzerland, but the lake-dwellers were perhaps the first inhabitants of any note. The site of the early Iron Age at La Tene near Neuchatel is well known. The lake-dwellers were almost certainly dominantly Alpine, and it may be that the Rhetians are in part descended from them. No doubt many of the K-speaking Keltae, who descended into Italy in prehistoric times, left representatives among the Alpine valleys, but the place-names show that large numbers of Welsh-speaking P Keltae lived in Switzerland at a relatively late date. In the writer's opinion the Helvetians were essentially Alpines of this migration. It is true that Welsh names are spread through parts of the Rhetian tribal country (Fig. 27) as around Lake Geneva. But this merely indicates the spread of the Welsh speech among the former Q-speaking Keltae, a procedure which we have noticed in many places (p. 114).

In our earlier study of Germany we noted the way in which the Teutonic (Nordic) tribes gradually drove out the southern Keltae, who were mainly Alpine. The Helvetians as stated moved into Switzerland about the 3rd century. They left much of the region north of Switzerland empty until it was occupied by the Alamanni and other German tribes about A.D. 400. So also the Burgundians moved into France by the Burgundy Gate (or Belfort Gap) along the River Doubs (Fig. 80), and were settled around Lake Geneva about 500 A.D. In the 8th century Germans migrated into the Swiss Downfold in large numbers, and also some way up the Reuss valley. The forests of the Juras were unoccupied until about the 10th century, and about this time small bands of Moors from Spain settled in the Alps and levied toll on the travellers crossing the Passes.

In B.C. 60 German tribes under Ariovistus marched west through the Gap at Belfort and conquered the Sequani and Aedui in Gaul (Fig. 84). In B.C. 58 the Helvetians also attempted to march to the west but were crushed by Caesar at Bibracte near Autun (Fig. 85). A little later he defeated Ariovistus near Basel and the German invasion was finally repulsed. In B.C. 25 the Salassi near Saint Bernard Pass were subjugated, and Aosta (Augusta) founded to overawe them. A decade later Drusus conquered the Eastern Alps near the Brenner Pass and Rhetia was organised as a Roman province with its chief towns at Chur and Augsburg (Fig. 85). The Latin dialects still spoken in the eastern Alps (called Romansh and Ladin) have changed little from those which developed in this period of early Roman culture.

A Roman road soon connected Milan with Chur by way of the Splugen Pass. Another ran from Aosta over the Saint Bernard Pass to Martigny, and thence to Nyon and France. From Vevey another went along the Downfold to Basel and the Rhine, while a branch led to Windisch (at the junction of the Aar and the Reuss near Hapsburg). The latter site was chosen as the military centre of Helvetia, and guarded both the Rhine and the Danube garrisons. About 265 the Alamanni first invaded Helvetia and destroyed Avenches, the Roman capital near Bern. In 406 as we have seen they again crossed the Rhine and settled in Switzerland. The Rhetians in the Grisons alone escaped from the Teuton domination, for the Alamanni practically destroyed the Roman civilisation and did not become Christians for 200 years. However, in 469 they came into conflict with the Franks and were largely conquered by them.

The province of Burgundy dates from the retreat of the Burgundians after their attack on Rome in A.D. 411. In 443 Aetius forced them to migrate to the Savoy, by which time they were already Christians. They made no attempt to uproot Roman customs, and in effect

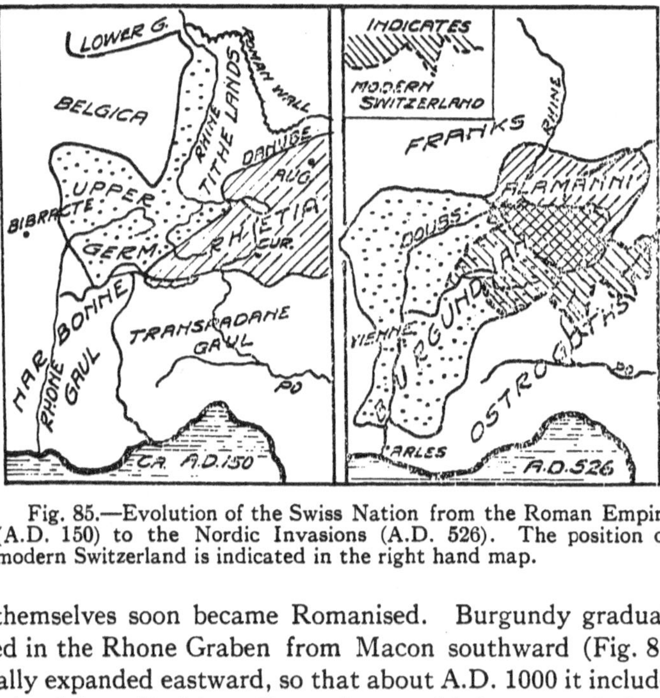

Fig. 85.—Evolution of the Swiss Nation from the Roman Empire (A.D. 150) to the Nordic Invasions (A.D. 526). The position of modern Switzerland is indicated in the right hand map.

they themselves soon became Romanised. Burgundy gradually developed in the Rhone Graben from Macon southward (Fig. 85). It gradually expanded eastward, so that about A.D. 1000 it included east Switzerland from Zurich to Sion. It was this long association of the western half of Switzerland with Burgundy which determined the French culture of this portion of the country.

The region occupied by the Helvetians and the Alamanni (in Roman Rhetia) later became linked to the allied tribes of south Germany, and formed part of the main group of the Alamanni or Suabians (Fig. 85). After the Partition of Mersen (870) Suabia became a Duchy under the Holy Roman Emperor, and Burgundy was also united to the Empire about 1034 (Fig. 85a).

D. The Historical Basis for Difference in Language and Religion

During the 13th century the Swiss living around Lake Lucerne were partly direct vassals of the Emperor, and partly under the control of the Hapsburgs, whose ancestral castle was only some 50 miles to the north (Fig. 85a). Uri, Schwyz and Unterwalden formed a League in 1291 primarily for protection against the Hapsburgs. In 1315 a large Austrian army under Leopold marched to attack the town of Schwyz, but was defeated by one-tenth of the number of Swiss at

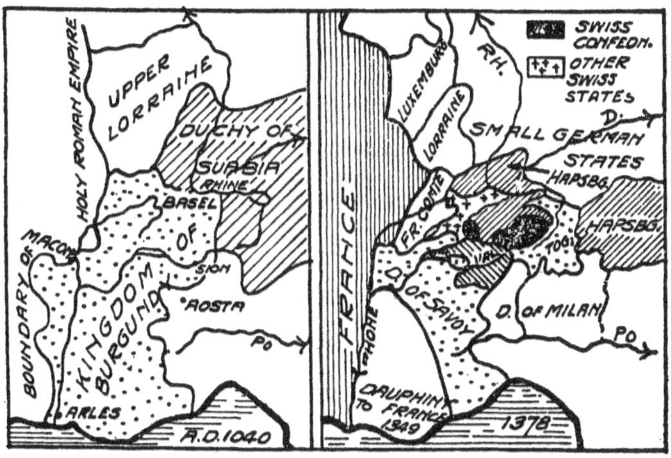

Fig. 85a—Switzerland in the Holy Roman Empire about 1040; and the Dawn of Independence about 1378.

Morgarten (Fig. 86). In 1320 the name Switzerland seems to have been first used for the whole region in honour of Schwyz, which was the leader of the struggle for independence. Neighbouring towns and districts joined the League, Lucerne in 1332, Glarus in 1352 and Bern in 1353. Thus the confederates obtained a strong footing in the lowlands, and approached the French culture-area of former Burgundy, near Bern. In 1386 the Austrians attacked Lucerne, but they were beaten by the Swiss League at Sempach, and two years later a similar invasion was repelled at Nafels (Fig. 86). It was becoming clear to the surrounding rulers that the Swiss would fight to the death to protect their barren territories, and as years went by they were accordingly left in peace. However, the Swiss themselves took the offensive in 1403, and advanced south along the great Alpine Route of the Reuss and Ticino. They occupied the Val Leventino on the south side of the Alps and levied taxes on the Italians. It was not, however, till 1440 that these lands became finally subject to Uri.[2]

In the north two areas, Aargau and Thurgau, were in the hands of Austria long after the rest of the country had joined the League. In 1415 the Emperor Sigismund quarrelled with Frederic of Austria and incited the Swiss Confederation to attack his lands. These new

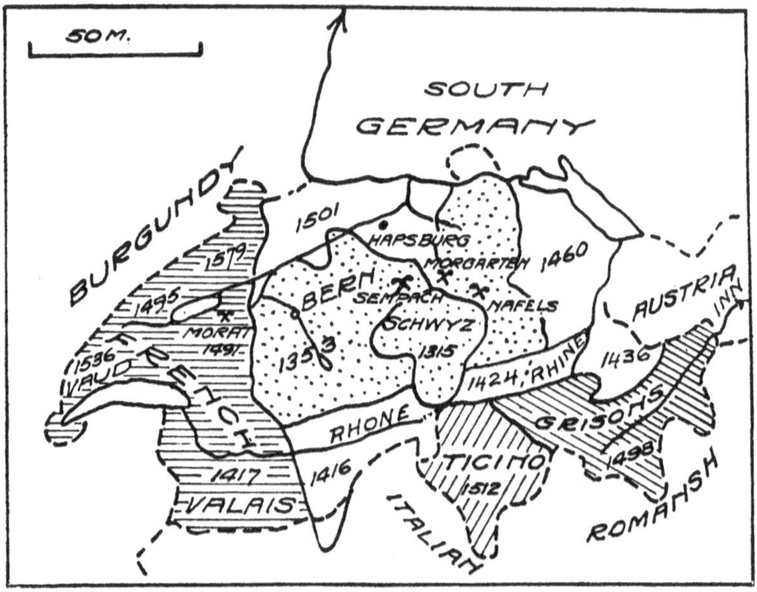

Fig. 86.—Languages in Switzerland. The Early German members of the Federation are shown by dots, the later German areas are left blank. The chief battles are shown also.

territories were accordingly occupied by the Swiss but were controlled by the League in common, a practice which led to much trouble.

Alliances and counter-alliances between the Swiss Confederation. France, Burgundy, the Empire and Savoy filled much of the 15th century. In 1475 Bern attacked the towns to the south in Savoy, and captured most of Valais. Charles of Burgundy hastened to help the Savoyards and was defeated by the Swiss at Grandson and Morat. Thus French districts were now linked to the Swiss Confederation, which had hitherto been wholly German. Charles was killed at Nancy in 1479, and much of Burgundy was occupied by Louis of France. A

[2]W. D. McCracken, "Rise of the Swiss Republic," 1901, New York.

year or two later Solothurn and Fribourg joined the League. In 1499 the Confederation virtually withdrew from the Holy Roman Empire, though their independence was not formally recognised till 1648. By 1513 Basel and Appenzell joined the Confederation and there were thirteen cantons in the League (Fig. 86).

We see that none of the French-speaking Cantons took any part in the struggle of the Swiss for independence until the latter had defeated Charles of Burgundy and won great renown for valour in the field of war. Entirely isolated from the northern German districts were Valais and Grisons, and for this reason they were late in joining the League. Moreover, Valais was half French and half German, the former occupying the western part of the enormous glacier-cut valley (Valais) of the Rhine below Sierre, while the German descendants of the Alamanni were in the upper half of the valley. They had bitter struggles with the House of Savoy, and in 1403 obtained help from the Confederation, but did not join it till later. Equally isolated from the Downfold and its main valleys was the district of Grisons. It was dominated in the 13th century by the Count of Toggenburg (Fig. 85a). But in 1436 he died without heirs, and the valleys in the south-east united to form a "League of Ten Districts" under democratic rule. They joined the Confederation as Allies in 1498, but were only admitted as a full canton in 1803.

In 1501, there were as stated 13 full members of the League. Of the remaining nine, who later joined the Swiss Republic, four including Valais and Grisons were Allies. Three including Aargau and Thurgau were Subject Lands, while Vaud and Geneva had not yet come into contact with the League in any way.

In 1512 Swiss troops poured down the southern slopes of the Alps and captured Lugano, Locarno and other towns. Not till 1803, however, were these Italian lands received into the League as Canton Ticino. Bern for a time held much of the territory of Savoy south of Lake Geneva, but lost it in 1564.

Meanwhile the Reformation was spreading apace. The rapid manner in which the Protestant ideas took hold in certain of the Swiss districts and left the others unaffected is well worthy of our attention. At first the Swiss were all faithful to the Pope, for like all secluded nations they were conservative. Indeed the typical Mountain Cantons never altered their faith. However, the travels of large numbers of Swiss mercenaries to Italy and France showed them many of the abuses of the Church of Rome. In 1523 Zwingli's attitude in

Zurich was strongly supported by the Town Council, and so this important district became Protestant (Fig. 87). His friend Vadianus preached in St. Gall and influenced many people in the northeast areas to swing over to the new religion. In Grisons the animosity against the Bishops was strong and this region also joined the Reformers. However, Basel was swayed the other way by Erasmus, who on the whole objected to the reform movement. In Bern Haller was an ardent convert; while Farel did not rest till Geneva and most of the region north of the Lake became Protestant.

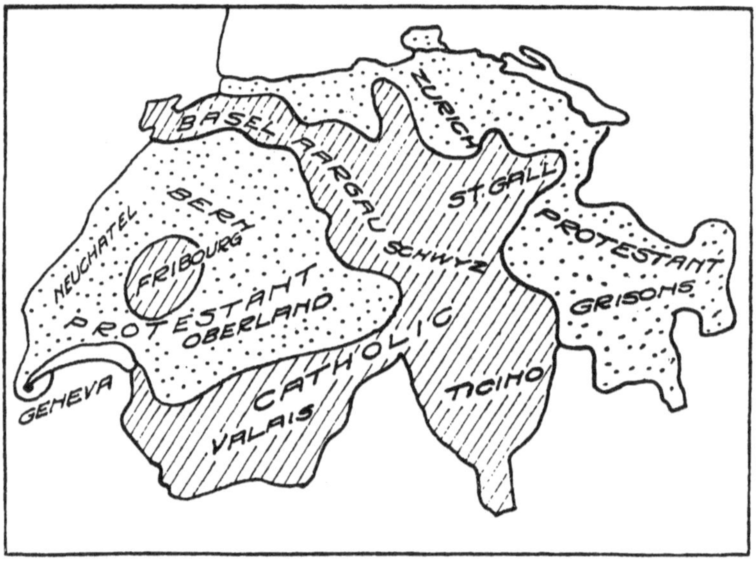

Fig. 87.—Distribution of Religions in Switzerland. They are not related to Race, Language or Density of Population.

In 1535 many of the Catholic clergy were driven from Geneva, and next year Calvin was persuaded to carry on the reformer's task in that city. However, Fribourg, Solothurn and the Forest Cantons held by the old religion (Fig. 87); partly because the reformers strongly opposed the hiring of Swiss mercenaries, who largely came from these cantons. Hence, as usual, there were economic factors which swayed people to join one or other religious party. A survey of the Reformation in Switzerland leads one to the conclusion that the people of the towns tended to become Protestant, while those in the hills were more likely to hold fast to the old church. Notice also that the *main corridor* from Basel to St. Gotthard and Italy remained Catholic. But outstanding personalities like Farel, Zwingli or Calvin—or strong local animosity to the Church—complicated the pattern (Fig. 87), so that no simple reasons suffice to explain the map of the Religions.

During the 18th century Switzerland was much under the control of France. Most of the conquered territories like Vaud or Ticino were governed harshly by their Swiss rulers, and there were several unsuccessful revolts. The French Revolution greatly affected the Swiss

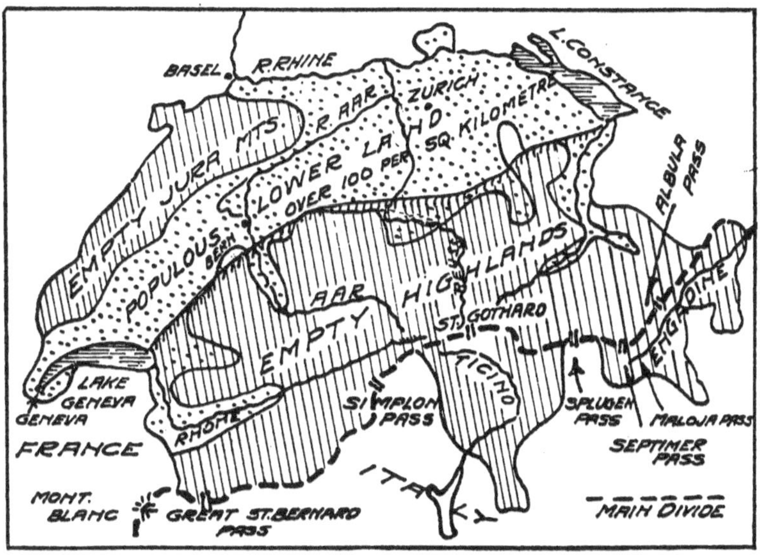

Fig. 88.—Density of Population in Switzerland, which is correlated closely with Topography. The Passes over the High Alps are also shown.

and the oppressed districts invited French aid. In 1798 the Helvetic Republic was proclaimed, and consisted of 19 Cantons (instead of the former 13 provinces with their subsidiary territories). After many struggles among the various Swiss parties Napoleon was accepted as mediator in 1803. His suggestions for the direction of the 19 Cantons were generally adopted. On his fall in 1815 three more regions were given full status as Cantons. These were the Protestant areas of Geneva, Valais and Neuchatel.

In conclusion a few words may be written on the present organisation of the Swiss nation. The Cantons are only loosely federated except as regards external affairs, so that each Canton is a law unto itself in many respects. The Federal Capital was naturally chosen in the Downfold, where almost all the population lives (Fig. 88), and near the linguistic boundary between the French and German speakers. Hence Bern was more suitable than either Geneva, Basel or Zurich, the other three large towns. All four are in the Downfold, but Basel and Geneva are right in the two Gates, one at each end of the Jura Folds. Being border towns they are greatly affected by foreign interests. Geneva is very French, and indeed its Protestant Reformation was Huguenot rather than Swiss. To-day its population has become largely Catholic by immigration from France. Basel, on the other hand, is in close touch with the Rhine Graben, and is naturally German in culture. The Downfold opens widely to the north (Fig. 88), passing without much change across the Rhine into Bavaria. This is the real reason why the speech and culture of Germany dominate Switzerland. Zurich has developed greatly in the last few decades. The great railways from Germany find it easier to run *via* Zurich and Schwyz up the Reuss valley to the Saint Gotthard Tunnel than to skirt the rocky cliffs of Lake Lucerne *via* Lucerne itself. The industrial area based on waterpower is also better favoured in the Zurich area than in any other portion of the Downfold.

Our greatest interest in the study of Switzerland is surely due to its representing in its diversity a little world in itself. With its two different religions, four commonly spoken languages, very varied environments and with no very natural centre, it is amazing that it has developed a vivid Nationality. It has been stated that the love of liberty and the love of mountains are the sole links uniting its many diverse peoples. Possibly a high level of education, a sense of the futility of war and a feeling that goodwill is the chief essential to solve most man-made troubles may also be more characteristic of the Swiss than of many continental nations. Their remarkable progress in education and in international co-operation would seem to support the hypothesis that what Switzerland has done a United Europe might imitate.

PART IV

THE CATHOLIC NATIONS OF THE SOUTHWEST AND CENTRE

CHAPTER XVII

ENVIRONMENT AND THE FRENCH NATION

A. The Structure of France

No region could be more suitable than that of France in which to study the relations with which we are here concerned. The *Nation* is peculiarly prominent in French culture. "La Patrie" is ever to the fore, and the French, perhaps in consequence, have less sympathy with the ideal of a World-State than is the case with most other nations. Yet France, though nationally a fairly well-defined unit, is by no means a simple structural, racial or cultural unit, as we shall see.

Structurally, France is best considered as consisting of three pairs of contrasted units. These are shown clearly in the following table:

TABLE T

STRUCTURAL UNITS IN FRANCE

Structure	Unit	Uses
1. Ancient Relic-Blocks....	Cevennes	Rather rugged, poor soils, but open and inhabited early.
2. " " "	Armorica	
3. Young Fold Mountains..	French Alps	Very rugged and always sparsely inhabited.
4. " " " ..	Pyrenees	
5. Flat Basins............	Paris Basin	Originally heavy forests; now, chief settlement.
6. " "	Aquitaine Basin	

Reference to Fig. 89 will show that these six areas comprise almost the whole of France. Almost the sole exceptions, which are of limited area but of great importance in the history of France, are the "Rhone Graben" and the "Lille Coalfield." We can best discuss these units as they affected history.

At the close of Paleozoic times about 150 million years ago extensive mountain-building occurred, as we have seen (p. 53). The worn-down stumps of these fold-mountains now survive as the rather high plateau of the Cevennes, which is largely over 3,000 feet, and the low uplands of Armorica (now Brittany), about 600 feet high. In the extreme northeast, the Ardennes Plateau of the same age, mostly above 1,000 feet, just enters the realm of France. These rather sterile, open expanses, where the ribs of ancient rocks often project above the thin mantle of soil, were peculiarly suitable for Neolithic settlement.

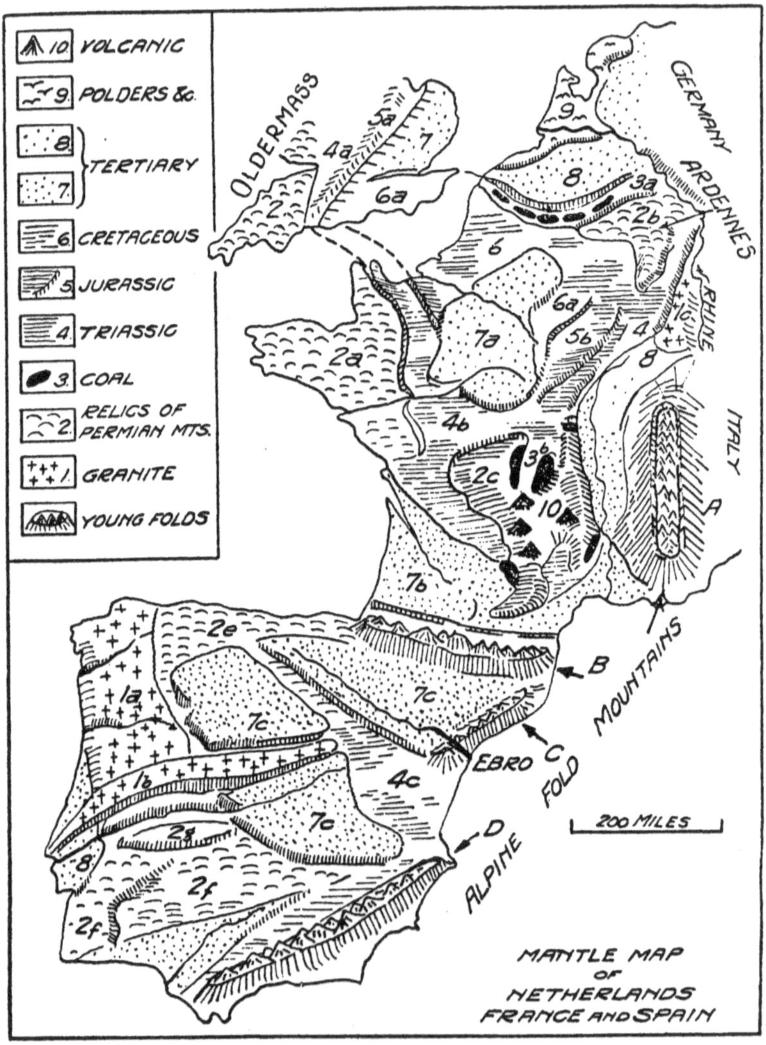

Fig. 89.—Mantle-Map of Netherlands, France and Spain. Notice the relations of the Paris Basin to the Cuestas of England. Fold Mountains of the Alpine Storm appear along the southeast edge of the map.

They agree in many ways with the "Older-Mass" of Wales and Cornwall, which had the same attraction for folk of early civilisation (p. 39).

While the Armorican Folds were being eroded to Relic-Stumps (or, as we should say in more precise language, were being "peneplained"), the debris from their slopes was being deposited in the shallow seas surrounding these mountains. Thus were built up a series of more or less level sediments which ultimately made up the rocks of the Paris and Aquitaine Basins. These agree again with the similar structure (around London) in England which was fairly fully described in Chapter II. In both lands are found more or less concentric cuestas and vales, which in France are centered near Paris. Indeed, it is this fact which has led to the "nodal point" of north France becoming the capital of the whole area. In Aquitaine the cuestas and vales are not so well developed. But the drainage of the Garonne (like the Seine) is towards the centre of the geological basin—so that Bordeaux is the natural centre of this basin (Fig. 89). It is significant that when the Germans closely menaced Paris in the Great War, it was to the centre of the other basin (Bordeaux) that the French Government proposed to move. Connecting these two basins is an "isthmus" of young rock around Poitiers. Little wonder that this has been the scene of many critical battles. We can only instance two. In 732 the Arabian Moslems made their most serious attack on Western Christendom. They were held here and then driven back to Spain by the Franks under "Charles of the Hammer." Later again, in 1355, when the English (who then controlled Aquitaine) marched against the French King, it was at Poitiers that the Black Prince crushed the defending army.

Towards the close of Tertiary times we may picture France as nearly level throughout. Probably the "Basins" of newer rock mantled more of the older "Relic-Blocks" than they do to-day. But now set in the more active periods of the Alpine Storm—say some ten million years ago. The chief result of this earth movement was the piling up of mighty ramparts along the south and southeast of France. These have since lost much (perhaps half) of their height by normal erosion, but still constitute the chief defence of France against invaders. But it is important to note that adjacent parts of France, although they did not yield to strain as did the weak sediments of the regions of the Alps and Pyrenees, were still considerably affected by the unrest to the south. To this same period of the Alpine Storm we can safely attribute much minor warping and uplift in other parts of France. Probably the Cuestas of the Paris Basin were exposed at a rapid rate after such warping or uplift. It is clear that the Cevennes were uplifted from their original peneplain (necessarily near sea level), because the numerous young volcanoes of the Auvergne (10 in Fig. 89) belong to this period. We know that volcanic action is one of the outward and visible signs of inward, invisible crustal movement.

This last period of great earth movement has, of course, produced the France of to-day. One of the chief minor topographic features lies between the Alps and the resistant Cevennes Block. This is the Rhone Graben, which is a fragment of the crust which has subsided in the same way as the Scottish Graben (pp. 45-48). In this fashion has been produced the "Way of Light," whereby first the *Pax Romana* and later Christianity passed from the Mediterranean to the northern peoples. Until relatively modern times this Rhone Graben has perhaps dominated French History. In earlier years it was the kernel of Burgundy—the dangerous rival of France. Later the Renaissance movement spread up the "Way of Light," as we have seen. It has always been the chief corridor of Man and Culture in the west of Europe.

But of equal though not such striking importance are the numberless other elements in the landscape produced by this general earth movement at the end of the Tertiary. All those flatter regions which were more or less elevated immediately entered upon a phase of valley deepening. These "juvenile valleys" (p. 60) are universal on the flanks of the moderately elevated parts of France. Nowhere are they more striking than in the Ardennes—where the Meuse and Moselle have cut deep trenches one or two hundred feet below the general level. Such a topography has always hindered human occupation and communication through history. We may well, therefore, contrast the advantages of the broad Rhone Graben with the disadvantages of the narrow juvenile valleys, though both were essentially due to the same earth forces.

The recent changes in topography are most marked in two regions, both in the south, namely, the Landes and the Rhone Delta (Fig. 90). The sandy plain of the Landes covers some 3,000 square miles, and is bordered by recent dunes allied in origin to those so common in the Netherlands. These are often 200 feet high, and retain on their inland sides a zone of lakes and swamps which are gradually being drained. The Rhone empties into the quiet waters of the Mediterranean and has built up a large delta of some 600 square miles. This has grown out to sea at a rate of about four miles in a century. St. Gilles, near Arles, was the port from which many crusaders proceeded to Palestine in the 12th century, but it is now 20 miles from the sea. The delta consists largely of fields of shingle and mud, and was formerly subject to fevers and almost uninhabited. Canals have been constructed at various times through the shingle (Crau) since the time of Marius, who dug the first in B.C. 104, during his campaign against the Teutons. Further results of the topography will be discussed in later paragraphs.

B. *Geographical Factors before A.D. 500*

We have seen that in England the limestone cuestas of the midlands and south were favoured habitats of early man. This was because they were relatively high and bare. But in France another feature of limestone topography led to the growth of what we may call the *Metropolis* of Paleolithic Man. Flowing into the River Dordogne (Fig. 90) from the Cevennes is the River Vezere, which cuts a gorge about 200 feet deep in the Mesozoic limestones a little south of Perigueux. The limestone dissolves readily to form shallow caves and deeper tunnels, and relics in these caves have taught us more about Paleolithic Man than those in any other site. Here within a mile or two of the village of Les Eyzies (*pron.* Laze-Aisy) are the Cromagnon, Madeleine, Moustier and Laugerie caves, all classic in the early history of Man.

But we shall do well to hurry on to the period of Julius Caesar. Everyone knows the opening phrases in his annals to the effect that "All Gaul is divided into three parts; the domains of the Keltae, Belgae, and Aquitanians." These three tribes which he described in B.C. 55 probably still build up the bulk of the French Nation of to-day. Moreover, each represents one of the major races of Europe. Indeed, Les Eyzies is near the meeting place of the three, at the most interesting point in the Racial Distribution of Europe—the tip of the "Alpine Wedge" (Fig. 22).

We may picture Cro-Magnon man some 12,000 years ago vainly resisting the great migrations of "Mediterranean" tribes who poured in from Africa. In time these in part merged to form the dark, short Aquitanian of whom Caesar speaks. "Aquitaine" has since been softened to Guienne. Then within the next few millenia began to appear bullet-headed Asiatics who had probably come up the Danube and along the flanks of the Alps, and so *via* the Gate of Belfort reached the east of France. These folk are the Keltae (or Alpines) of Caesar's annals. Throughout the last 10,000 years the same irresistible movement of Alpine Waves has pressed into Western Europe, as we shall have occasion to see.

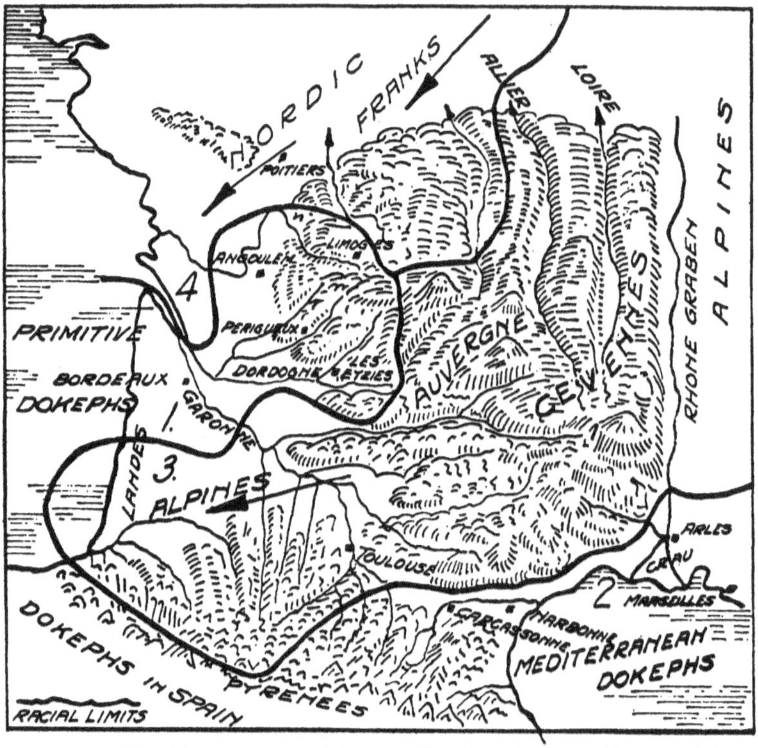

Fig. 90.—Topography and Race in Southern France. The earliest Paleolithic Dokephs still cluster about Les Eyzies. Other Mediterraneans inhabit the south coast. Alpines in the mountains came from the east, and Nordics in the plains from the north.

Many centuries later, perhaps not much before Caesar's time, some small bands of tall, fair dokephs (narrow-heads) began to appear in France. They also pressed down the gap of Poitiers towards the Dordogne, though they seem not to have reached Perigueux in any large numbers (Fig. 90). In Caesar's time we may be sure many of the "Belgae" were of Nordic stock, though many were Alpine. But it was not until the 5th century A.D. that the invasion of the Franks altered the whole racial composition of northern France.

We may here consider briefly the way in which the *Pax Romana* affected France. Naturally this to some extent depended on the ease of communications with Rome. The Rhone Graben was visited very early by civilised folk—long before Rome was of any importance—and

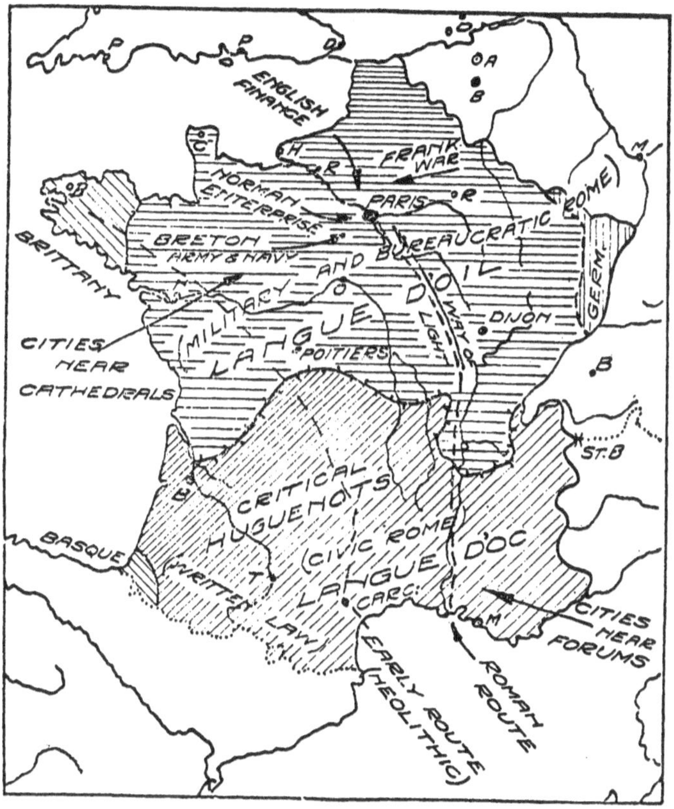

Fig. 91.—Correlations of Langue d'oil and Langue d'oc with various culture-facts in France. Influences affecting Paris are indicated. (Based on Fleure.)

Marseilles (Massilia) is supposed to go back to B.C. 600. The Narbonne Gate (see p. 64) was also well known to ancient traders. Hence the Rhone Graben and Aquitaine were in touch with the eastern civilisation long before the Paris Basin. Indeed, most of this southern area became part of the Roman Territory about B.C. 100. The northern half of France was not much affected by Rome until the time of the first Emperors. Thus we may follow Fleure in considering that southern France derived its civilisation from Rome in its simpler *Republican* period, while northern France obtained her Roman culture during the more elaborate days of the *Empire.* In Fig. 91 some of the consequences of this difference are indicated. For instance, the French dialect differs considerably north and south of the Poitiers-Lyons line. To the south is the older harsher form of Langue-d'oc (Provençal). This name is derived from the common early Latin form of assent, which included the word "*Hoc.*" Indeed, the large province centered about Toulouse was long known as Languedoc (Fleure, *loc. cit.*, 1928).

The north of France, around Paris, developed the standard language, where the similar form of assent was "*Hoc illud,*" softened to "*Oil,*" and finally to "*Oui*" (Yes). We seem to see on the map (Fig. 91) how the dominant speech is ousting the dialect down the two corridors leading from Paris *via* Poitiers and Lyons, whereas the Provençal holds its own in the rugged Cevennes between. So also in the south the cities grew up near Roman market-places (forums), while the later cities of the north are more definitely associated with the ecclesiastical centres of the close of the Roman Empire. Of course the later invasions of the Franks very largely altered the character of the culture of the north, whereas they settled to a much less degree in the area of the Langue d'oc. Thus to-day some of the laws of southern France show a closer relation to Roman Law, while in the north the customs of the Franks can be shown to have left their mark. It is logical to apply this idea of the essential difference between early and later Roman culture even further afield. The early civilisation of Germany derived from north France, and was more military and bureaucratic than that which obtained in southern France. No doubt also these differences and the relative isolation of Aquitaine and Languedoc led to Toulouse and its vicinity often standing in opposition to the views and wishes of Paris. The early Albigenses and the later Huguenots lived mainly in the west and southwest of France, and were always critical, if not hostile, as regards the official proceedings in Paris.

C. *Geographical Factors during Frankish Times*

When the most terrible of the Eastern Invasions under Attila the Hun reached France in 451, they found it disorganised like almost all the other parts of the Empire. However, Aetius the Roman leader collected a mixed force of Gauls, Franks and Visogoths and was able to throw back the Huns at the battle south of Chalons. It is no mere coincidence that the battle was fought here, for the safety of the heart of the Paris Basin turns on the defence of the cuestas and vales of the region of Champagne. Chalons is but a few miles from Chateau Thierry and Rheims, where the decisive battles of a later and greater war occurred (pp. 331-4). Franks had first settled in Gaul about 310, but it was not till 486 that their leader, Chlodwig (*i.e.*, Ludwig, Clovis or Louis), attacked Syagrius, the Roman Governor of Gaul, at Soissons and replaced the decadent Roman rule by Frankish barbarism (Fig. 41). However, Chlodwig became a Christian about 496, and as he was a convert of Rome he strongly objected to the Arians, who had originally been converted from Byzantium. He therefore attacked the Visigoths of southern France at Poitiers, the gate to their territory, in 507. His successors conquered the Burgundians; and as we have seen Charles Martel and his Franks saved France from Moslem domination by another battle in the same "Poitiers Gate" in 732.

For many centuries the northern tribes had been progressing towards a higher civilisation, chiefly under the guidance of Rome. Now, however, after nearly 300 years the Franks were become much more powerful than any of the various rival powers in Italy; of which the Pope and the Lombards were the chief (p. 173). Hence 755 is an important date in European history, for in that year the independent Frankish king, Pippin (or Pepin), returned some of the help which the north owed to Rome. He vanquished the Lombards and commenced that real association between France and the Papal ruler which endured in some form or other until 1870. Pippin's son was the great Charlemagne, whose conquests for France had no parallel until the time of Napoleon. It is important to remember that the Franks had not yet identified themselves with modern France. Indeed, Austrasia, the eastern half of their kingdom, extended from the Meuse to the Weser, *i.e.*, in German lands (Fig. 92). Charlemagne extended his dominions far to the east and south, for he conquered the pagan Saxons, the fierce Avars (of the Turkish stock) as well as the Lombards in Italy and Moors in Spain.

It is interesting to compare Charlemagne's eastern boundary with that of the Roman Empire. The latter lay along the Rhine-Danube line for reasons that have been discussed already (p. 158). But Charlemagne was naturally familiar with the cool, wet climate of the northern forests, and so extended his empire far to the east of the Rhine. At the same time he did not subjugate the northern Slavs, nor did he occupy the plains of the Alfold (Hungary). From this latter area the Avars were soon expelled by the Magyars coming from the east. Charlemagne gained, however, some territory from the Croats between the Drava and the Sava (and his son subjugated many northern Slavs). In Italy his rule extended but little south of Rome, while in Spain he only held the march (just south of the Pyrenees) as a defence against the Moors (Fig. 92).

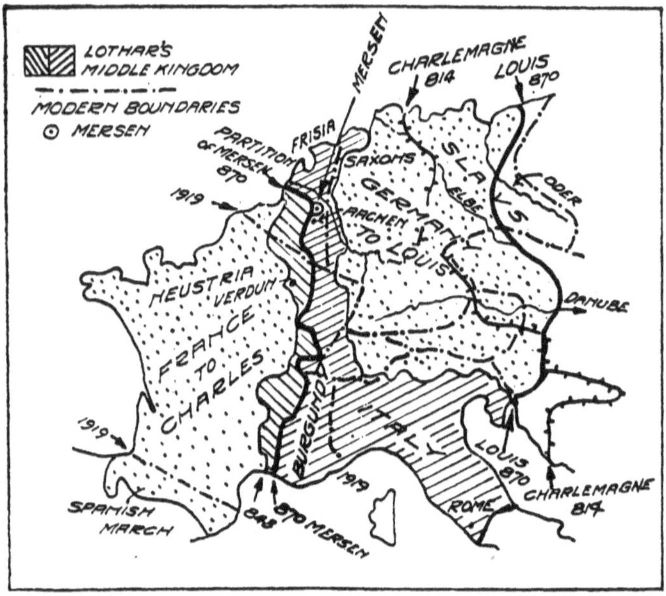

Fig. 92.—The Partitions of Charlemagne's Empire. They marked the birth of France and Germany, and to a less definite degree of Italy, the Netherlands and Switzerland.

There was not much unity either in structure, race, language or religion in Charlemagne's Empire. It was inevitable that it should fall to pieces, unless his successors were men of his own calibre, which was not the case. The Alps separated Italy, both structurally and linguistically, from the rest of his Empire. The Rhine was a potent cultural boundary separating the Franks of Gaul with their fairly strong Roman culture from the allied Teutons east of the Rhine without that association. Burgundy had been settled for a time by Teutonic rivals of the Franks; while Aquitaine, though more closely in touch with the north of France, as we have seen, had many strands in her culture-pattern not found in the Paris Basin. Charlemagne placed his capital at Aachen (*i.e.*, Aix), which lies almost exactly where two of the great routes of his empire crossed. Here the Rhine Trade met the "Route of the Foothills," which linked Paris and Flanders with the Plains of Germany and with the barbarians of the East. Aachen also was within striking distance of the Saxons and Avars, who were his most turbulent enemies. (The geographical reasons for the site of his capital recall those influencing in other days Constantine and Peter the Great.)

The events following Charlemagne's death might easily have been foretold. While his son Louis was reigning many factions arose in various parts of the Empire. On his death his three sons fought for the succession and finally the empire was divided among them in 843 (Fig. 92). The two younger sons received fairly definite geographical units, which in time evolved into France and Germany. But to the eldest, Lothar, was allotted an elongated strip of territory for whose boundaries at first glance no reasons can be assigned. This Middle Kingdom (Fig. 92) extended from the North Sea to Rome, and included Frisia, most of Austrasia and Burgundy, as well as the conquests in North Italy. But Lothar's chief advantage was that he controlled the trade along one of the main roads of western Europe, that leading up the Rhone Graben and down the Meuse to Aachen and the Netherlands. He also ruled over the mouths of the Rhone and the rich lands of North Italy, including Rome itself. These assets, together with the title of Emperor, seem to have prevented Lothar from realising that his realm had practically no defensible boundaries. At any rate the Middle Kingdom vanished on Lothar's death and nearly all his realm was given to his brother Ludwig (or Louis), ruler of the German lands.

The half century following the death of Charlemagne is of the greatest importance in our study of National Evolution. Almost all the troubles of Western Europe have their roots in these partitions of Charlemagne's Empire. The rulers of France fought almost continuously until 1919 to incorporate all those portions of Lothar's Kingdom which lay *west* of the Rhine and the Alps. In medieval times the rulers of Germany and Italy were so busy trying to link these two very different environments to form the Holy Roman Empire that neither region could flourish. The evils which resulted from their vain efforts to some degree determined the fact that neither Italy nor Germany achieved real unity until the latter half of the 19th century. Nevertheless, the treaties of Verdun (843) and Mersen (870) do mark the birth of France and Germany, though it was not till the end of the 12th century that the Italian regions for all practical purposes broke away from the so-called Holy Roman Empire.

D. The Growth of France in Medieval Times

In medieval times the social environment in France was very different from its condition under the aegis of Rome. We may here consider three such differences. Attacks by private or public aggressors were common throughout the country. The northern pirates made raids on most of the northern coast, while Arab corsairs descended on the Mediterranean shores. For many centuries after Charlemagne France was split up among powerful barons who had little or no respect for any authority but their own. This unhappy state of general insecurity led to the development of the Feudal system, whereby each individual was, nominally at any rate, under the protection of some higher power until the Feudal Head was reached. Theoretically, the King was the supreme controller, but in practice the chiefs of the second rank, Dukes and Barons and the like, generally used their large powers to attack each other, and to increase their private possessions. Indeed, about A.D. 1000 the Royal Domain extended hardly 100 miles from Paris, and the other topographic units, such as Provence, Languedoc, Gascony, Aquitaine (Guienne), Brittany, Flanders, etc., were so isolated from Paris that they were often independent of the King. In the third place most of the good roads and bridges of Roman times had so deteriorated that they had fallen out of use, and communications across any length of territory were difficult and, indeed, dangerous.

In this brief account there is no space for a lengthy discussion of the relation of the Feudal System to the growth of France. Only two or three salient features can be noticed. Clearly the King could exert little or no authority as long as his nobles were as powerful as himself. Indeed, one of his vassals (the King of England about 1180) was much more powerful. Moreover, these huge feudal territories passed from family to family—often as the result of a marriage—in a way that is hardly understandable to-day. We may consider very briefly the results of some of these transfers, which were of tremendous importance to France from the eleventh to the fifteenth centuries.

The Norman King Henry I of England, married his only child, Margaret, to Geoffrey Plantagenet, Count of Anjou. Their child (Henry II) in turn married Eleanor of Guienne, and so her vast possessions in southern France were added to those belonging to Normandy and Anjou. All of this territory belonged to Henry II by feudal law; but as a fief of the French crown. (It is denoted in Fig. 93 by his date, 1154.) The French King at this time, therefore, ruled merely a thin strip, of somewhat less area than Henry's possessions, between these latter lands and the Holy Roman Empire. Yet this

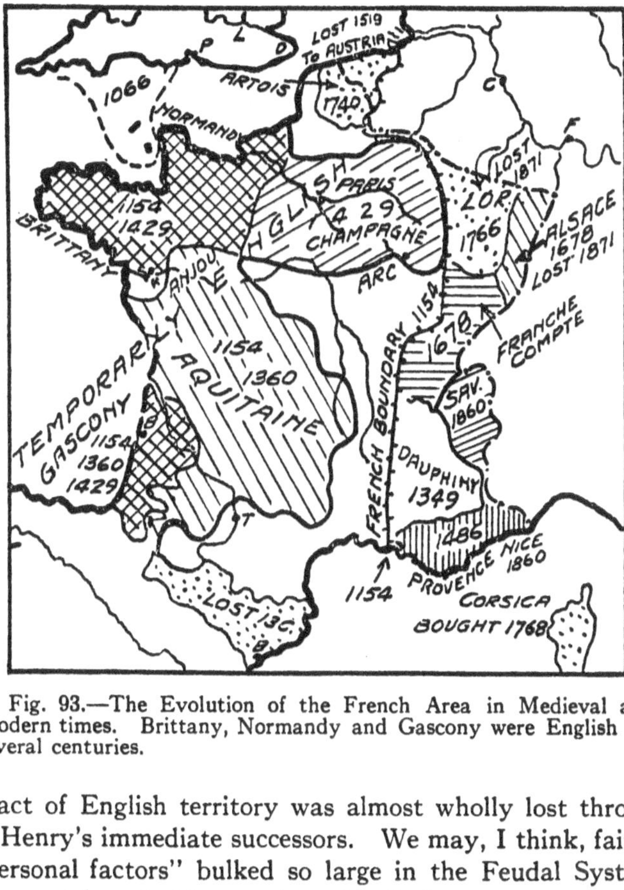

Fig. 93.—The Evolution of the French Area in Medieval and Modern times. Brittany, Normandy and Gascony were English for several centuries.

huge tract of English territory was almost wholly lost through the folly of Henry's immediate successors. We may, I think, fairly state that "personal factors" bulked so large in the Feudal System that "environmental control" seemed to play little part.

The Norman Kings of England were indeed more at home in Western France than in England itself. Conceivably if they had continued to hold their French domain (which was as large in the middle of the 12th century as that in Britain) England might have become merely an appanage of France, instead of a separate flourishing nation. Howbeit, Philip Augustus of France soon acquired from John of England most of Normandy and Anjou; and a little later most of Guienne passed to France. Thus in 1337 only the coast near Bordeaux belonged to England.

We may here summarise the result of the Hundred Years' War waged from 1337 to 1450. Edward II of England (the grandson of Philip of France) claimed the throne of France in 1328. Some years later he invaded France, and increased the British holdings until they comprised the region labelled "1360" in Fig. 93. After an interval of deadlock Henry V in turn invaded France, and in 1429 had won all the region (labelled "1429") north of the Loire. But the heroic leadership of Joan of Arc inspired the French troops to new efforts; and by 1453 the English had abandoned their unwarranted attempt to conquer a land with such a different people and culture from their own. Never again was France invaded for lengthy periods like these.

These wars with England naturally intensified the French national feeling, but two other factors also largely contributed to this result. The Crusades were undertaken to win back the holy places of Palestine from the Moslems, and they were largely manned by French troops. Thus the First Crusade (1097) was almost entirely French; the Second (1147) they divided with the Germans; the third (1190) with the English; and the fourth (1202) with the Venetians. The last two, in 1248 and 1270, were again almost entirely French. From our present point of view we may note that these Crusades helped indirectly to develop the French Nation in several directions. Many turbulent barons with their most enterprising followers went to Palestine, and there succumbed to the vicissitudes of war and disease. This inevitably made things easier for the Kings of France, who in general stayed at home. Secondly, the Crusades greatly stimulated trade, and so contributed to the growth of a class of merchants and town-dwellers, which had almost dwindled out of sight in the anarchy after the Fall of Rome.

During the twelfth century the growth of the Town Guilds was a very important phase of the nation's development. Philip Augustus encouraged the guilds in many directions; and was, in return, assisted by them in his struggles against the German Emperor, Otto IV, King John of England, and the Counts of Flanders. The Battle of Bouvines, fought in 1214 just on the edge of the "Cockpit of Europe," was indeed a notable event. "The victory [says Guizot] marked the commencement of the time when men might speak by one single name of *the French* . . . . The nation in France on that day rose above the feudal system." Philip III (1270-1285) inherited a large part of southern France from his uncle Alphonso, and his son Philip married the heiress of Navarre and Champagne.

In 1349 Dauphiny was bought by Charles V from its former ruler. This purchase marked an important movement of the French Kings in their advance to the Rhine-Alps boundary. About a century later in 1486 another large territory, Provence (the "Provincia" of the Roman Empire), was left to the crown of France by its last independent ruler. There were few changes of territory for us to note for nearly three centuries.

E. The Growth of France in Modern Times

Our account now takes up the story in the time of Louis XIV, who ruled from 1643 to 1715. France at this time undoubtedly led Europe in arms as in arts. From 1667 onwards Louis waged a continuous series of wars, but of even greater importance to the nation was the revocation of the Edict of Nantes in 1685. By this unjust law the Protestants (Huguenots) of France were driven out in thousands, and so carried the best of the French arts and crafts to almost every country in Europe. Moreover, this cruel treatment of the Huguenots alienated all the Protestant Powers in Europe. In the rugged fastnesses of the Cevennes (Fig. 90) the Protestant Camisards fought the royal troops from 1703 to 1711, but in the end the Huguenot cause was lost. In his foreign policy Louis XIV definitely desired to eclipse the power of the Hapsburgs, and his well-disciplined troops were victorious in Flanders in 1667 and 1678. He thus won from the Spanish half of Flanders, Franche Comte and part of Alsace.

Some attention should be given to Alsace, for it is the sole region of German-speaking people in the French domain. Its situation is mainly in the western half of the Rhine Graben (Fig. 69), so that it is a rich land of deep soil and large water supply. Moreover, in the ancient hills of the Vosges nearby are valuable mineral deposits. It had early passed under Roman sway, so that it shared with Gaul the benefits of Roman culture. During much of medieval times it was part of a Rhineland Confederation which owed allegiance neither to Germany nor France. Later it became an outlying portion of the Hapsburg dominions; but most of it was ceded to France, as we have seen, in 1678. It is surprising that a region first settled by German tribes and speaking a German tongue should prefer French rule to German, but such is the case. In 1870-71 the Alsatians fiercely resisted the German armies. After the war Alsace was given to the newly formed German Empire, but 45,000 Alsatians elected to leave their homes and migrate into France. Since that date the Germans have not been successful in gaining the goodwill of the Alsatians; and in 1919 the province was once more restored to France.

Lorraine preserves the ancient name of Lotharingia of which it formed part in the ninth century. It passed in turn into the hands of many of the ruling houses of Europe, including those of Burgundy (Fig. 79) and Hapsburg. In 1735 Prince Francis of Lorraine exchanged his territory for Tuscany, while Lorraine was given to Stanislaus, the father-in-law of Louis XV. On the death of Stanislaus in 1766 Lorraine was annexed to France. Thus by the addition of Lorraine the long-held hopes of the French rulers were fulfilled, and only the small region of Savoy (south of Lake Geneva) remain unannexed on the French side of the Rhine-Alps line. This was the ancient domain of the Italian House of Savoy—which later became the House of Sardinia. As the price of French help in the Italian War of Liberation (p. 369), Savoy was handed over to the French in 1860 (Fig. 104).

It remains to add a few words on Corsica. This rugged island, inhabited by some of the earliest "Mediterranean" tribes in Europe, had long been nominally part of the Genoese Republic. Frequent rebellions broke out in which the Corsicans were often aided by British ships. In 1768 the Genoese were glad to sell their "rights" to the French, who put down the Corsicans in 1769, and thereafter assumed control of the island.

F. Topography as a Factor in Defence of France

The outline of France is that of a rough square, of which the north, west and south are well protected by nature either with the open sea or with the Pyrenees (Fig. 94). In the latter region it is difficult to imagine France having much fear of her Spanish neighbour, but it is, of course, along the lengthy eastern frontier that France is less adequately protected. The boundary between France and Italy consists for the most part of the summit ridge of the Alps. In general the region which is over 3,000 feet high is about 90 miles wide, though the section through Mt. Pelvoux (13,400 feet) in the widest part is no less than 120 miles across. Yet throughout, the divide (and frontier) is usually within 30 miles of the plains of Lombardy. Hence it is obvious that the Western Alps have a short steep face on the Italian side, and a much wider and less steep slope on the French side. As stated earlier, the migrations in historic times have almost always been from the cooler north and northwest to the richer and warmer Italian territories. Indeed, considerable French-speaking communities live in Italy around the headwaters of the Maira and Dora Baltea, where

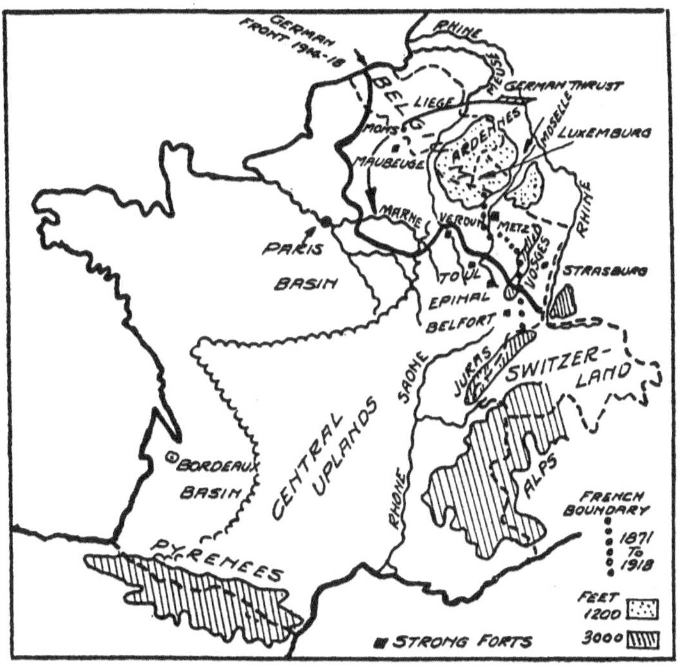

Fig. 94.—Topography and French Defence. In 1914 the Germans attacked through the lowland of the "Cockpit of Europe." Note the importance of the four great Forts in the southern line of defence in 1914-18.

they have "spilled over" the natural and political frontier. However, this rock barrier almost wholly above 6,000 feet high has made the defence of this frontier rather easy for the French.

The boundary with Switzerland begins at Mount Dolent, ten miles north of Mont Blanc, which is by so much entirely outside of Switzerland, though it is often stated to be in that country. The Swiss-French boundary runs across the River Trient at Valorcine and through the peaks of the Dents du Midi to the south shore of Lake

Geneva. From this point northward to the North Sea there are many breaks in the physical barriers, which, indeed, become non-existent at the northern end.

The Gap at Geneva is due to the "Downfold" in front of the Alpine folds (p. 294). Here has grown up the famous city of Geneva—often a "thorn in the flesh" of French policies. Here, for instance, Calvin received a refuge when his religious views led to his being driven from France. However, the deeply entrenched gorges of the Rhone make the Geneva Gate less easy to negotiate than many of the "gates" to the north. Moreover, Switzerland has been free from foreign aggression, whether French or otherwise, because it is hard to attack her fastnesses, impossible to conquer them, and because the country as a whole is poor in those resources which have almost always led to wars of aggression. Switzerland is a very useful "buffer state" between the rival nations of France, Germany, Austria and Italy—but it has had a very different history from Belgium (its companion buffer) further to the north. This difference is in no small degree due to the different environments—which, indeed, could hardly be more sharply contrasted. The frontier, then, passes north from the Geneva Gate along the low crests of the Jura Mountains (Fig. 94). These crests rarely rise above 4,000 feet—but they are difficult to cross since they consist of parallel "earth ridges" separated by deep and narrow fold troughs. North of Pontarlier the boundary leaves the Juras and crosses fairly level country to join the Rhine at Basle. From this point northward begins the real problem of French defence.

Before 1871 and again since 1919 the French boundary from Basle to Karlsruhe has been the Rhine. This river, though flowing over the relatively flat floor of a wide graben, is so wide, deep and swift that it forms a fairly strong line of defence, especially as it is fringed by wide strips of marshy country. Above the flats is the steep scarp of the Vosges—a fault-plane facing Germany and eminently suitable for defence, which the French are fortifying strongly. The gap where the Juras die out is the famous Gate of Burgundy held by the fortress of Belfort—which successfully resisted the Germans both in 1871 and 1914-18. A number of forts surround the older fortifications forming a circle of about 25 miles in circumference. Similar rings of forts have been built at Epinal, Toul and Verdun, and were of vital importance in the Great War (Fig. 94). Now, however, the Vosges and regions around Metz form the outer defence against future German aggression, and here immense sums are being expended to strengthen the frontier.

Just opposite Strasbourg (which is strongly fortified) the Vosges drop to the Pass of Saverne (the Roman "Tavern"). Thence the frontier crosses undulating uplands about 1,000 feet high which soon rise to the somewhat higher plateau of the Ardennes. The latter plateau is crossed by the deep gorges of the Moselle and Meuse; and these gorges furnish the main routes through their basins. The Moselle is defended by the fortresses around Metz, while the Meuse is similarly defended at Namur and Dinant in Belgium and Givet in France. The River Sambre also cuts a gorge through the northern portion of the plateau and is defended at Maubeuge (Fig. 94). Westward of Maubeuge is the lowland of Flanders; where nature has provided no defences in the way of mountains, but where canals, sluices and swamps have been utilised by soldiers in the frequent wars in the "Cockpit of Europe." The Germans in the Great War planned their main attack in direct relation to the topography. They marched along the line from Cologne to Liege, Mons and Maubeuge for the same reasons that have made this the chief artery of northern Europe. This route lies between the rugged Ardennes and the swampy lands of Flanders. Moreover, the Belgians had not erected extensive defences against Germany, since the latter had solemnly sworn not to invade her territory. The manner in which the topography of northeast France helped to defend Paris against the invaders may well form the concluding paragraphs of this chapter.

G. *Topography in the Great War*

It has somewhere been stated that the cuestas of the Paris Basin won the war for France. This is, of course, a picturesque exaggeration, but no one can doubt that the warfare on the Western Front was largely determined by the build of the country. This is shown in a slightly generalised form in Fig. 95. Examining the front edge of the Block Diagram we see a series of six hard layers alternating with softer layers and together building the Paris Basin. Beneath the soft rocks of the Basin (and seen on the right) are the ancient rocks of the Vosges. (These are of much the same origin as those of the Ardennes to the north of our map.) On the surface of the block diagram the hard rocks appear as cuestas and the soft rocks as vales. These have so much individuality that they have been given names. Thus La Brie and Argonne refer to cuestas, while "Wet" and "Dry" Champagne (Latin, *Campania*, a plain) mean merely "Clay Plains" and "Sandy

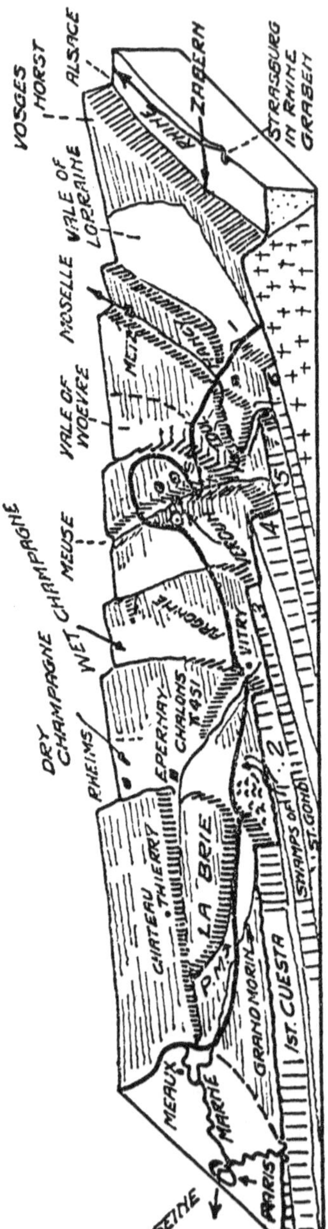

Fig. 95.—Block Diagram of the Five Cuestas between Paris and the Rhine, which largely controlled field operations on the Western Front in 1914. The heavy black line is the furthest advance of the Germans. P.M. is the juvenile gorge of the Petit Morin. S is Saint Mihiel on the Meuse. (Mainly after W. M. Davis.)

Plains," and refer to two of the vales. Woevre and Lorraine are other vales to the east.

The German plan of August, 1914, was to overwhelm the Belgians and the "contemptible" British Army, and then march rapidly on Paris along the Sambre-Oise route used by the main Berlin-Paris Railway. It is essentially the "Route of the Foothills" of medieval times. It is a route (as D. W. Johnson writes) which enables an invader to take in the flank the entire series of plateau barriers (cuestas) which lie to the south. On the other hand, it is 50 per cent. longer than the route (*via* Metz) successfully followed by the Germans in 1870; but the easier topography determined the change of plan in 1914.

In August, 1914, the British and French forces tried to hold the gorges back of Namur, but were beaten back within twenty-four hours (Fig. 78). The Allies retreated quickly to the northern ends of the cuestas and there slowly retired from a line from Noyon through La Fere to Rethel and Verdun. (This line is about 40 miles north of the Marne.) Now followed a battle which, it has been said, may determine the world's history for many centuries to come. Von Kluck hoped to envelop the western end of the allied line and crush it before he attacked Paris—which lay only 50 miles to the west. He crossed the Marne and reached the trench of the Grand Morin on the cuesta of La Brie (Fig. 95). Now the French attacked his right wing at Meaux, while the British drove forward at Grand Morin. The Petit Morin has cut a deep gorge in the First Cuesta (La Brie), and it actually rises in the vale of "Dry Champagne" in the swamps of Saint Gond. The French line was protected by these marshes and by the gorge to the east, and it was along this barrier that the issue of the war was finally decided. The French launched their supreme attack just to the east of the swamp where the ground became firmer. They pierced the German line and caused the retreat of the whole German Front northward to the valley of the Aisne (9-12 Sept., 1914). Only for a short time in the summer of 1918 did the Germans ever again reach the Marne.

Thus the First Cuesta played its part in the defence of Paris, and in similar fashion the other cuestas formed admirable lines of defence; for their gentle western slopes were in favour of the French, while their steep slopes faced east, and were found too difficult for the movement of large masses of German material and men. Perhaps the locality which stands only second to St. Gond in topographic interest on the Western Front is the Fortress of Verdun. Near here are the Fourth, Fifth and Sixth Cuestas (see Fig. 95). In their first attack on the scarps near Nancy (in September, 1914) the Germans lost 40,000 men. They then attempted to seize Verdun from the Vale of Woevre. They actually captured St. Mihiel (S. in Fig. 95) near the Moselle, but pushed their main attacks along the Meuse Valley from the north. The swampy valley of the Woevre and the great scarp protected Verdun on the east. The deep gorge of the Meuse (an "entrenched meander" in topographic language) proved impassable to the grey waves of the Germans. During March, April, May and June, 1916, they attacked the ring of forts around Verdun almost continuously. In the words of D. W. Johnson:[1] "Verdun saw half a million of Germany's finest troops sacrificed in the fruitless endeavour to conquer the easternmost belt of the natural defenses of Paris."

[1]D. W. Johnson, "Topography and Strategy in the War," 1917, New York.

CHAPTER XVIII

SPAIN—A NATION FOUNDED ON A DRY PLATEAU

A. The Build of Spain and its Historical Implications

Europe is sometimes described as the peninsular portion of Eurasia, and it is indeed more peninsular than any other continent. Peninsulas tend to produce national isolation, though not to the same extent as do islands. We may compare for a moment the four main peninsular regions of Europe, *i.e.*, Scandinavia, Spain, Italy and the Balkans. Each has its special pattern with historical consequences. In Scandinavia the peninsula is largely a bleak highland, which permanently isolates the two main coasts, and so has kept Norway distinct from Sweden. The land connection with the continent is with the far north in an inhospitable climate, and so has never been of much consequence. Scandinavia is in essence an island, and much like Britain, for it is removed only a short distance from the mainland. Italy is the most typical peninsula, being so long and narrow. As a result it has no plateau country of importance, almost all the settlement being in the coastal plains. It would serve as the land bridge to Africa, but for the tremendous barrier of the Alps. For the same reason its culture and history have developed on somewhat different lines from those of the rest of Europe. The Balkans are so broad that the peninsula shape is somewhat masked. It is, moreover, unlike Spain in that it is rather open to the north, mainly along the Morava-Vardar Corridor, but it is so mountainous that a number of small national groups developed in the several corners of the Balkans. On the whole it offers perhaps a better geographic parallel to the Spanish Peninsula than do either of the other two.

The relation of Spain to France and the build of both are given in the Mantle-map (Fig. 89). The boundary between the two countries is the rugged fold-range of the Pyrenees, which rises to 11,000 feet. A very large proportion of Spain consists of an ancient relic-block, almost all the western and central area in fact. Only in the eastern third is there a series of fold-mountains and elevated Triassic areas (4) which are as young as the greater part of France. The present topography seems to have developed as follows. In Middle Tertiary times the crust began to buckle, possibly in the vicinity of Morocco. The

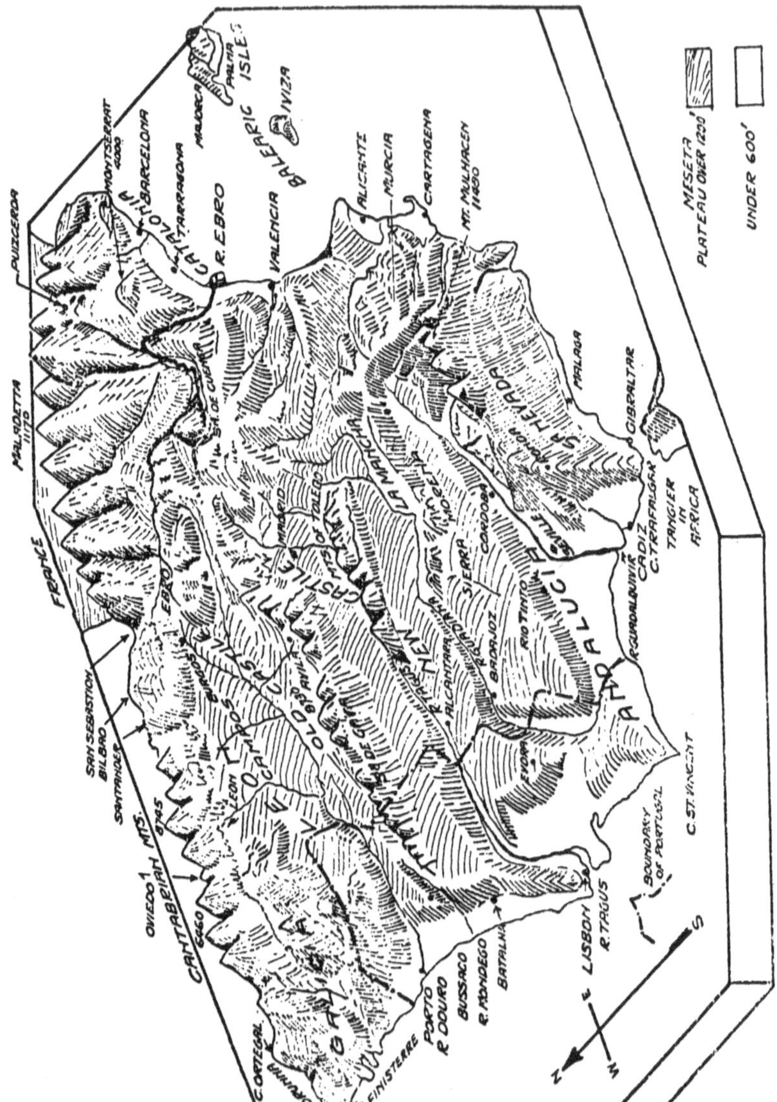

Fig. 96.—Block Diagram of Spain and Portugal showing the Meseta bordered by young Fold-Mountains on the north and south. (See also Fig. 89.)

two relic-blocks of the Meseta and the Cevennes (p. 57) resisted folding, but the softer rocks (largely of Jurassic age) between them were strongly buckled, forming the Sierra Nevada, the Catalonian Folds and the Pyrenees. At the same time the relic-blocks themselves were lifted *en masse*, and probably cracked across to form Horsts. We judge that this is so because the edges of these blocks are gashed by the deep juvenile gorges of the various rivers. Such gorges invariably indicate recent uplift. Tertiary deposits (7c) shared in the elevation and some of these seem to have been laid down in vast fresh-water lakes. The build of the Cantabrian Mountains along the north coast is somewhat complicated. Some geologists believe that the Meseta was actually folded or faulted to form the height of the Penas de Europa (8,740 feet).

The three main rivers flowing to the west, the Douro, Tagus and Gaudiana, traverse the 2,000 feet plateau of the ancient Meseta (Fig. 96). They are separated by two parallel ridges of granite, etc., which are the horsts (*i.e.*, raised blocks) called the Sierra de Guadarrama (7,900 feet) and the Mountains of Toledo. These rivers have cut deep narrow valleys in their lower courses, which render communication difficult and help to isolate Portugal. Their upper waters, however, flow across the undissected plateau, built largely of Miocene deposits (7). These are given various names such as the Campos (2,500 feet) of Old Castile at the head of the Douro, the Alcarria (3,000 feet) at the head of the Tagus, and La Mancha (2,500 feet) at the head of the Guadiana. There are also four low-lying valleys, probably graben (fault troughs) in part, and bounded by fault-scarps on the side next the Meseta. Two of these are drained by the Ebro and the Guadalquivir, while two are smaller coastal plains around Lisbon and Valencia. The last three, blessed with adequate rains, are among the most fertile districts in Europe. The Ebro Basin, however, is very dry, and much of this region has a semi-desert appearance during most of the year.

Spain, therefore, consists essentially of the square block of the Meseta, a plateau of two or three thousand feet, surrounded by rather steep scarps which are trenched by juvenile canyons. The plain of Andalusia is the chief exception to this general plan (Fig. 96). It is occupied by the "Great River" of the Moors, the Wad-el-Kebir which the Spanish call the Guadalquivir. The Ebro Basin is shut in by the coastal folds, and the other rivers flow along difficult gorges in their lower courses. All these factors make for inaccessibility, which is greatly accentuated by the distribution of the rainfall.

In general, in Europe there is a sufficiency of rain, as pointed out in Chapter IV. But Spain is one of the rare exceptions, so that this aspect of the environment must be given more space than usual in our description. It lies on the border of the Mediterranean type of climate, and receives its rains rather in winter than in summer, because in the former season the track of the rainy northern cyclones lies further south. However, the west winds from the ocean blow fairly

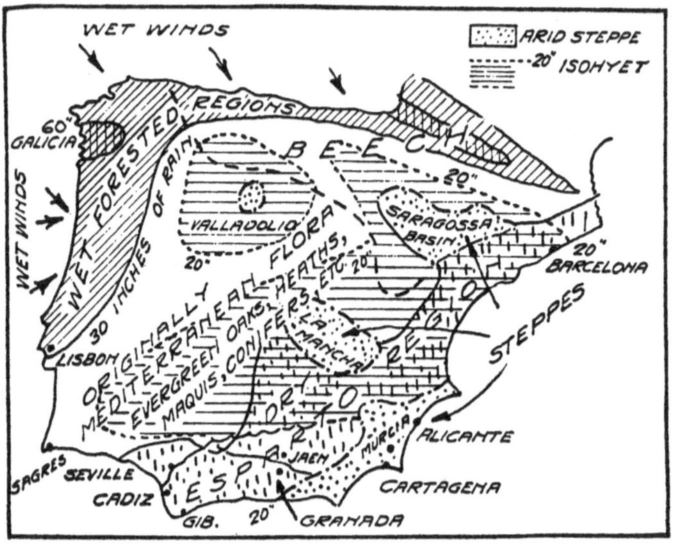

Fig. 97.—Rainfall and Vegetation in Spain.

steadily on to the northwest coast throughout the year, so that this region has rather constant "topographic rains" in addition to the cyclonic rains in winter. Since both these rain-bringers come from the west, there is a marked falling-off in the total towards the east, which is only modified where there are high mountains. The rainfall map is shown in a simplified form in Fig. 97.

The rugged hills of Galicia and the Western Pyrenees receive over 60 inches of rain in the year. Most of Portugal and the northern coastal ranges have over 30 inches, while more than three-quarters of the Peninsula has less than 30 inches. Most of Spain indeed has less than 20 inches, and only the north coasts and the Horsts in the centre exceed this amount. The high Sierra Nevada in the southeast is in an unfavourable position for rain, but it carries snow most of the year, as its name implies. Indeed the River Genil rises in a small glacier, the most southern in Europe. The worst areas in Spain are the centres of the basins into which the plateau is divided. Here are veritable Steppes or semi-deserts, such as near Valladolid (*pron.* Ballad-o-leèd), La Mancha and Saragossa. The rain-shadow in Murcia to the lee of the Sierra Nevada is also a steppe region (Fig. 97).

Owing to the rapid changes in rainfall the vegetation changes greatly between Galicia and Murcia. In the former are still some thick forests of beech and pine with numerous oaks, walnuts and chestnuts. As we proceed to the southeast the open plateaus are soon reached in which there is very little of the original vegetation remaining. All has been removed, partly for fuel, partly to make way for crops; while any seedlings are cleared off by the goats. Originally a Mediterranean flora of heaths, evergreen oaks and scattered pines flourished here with large tracts of scrub or *maquis*. Extensive areas in the east carry Broom and Esparto Grass. However, there are still woods and thickets on the mountain sides, containing oaks, chestnuts and pines, while the slopes of the Sierra Morena in the spring are covered with "Rock Roses" and other flowers.

There are veritable salt steppes along the southern bank of the River Genil below Granada and also near Jaen. Similar steppes are found for sixty miles north of Murcia. Many of the rocks involved in the Tertiary folding contain salt deposits, and these give rise to Brine Springs which carry the salt into small lakes as near Alicante. The coastal steppes have, however, been irrigated from the rivers, and oases of palms and "huertas" (gardens) of great fertility have resulted in favourable localities. These irrigation works were initiated by the Moors.

It is, therefore, not surprising that the Peninsula should have developed as a series of isolated small "nations" rather than as a united Spain. It is indeed much more broken up than Italy, but as we shall see reached unity four hundred years before the land of the ancient Romans. It might at first be thought likely that a central authority would develop on the plateau, but the latter is so much drier than most of the coastlands that the chief settlement has always been in the latter (Fig. 98). These coastal nuclei are clearly separated from each other by the steep coastal scarps and rocky promontories. The latter are indeed characteristic of Spain, as witness the series of seven along the south coast from Cape Nao near Alicante around to Cape Roca near Lisbon. Even to-day there are no important centres of population inland except Vallodolid and Madrid, and the latter is a city of almost wholly political growth.

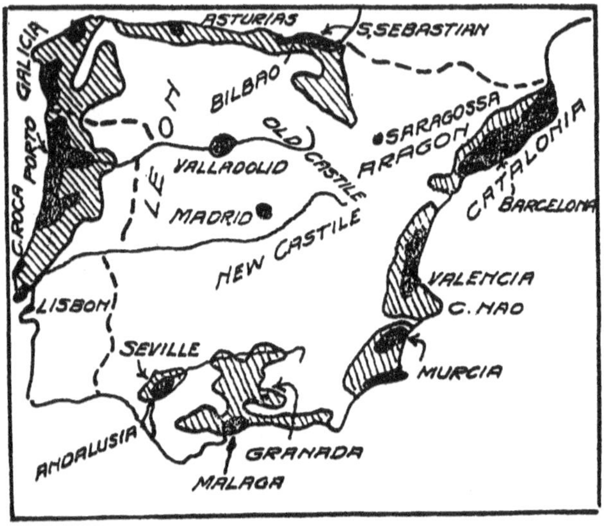

Fig. 98.—The Population of Spain is distributed in isolated provinces. The black areas have over 250 per square mile, and the ruled over 135. (Based on Von Seydlitz.)

B. The History of the Early Peoples of the Peninsula

The Archeology of Spain is fairly well known, but it is not necessary for us to delve into pre-history further back than those tribes called Iberians. These were typical dark "dokeph" Mediterraneans who doubtless formed part of the great migrations which left Africa about the beginnings of Neolithic times. It seems probable that they spoke Hamitic tongues akin to those of the dokeph tribes still in North Africa. Perhaps Basque is a still more ancient tongue as was suggested on page 121. Judging from place-names the Q-speaking Keltae did not migrate into Spain in noticeable numbers. But the next wave of Alpines, the P-speaking Brythons (*i.e.*, Welsh), colonised much of North Spain. They are said to have arrived about the 5th or 4th century B.C. but later were entirely merged with the Iberians (Fig. 99). Meanwhile many traders had visited the mines of Spain, where indeed Tarshish of the Bible is placed. Probably before B.C. 1000 the Phoenicians were at Cadiz, Malaga and Algeciras. Indeed "Spain" is said to be a Phoenician name meaning the "Remote Land." Their great colony of Carthage sent emigrants to Spain, to Cartagena

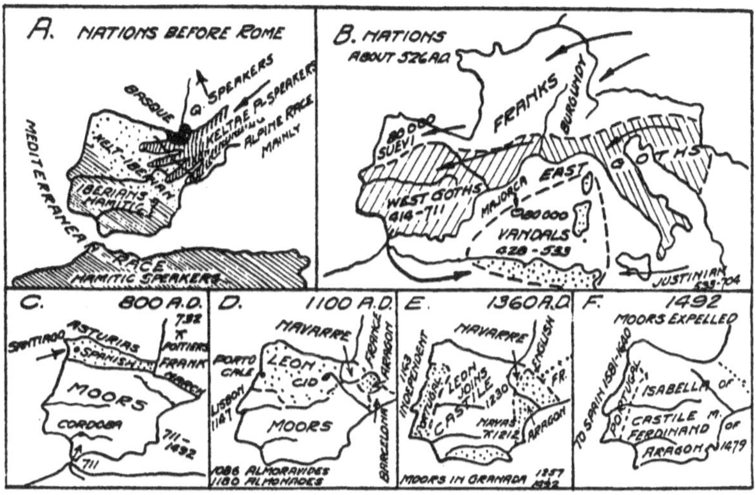

Fig. 99.—The Evolution of the Spanish Nation. The Race is essentially Mediterranean, but invasions of Keltae, Romans, Goths and Moors have modified it to some extent.

among other places, about B.C. 600. This was about the same time that the Greek traders were introducing the vine and the olive into northeast Spain. Emporium near Barcelona was the chief Greek trading post. In an earlier chapter (p. 154) the struggles of Rome and Carthage have been described. Hamilcar is said to have named Barcelona in honour of his family name of Barca, but by B.C. 206 Carthage had abandoned Spain to her Roman victors (Fig. 38).

The Romans naturally expanded first into the Eastern part of Spain, and it was not till about the birth of Christ that Agrippa conquered the Iberians and the mixed Keltiberian people of the northwest mountains. Further invasions from the south continued in spite of Roman authority. Thus the Berbers of Algeria made repeated raids from A.D. 170 to 180. But many Roman soldiers married local women and settled in the south of Spain, so that Andalusia was thoroughly Roman in culture by the second century. The camps of the Roman garrisons became centres of civilisation. Thus Leon dates from the occupation by the Roman "Legion," Badajoz is Camp "*Pax Augusta*," Saragossa is Camp "*Caesar Augusta*," etc. Their roads, bridges and aqueducts were so well made that some of them are in use to this day. Many famous Emperors came from Spain, three of them, Trajan, Hadrian and Theodosius, from Seville. The Roman Church followed the Roman armies, and Spain was the first and one of the richest of Rome's external provinces. It is to be noted that Spain has almost the same environment as Italy, save for the Steppes in the southeast. Hence the Romans found themselves soon very much at home in the Spanish Province.

When Rome fell upon evil times just after A.D. 400 Spain was one of the first provinces to be affected. The Vandals who had attacked the Imperial City, immediately afterwards marched west, and in 409 entered Spain. There does not seem to have been much local resistance to their invasion, and those areas which did not yield to the Vandals set up semi-independent rulers of their own. The Visigoths followed the Vandals in 414, and Ataulf at once occupied Barcelona (Fig. 99). For over a century, however, the Visigoths made their headquarters in South France. They conquered the Alans (who with the Suevi had accompanied the Vandals) and penned the Suevi in the rugged northwest. The Vandals were soon driven right out of the Peninsula into North Africa (p. 170). In 485 Euric ruled from the Loire to Gibraltar, except for the realm of the Suevi; but in 508 Clovis drove the Visigoths southward and they made their capital at Narbonne. In 560 further Frank victories again displaced the Visigoths, and Toledo became their first Spanish capital. Constant religious conflicts between the Orthodox Christians and the ruling Visigoths led King Reccared in 587 to give up his Arian tenets, and it is worthy of note that the Jews were fairly treated by the Visigoths. Later, however, Spain was filled with rival factions, and it is stated that the Jews in 711 took part in inviting the Moslems from Africa to help their party against King Roderick. As a result Tarik the Moor smashed the Gothic Army at Medina Sidonia near Gibraltar.

Let us, however, digress for a short time to study the development of the Moors, who were a new force in European history. This is of especial interest because each of the great nations, France, Germany and Spain diverged about this time. France adopted absolute monarchy, while Germany split into innumerable small independent states. Both, however, made use of the Feudal System. In Spain the Moslem invasion delayed the Feudal System, and substituted for it a partial fusion of Eastern and Western civilisations, which made Moorish Spain a leader in European progress.

C. The Moorish Conquest of Spain

Three times in medieval history the followers of Mohammed came into lengthy conflict with the Europeans. First in Spain from the eighth to the fifteenth centuries; second in the Levant from the eleventh to the thirteenth centuries; and third in the whole of South-east Europe from the fourteenth to the twentieth centuries. These Moslem invaders are better known to us as the Moors, Saracens and Turks respectively. It was their religion which united them, for racially they were different. The Moors were almost wholly Mediterraneans, the Saracens were a hybrid group, probably mainly Mediterranean but with Nordic (Kurd?) and Alpine (Armenian) components. The Turks were mainly Alpine, but no doubt had many Nordic allies.

Mohammed was born in Mecca near the coast of the Red Sea about A.D. 570. This part of Arabia had been influenced by Jewish, Christian and Zoroastrian teaching; and it was also a stronghold of the primitive animistic cult centered about the Holy Black Stone of the Kaaba. In 622 Mohammed was driven from Mecca to the more enlightened Medina. This flight is the Hegira (better Hijra), from which the Moslem dates his chronology. In 632 when the leader died he was in control of all Arabia. While Mohammed seems to have been endowed with a sufficient supply of common human failings, the Moslem religion had many attributes which made it a successful rival of the debased Christianity of the eighth century. It preached equality of races, making the negro equal to the white man, something that Christianity has not yet really done. It insisted on temperance. It was devoid of the dogmas of the theologian. It had no organised priesthood. As H. G. Wells writes:—"Against it were pitted Judaism, which had made a racial hoard of God, Christianity preaching endlessly of trinities and doctrines no ordinary man could make head or tail of, and Mazdaism, the cult of the Zoroastrian Magi who had inspired the crucifixation of Mani" ("Outline of History").

We must traverse very rapidly the expansion of Islam, which was due mainly to three men. Abu Bekr the lifelong friend of Mohammed succeeded him as Caliph. His general Khalid conquered the armies of Byzantium on the Jordan in 634, and three years later in a three days' battle he defeated Rustam the Persian at Kadessia near Bagdad.

Omar succeeded as Caliph in 634, and in 638 he occupied Jerusalem. Struggles between Othman the third Caliph and Ali (son-in-law of Mohammed) rent the Moslem world. Othman was murdered in 656 and Ali in 661. His sons Hasan and Husein in turn were killed and became the martyrs of the Shiite sect. Their opponents the Sunnites affirm their right to appoint their own leaders, whether descended from Mohammed or not. However, the Omayyad (Sunnite) family of Othman controlled Islam for many decades. Another group the Abbasids (named from Abbas the uncle of Mohammed) took up the quarrel on the Shiite side, and all the strife of the Christians was reproduced among the Moslems. In 673 the Crescent reached the Bosporus, but Constantinople was besieged without success. Again in 718 they attacked the heart of the Eastern Empire with a powerful fleet, but were driven off by the Emperor Leo, largely by the use of a primitive gunpowder called Greek Fire. In 749 the Abbas leader massacred all the chief Omayyads at Damascus, and legend states that he actually feasted on a carpet laid over their dead bodies.

Let us now turn to the Moslem conquest of Africa. In 639 Omar's troops invaded Egypt and found the Greek Christians bitterly opposed to the native Christians (Copts). The latter welcomed the Arabs, and so the occupation of Egypt was easily accomplished. The Arabs rapidly advanced through Tripoli; but the Eastern Roman Empire with command of the sea could attack at will. Thus in 683 the great Arab general Okbar was defeated near Carthage by the Byzantines from the sea and the Berbers from the Atlas Mountains. However, by 712 the Arabs had conquered the whole of North Africa and were ready to attack Spain. It is obvious that the Arabs were specially fitted to civilise the Moors and other native peoples of North Africa. The Arab religion appealed to warlike tribes; their desert and mountain environment was the same. Moreover, the Arabs already had a considerable knowledge of literature, especially in poetry, and their leaders had been for centuries in touch with all the philosophies of the east. For a century or so the Moslem rulers at Bagdad controlled North Africa, and the wild Berbers of the mountains gradually became devout followers of Mohammed.

In 710 the Visigoth King Witiza was overthrown by Roderick of Cordoba, who took his place. Rival factions sought aid from Musa, the Arab leader in North Africa, and he sent his general Tarik into Andalusia in 711 with an army of Berbers. Roderick was completely defeated, and Musa soon subjugated most of Spain. The study of the contacts of the various racial groups in Spain, which produced the modern Spaniard, is one of the most interesting in Europe. We have seen that the Nordic tribes merged in the essentially Mediterranean Roman-Keltiberian peoples. Probably the number of Nordics was rather small. It is stated that the Vandal horde was only about 80,000 in numbers, and the Suevi were about the same. The Visigoths were no doubt more numerous, but would not greatly alter an aboriginal population of several millions. Jews in considerable numbers had arrived, but they were Sephardim, *i.e.*, of Mediterranean race (p. 416). There were even groups of Slav warriors who had been captured in East Germany, and sold as "slaves" in Spain by Jewish merchants. The new Arab groups were also dominantly Mediterranean, though consisting of two different cultures. There were the Arab leaders, the original Semitic-speakers, and the African Berbers who spoke Hamitic. These Moslems at first formed an aristocracy as did the former Goths, and indeed Gothic landowners were still allowed to control affairs in many parts of Spain. It seems clear that it was the merging of cultures rather than of races, which produced the Moorish Renaissance in Spain in the 12th century.

For a century Spain was an outlying portion of the great Saracen Empire (Fig. 99 at C). The Arab leaders settled as landowners mainly in the fertile south, while the Berber soldiers were given territories in the less attractive north. Civil war among the Moors was of constant occurrence, and enabled the weak Christian states to maintain their independence in the Cantabrian Mountains. Abderrahman I (an Omayyad) escaped from the massacre at Damascus to Spain and defeated the rival Abbasids. It was about this time that Charlemagne invaded the north of Spain, but his defeat at Roncesvalles was the work of the Christian Basques as much as the Moors. Renegades (Mozarabes) and Jews formed important communities among the Moors, and in the time of Abderrahman II (912-961) Cordoba reached its zenith with half a million inhabitants.

D. *The Christians Reconquer Spain*

It is quite incorrect to picture a holy war waged by the small Christian states against Islam. They made attacks to regain their ancient homes, but the various Basque, Aragonese and Galician states fought quite independently. However, the troubles of the Moors naturally helped the Christian raids. In 874 Catalonia broke away from the Franks, and in the tenth century Leon and Castile were growing in power. Early in the eleventh century Sancho of Navarre conquered much of Moslem Spain, thus arousing the national spirit; but he nullified his gains by splitting them among his three sons. About 1086 the Moors in Spain invited the Almorabides, a powerful group of African Berbers, to come over and help them. They drove the Christians north, but in less than a century their power waned, and they in turn were overthrown by a fresh African migration of Almohades about 1146 (Fig. 99).

Meanwhile Ferdinand of Castile conquered Leon, and his successor Alphonso captured Toledo in 1085. Ruy Diaz "The Cid" (*i.e.*, Chief) was a famous mercenary of the times who fought impartially in turn for Castile, Aragon or the Moslems. In 1086 he conquered Valencia on behalf of a Moslem ruler, and on the death of the latter the Cid ruled there till 1099. In 1095 Alphonso gave the northern province near Oporto to a foreigner Henry of Lorraine, who had married a Spanish princess. His successor Alfonso drove the Moslems south, and in 1147 the Pope recognised him as King of Portugal. The name of the country is derived from "Portus Cale," now a suburb of Oporto. Thus the Moors first relinquished the wet forested region of the west which was least like their home environment (Fig. 99).

The two greatest periods of intellectual activity in Spain were the 12th century and the 17th century. The first occurred in Moslem Spain during the rule of the Almohades, when such teachers as Averroes (1126-1198) flourished at Cordoba. But these intolerant Moslems drove out most of the Jews, who for a time found refuge in Castile. They played a large part in introducing Moslem and especially Greek knowledge into Christian Spain. Thousands of Mozarabes were brought from the conquered southern lands and settled in Aragon in the 12th century, where they introduced better methods of agriculture.

Alphonso VIII broke the Almohade power at the battle of Navas de Tolosa in 1212 (Fig. 99 at E), and his grandson Fernando captured Cordoba and Seville (1248). But after the death of this warlike monarch no effort was made to drive the Moors from the mountains of Granada, where they had taken refuge, for more than two centuries. The greatly enlarged kingdoms of Castile and Aragon were busy with developments of their own. Towns sprang up and acquired considerable degree of self-government owing to their wealth. In Spain as in other lands the nobles fought for their privileges, while the stronger Kings aimed at absolute rule. However, Henry IV of Castile, a very weak ruler, died in 1474 without having nominated a successor. His sister Isabella had married Ferdinand of Aragon in 1469, and she was now proclaimed Queen of Castile. In 1479 Ferdinand succeeded to the crown of Aragon. This union of most of Spain was destined to remain permanent (Fig. 99).

Transfers of peoples continued in Spain. Thousands of Moslems emigrated from Granada into Christian Spain in the 14th century, and were allowed relative freedom of worship; but the Jews suffered much persecution, and there were many massacres about 1391. Barcelona and Santiago were early centres of manufacture, but the internal trade was small, partly because the nobles prevented the building of roads and bridges, which might reduce their fees for ferries, etc. Castile still maintained the lead in the peninsula, in spite of its general poverty of resources, since it had been the most energetic in the war of liberation. The Castilian dialect was used at the Courts of Castile and of Aragon and ousted Latin and Catalan, so that it gradually became the national language.

The two rulers, Ferdinand and Isabella, raised Spain from relative insignificance to a position of world power. Civil wars ceased and Granada was conquered by 1492. In the same year Columbus led the Spanish fleet to the West Indies, and in this remarkable year the Jews were expelled from Spain. All three events were of the greatest importance. About 165,000 Jews left the country, some 50,000 became Christian, while 20,000 are said to have been killed. There can be no doubt that this expulsion was a great loss to Spain. Moreover, in 1502 all the Moors were ordered to become Christians or leave the country. Thus the Jews, the most energetic merchant class, and the Moors, the chief source of agricultural labour in the south, were in large part removed from Spain. Also the rulers did not realise that crops paid better than grazing, and so a great Sheep Corporation known as the "Mesta" was allowed to pasture its flocks almost where desired in Castile. The Mediterranean ports declined greatly in importance, partly due to the growing power of the Turkish pirates, partly because the main sea-lanes had shifted to the Atlantic.

E. Charles V (1516-56) and the Gradual Decline of Spain

Ferdinand of Aragon married his daughter Juana to Philip of Burgundy, and their child Charles V was born in Ghent (Fig. 108). The enormous territory which Charles inherited from his Spanish mother, his Burgundian father, and later from the Austrian Maximilian were in the long run the chief causes of the decline in the power of Spain. An Empire as unwieldy as that of Charles could only survive if it were wisely administered and supported by constantly renewed and enormous resources. The British Empire is greatly scattered, but it is based upon mutual trade-relations between the mother country and her dependents. Spain made no attempt to use this plan, which indeed was a growth of later centuries. The colonisation of America was in the long run good for the colonies, but it was not to the advantage of Spain. If Charles had succeeded only to his mother's realm Spain might have been strengthened by her new colonial possessions. But the treasures of America were squandered by Charles and his son Philip in a vain effort to maintain and even to expand their Empire in Europe. France was a stronger and more united nation than Spain, and was bitterly opposed to the growth of either Austria, Burgundy or Spain. Hence France continually attacked Charles, chiefly in the latter's distant northern possessions. At the end of his reign Charles had spent the riches of America, and had nothing to show for them except some small gains in Italy. Moreover, the Reformation developed just at this critical period and added immensely to Charles's difficulties.

Philip (1556-98) who succeeded Charles was an able King like his father, and probably better trained for government. He was, however, a bigoted Catholic, and his religious views led him to treat the Dutch with such harshness that they fought for nearly a century to gain their independence (p. 291). He had better fortune in Portugal whose King died in 1580 leaving no near heirs. Philip's mother was a Portuguese princess, and his occupation of Portugal was effected fairly peacefully. Since he was dealing with a Catholic country he governed wisely, but many Portuguese nobles were dissatisfied, and the union lasted only for sixty years.

In 1588 Philip sent the Invincible Armada to conquer England with a view to restoring the Catholic religion. His admiral was incompetent, and British ships and British storms destroyed the Spanish fleet. Philip's successors on the throne proved men of no character, indeed Charles II was an epileptic. Hence the power of Spain quickly diminished, and this rapid decline in a nation's importance is well worth our study.[1]

It is easy to pick out many reasons for this decline, but hard to arrange them in order of importance. The Spanish historian Altamira

[1]These sections are founded on C. E. Chapman, "A History of Spain," New York.

thinks that they were mainly economic. In effect he writes that Spain entered the modern age with weak economic energies. Its governments were led into an expansive and much too expensive imperialistic policy, but neglected those measures which could better the productive power and well-being of the country. For instance, war with Portugal was due to the imposition of heavier taxes, and led to the Duke of Braganza gaining the crown in 1640. France attacked Spain repeatedly from 1667 to 1697, and whittled away her northern territories. Moreover, Spain expelled over half a million Moors in 1609-14 who would not become Catholics. Popular objection to them was that they ate too little (and so paid few taxes for food), worked too hard and had too many children! Naturally agriculture declined still further. Of course many of the most energetic of the Spanish went to America and few returned. The same, however, could be said about the English about the same time, and yet England increased in power. Many of the Spanish disdained hard manual labour, which was in their opinion the work of slaves or Moriscoes. They preferred the more honourable profession of begging. By 1650 poverty was general in spite of the enormous conquests in America. Yet the "Golden Age" in Spain (*Siglo de Oro*) occurred just about this half century. The mental stimulus due to the glorious deeds in the New World and to the vast possessions in Europe persisted during the early stages of economic decline. Cervantes flourished late in the 16th century, while three of the greatest European artists, Velasquez, El Greco and Murillo, painted in Spain during the middle of the 17th century.

We can now summarise the major factors which have produced the Spanish nation. Isolation due to the Pyrenees has given her a Romance culture and language different from that of France. Her marginal position far from the original centres of civilisation did not much handicap her, since the Mediterranean climate and environment linked her so closely to Rome that Rome occupied Spain even before France. The distance from the centre of the Roman Empire, however, resulted in the country being occupied by Rome's Nordic enemies sooner than Britain or France. Her proximity to Africa led to the counter-current of Moslem invaders, who used the Gibraltar "Bridge" into Europe, just as later they used the Sicilian "Bridge" with less success and the Anatolian "Bridge" with greater success. We have seen that not much racial change occurred. Possibly the negro blood in South Portugal is more prevalent than any other non-Mediterranean strain.

The two rival cultures which developed in rugged Asturias and fertile Andalusia were kept from complete fusion by the semi-desert character of the Eastern Meseta. Moreover, the edges of the Meseta are difficult to negotiate even to-day, and trade and war entered Spain by well-defined routes. Roncesvalles was the key to Navarre, and its possession gave power to that small principality for several centuries. Gerona at the other end of the Pyrenees also commanded trade in and out of Catalonia. Nearer the Meseta the command of the Burgos Gate in the north led to many battles, and so also did control of the "Dog's Pass" near Linares in the south. The Moors, while they added to the intellectual growth of Spain, certainly hindered the national development for several centuries. But France was similarly compelled to spend much of her energy and resources in driving out the English during the 14th century, after the Moors had retired to Granada. Hence we must not lay too much stress on the Moorish occupation in this connection.

The Peninsula had an unrivalled position to take advantage of the new Atlantic trade. Prince Henry (1394-1460) placed his College of Seamanship on Cape St. Vincent at Sagres, where the Meseta juts out towards Brazil and West Africa. For nearly two centuries Spain and Portugal maintained their lead, but lost it owing to their lack of trading capacity. This in turn resulted from the lack of a middle class, due partly to a poor environment and partly to the lack of adequate encouragement of industry.

The separation of Portugal from Spain seems to have been rather accidental than inevitable. In the 12th century Portugal was no more different from Castile than was Catalonia, which spoke Catalan akin to Southern French. There was indeed not so much animosity between them, and Galicia still speaks Portuguese although part of Spain. The change of environment from the arid Meseta of Castile to Valencia, Murcia and Andalusia is less than that towards Portugal with its wetter lands, its forests and deep canyons. This in part explains why Castile turned to the southeast first, but the Portuguese had learnt to dislike the Spanish before the brief union under Philip. It seems obvious that but for the ambition of the Duchess of Braganza the two countries might easily have come together finally, since Philip and his successors treated the Portuguese well enough considering the times. Luckily the Spanish Kings did not persist in the practice of splitting their kingdoms among several heirs. This in itself would not necessarily have been evil, had not their heirs almost invariably proceeded to war to extend their small territories.

One outstanding feature of Spain is the unfavourable environment of Madrid. It has no natural features to commend it, for it is in a fairly level arid district without a river of any note. Moreover, the climate is continental: very cold in winter, and dry, dusty and hot in summer. It was used as a hunting resort by the early Kings, but as a town it dates from 1560 when Philip II placed his court there. Its position in the centre of Spain makes it an important railway junction to-day, but few manufactures (except perhaps leather goods) seem to be naturally sited there. Hence we may say that Spain developed in spite of the absence of a natural centre.

The actual date of the emergence of the Spanish Nation may be placed at 1479, and although this was far later than France or England, it was ahead of the union of many European nations, notably Germany and Italy. Spain had a lively history during the Napoleonic wars, but since these made little difference to the cultural or political distributions we cannot give them space in this necessarily rapid survey.

CHAPTER XIX

ITALY—DISUNION IN THE APENNINE PENINSULA

A. Mountain and Plain in Italy

In Chapter VIII the build of Italy has been described, since it has a considerable bearing on the rise of the Roman Republic. But the topography was not discussed in any detail, and no student of Italian history can ignore the important control exercised by the varied landscape on the evolution of the nation. The broader features are indicated in Fig. 100. The great arc of the Alps bounds the Lombard Plain on the west and north. Its structure is fully described in the chapter on Switzerland. The slopes of the Alps above Turin are very steep on the Italian side but descend much more gradually to France. Hence the migration of French into the Italian valleys (p. 329), and the ease with which French armies invaded the plains during the 10th and 19th centuries. In Fig. 101 the great barrier is shown as if stretched out in a straight line. We see at a glance the importance of the Rhone Graben on the west and the Pear Tree Pass on the east. They are far lower than any of the passes through the Alps. The chief passes are shown in the following list with their heights and routes.

TABLE U

PASSES OVER THE ALPS FROM ITALY LEADING WEST AND NORTH

Pass	Height	Route
Larche	6540	Cuneo to Barcellonette.
Genevra	6100	Susa to Briancon.
Cenis	6800	Susa to Modane.
Little St. Bernard	7060	Aosta to Moutiers.
Great St. Bernard	8120	Aosta to Martigny.
Simplon	6600	Domo Dossola to Brigue.
St. Gotthard	6890	Airolo to Andermatt.
Splugen	6950	Chiavenna to Chur.
Maloja	5940	Chiavenna to Engadine.
Reschen	4890	Bolzano (Bozen) to Engadine.
Brenner	4470	Bolzano to Innsbruck.
Pusterthal	4000	Bolzano to Lienz.
Tarvisio	2640	Udine to Villach.
Pear Tree	2900	Trieste to Laibach.

In the history of Italy these Alpine Passes have been of vital importance, from the crossing of Genevra in B.C. 218 by Hannibal to the many battles fought around the eastern end of the Alps in the Great War. The Brenner Area is so important that a fuller description is given on page 356.

The Apennine Ranges belong to the Dinaric series of Folds (see inset in Fig. 13) and commence near Savona. Possibly their roots may pass north and under the Lombard Plain to link up with the Dinaric Region east of Ivrea (Fig. 80). The overfolds of the main Apennines have moved across Italy from the west. Hence the ranges consist of an eastern not much folded series covered by two overfolds

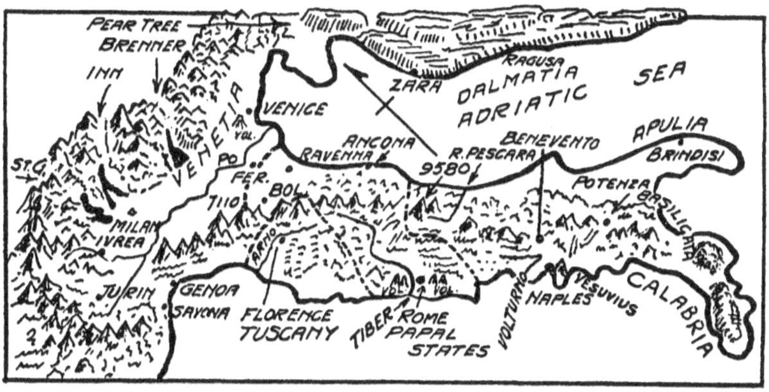

Fig. 100.—A Topographic Sketch of Italy showing the five Plains. Notice the Papal States astride the Peninsula, and Medieval Tuscany.

(Nappes). The Upper Ligurian Nappe runs along the coast from Genoa to Rome, but is quite low in the Arno Basin. The earlier Tuscan Nappe is more prominent and extends from Florence to Naples. In the far south in the "toe" of Italy are the ancient relic-blocks of the former Tyrrhenian Land (Fig. 35). They differ entirely in age and structure from the lately-folded Apennines. It should be remembered that the southern folds of the Apennines are perhaps the youngest in Europe, and are far later than the Alps proper (Fig. 36). Accompanying the folding are the volcanic rocks north and south of Rome, and the remarkable group of cones including Vesuvius near Naples. These have greatly affected settlement, since the lavas have usually improved the fertility of the soils.

The main divide of the Apennines runs from Genoa across the Peninsula towards Ancona (Fig. 100) in a fairly narrow belt with passes from 3,000 to 4,000 feet high. The range rises to 7,100 feet at Cimone, east of which is the main pass connecting Bologna to Florence. It is 2,000 feet above the sea and was the scene of several battles in the the 15th century. Indeed Pistoia on the southern slope near Florence, gave its name to the *pistol*, first made in this little iron-working town. One of the western folds of the ranges contains an altered limestone which has furnished the wonderful Carrara marble, the basis of the best statuary in Italy.

The central portion of the Apennines is much broader. The highest ridges keep near the Adriatic coast, and the upper portions of the Tiber and the Pescara flow in longitudinal valleys parallel to the folds. A broad plateau-like fold rises well above 3,000 feet and constitutes the

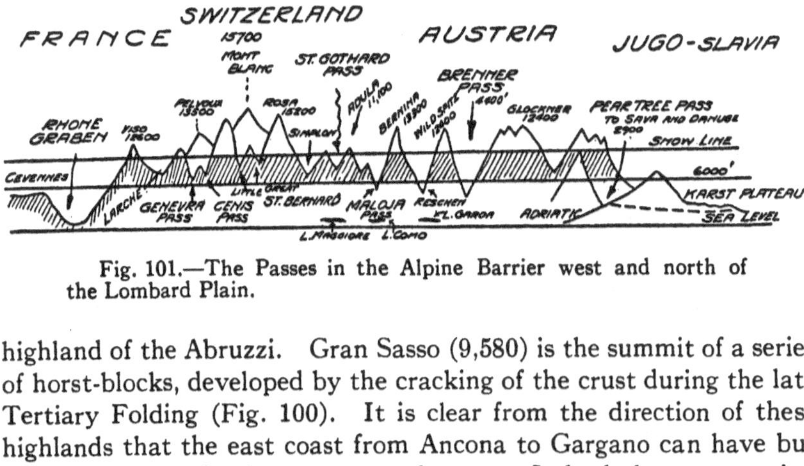

Fig. 101.—The Passes in the Alpine Barrier west and north of the Lombard Plain.

highland of the Abruzzi. Gran Sasso (9,580) is the summit of a series of horst-blocks, developed by the cracking of the crust during the late Tertiary Folding (Fig. 100). It is clear from the direction of these highlands that the east coast from Ancona to Gargano can have but few possibilities for important settlement. Indeed the construction of roads in the Mark of Ancona has always been difficult across the numberless steep gullies which seam the Apennines.

The west coast, apart from the lower valleys of the Arno and the Tiber, is also somewhat difficult as regards communications. Great flows of lava have blocked many of the rivers in the past, and they zigzag from basin to basin, now longitudinal, now transverse, in a fashion which shows that they have developed quite recently. Soil-wash from the hills, in part due to the ruthless cutting of the ancient forests, has filled in hollows and produced marshes in these coastal plains. Hence the wide prevalence of malaria in the Middle Ages, which contributed in no small degree to the decline in the prosperity of Italy.

The Arno is cut in somewhat soft Tertiary rocks, and so the plain around Florence is much larger than that occupied by the middle Tiber. Moreover, the folding and uplift on the west coast turned former marine gulfs into rock-rimmed plains, and provided a very varied series of capes and bays all down the west coast from Leghorn to Calabria. On the eastern side of the Apennines there is a linear coast formed of young Tertiary slopes the whole length, except for the fragment of Gargano which is really a bit of Dalmatia isolated by the foundering of the Adriatic Graben (Fig. 35).

There is a considerable change in the topography of the Apennines between Rome and Naples. They become lower and narrower just where the older fold mountains ended before the latest uplift (Fig. 36). Much of the peninsula northeast of Naples is either built of volcanic lava or of silts brought down by the Volturno River. Here also it was naturally relatively easy to cross the peninsula, and the town of Benevento commands this highway and was of great importance in Italian history. Just near here are marine Pliocene beds nearly 4,000 feet above the sea (Newbigin). Clearly this was all sea-floor just about the time that ape-man was in existence, in spite of its height to-day. A still lower area than that at Benevento is commanded by Potenza on the route from Naples to Taranto and Brindisi (Fig. 100). The Apennine folds die out south of the Basilicata, and the "toe" is made up of square relic-blocks which rise in Aspromonte to 6,420 feet.

The regions of Lombardy and Venetia are so important that we must now give them some attention. The general features of the topography are shown in Fig. 102, where the various changes in the national boundaries can also be compared with the main routes through the Alps. Generally speaking the Lombard Plain contains much the same area as the *former* Austrian Alps; but now, by the transfer of the Trentino (the central triangular area in the map) the Italians own not only all the plain but most of the mountains also. The outstanding feature of the Eastern Alps (in Fig. 106) is the oblique valley of the Inn, which has never been Italian. To the east is the branching valley of the Adige leading up to the three important passes, Reschen, Brenner and Pustertal. To the east are the longitudinal valleys of the Salzach and Drava (Fig. 102). Great mountain groups lie between these main routes, clustering around Adamello (11,600), Bernina (13,500), Gross Glockner (13,460), Marmolata (11,000), Ortler (12,800) and Wildspitz (12,400). (These are represented by their initials in Fig. 102.)

The waste torn from the Alps by the bygone glaciers and by modern rivers has been piled up in giant moraines at the mouths of the mountain valleys. Typical terminal moraines, sometimes 1,500 feet high, are shown in Fig. 102 south of Lakes Como and Garda. These lakes are in part cut out by the glaciers, and in part dammed by their

Fig. 102.—Topography, Boundaries, Soils, etc., in North Italy and the Tyrol. The central triangle (broken line) round Trent is the Trentino. The black branching valleys show where 230,000 Austrians live in Italy. The initials indicate the chief Peaks; P.T. is Pustertal. (Data partly from Newbigin.)

moraines. Surrounding the foothills is a broad zone of fluvio-glacial terraces, consisting of glacial debris re-distributed by the hundreds of rivers and streams draining the Alpine slopes. This coarse material does not produce a fertile soil, and is dissected into deep gullies. But along the outer edge the soil becomes finer and more fertile and is watered by many springs. Here (as Newbigin points out), the sites for settlement are much more favourable, so that a belt of ancient towns follows this line of springs and fertility. Milan was founded in such a position by ancient Keltic tribes, while Turin, Novara, Brescia, Verona, Vicenza and Udine have exactly similar reasons for their development. The central part of the Lombard Plain is covered with alluvium carried down to an enormous extent by every stream which enters the Po. Hence the plain is rapidly building its way into the Adriatic, and Adria and Ravenna (which were ports during historic times) are now miles back from the sea. Large boats can navigate the Po as high as Pavia, but the senile river is often changing its course, and in many places flows at a higher level than its flood-plain.

There is naturally a belt of towns just where the routes reach the plains, *i.e.*, well north of the line of springs mentioned previously. These towns were key positions, and have always been of importance in Italian history. Susa is at a height of 1,650 feet at the foot of the Genevra Pass and has thrice been destroyed, by Romans (312), Saracens (906) and Germans (1179). Ivrea was the capital of Berengar and Audouin, who played an important part in Italian history at the close of the tenth century (Fig. 102). Behind it lies the Val d'Aosta, where French is spoken owing to its early connection with Savoy. Curiously enough German has spilled over into the village of Gressoney, just south of Monte Rosa. Napoleon led his army down the Val d'Aosta from the Saint Bernard Pass in 1800. Lugano near the outlet of the Ticino from the Alps was the scene of much warfare early in the 16th century (p. 307). Again from 1846 to 1866 it was the headquarters of Mazzini during the Italian War of Independence. Como is built among the high moraines right at the foot of the lake of the same name. It has been engaged in war through all its history since it was founded by Rome in the 2nd century B.C. Gauls ravaged it till B.C. 50, the Milanese destroyed it in 1155, the Visconti won it in 1335, and the French armies attacked it in the 16th century.

Near Brescia were fought Lodi and Solferino (1859) together with many another battle, which give to this plain the name of the "Cockpit of Southern Europe." The famous Quadrilateral consisted of the four fortresses of Verona, Peschiera, Mantua and Legnano, all south of Lake Garda (Fig. 102). Austria thus controlled Lombardy, since Mantua is naturally defended by the broad Mincio River, while Verona commanded the main route from Austria *via* the Brenner Pass. Feltre and Udine are similar "outlet towns" to the east with warlike histories.

On the southern side of the Lombard Plain the mountains are much lower, the rivers much smaller and the rainfall lighter. Hence the gravel terraces and belt of alluvium are much narrower. Naturally, however, the main roads and railways traverse this belt, just where the gravels grade into the alluvium. In Roman times the Emilian Way led from Placentia (Piacenza) in an almost straight line to Ariminum (Rimini) on the Adriatic (Fig. 102). Parma and Bononia (Bologna) were even then important towns as they are to-day, but the main Flaminium Road to Rome led south from Rimini through Umbria and Spoleto, and not *via* Florence. The silts of the Po settle as soon as they reach the salt waters of the Adriatic. Hence a wide delta has developed, so much so that the old port of Adria is now 12 miles from the sea. Venice lost her trade when the Atlantic became the chief seaway, and its channels are too shallow to enable it to rival Trieste, now that the Suez Canal has resurrected Mediterranean trade.

B. Ethnical Distribution in Italy

In Chapter VIII some attention was given to the relations of the Gauls to the "Latins" and Etruscans, but the ethnical composition of modern Italy was not discussed. Let us follow Dixon in his survey of the migrations into Italy.[1] By the end of the Bronze Period the population of the northern half of Italy had become very much mixed. The old Mediterranean "dokephs" had been overlaid by and blended with Alpine peoples who had come into the Po valley from the north. The Iron Age was apparently marked by a recrudescence of dokeph types, perhaps partly Nordic. By the 6th century B.C. the Alpines had become the kernel of the Etruscans, and the allied Sabines entered into that fusion of tribes which took part in the founding of Rome; where, as the Patrician Class, they formed the aristocracy of the Roman people. There was thus a racial factor in common between the early Romans and the Etruscans. There seems to have been a complete change by the first century A.D., for the Roman crania from Pompeii are dominantly Alpine. This change is to be explained by immigration of plebs from the north. Even the later Etruscans had become brakeph also.

The invasions of the Goths and other Nordics seem for a time to have affected the Italian population; for a series of skulls from Aquilea in Venetia dated between the 8th and the 12th centuries, shows dokeph values as high as 40 per cent. But by the 14th century a similar series only shows 15 per cent. as belonging to the (possibly)

[1]R. Dixon, "Racial History of Man," New York, 1923.

Nordic class. In modern times the crania from Venetia, Umbria and Tuscany exhibit marked Alpine characters. It is worth noting that these folk have lived for centuries in the low-lying plain of the Po, though some writers state that Alpines are brakeph because they live in *mountain* regions!

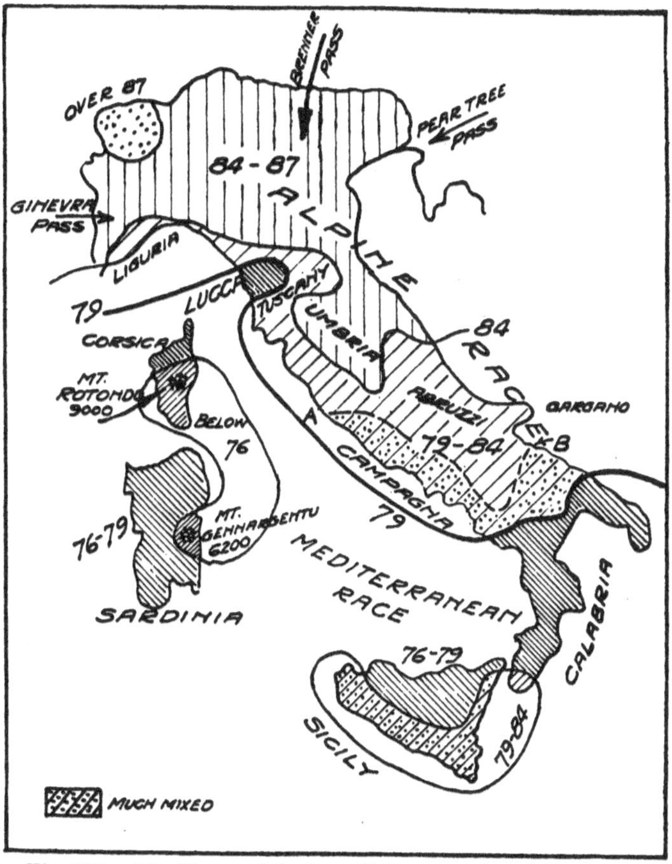

Fig. 103.—The Races of Italy. Alpines are found in the north and Mediterraneans in the south. More than 54 per cent. are brunette south of the line *A B*. (Based on Ripley.)

Ripley was unable to find any clear indication in the present population of the Nordic migrations. It has been stated that the Lombard Conquest was due to 60,000 Lombards assisted by 20,000 Saxons. If this be true, and the writer rather doubts this low number, then one would hardly expect much effect on the millions of Romans. In

Lombardy (Fig. 103) and Milan the people are rather tall for Western Alpines, but Ripley does not think that this indicates Gothic or Lombard blood so much as the presence of an Illyrian strain from Dalmatia. The latter are unusually tall for Alpines. In Savoy the hair is rather lighter than in the rest of North Italy. Perhaps this indicates a survival of the Gothic strain, though it is not apparent in the shape of the head for they are unusually broad (Fig. 103). The northern Alpines have spread through the passes from the Lombard Plain into Umbria, as the map shows, and here the average head index is over 84.

A long transition belt extends from Tuscany to Gargano, where Alpines seem to have overwhelmed the earlier Mediterranean stock, but in the hills north of Lucca, protected by the higher mountains of the Apennines around Cimone, is a nest of short dark dokephs of the primitive Mediterranean type. These folk also survive in Calabria and Apulia and in the north of Sicily. In the south of Sicily, perhaps the most conquered piece of Europe, where Carthaginian, Greek, Goth, Vandal, Saracen, Lombard, Norman, Frenchman and Spaniard have all had their day, it is natural to find another transition area, neither typically Alpine nor Mediterranean.

In the two rather distant islands of Sardinia and Corsica we should expect to find the Mediterranean type less altered by mixture than in any other part of Europe (Fig. 103). Thus the Sardinian speech has a strong resemblance to Latin, of which the Sardinians have preserved to some extent the endings and the vowels. In the mountain districts they still wear the goat-skin mantle described by Cicero. Their legs are covered by a short kilt, and the calves by leather gaiters (*vide* Deecke). Stature is remarkably low; but in Ripley's opinion this is due to emigration having removed the larger people. There are also many negroid characteristics, such as flattened nose, open nostrils, thick lips, retreating forehead and a long narrow face. "Imagine the black hair and eyes, with a stature scarcely above five feet, and a very un-European appearance is presented."

C. *Disunion in Italy from 900 to 1500*

Few stories of national growth are so difficult to interpret as that of medieval and modern Italy. With the glorious tradition of Rome to inspire her one might have imagined that Italy would have reached unity earlier than most European nations. Yet it was practically the last of all, and was subjected to more foreign invasions and to greater dismemberment than any other of the important nations of Europe. In earlier chapters we have read how the Ostrogoths succeeded the old Roman Empire. In 552 Byzantium conquered Italy; and Ravenna, well-protected by the marshes of the Po, became the capital for two centuries (Fig. 41). Then the Lombards overthrew Byzantine rule in most of Italy, but the Bishop of Rome successfully withstood them, until at his request the Franks conquered the Lombards (p. 174). Charlemagne and his descendants have been described in earlier pages. The crowning of the Saxon Emperor Otto in Rome (962) again linked north Italy closely with Germany (Fig. 92).

In the south, however, developments were different throughout Italian history, and indeed it is only by realising what are the "natural regions" of Italy that any clear picture of the evolution of the nation can be gained. Throughout later centuries there have been five main historical regions in Italy, which agree in part with the natural regions. These five are Naples, the Papal States, Tuscany, Venetia and Piedmont (Turin). Considering these in terms of topography, two belong to the basin of the Po, and three to the Apennine Peninsula (Fig. 100). The latter region may be divided physically into sub-regions as follows: the basin of the Arno, constituting Tuscany, the basin of the Tiber around Rome, and the two southern plains, those of Naples and Apulia. The mountain backbone and the "toe" of Calabria are obviously of little value for important settlement.

The geographer, considering only the build of Italy, would expect to see a homogeneous nation gradually develop in the basin of the Po with a capital at Milan (somewhat as Mazzini suggested), and possibly a second nation in the Peninsula with a capital at Rome. These if left to normal developments would in turn have united to form the Italian nation. But there has always been a unique factor in Italian history which has blocked normal development. Although as we shall see there were many factors which tended to keep Italy a group of independent states, yet (in the words of H. M. Vernon)[2] "these difficulties would not have been insuperable had it not been for the existence in the midst of Italy of the Papacy, which could not absorb and would not be absorbed. Indeed Machiavelli went to the root of the matter when he declared that it was the Papacy which had always kept Italy divided and still continued to do so."

Keeping this barrier of the Papal States in mind (Fig. 100) it is easier to understand the piecemeal history of Italy from the time of

[2] "Italy, 1494-1790," London, 1909.

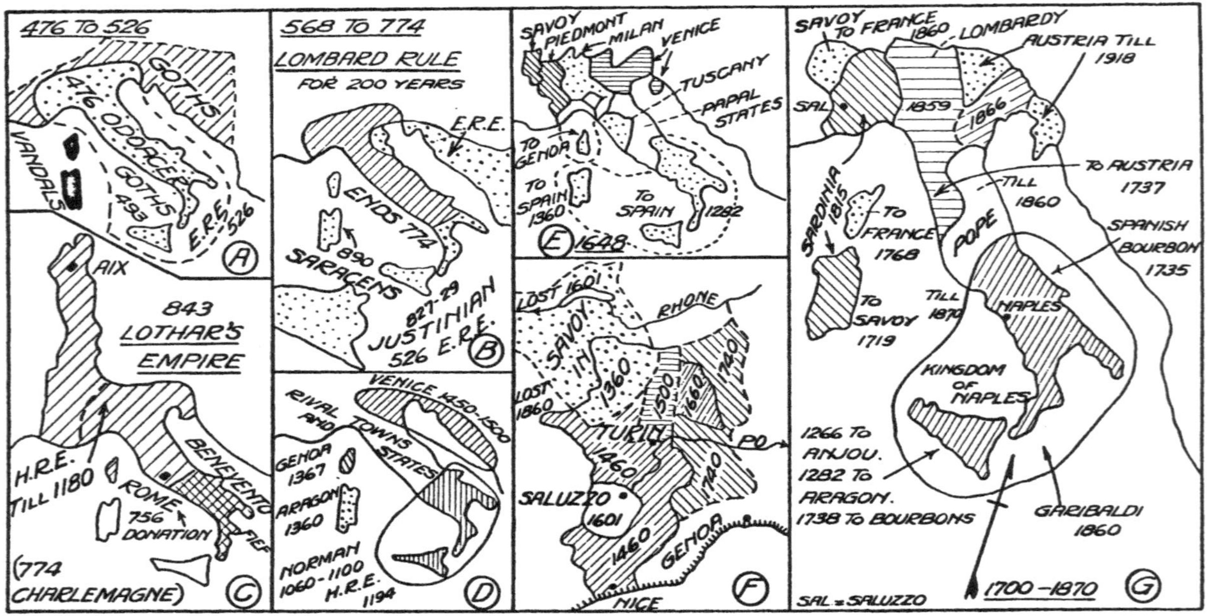

Fig. 104.—Political changes in Italy from 476 to 1870. Lombard rule ended in 774. The gradual expansion of Savoy is shown at *F*. The final Unification is illustrated at *G*.

Charlemagne onwards. We may consider first the course of events in Naples and Sicily. During the ninth century the Saracens (Arabs) from Africa gradually conquered Sicily, Sardinia and part of South Italy (Fig. 104). But the Byzantines still maintained control of parts of Apulia, and they drove out the Saracens from Calabria about 890. All this region became so Byzantine in culture that part of it is known as Basilicata (Greek for "Imperial") to this day.

Norman raiders appeared in southern Italy about 1017 (Fig. 104), and in 1035 the Duchess of Amalfi married the Norman Rainulf. The Normans gradually obtained a strong footing in Italy, and they fought and captured Pope Leo IX in 1053. Robert Guiscard became Count of Apulia in 1057, and a year or two later he captured Bari from the Byzantines who had held it since 876. In 1077 the Normans occupied the last of the Lombard territory, which had lasted here 300 years after the Frankish victory in north Italy.

In Sicily Roger Guiscard drove out the Arabs and founded a Norman dynasty which ruled successfully for several generations. Naples and Sicily were united in 1061. Roger II was a particularly enlightened monarch who ruled from 1112 to 1154. But the Nordics seemed to deteriorate under the southern skies and the last of the Normans was Constance who married Henry VI of the Hohenstaufen line in 1195.

Frederic II was the son of Constance and his struggles with the Pope have already been recounted. On his death the Pope, following the practice of his predecessors, invoked French aid to get rid of his enemies. Charles of Anjou was successful in defeating Manfred (son of Frederic II), and Angevin rule on a feudal basis was established in Naples and Sicily in 1266. Sicily, however, revolted in 1282 against the French, and the King of Aragon took advantage of the conflict to add Sicily to his dominions. The history of Angevin Naples has little of interest, save the alliance by marriage with Hungary. For a time the enlightened Ladislaus (1390 to 1414) seemed likely to further the union of Italy, but the country was conquered by Alphonso of Aragon (and Sicily) early in the 15th century. In 1516 the whole realm was inherited by the Hapsburgs.

The development of the Papal States forms a very interesting chapter of political history. The transfer of the Roman capital to Constantinople tended to separate the Christian Church into two divisions, and this has been discussed in Chapter X. Pepin and Charlemagne confirmed the authority of the Pope over the lands around Rome. During the 10th and 11th centuries many of the Popes led most immoral lives, which greatly reduced their spiritual authority. In 1278 Rudolf of Hapsburg, in return for the support of the Pope, acknowledged the latter's rights over Romagna, Umbria and Ancona, as well as the original Patrimony near Rome. Reference to Fig. 100 will show that the limits of the Papal States have no relation to the topography. They really express the desire of the Pope to maintain his hold on the old Flaminian Way and on Ravenna, the old capital and the headquarters of the Eastern Roman Empire in Italy. However, no ruler dared to dispossess the Pope until the time of Napoleon, while the Pope had not enough power in arms or resources to undertake the unification of Italy himself. As regards the Unity of Italy, therefore, the early Popes remind one of the "Dog in the Manger" of the fable.

In 1294 Boniface VIII became Pope and his occupation of the Papal throne was marked by constant conflict with Philip of France. In 1302 he threatened France with excommunication. In return French soldiers took him prisoner, and a few months later he died. A French Archbishop was chosen to succeed him, and he decided to stay in France. Thus for seventy years after 1309 the Popes lived at Avignon in the south of France. This transfer from Rome greatly damaged the authority of the Pope in the eyes of the common people. Gregory VII returned to Rome in 1377, but in 1378 the French Cardinals elected a French Pope who dwelt at Avignon, while another Pope was appointed at Rome. In 1409 there were actually three Popes at once, but in 1414 the Emperor summoned a Council at Constance which put an end to this absurdity.

The various regions north of the Papal States had for a time a common development. In the warlike times of the 10th century, the Marches of the Empire tended to become independent as has been narrated. Tuscany and Piedmont were two such areas whose Margraves paid little attention to the desires of the Emperor. But even more significant was the growth in power of the burghers, which followed along the lines already discussed regarding Flanders. The wealth of towns like Milan, Venice, Verona and Florence gradually became very great, as trade spread throughout the Mediterranean world. The exactions of the Emperor Barbarossa led to the formation of the Lombard League. Helped by the Pope the towns succeeded in defeating the Emperor in 1176. The Tuscan towns soon formed a similar league and set up various communes, in spite of the fact that the Duchess of Tuscany in 1115 left her lands to the Pope.

The 13th century in Italy was marked by almost continuous strife. Burghers fought nobles in each town, while the towns as a whole fought each other. At last the state of affairs was so evil that the towns placed themselves in the hands of capable leaders who soon developed into the so-called Tyrants. Thus the family of Este obtained control in Ferrara, the Visconti ruled Milan and the Della Scala were in Verona (Fig. 104). By 1400 the Visconti had succeeded in conquering almost one-third of Italy. Possibly they might have gone further and brought about a United Italy under a Visconti Dynasty, but for the violent jealousies of the various towns and districts, which nothing could bridle. Sforza a famous Condottiere married a Visconti and gave prosperity to Milan during the latter half of the 13th century. Ludovic Sforza married an Este, and for a time controlled politics in North Italy. But he antagonised too many powerful enemies, and his plans to increase the power of Milan came to nought.

Tuscany offers a specially interesting study about this time, since it illustrates how financial ability acquires power, even in medieval times. By degrees Florence, the natural centre of the basin of the Arno, obtained the leadership in Tuscany. Ostensibly it was a Republic, which controlled rather arbitrarily a number of other smaller Town-Republics in Tuscany. Little by little the power came into the hands of the Medici family. These were bankers who had financial dealings throughout much of western Europe. They had no official position, but like a "Boss" in a modern American district, had most of the authority. The Medicis greatly encouraged Art and Literature, and were unrivalled diplomatists. The Medici Family flourished from 1400 till 1737; for though they were expelled in 1494, they were reinstated and finally became Dukes of Tuscany. Two Popes, Leo X and Clement VII, were members of the same family.

Venice is, however, the most remarkable of the independent units into which Italy was divided at this period. The town was founded in the 6th century in the remote marshes at the edge of the Lombard Plain, by refugees from the Nordic invaders. It was of little importance till the time of the Crusades, when its situation gave it an advantage as a port for the troops from Germany and northwest Europe. Venice naturally participated in the trade with the Levant; and the conflicts in the south of Italy between the Normans, Arabs and Byzantines benefited indirectly northern ports like Venice, Pisa and Genoa.

In 1154 she was recognised as Mistress of the Adriatic, and had developed a powerful navy. Her rival Genoa was an ally of the Byzantines at Constantinople. Hence Venice played an important part in the so-called Fourth Crusade, which resulted chiefly in the capture of *Christian* Constantinople in 1204 for the benefit of Venice! She placed her soldiers in Crete, Cyprus, the Morea (Greece) and all round the Aegean. Her government was remarkably successful, being an oligarchy of talented nobles. A secret Council of Ten carried out the plans of the Republic swiftly and ruthlessly. There was no hereditary ruler and no single family gained all the power as in other Italian States. The cities under Venetian rule were prosperous and in general peaceful. Yet the jealousy of the neighbouring states led to the formation of a League against Venice, which stripped her of many of her possessions in 1509.

We can now sum up some of the factors which prevented any developments of Italian unity in the period 962 to 1500. At the earlier date two-thirds of Italy was under the control of the "Holy Roman Emperor." The latter, however, could not control German Nobles and Italian communes in the face of the constant opposition of the Pope. The real separation of Germany and Italy occurred after Legnano in 1176. But there seems no very good reason why the Italian States should not have grown closer together as did the Swiss provinces after their separation from the Empire. If we grant that the Pope prevented any union between north and south Italy, which is probably true, then it is not obvious why north Italy did not come together. The Lombard plain is a natural unit with a centre at Milan or Mantua, from which Piedmont, Lombardy, Venetia and Tuscany might well have been administered. No doubt the Venetian interests in trade with distant lands were different from those of Piedmont, which had no coastal trade. Genoa's lands for the most part lay along the coast in a narrow strip south of the Apennines, and she was not politically associated with Milan or Turin. Religion, Language, Common Race and Tradition all linked these northern provinces. Possibly (as Vernon states) the decline of a moral sense, of a love of freedom, and of a love of arms which marked the dawn of the 16th century accounts for their failure as a nation. Luxury and pleasure were counted the chief assets and military prowess was left to mercenaries. During the next four centuries Italy paid dearly for her neglect of the primitive virtues mentioned above.

D. *Foreign Domination from 1500 Onwards*

Much of the national story about 1500 has been indicated in the chapters on the Renaissance and the Reformation. The end of this great period of intellectual activity was mainly due to the prevalence of foreign invasions from 1527 onwards. It is true that there was much civil strife in the earlier period, but many cities kept aloof from such struggles, and the small Italian armies did not ravage and destroy whole provinces as was the case with the later French, Spanish, German and Austrian invasions. "Men who walk daily in distress and danger cannot give themselves freely to Art" (Vernon).

After Venice was crushed in 1509 by a league of envious States aided by France, struggles for domination continued for decades between the French and the Spanish. In 1512 the Pope drove the French out of Italy by the help of the Swiss, and Spain extended her influence by restoring the Medici in Tuscany. In 1515 Francis of France defeated the Swiss and invaded Lombardy. However, in 1525 Francis was captured by Charles V of Spain after a fruitless attack at Pavia. In 1527 occurred the disgraceful sack of Rome by the German troops of Charles. Once again Francis invaded unhappy Lombardy now aided by the Pope and Genoan ships. He was again unsuccessful, and the chief result was that Charles was crowned King of Lombardy. Henceforth France had little to do with any Italian affairs (except in Savoy) for nearly 300 years.

It was in Savoy that the chief change in the national parties developed. This province dated back to Humbert of Savoy who controlled the three chief passes near Mont Blanc as early as 1034. His descendants now rule Italy. By 1060 Savoy had obtained Turin by marriage and soon controlled most of Piedmont. Savoy lay between Spanish Milan, France and Spanish Burgundy. The Duke of Savoy built fortresses in the Alps and obtained allies among the Swiss. Gradually he extended his lands in the Italian sphere, sometimes as in 1601 by exchanging districts near the Rhone for Saluzzo south of Turin (Fig. 104).

The conflict between Venice and the Turks became more bitter as the Turkish power increased. By 1500 they owned all the Venetian possessions except Crete, Cyprus and some towns in Dalmatia. In 1570 Turkey had a navy of 300 vessels and could readily replace what she lost, which was not the case with Venice. In 1570 the Sultan captured Cyprus, but this was balanced by his defeat off Greece at

Lepanto, where Don John of Austria and the Venetians crushed the Turkish Fleet in 1571. Crete, however, was now attacked and Venice was unable to enlist help, for England and Holland were her trade rivals. In 1648 Mocenigo was Governor of Candia in Crete, and he held the town against the Turks during a blockade of over twenty years. In the last three years of the struggle 100,000 Turks were killed in abortive attacks. In 1687 Greece was recaptured by Venice, but the Turks had gained all but Corfu by 1718. It was, however, the courage shown by Venice in these gallant fights against the enormous resources of the Turks which gained for Venice the admiration of Europe. Partly for this reason the Venetian Republic lasted long after most of her early rivals in Italy.

On the whole Italy had been fairly peaceful from 1590 to 1690, but during the next 50 years she again became a theatre of war. Venice lost more and more power, and the Papal States became of little significance, but Savoy now came quickly to the front. The Duke used his position athwart the Alps to bargain with France and with Austria. After a number of defeats Eugene of Austria (and Savoy) attacked Milan in 1706 and thus ended 150 years of Spanish misrule. In 1712 the Duke of Savoy was given a place among the Great Powers at the Peace of Utrecht, at which he was granted the Kingdom of Sicily. A little later Austria received Sardinia, and from this time the Austrian Hapsburgs replaced the Spanish as the controlling factor in Italy. A few years later Savoy was compelled to exchange Sicily for Sardinia, and the Duke of Savoy now became King of Sardinia (Fig. 104). A Spanish army placed the Spanish Bourbons on the throne of Naples in 1735, but they remained independent of Spain. Corsica was purchased from Genoa by the French in 1768.

The Revolution in France antagonised the Pope, who formed a coalition against the French, but in 1798 a French army carried him off to France. Napoleon entirely altered the map of Italy for a short time. He placed his favourites on the thrones of Milan and Naples and annexed the States of the Church to France in 1809. The Treaty of Vienna in 1815 returned the regions to their original rulers for the most part, and paid no attention to the desires of the people. The Pope was dependent on Austrian help at first, and later, on French regiments stationed at Rome. The revolutions of 1830 and 1848, chiefly directed against Austrian control, were abortive. Indeed the Sardinian army was crushed by Radetsky in 1849.

In the Crimean War in 1855 Sardinia, on Cavour's advice, gave important help to the French, and in return the French agreed to help

Sardinia against Austria. The Austrians were defeated at Solferino and Magenta in 1859 and finally driven out of Lombardy. However, Napoleon III gained Savoy in return for his military assistance. Thus Sardinia had now lost the ancestral lands of the early Dukes of Savoy. The Tuscans soon revolted against their rulers and joined Sardinia, while Garibaldi successfully aided the rebels in Sicily and Naples, which followed suit in 1860. Soon only Venetia and the Papal Patrimony remained outside United Italy. When Prussia declared war on Austria in 1866 Italy joined the former. Though the Italians lost their battles, Austria was crushed in the north at Sadowa (p. 380), and Venetia was given to Italy. King Victor Emanuel waited until Napoleon III withdrew his army from Rome to fight Prussia, and then he occupied the city. Rome became the capital of a fully united Italy in 1870.

We have no space to discuss the part Italy played in the Great War. However, the control of the Trentino was one of the aims of Italy, and the national problems here are so interesting that they deserve a paragraph in this brief survey. The Trentino is the Italian portion of the main highway through the Alps from Italy to the Brenner and Reschen Passes (Fig. 102). It was the home of the Tridentini in Roman times, hence the name (Trent) of the only large town. In 1027 the German Emperor Conrad II gave it to the Bishop of Trent, who held it till 1803, in spite of many attacks by the Venetians. Then it passed to Austria and it was annexed to the Tyrol in 1814. Hence it is difficult to see that Italy has any good *political* claim to this region. However, south of Bolzano (Bozen) the population in the Austrian area before the war was exclusively Italian, and this has always been true of the town of Trent itself.

As a result of the war the boundary was placed at the Brenner, so that the Italians now hold command of the main divide. By this transfer a large population of some 230,000 Austrians is now compelled to live under the Italian flag. Their villages are in the three main route valleys leading to the Brenner, Reschen and Pustertal Passes. Thus has been produced an Austrian Irredenta which is likely to be a danger-spot for many decades. With Austria so reduced in power as to be almost negligible it seems unnecessary to confine the western Tyrol to the Lower Inn valley. A line from the Ortler through the village of Salzano to the Marmolata would sharply separate the Italian and German speakers (Fig. 102). North Italy was harshly treated by Austria for generations, and like most other nations Italy has not yet learnt the practical lesson that it *pays* to be magnanimous!

CHAPTER XX

AUSTRIA—THE CRUMBLING OF AN EMPIRE

A. The Build of Present-Day Austria

It is somewhat difficult to know what geographical limits to assign to our discussion of the Rise and Fall of the Austrian Empire. Austria of to-day is so insignificant in comparison with its proud position up till 1914 that it seems absurd to include Transylvanian topography or the Dalmatian coastline in connection with the "Build" of Austria. But there is, as we shall see, a certain geographic unity about the old Empire which helped to bring the areas under the Hapsburg rule—and which may again bring about an economic union, if not something closer, among the former members of the Empire.

The Block Diagram shown in Fig. 105 gives us a bird's-eye view of the "Young Folds" of the Carpathian Arc, together with the Bohemian Basin and lands adjacent to these. The dominant line of communication is clearly the River Danube, which in our sketch rises at the top of the drawing in Bavaria and flows into the Black Sea at the foot of the drawing. It is very significant that the former Austrian Empire included almost the same area, *i.e.*, the Danube from Passau to Orshova, the Bohemian Basin and all the inner (southern) slopes of the Carpathians. Not shown in the diagram, and of much less importance, are the southeast Alps and Istria (now Italian), and Bosnia and Dalmatia which are now Yugo-Slav. Since succeeding chapters will deal with the growth of Bohemia, Hungary, Rumania and Yugo-Slavia, only the general geographical aspects of these countries will be here discussed.

Austria to-day consists of two very different regions, the Alpine section and the Danube section. The build of the former has been referred to in describing Fig. 80. Essentially Alpine Austria consists of the three overfolds of the eastern Alps, while underneath all three are the Pennine Nappes. The latter show through the eroded coverings of later folds in the Engadine and Tauern "Windows" (Fig. 80). Flung over all the others and furthest to the north are the Calcareous Alps which appear in a zone north of Innsbruck (Fig. 106). The Schistose Alps are below these in order, but on the surface (owing to erosion) occupy most of the area between the middle Inn and the upper Salzach.

The central part of the overfolds consists of the Crystalline Alps, which carry the high glacier-bearing peaks east of the Brenner. On the southern side of the divide the folds appear in reverse order, but since the boundary of Austria runs along the Schistose Alps, the Calcareous Alps of the south slopes are all in Italy. (See Inset in Fig. 106.)

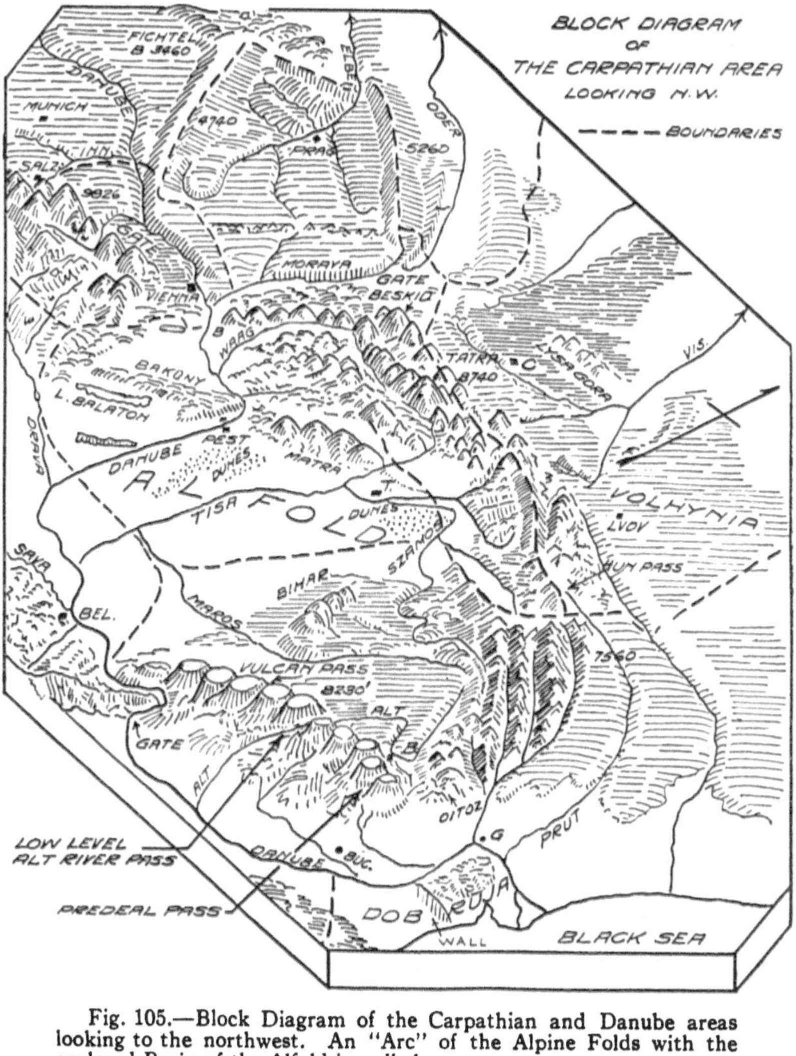

Fig. 105.—Block Diagram of the Carpathian and Danube areas looking to the northwest. An "Arc" of the Alpine Folds with the enclosed Basin of the Alfold is well shown.

According to De Martonne's map[1] the main rivers have eroded their valleys along the boundaries of these major fold-regions, while the passes and lakes are also largely determined by the build. Thus the moraine hills of the Bavarian Downfold (p. 260) contain the large Chiem Lake. The Alter and Trauen Lakes (both east of Salzburg) are cut out of the softer Flysch deposits. The small but very picturesque lakes of Hallstatt and Ischl are eroded in the Calcareous Alps. The main central mass of schistose and crystalline Alps is free from lakes, but carries many cirque-cut crests which become higher (Gross Glockner, 12,464) to the east of the Brenner and are fringed by fairly

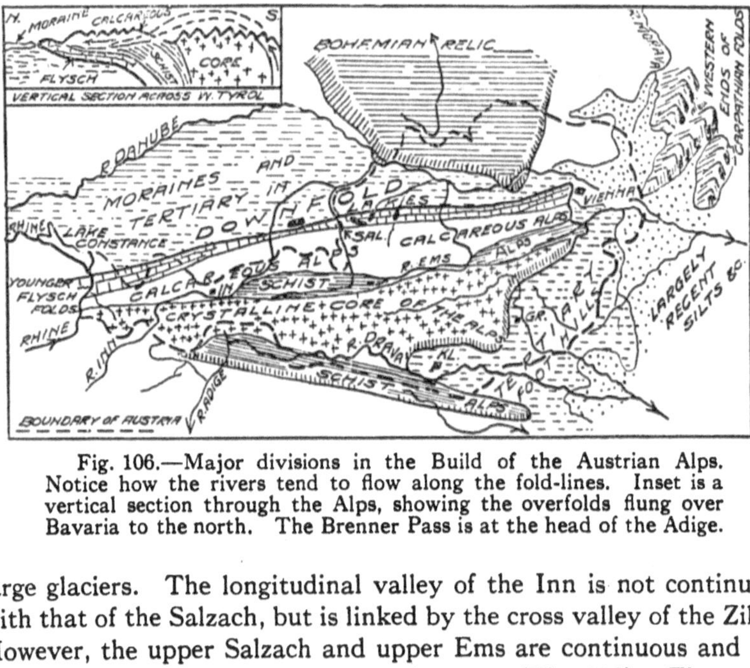

Fig. 106.—Major divisions in the Build of the Austrian Alps. Notice how the rivers tend to flow along the fold-lines. Inset is a vertical section through the Alps, showing the overfolds flung over Bavaria to the north. The Brenner Pass is at the head of the Adige.

large glaciers. The longitudinal valley of the Inn is not continuous with that of the Salzach, but is linked by the cross valley of the Ziller. However, the upper Salzach and upper Ems are continuous and run along the north edge of the crystalline core (Fig. 106). The upper Mur and the Drava, while longitudinal, are wholly contained in the crystallines. On the south the border of the latter is marked by the Pustertal and its extension (to the east) in the Gail, a tributary of the Drava.

The main passes, along the upper Inn to the Maloja Pass, and over the Brenner Pass, are seen in Fig. 106 to be related to the presence of

[1] "Europe Central," Paris, 1931, p. 465.

the schistose fold-rocks in those regions which are more easily eroded than the crystallines. The Brenner Pass is just where the hard crystallines are at their narrowest. Glacial erosion has, of course, also helped to lower the divide in both these passes. In the Alps of Austria the upper slopes of the rugged mountains are only of use for pastures in the summer. The population lives mainly in the great longitudinal valleys and in the main cross valleys. Very little, therefore, of Alpine Austria has over 25 people per square kilometre except around Bregenz near Lake Constance, near Innsbruck, and between Klagenfurt and Graz. The habitable Tyrol is now limited almost to the longitudinal valley from Bregenz (*via* the Arlberg Pass) to the Inn Valley and so eastward along the Salzach and Ems. The upper Drava and upper Mur are less important regions of settlement in the eastern Alps. The eastern Alps are fairly rich in salt, especially near Salzburg and Hallstatt—as these names imply. Much iron ore has long been mined both northwest and southwest of Graz.

A second fairly well-defined area in Austria extends along the Danube between Passau and Vienna. Two smaller intermediate towns are Linz and Krems. This is the Vienna Gate (Fig. 112), between the Alps on the south and the Bohemian Horst (p. 394) on the north. Much of it is filled with late Tertiary or recent deposits laid down in the depression just at the foot of the Alpine folds. But a corner of the Bohemian Relic-Block is included in Austria, and, indeed, the Danube has cut a number of small gorges in this hard block—whereas it flows in a wide valley over the Tertiaries. No portion of this northern area is mountainous, and most of it west of Vienna is between one and three thousand feet. It is rolling woodland, occupied with rich farmlands. There is a good deal of water power to-day—and a few textile factories.

There is no natural feature to define the eastern boundary of Austria. As we shall see later, the German defenders of the Mark occupied the slopes of the Alps leading down to the great alluvial plains of Hungary. The Magyars never settled in the uplands, and the present boundary is based on ethnic grounds. Indeed, Austria was allotted Burgenland—a tract containing 218,000 people, mostly German-speakers—by the Treaty of Versailles. This gift from the Allies to an enemy country, it is hardly necessary to say, was made at the expense of another enemy, Hungary.

B. *Major Structural Features of the Former Austrian Empire*

To understand the main topographic features of the structure of the former Austrian Empire we must consult Fig. 35. Here the Danube is seen to enter the oval basin of the Alfold in the northwest and leave it at the southeast. The surface of the basin consists of silts, clays, sands and gravels brought by the numerous tributaries of the Danube. The margins of the basin consist of sloping beds of Mesozoic (4) or Tertiary (5) rocks, or else of ancient shield and core rocks as in the southeast. On the west of the Alfold Basin are the rocks of the Eastern Alps, already described. On the north and east is the fold of the Carpathians, which is like that of the Alps, but is less high and less eroded. For the most part the Tertiary rocks—which have been stripped from the Alps (if they ever covered them)—are still present in the Carpathians. They have been removed, however, in the southern part of the Carpathians—where the core is exposed just as in the Alps (see p. 436). The southwest side of the Alfold Basin consists of a similar fold system—the Dinarics. These consist of folded chalk and Trias rocks, from which apparently the Tertiary "mantle" (again, if it were ever present) has been eroded.

There is thus a very interesting series of four fold-ranges in the vicinity of the Austrian Empire, each differing from the others (Fig. 35). The oldest type is that shown in the Swiss Alps—where the granitic cores (2) are widely exposed and weather to gigantic peaks and pyramids. The surface rocks of the Austrian Alps are somewhat further from the *central* "core" of the overfolds—but here also no later mantle of secondary rocks covers them. The Dinarics do not show much of the core, and are fairly uniformly covered by a Mesozoic mantle (3). The Carpathians show the mantle of Tertiary rocks still fairly complete over most of the arc of buckled crust. Lastly, in the Apennines the latest Tertiary (Pliocene) rocks still mantle the folded crust in the south, though most of the mantle rocks are *early* Tertiary (5) in age. Since the Carpathians and Apennines are built up of much later rocks, their "profiles" are softer than those of the Alps, and the hill slopes erode much more readily than do the ancient granites of the Swiss Alps.

The heart of this large region is clearly the Alfold Basin, and we shall try to interpret its history in terms of this marked topography. The Austrian Mark controlled the Austrian Alps between the Vienna Gate and the Pear Tree Pass (p. 266). The Bohemians, after the 5th century, were firmly established in the Bohemian Basin to the west of the Yablunka Pass. The Serbs about the 6th century occupied Serbia (Fig. 111), which consisted mainly of the eastern flanks of the Dinarics and the Valley of the Morava and adjacent streams. But elsewhere along the whole Carpathian arc there was no permanent occupancy apart from certain small mountain districts where the Romano-Dacian culture survived. The Carpathians are probably younger and certainly less massive ranges than the Alps. Certain ancient rivers such as the Alt (and, of course, the Danube) have not been blocked by the Carpathian uplift, but have kept passes open through the young ranges. Hence the Carpathians did not act as a barrier to protect the Alfold Plain from invasions from the east and southeast (Fig. 12 at F).

In our present chapter we are dealing with an area of 264,000 square miles, which includes the following fairly distinct provinces inhabited by nearly 50 million people in 1910; Austria Proper (including Tyrol, Carniola, etc.), Bohemia, Moravia, Galicia, Bukovina and Dalmatia. The foregoing belonged to the Austrian Crown. Hungary, Transylvania, Croatia, Slavonia belonged to the Hungarian Crown; while Bosnia and Herzegovina were held in common. In addition to the regional diversity, there was an extraordinary national variety, which is indicated by the number of languages in use. About 20 million spoke Slav languages (Czech (6), Polish (4), Russian (4), Slovak (2), Serb (2), Croat (2), etc.), about eleven millions spoke German, nine millions Magyar, and three millions Rumanian. There were about two million Jews. It seems advisable, therefore, to study the development of the Hapsburg Territories first, and to show how they gradually dominated the Czechs, Magyars and other peoples. Then in later chapters the early history of these non-Hapsburg lands will be described, and related to the local environments.

C. *The Development of the Austrian Nation*

The Roman conquest of the Austrian Alps and the upper Danube was briefly described in Chapter VIII. In B.C. 15 Noricum (S.W. of Vienna) and Rhetia (Tyrol) were added to the Empire, and about 10 A.D. Octavius conquered Pannonia, *i.e.*, the western Alfold. The first two centuries of our era were the most prosperous in the Empire, but about A.D. 200 Suabians from Bavaria, etc., devastated Pannonia. Other Germans later attacked Milan, and Gallienus allowed them to settle in Pannonia. During the Folk-Wandering, Huns, Gepids, Avars and Lombards fought for mastery in the western part of the Alfold (p. 173). The Avars attacked Noricum about 563, and it is said they almost rooted out Christianity, so that it remained pagan for 250 years. About 700 many of the Bavarians became Christians, and Bishop Rupert founded the first church at Salzburg. During this century the Bavarians tried to break away from the Frankish rule, and in 766 they gained the Tyrol. In 768 Charlemagne began his victorious career. In 45 years he was engaged in 53 wars—during which he widely extended his dominions until he controlled all of Austria Proper (Fig. 92). In 795 he captured the Avar capital in the Alfold, and by 803 the Avar nation was completely disintegrated.

In 895 the Magyars, another fierce eastern nation, entered the Alfold from the northeast by the Vereczka (or Hun) Pass near the source of the Dniester (Fig. 105). During the next decade they raided all central Europe, from Bremen to Vesuvius; but in 955 they were crushed at Lechfeld by Otto the Great. About 1000 the name Ostarrichi came into use for the eastern part of Bavaria.[2]

During the next two centuries Austria passed into various hands, usually as a fief of the Holy Roman Empire. In 1136 Austria, Styria and Carniola were united to form a Duchy, and about this time Vienna, the old Roman Vindobona, began to become an important centre. In 1253 Ottakar of Bohemia bought Carinthia, and for a time it seemed likely that he would dominate the Holy Roman Empire (Fig. 113a). But a new dynasty was arising in the Castle of Hapsburg near the junction of the Aar and the Rhine (Fig. 107). Here had been the Roman military centre of Windisch (p. 303), and here in 1218 was born Rudolph of Hapsburg. The Hapsburgs also owned lands in Swabia (Bavaria), but their scions served in many fields. Thus, Rudolf was for a time under Frederic II in Sicily and later with Ottakar. He was a brilliant soldier and fought in many of the local wars, e.g. against the rulers of Strasbourg and Regensburg. He was a rival with Ottakar of Bohemia for the title of King of the Romans and to Ottakar's disgust Rudolf was elected. In 1276 Ottakar was compelled to yield the four Austrian provinces which he had seized. In 1278 Rudolf defeated Ottakar on the Marchfeld near Vienna and thereafter reigned successfully in Germany until his death in 1291. His son Albrecht became Emperor in 1298 and had a warlike career fighting against the Bohemians, Magyars and Swiss. His sons were given Austria,

[2]Sidney Whitman, "Austria," New York, 1898.

Styria and Carniola (Fig. 107). He was assassinated at Windisch and thereafter the Empire passed into other hands.

Albrecht II of Hapsburg inherited the feuds against Bohemia and Switzerland (p. 305), and acquired Carinthia and part of Carniola in 1335 for aid rendered to the Emperor Louis. His son Rudolf IV persuaded Margaret (of the Wide Mouth) to cede Tyrol to him in 1361 (Fig. 108). A later ruler, Albrecht V, became a strong friend of the Emperor Sigismund of Bohemia, helped him against the Hussites and

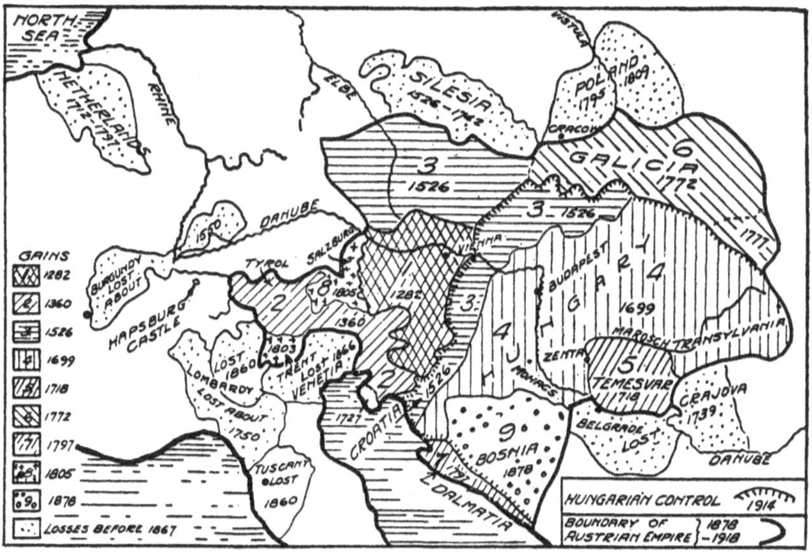

Fig. 107.—The Growth of the Hapsburg Empire. The Austrian March came into their hands about 1276. After the defeat of Mohacs in 1526, Bohemia and West Hungary joined Austria. The Turks were driven out in 1699.

in 1422 married the latter's daughter. Sigismund advised his nobles to make Albrecht his heir, and accordingly in 1438 Albrecht "the Magnanimous" became Emperor, and for a time the lands of Austria, Bohemia and Hungary were united. Thus the Imperial Crown came back to Hapsburg after 130 years. By this time the Turks were threatening southern Hungary, and Albrecht died in 1439 while marching to attack them.

It is impossible to explain clearly in such a brief survey as this the ramifications of the Hapsburg family (Fig. 108). Frederick (grandson of Albrecht II) had inherited the Tyrol, and his struggles with the

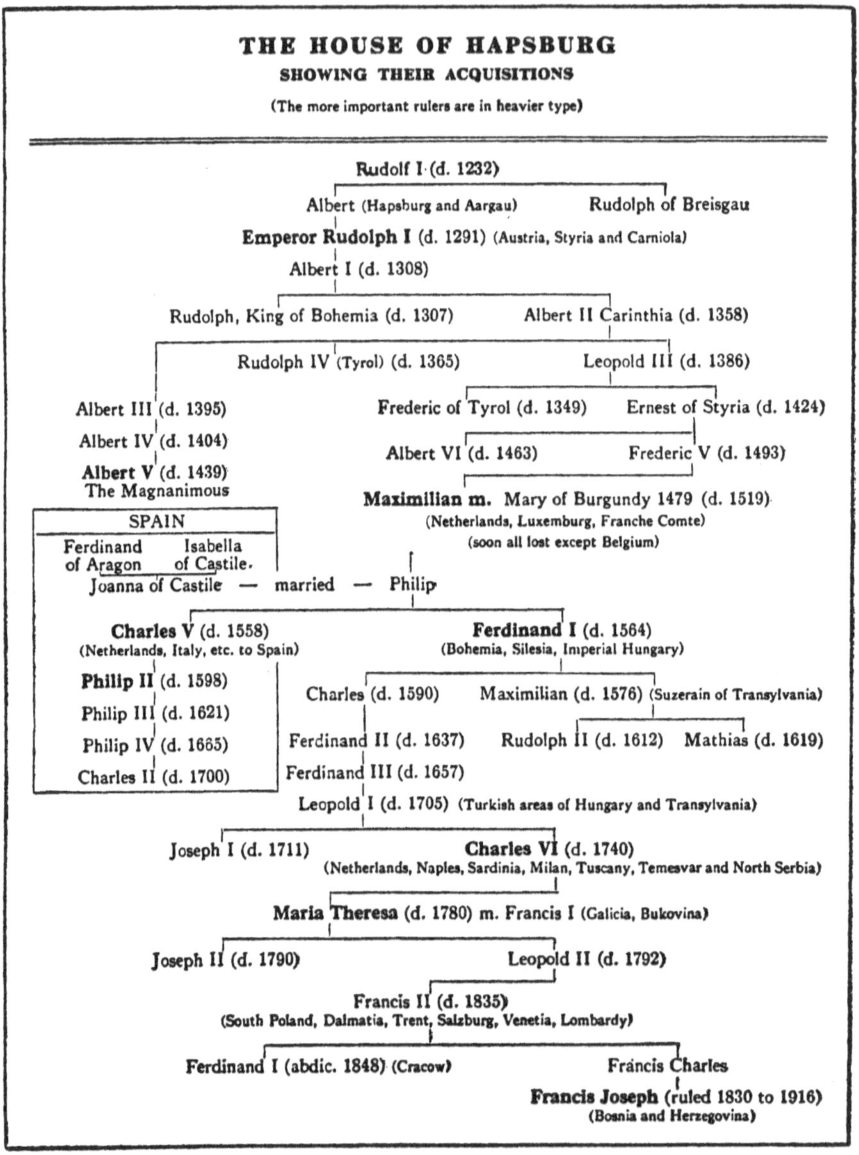

Fig. 108.—A Table showing the ramifications of the Hapsburg Dynasty, together with the territories added to Austria by the various rulers.

rising Swiss Confederation have been referred to (p. 302). In 1411 he also inherited his brother's lands in Swabia, but his line became extinct in 1496, while that of Albrecht had died out some 40 years earlier. Thus the sole Hapsburgs left were the Styrian line, and the leader thereof, Frederick, became Emperor in 1442. His was an unlucky reign spent fighting the Swiss, the Bohemians, the Hungarians, and his own brother Albert. Moreover, in 1473 the Turks invaded Carinthia, and in 1475 they actually passed the Alps and reached Salzburg. After 1491 he left the government largely to his energetic son, Maximilian—one of the most illustrious of the Hapsburg line. Frederick is credited with the epigram "Austria est imperare orbi universo;" and it was of him that the King of Hungary said "*Bella gerant alii; Tu, felix Austria, nube*" (Happy Austria, expanding by marriages rather than by wars).

At this stage we may pause for a moment to realise that the Hapsburgs during Frederick's reign owned much the same area as is now included in Austria (Fig. 107). His realm comprised the German-speaking parts of the Eastern Alps as well as the lower lands of Klagenfurt, Graz and of the Vienna Gate (cf. fig. 122). The chief difference was that Salzburg was under the control of the Church till 1805. All the remaining huge areas which were gradually acquired by the Hapsburgs were inhabited by "foreigners," by Czechs, Magyars, Rumanians, etc. All these lands fell from Austria at one stroke, when the Empire disintegrated in 1918.

It was during the lives of Maximilian and his descendants, Philip and Charles V, that the Hapsburg marriages brought them the largest accessions of territory. In 1477 Maximilian married Mary of Burgundy, daughter of Charles the Bold. This gained for Austria most of the Netherlands, Luxemburg and the eastern part of Burgundy (Franche Comte). But his later attempts to combat the aggressions of France were for the most part unsuccessful, and later France managed to acquire Luxemburg and Franche Comte. He made war on Bavaria to support his candidate for the succession with better success. He married his son Philip to Joanna of Castile and his grandson Ferdinand to Anne (of Hungary and Bohemia). Maximilian died in 1519 and his grandson, Charles V, became Emperor of the Holy Roman Empire, as well as King of Spain (Fig. 108). He did not, however, obtain the Austrian possessions, which were held by his brother Ferdinand. For a time, therefore, there was a brief break between Austria and the Empire in that they were not controlled by the same ruler. But on Charles's retirement Ferdinand in 1556 became Head of the Holy Roman Empire, while Spain now remained entirely dissociated under Philip II. Charles V spent his time trying to control his vast but weakly-united Empire. Troubles in Spain and the Netherlands prevented him from crushing the Protestant forces in Germany (p. 291).

D. The Expansion of the Empire into Magyar, Slav, Rumanian and Polish Lands

During Ferdinand's reign over Austria enormous areas were added to the Hapsburg territories. This was due to the death of Louis (King of Bohemia and Hungary) at the battle of Mohacs (1526) in a vain attempt to prevent the Turkish invasion. Ferdinand claimed both kingdoms as husband of Louis' sister. He succeeded in Bohemia, Silesia and Moravia, but a rival claimant, John of Zapolya (in Transylvania) gained most of Hungary as a vassal of the Sultan (Fig. 107). For nearly two centuries Hungary was divided into an elongated Imperial region on the west, which was controlled by Austria, and the main portion of the Alfold mostly under control of the Turks. The line of division during the early 17th century is an interesting one, for it runs from the highlands of Zagreb along the Bakony ridge (Fig. 105) to the Matra Mountains across the Danube. Thus the well-defined Hungarian plain between the Bakony ridge and Pressburg was thenceforward in Austrian hands—though the Turks attacked Vienna itself in 1529. Transylvania remained Hungarian and for the most part successfully resisted the Turks.

Ferdinand died in 1564 and his successor Maximilian II was an enlightened prince, who successfully curbed the religious struggles in his Empire. He came to terms with John Sigismund of Hungary, and on the latter's death in 1571 the new ruler of Transylvania, Stephen Bathori, accepted the Emperor as his overlord. Rudolf II of Austria was crowned King of Hungary in 1572 and in 1575 Bohemia also accepted Rudolf as King. Maximilian's efforts to acquire Poland failed, and he died in 1576. Rudolf II was an incapable ruler who persecuted the Protestants, and alienated many powerful nobles. Revolts broke out in Hungary, and in 1608 his brother Mathias displaced him in Austria, Hungary and Moravia. A year or two later Matthias gained Bohemia also. He appointed Ferdinand of Styria as his successor, but revolt against the latter broke out in 1618, and started the terrible Thirty Years' War (p. 197). The struggles of the Catholic Emperors, Ferdinand II and Ferdinand III, against

Bohemians, Germans, Danes, Swedes and French have received attention elsewhere (p. 197), and they did not affect territorial changes in Austria materially. In spite of strong opposition from the powerful French party, Leopold I became Holy Roman Emperor in 1657. During the period 1660-1664 he was fighting the Turks who threatened Vienna in 1663. The War of the Spanish Succession occupied him during a large part of his reign but had no great effect on Austrian territories.

Hungary revolted against Leopold in 1678, and obtained help from the independent Prince of Transylvania. The combined forces under Tekeli invaded Austria, while 200,000 Turks besieged Vienna in 1683. John Sobieski of Poland came to Austria's assistance and speedily drove the Turks southward as far as Belgrade. At a second battle of Mohacs in 1687 the Austrians decisively defeated the Turks. The Turks returned with a large army in 1691, but were defeated again by Austria at Szalankamen. Another invasion in 1697 was driven back by Prince Eugene of Savoy after the battle of Zenta on the Theiss. The Turks now gave up all control in Transylvania and Hungary north of the Marosch (Fig. 107). After a long and successful reign Leopold died in 1705.

Joseph I aided by England continued the long struggle with France, Eugene and Marlborough being the chief generals. He also pacified Hungary and Transylvania, granting the Protestant religions freedom in 1711. His death in the same year gave the crown to his brother Charles VI, who became King of Hungary and Bohemia, Archduke of Austria, and a little later Holy Roman Emperor. In 1713 the long struggle with France ended and Austria obtained Milan, Sardinia, Naples and the Spanish Netherlands. Eugene of Savoy led Austrian armies against the Turks, and won a notable victory beneath the walls of Belgrade in 1717. As a result Temesvar, Crajova and much of Serbia were handed over to Austria (Fig. 107).

The later years of Charles's reign—after the death of his great general Eugene of Savoy in 1736—were marked by many reverses. He had no son, and his great ambition was to be succeeded by his high-minded daughter Maria Theresa (Fig. 108), who married Francis of Lorraine in 1736. The Elector of Bavaria, a descendant of Ferdinand I, claimed the throne in spite of the "Pragmatic Sanction." Frederick of Prussia invaded and conquered Silesia with Prussian disregard of territorial rights. In 1742 through French influence the Elector of Bavaria was crowned Holy Roman Emperor as Charles VII, almost the sole exception to the Hapsburg elections for centuries. However, he died in 1745 and Francis, the husband of Maria Theresa, was elected. The Seven Years' War with Prussia lasted from 1756 to 1763, but had little geographical effect in spite of the waste in men and resources.

The first partition of Poland occurred in 1772, and the reasons leading to it are somewhat involved (Fig. 118). Poland was torn by factions and Maria Theresa supported one, Russia the other. Meanwhile, Russia defeated Turkey and was persuaded by Maria Theresa and her ancient foe, Frederick of Prussia, to annex a portion of Poland instead of the Rumanian lands (held by Turkey) which Russia desired to attach to its Empire. Prussia obtained West Prussia and Austria Galicia (Fig. 107). The Sultan of Turkey gave Bukovina to Austria to show his gratitude! It is more easy to see some justification for this barefaced "grab" on the part of Russia and Prussia than of Austria, which now added Poles and Ruthenians to her already very varied national medley. Maria Theresa died in 1780 and her son, Joseph II, made many useful reforms in his Empire, but saw most of his foreign policies frustrated. Thus the Netherlands revolted in 1787 and succeeded for a time in becoming independent. Joseph was succeeded by his brother Leopold in 1790, who, however, only reigned two years. His son, Francis II, was the next emperor and he lived during the stirring times of the French Revolution. Austria was naturally antagonistic to the new Republic, and much indecisive fighting occurred in the Netherlands. In the second partition of Poland (1792) Austria did not participate except to express approval of Prussia's action. Kosciusko led a vigorous revolt in Poland in 1794 but was crushed by the three absolute monarchs of Austria, Russia and Prussia. Austria as her share received the ancient capital, Cracow, and much territory to the northeast (Fig. 107). Most of this, however, was lost to Russia in 1809.

E. The Fairly Stable Boundaries of the Nineteenth Century

In the first decade of the nineteenth century the Austrian Empire formed for the most part a well-rounded geographic unit. It consisted of the original Austrian lands plus those included within the great Carpathian Arc, and was bounded on the south by the river Sava which limited the great Alfold Plain. Economically, it was to function well for a century. The salt and iron mines of the Austrian Alps, the coal and iron of Bohemia were supplemented by the gold, copper and silver of the Carpathians. The growing industries of the Vienna Gate, industrially the most advanced portion of the Empire, could be exchanged with the pastoral products of the Alfold where grazed some of the largest flocks and herds in Europe. Hungary also is one of the largest grain producers, and Moravia a region of intensive farming. Only Galicia, the fruit of an inexcusable annexation, lay outside the borders outlined above. Of course, culturally the region could hardly have been more diverse. Possibly had Austria made no attempt to hold her recent Italian conquests and relinquished her Polish territories she might have succeeded better in "driving the rest of her unruly team."

War with the French continued, and Bonaparte was now rising to fame. In Italy Austria was defeated at Castiglione (1796) and Rivoli (1797), both at the southern outlet of the Brenner route. As a result, the Netherlands were transferred to France, but Salzburg, Dalmatia and Venice were given to Austria. The bishopric of Trent, so vital a region in the Great War, was annexed to Austria in 1803. In May, 1804, Napoleon assumed the title Emperor of the French, and Francis, not to be outdone, was declared hereditary *Emperor of Austria* in August of the same year. Fierce struggles with the French under Napoleon followed. In 1805 the French occupied Vienna and in 1806 Berlin. In 1806 the North German States formed the Rheinbund, and Francis thereupon resigned the Imperial Crown. Such was the end of the Holy Roman Empire. Further defeats of the Austrians followed, notably Wagram on the plains near Vienna in 1809. However, the French were defeated by the Allies in 1814, and Napoleon was finally crushed in 1815 at Waterloo. Lombardy, Venetia and Tuscany were assigned to Austria at the Congress of Vienna in 1814. Francis II, the first hereditary Emperor of Austria, died in 1835 (Fig. 108).

During Ferdinand's reign the free city of Cracow was a centre of Polish disaffection and was deprived of its rights by Austria in 1846. In 1848 revolts broke out in Milan and in Hungary, which led to the abdication of Ferdinand. Francis Joseph adopted a more constitutional government which in time pacified the disaffected provinces, though not until Kossuth's army had been defeated with the aid of Russia. The loss of Lombardy to Sardinia has already been described (p. 369).

We have seen that Austria in the 17th century had been foiled by the French in her desire to crush the Protestant Princes of Germany.

The Seven Years' War with Prussia was another trial of strength. The Rheinbund in 1806 was a further challenge; but the final struggle for the leadership of the German states occurred in 1866. The salient features have been described in connection with the chapters on Germany and Italy, but may briefly be repeated. The quarrel was ostensibly as to the control of Schleswig-Holstein (p. 279). The smaller German states such as Saxony, Hanover and Bavaria were all jealous of Prussia, and practically all of them joined Austria. Prussia,

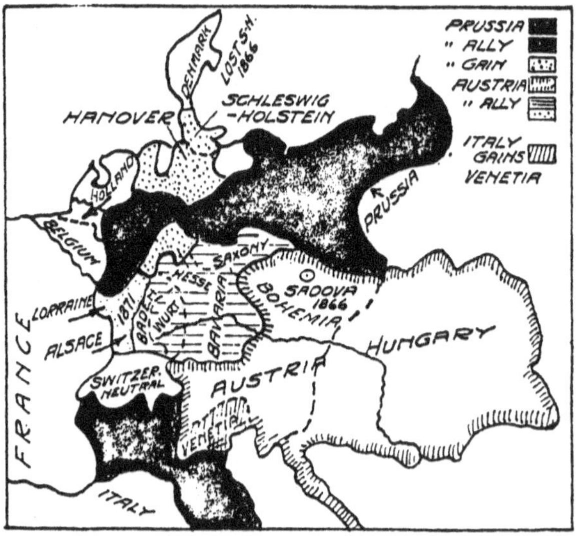

Fig. 109.—The final struggle between Austria and Prussia in 1866. The dotted areas are those gained by Prussia.

however, had made a close alliance with Italy, which appeared to make the contest a very equal one (Fig. 109). Austria had to divide her forces to resist Italy as well as Prussia, while the latter also had to divide in order to attack the South German armies. As on so many previous occasions Bohemia, between the rivals, was the chief battleground. The Hanoverians were defeated near Langensalza and two other armies advanced into Northern Bohemia. Here Benedek awaited them at Sadowa on the upper Elbe, where he was completely defeated, losing 44,000 men. As a result Austria gave up Venetia to Italy; and the German allies of Austria lost Hanover and Hesse, which were incorporated in Prussia.

Hungary had been deprived of most of her local authority after the revolt of 1849, but now took advantage of Austria's defeat to gain almost complete independence under a dual system of control by Austria-Hungary. Constitutional government in a liberal form was granted by Austria so that "all the blessings of a modern state were bestowed at one blow on a people which a few months before had been governed like a herd of cattle." Clearly the dual control was at the expense of the Slavs, for in "Hungary" the Magyars did not constitute one-half of the population, while in "Austria" the Germans also formed a minority. The "cement" holding the Empire was loyalty to Francis Joseph, and the fear that the Emperor could not live much longer no doubt contributed to the Great War starting in 1914. With northern aspirations removed from Austrian policies she turned to the southeast and commenced that "Thrust to the East" which was another important factor leading to the Great War. In 1878, after the victory of Russia over the Turks, the Powers met at Berlin, and Bosnia and Herzegovina were handed over to Austria. This marked the zenith of the Hapsburg expansions (Fig. 107).

We may here pause a short time to consider the various reasons for the growth of this unwieldy Empire, which held together till 1918 largely through the personal popularity of one aged man. The Austrian Alps were occupied far back in Roman times by German and allied tribes who have preserved their independence under native leaders practically ever since. The proximity of this mountain region to the Vienna Gate and to Vienna, where the ruler dwelt, may help to account for the fact that Austrian mountaineers remained loyal to the Emperors—unlike the Swiss in similar conditions.

Bohemia is a well-defined environment surrounded by horsts on three sides (p. 57) and inhabited by the western wedge of the Slavs. Doubtless the friendship of Albert the Magnanimous with the Emperor Sigismund drew the Bohemians towards Austria rather than to the north. From the rapid growth in numbers of the Hussites and Calvinists it is obvious that Bohemia might readily have joined the northern Protestant powers in the later religious struggles, yet it seems likely that the environment linked Bohemia on the south with Vienna, for only on the southeast is the "Wall of Horsts" absent (Fig. 112).

Perhaps we can say that the Danube was the chief factor in determining later expansions. The Danube filled the Carpathian Arc with the silts of the Alfold, and so made an environment of dry plains (p. 56) which determined the settlement of the nomad Magyars. It was the Danube valley which led the Turks into the heart of Europe, so that when the Magyars had to give way they naturally joined with Austria to resist the common foe. The final defeat of the Turks mainly by Austrian armies accounts for the union of Austria and Hungary as narrated.

F. Pan-Germanism and the Plight of Austria

This is not the place to discuss the moral aspects of the Great War, but the economic pressure exerted on the central powers is patent to every student. In Germany and Austria industrial growth had been delayed until there was little empty land left in the world for them to colonise and develop under their own flags. To quote R. G. Usher: "The Germans and Austrians of the twentieth century were filled with an uncontrollable determination to establish their economic well-being. Few nations have refrained because of moral scruples from advancing their economic or national welfare by any means they could. The substitution of machines for labour and the use of new lands for food has literally made possible the growth of standing armies of hundreds of thousands of men who can thus be spared from the task of keeping the community alive. Pan-Germanism has been made possible by the economic growth of the nineteenth century. Further, the unprecedented growth of population has produced in Germany, as elsewhere, a sense of intolerable crowding—since Germany's effective area is not much larger now than it was several centuries ago. The intensification of national feeling is also due in large part to the inventions of newspapers, wireless, etc., and these have helped to create the present acute crises. The Germans claim with much justice that England and France secured their empires by precisely the methods Germany employed against them."[3] In the present writer's opinion the crime of Germany was mainly due to the usual cause, *i.e.*, Ignorance. She did not and does not yet realise that war under modern conditions is deplorable because it curses the victors as well as the victims.

The motif behind the German-Austrian Plan is often called Pan-Germanism. The central Powers hoped to construct a great confederation of states, including Germany, Austria, Hungary, the Balkan States and Turkey, which would control a large area of territory stretching southeast from the North Sea to the Persian Gulf. A railway from Constantinople to Bagdad would effectually tie the great trunk lines leading from the Rhine and Danube valleys eastwards, and

[3]Roland G. Usher, "Pan-Germanism," New York, 1914.

so establish a shorter route to India than that *via* Suez. Pan-Germanism is, therefore, in the first place a defensive movement for self-preservation, for escaping the pressure of France and Russia. Germany (and Austria) hoped thus to obtain an outlet for their surplus population and manufactures and to create an empire as little vulnerable politically, economically or strategically as any the world has yet seen. This World Plan obviously developed from the policy which led Austria to annex Bosnia, Herzegovina and to occupy Novibazar.

At the end of the Great War in November, 1918, the components of the long-lived Austrian Empire fell apart immediately. A former Empire of some 50 millions was reduced to about six millions, living in an area of 32,000 square miles (Fig. 106). Especially hard is the plight of Vienna—a city of two millions; a reasonable figure for the capital of a huge empire with the diversified resources mentioned above, but entirely disproportionate when the seven-eighths of the population joined separate states. Austria has lost her ports, notably Trieste and Pola, as well as the Dalmatian coast. She has now only the Danube, with its outlet upon the Black Sea—far from ocean routes, in place of her basis for important commerce *via* the Adriatic. The loss of many nationals in Trentino has been described. Somewhat different is the case of the adjacent peninsula of Istria—also lost by Austria; for it contained few German-speakers. It was handed to Italy in spite of the fact that half the area is settled by Yugo-Slavs.

In view of these facts one can readily understand that many Austrians would prefer to join the German Reich for economic reasons (the "Anschluss"), even though they would undoubtedly be submerged in a huge population of 70 millions dominated either by Nazi or Prussian ideals. It will not be out of place to repeat that the dissensions in the former Empire were entirely due to man-made, cultural differences and are not in the least due to *racial* differences, such as are so obvious in France or even in Britain and Germany. The whole of the old Austrian Empire is Alpine (Fig. 110). Even the differences of religion and language are not much more marked than in Switzerland and Belgium, where they have been accepted without much friction for several centuries. Time and goodwill could cure the cultural differences of the "Danubian Economic Unit" (as the old Austrian Empire might be termed), but unfortunately there seems little chance at present of man following the teachings of history. In several later chapters we take up in more detail the problems of the disintegration of the Empire.

CHAPTER XXI

THE WESTERN SLAVS AND THE HUNGARIANS

A. The Origin of the Slavs

We have hitherto mainly been studying nations of Teutonic or Romance culture, but we now turn to a very different group of peoples. The Slavs in Europe number about 150 millions and occupy all the eastern and much of the central part of the continent. It is customary to divide them into the Western, Southern and Eastern Slavs, but this is not very satisfactory, for the Slovenes (South Slavs in Istria) extend as far to the West as do the Czechs in Bohemia. The significant feature of the distribution is that the Magyars and Rumanians occupy the central area, breaking the continuity between the West and South Slavs. This division of the Slavs into separate groups occurred in historic times and is of the greatest importance in our study.

The relation of the Slav languages to the Alpine Race is illustrated in Fig. 110. Here we see that almost all the Slav languages are included in the "Alpine Wedge," as also are the Magyar and the Rumanian peoples who isolate the South Slavs from their fellows. It seems likely that the Slav represents the Aryan language nearest to the original home of the Alpine peoples, which does not necessarily mean that it represents the "purest" Aryan speech. Of course, when a language spreads far from its cradle it becomes modified by contact with other speeches. Sometimes, however, owing to isolation an early language like Icelandic shows very little change from old times, because it has never displaced another language.

Slav has a good many points in common with Teutonic, but judging from its position in the language zones one would expect it to be a later development of Aryan than Teutonic (Fig. 25). The map (Fig. 110) further illustrates that the Old Austrian Empire (which approximately filled the circle shown in the map) was exclusively Alpine in race. No doubt the Asiatic Magyars differed somewhat from the European Alpines when these nomads finally arrived in their final homes. But long centuries of mixing with the much larger number of European Alpines have made it difficult to distinguish anything characteristic in the physical anthropology of any of these nations.

The cradleland of the Slavs has been placed in many areas, of which Poland, Polesia and the Lower Volga are perhaps the most usually quoted. The present writer believes that the Slavs inhabited all three at various periods in their race history. We must go far back of any written history, far earlier than Herodotus, for our clues as to the origin of the Slavs. They are mainly of Alpine Race, but in the north are dominantly Nordic. Hence the Slav culture (which is essentially based on a common type of language) has clearly spread from one race to the other. We do not know which race originated Slav, but as is suggested earlier, since the Alpines certainly came from

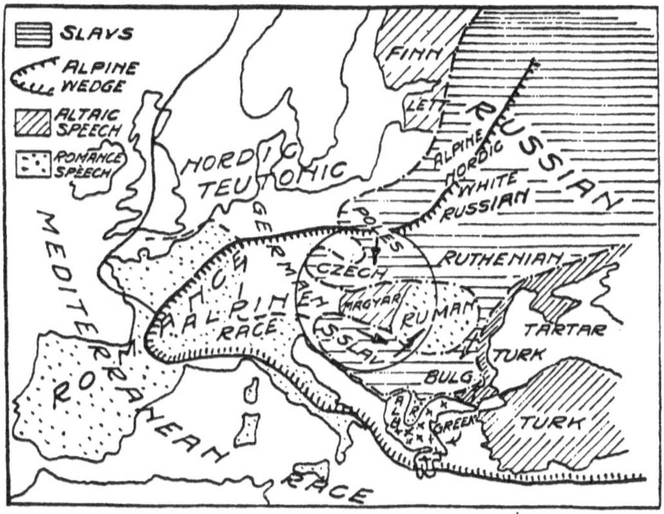

Fig. 110.—The Relation of the Slav Languages to the "Alpine Wedge." The North Russians and Northern Poles have a strong Nordic strain. The circle shows approximately the Austrian Empire in 1914. It was wholly Alpine in Race, but very varied in Nation.

Central Asia and the Nordics almost certainly came from Southern Siberia, it seems equally likely that the ancestors of the Slavs also reached Europe from the East. Another hypothesis which is more generally held is that the language has developed after the arrival of the Aryan-speakers in Russia or Poland. (This has been discussed on page 111.)

As regards the position of the Slav-speakers at the dawn of history there is little doubt that they had by this time arrived in Europe. Some writers place them in the Vistula Basin, others in the Pripet Marshes (Polesia), truly not far to the east. Rostafinska believes that, since a word for the "Hornbeam" tree is found in Slav and not in the other main Aryan languages, the Slavs originated in the marshy lands of the Pripet, which is the main habitat of that tree (Fig. 19). As mentioned elsewhere names of trees change very readily, but certainly when we first hear of the Slavs they were living in a region extending westward from the Pripet Marshes to the Elbe, and eastward for some considerable distance into Russia (Fig. 111).

According to Peisker[1] the Proto-Slavs of the marshes were a timid hunted folk, who had probably deteriorated in physique and character

Fig. 111.—Migrations of the Slavs from the Vistula region. Migrations of the Illyrians, Bulgars and Magyars are also indicated.

owing to the disadvantages of their environment. They had no cattle and dwelt in isolated groups almost without tribal consciousness. They lived on fish, honey and a little grain grown on the scattered islets in the marshes. They kept away from the forests to the west where dwelt the Germans, and from the steppes to the southeast where lived the fierce Asiatic nomads. During the time of Herodotus (B.C. 400) it seems probable that Scythians occupied the steppes west of the

[1] "Cambridge Medieval History," Vol. II, Cambridge, 1913, p. 418.

Don, and Sarmatians those east of the Don. For several centuries the steppes seem to have been peacefully occupied, so that the Scythians encouraged Greek traders and Slav farmers to live among them. Then about 100 B.C. evil times began. The Sarmatians moved against the Scythians and overwhelmed them, while the Greeks abandoned their trading posts, and the Slavs presumably retreated to their marshes.

Something the same kind of social relationship seems to have existed between the Germans and the Slavs. About B.C. 200 the Bastarnae were over-lords of the Slavs in Galicia and thereabouts (Fig. 38). The Venedae (who may originally have been Keltae) merged with the Slav stocks and gave rise to the Wends, who still live south of Berlin in Lusatia (Fig. 70). Some tribes called Antae lived in South Russia, and they fought bravely against their oppressors. They were probably Slav peasants dominated by over-lords of other nations. The Goths, whose movements have been discussed earlier, also made serfs of the Slavs during their wanderings through Eastern Europe, and these in turn gave way to the Huns before A.D. 400. About this time the Slavs extended westward to the Elbe, into the lands abandoned by the Teutonic tribes.

In Bohemia the Keltae known as Boii seem to have been overcome by Marbod and his Marcomanni (a German tribe) about B.C. 12. Slav pressure on the north in turn drove the Marcomanni upon the Roman frontiers, so that they attacked towns in Italy on several occasions, notably in the time of Aurelian. They were crushed by the Hun invasions about 450, and Bohemia for a time seems to have been left almost empty. The Slavs from the northern plains thereupon occupied the clearings among the great forests almost without opposition about the middle of the 5th century. Such was the origin of the Czech nation.

The invasion of the Avars is given by Peisker as the main reason for the migrations of the Slavs to the southeast. During the 6th century the Suabians abandoned much of north Germany, the Lombards entered Italy and the Avars occupied the Alfold Plain (Fig. 121). Here they enslaved the peaceful Slavs; living among them in the winter, but carrying out lengthy raids during the summer. Their leader Bajan placed the Slav tribes in various outlying parts of the region which he dominated. Some Sorabs were taken from Germany and planted in the south, where they gave rise to the Sorab (Serb) nation. Another northern tribe, the Chrobates, was brought into the Alfold and finally settled to the south in Croatia (Chrobatia). The

Avars ravaged the Byzantine Empire, and their Slav dependants moved south into the lands which the Avars depopulated during the sixth century A.D. (Fig. 111).

It seems likely that the Slavs improved in fighting qualities as they mixed with the Nordic or Altaic tribes. It is interesting to learn that the leader of the Croat warriors was long called Ban, in memory of the Avar Bajan. (Compare the use of "Caesar" as Kaiser or Czar.) The Slavs who entered Bulgaria in large numbers about 650 were dominated by Bulgar nomads (allied to the Finns) who reached Bulgaria from the Volga area about the same time (Fig. 111). The overlords soon merged in the Slav populace, and gave rise to a vigorous and warlike nation, the modern Bulgarians.

Part of the same movement of the Slavs southward from the basin of the Vistula brought the Slovaks south of the Carpathians. No doubt like the Czechs they made use of the Moravian Gate at the head of the Oder. They joined with the Czechs to form part of the empire of Samo about 650. The latter is said to have been a Frank who led the Slavs to victory against Dagobert, King of the Franks. His empire did not, however, persist, and the Slovaks were conquered by the Magyars in 907 and remained subject to them for most of historic times. So also the Slovenes (who now live around Laibach and Klagenfurt) were Slavs who migrated south in the 7th century. They also formed part of Samo's rather legendary empire, but were soon dominated by the Franks. They have naturally been in closer contact with early Italian culture than has any other group of the Slavs.

B. *The Build of Czecho-Slovakia*

The modern state of Czecho-Slovakia is built up of three fairly well-defined units which are indicated in Fig. 112. Czecho-Slovakia has an awkward shape, like a carrot with the root to the east. It occupies an area of 55,000 square miles. The west is a peneplain, and the east consists essentially of the western Carpathian Arc. Only near Bratislava and Munkacs are there small low-level plains. As Bowman points out, five nations, all at one time or another hostile, press upon this dangerously elongated state.

The western portion consists of the Bohemian Relic-Block, which differs from most other examples in Europe because it is bounded on three sides by well-marked Horsts. The latter rise to four or five thousand feet, while the peneplain which forms most of Bohemia has an elevation of about 2,000 feet above the sea. Moravia to the east is of much the same structure, consisting of an undulating peneplain about 2,000 feet high, but it is only shut in by the Sudetes Horst on the north, and on the east by the rather low western end of the Carpathian Arc. To the south Moravia consists of the wide low plains of the river March (Morava) under 600 feet above sea level.

Interesting and important details in these topographies must now be mentioned. The Bohemian peneplain on the south is deeply trenched by the Ultawa (Moldau) and Sazawa Rivers, and the northern part of the peneplain is still covered by level deposits of Cretaceous

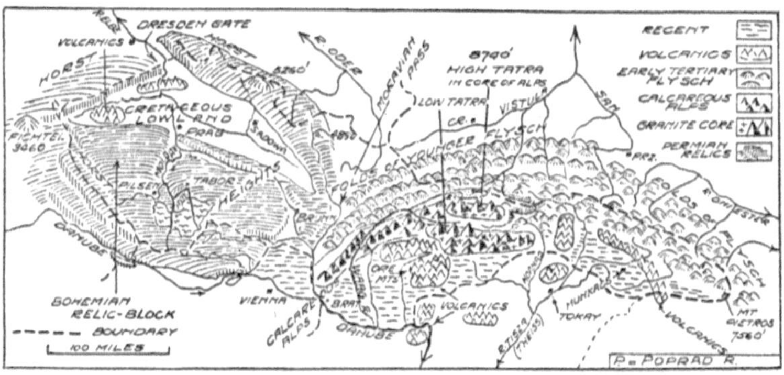

Fig. 112.—The Build of Czecho-Slovakia. Notice that Bohemia is a Relic-Block, while Slovakia is built up of the young Folds of the Alps. P. indicates the "antecedent" Poprad R.

age. Through these rocks flows the Upper Elbe from Sadowa almost to the border. On the inner side of the arc of Horsts (*i.e.*, east of the Ore Mountains) there have been extensive flows of volcanic lava, through which the River Ohre has cut its way in deep gorges.

The Elbe escapes from Bohemia by the Dresden Gate, which has always been a main line of access into Bohemia. But the Horst Mountains are little like the Alps in extent and height, and have not proved a barrier in the same sense as have the Alps. A number of roads cross them especially in the western corner near Eger, and between the Giant and Sudetes Mountains. The most important gap, however, along the whole line of Horsts and Carpathians is the Moravian Gate (1,000 feet) east of Olmutz just where the Oder rises. The river March flows through Olmutz and so leads southwards to Pressburg (Bratislava) and Vienna. Right in this important gap is the coalfield of Ostrava (Fig. 112). Immediately to the east are the very much higher mountains of the Beskids in the Carpathians, and we shall not find so low a gap in this great Arc until we reach the Red Tower Pass, 600 miles to the southeast. To the south, as already mentioned, both Bohemia and Moravia descend fairly gradually to the Danube; but between the two regions the Moravian Hills rise to 2,000 feet or so as rounded summits formed of the ancient rocks of the Relic-Block.

This Bohemian Basin, as stated, is unlike any other structure in Europe. It is a Horst-rimmed peneplain, where the central area is depressed relatively to the margins. Unlike the other peneplains, Norway, Finland and Spain, it is thus fairly well-protected against invasion, and this fact has clearly enabled the Bohemians to maintain themselves so successfully in the past against their German foes. Since the Bohemian peneplain has a rim it is not a region subject to so much erosion as are the surfaces of other elevated peneplains. Hence it has good agricultural lands, and also has preserved much of its coal. We have seen something of the same sort of result in the relatively depressed Scottish Lowlands (Fig. 8), to which indeed Bohemia has a slight structural resemblance.

The second structural unit in Czecho-Slovakia consists of the rather complicated Western Carpathian Folds, which build up the land of the Slovaks (Fig. 112). Here the Crystalline Core is exposed and is flanked on the north by the early Tertiary "Flysch" which also participated in the folding. On the inner side of the Arc (just as in West Bohemia) there has been an outpouring of volcanic lava which has been cut through by the river Gran. The Poprad, a tributary of the Dunayetz River, rises south of the main Carpathians, and flows *across the folds* to the Vistula. The head-waters of the Waag and the Gran on the west, and of the Poprad and the Hernad on the east flow in east-west valleys, and thus divide the Carpathians into four parallel ridges. The first of these is the White Mountain-Beskid ridge which is built up of "Flysch;" the second ridge consists of the Little Carpathians (which reach the Danube at Bratislava), the Fatra and the High Tatra, all of which are mainly Crystalline rocks; then comes the Lower Tatra; and lastly the Ore Mountains which are mostly volcanic. It is to be noted, however, that the lower portions of three of these rivers all lead directly to the Alfold Plain. Hence the contacts of the Slovaks have been rather with the Magyars of the Plain than with their compatriots the Czechs on the west or the Ruthenians on the east. In the early days the Slovaks occupied the region round Buda-

Pest and the Matra Mountains (Fig. 105), but the Magyars gradually drove them up into the Carpathians. Hence they became rather isolated and never developed so far as did the Czechs in the Bohemian Peneplain.

The third structural unit in Czecho-Slovakia lies east of the Dukla Pass (Fig. 112), and consists mainly of the region called Ruthenia. Here the Carpathians are much lower and are built up very largely of the soft Flysch. The Dukla Pass is only 1,650 feet high, and leads from the Middle Vistula to the Upper Theiss. Further to the east the Carpathians rise to 7,560 feet (Mount Pietros), alongside the high "Pass of the Magyars" (or Hun Pass, see Fig. 105), where many of those fierce nomads crossed the ranges about 900 A.D. The hard crystalline rocks are absent in Ruthenia, but there is a zone of volcanic lavas on the southern inner side of the arc akin to those farther west.

C. *The Growth of the Bohemian Nations*

Only for a few relatively short periods did the Bohemian territory extend far from its cradleland in the Bohemian Relic-Block. Hence not much space can be given to its history up to 1526, when Ferdinand Archduke of Austria also became King of Bohemia. While the Franks were building up a vast Christian Empire in western Europe the pagan Moravians were extending their territory in the centre of the continent. Rastilaus in 850 formed an alliance with the Bulgarian and Byzantine rulers in the east, and Svatopluk in 894 ruled the Danube Plain as far south as Lake Balaton (Fig. 105). However, the Magyars entered the Alfold Plain about this time, and were encouraged to attack the Moravians by their German rivals. The Magyars in 907 conquered the region along the river Morava, and for a century the Marchland was a bone of contention between Bohemia, Hungary and Poland.

Meanwhile Bohemia had asked aid of the Germans against the Magyars, and they paid tribute to Henry the Fowler in 928. Gradually the Bohemians became Christians (p. 189), and the conflicts with Stephen of Hungary (997-1038) were settled by giving up Slovakia and southern Moravia to the Magyar kingdom. The Magyars made no attempts to colonise the hills surrounding the plain, which resemble so closely their ancestral steppes. The Bohemians, however, obtained from the German Emperor the right to elect their own dukes, and about 1050 their ruler became one of the Electors of the Emperor.

Later Bohemia was promoted to a kingdom, the first King being Ottakar (1197). His grandson, Ottakar II, has already appeared in connection with Austria (p. 376). He defeated the Magyars in 1260 and his kingdom extended to the Adriatic (Fig. 113). Hence for a short time Shakespeare's reference to the "coasts of Bohemia" was

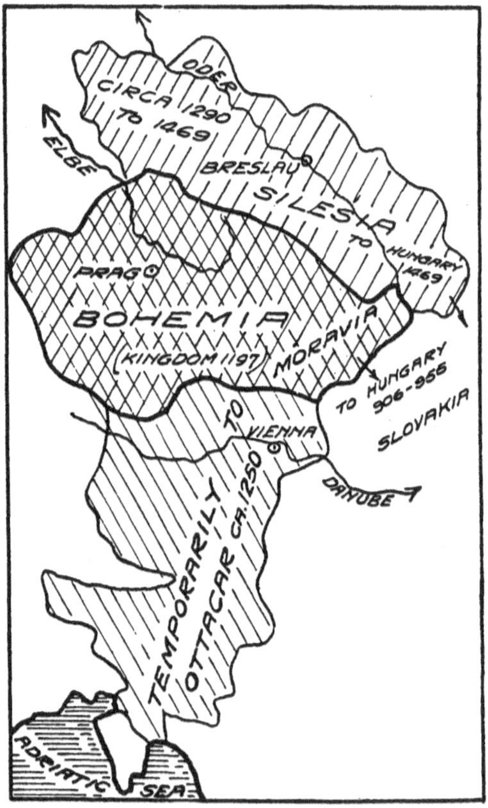

Fig. 113.—Ottakar's Empire of Bohemia about 1250.

correct. Ottakar came into conflict with Rudolf of Hapsburg, and lost most of his territory and was killed in battle in 1278.

During the Mongol invasion of 1241 much of Slovakia was depopulated, and large numbers of Germans were invited to settle in the empty lands by the King of Hungary. Thus developed the Zips towns, in the region of the Ore Mountains betwen the Gran and Hernad Rivers

(Fig. 112). The last of Ottakar's family died in 1306, and in 1310 the crown passed to John of Luxemburg who had married a Bohemian princess. He made a poor ruler, spending most of his time fighting for other princes, and died at Crecy in 1346. His son Charles next became Emperor and devoted his best energies to the Bohemian nation. He founded the University at Prague, which became perhaps the foremost city in Europe about this time.

The next period in Bohemian history is largely concerned with the activities of John Huss, who was a strong Czech nationalist as well as a vigorous opponent of the Pope. As narrated elsewhere he was burned in 1415. In 1419 Sigismund, the son of Charles, became the King of Bohemia and also Emperor of the Holy Roman Empire. His reign was followed by many years of civil war between the moderate Hussites (the Utraquists) and the extreme Hussites (the Taborites). George Podebrad, a Czech noble, was elected King in 1457, and was the sole Protestant ruler of Bohemia. George was involved with wars with Hungary which led him to seek aid from the King of Poland. On his death in 1477, Vladislav a Polish prince, was offered the throne. In 1490 he was also elected King of Hungary, and his two children were married to children of Ferdinand of Austria (p. 379). Vladislav died in 1516, while his son Louis was a minor. In 1526 Louis led an army to repel the Turks in the south of Hungary. He died in the disastrous battle of Mohacs, and as narrated elsewhere the Archduke Ferdinand of Austria was offered the two thrones, partly because his wife was the daughter of Vladislav and partly to enlist his aid against the Turks (p. 380).

During the next century the Bohemians made several revolts against the rule of the Hapsburgs. The Protestants, however, were unwilling to link themselves closely with the Lutherans of Germany. Moreover, the peasants were not much above serfs, and did not unite heartily in prosecuting a war against the Hapsburgs. The rebels chose a Bavarian prince as their King in 1619, but he was easily defeated by the Imperial army at the Battle of White Mountain in 1620. Thereafter the Protestant religion practically vanished, and for a time the famous University of Prague actually became a Jesuit College. Although most of the later Hapsburg monarchs were crowned separately as Kings of Bohemia, yet Bohemia was ruled most autocratically, and did not receive self-government when Hungary received that privilege in 1866. It was indeed the Great War which enabled Bohemia to obtain those rights which she lost in the 16th century.

(It is to be noted that Slovakia and Ruthenia had been linked to Hungary during most of their history.)

Modern Czecho-Slovakia in spite of its wise rulers would, therefore, appear to suffer from many disabilities as a nation. Among its population of 14 million over a third are, in general, hostile to the Czechs. There are over three million Germans, nearly a million of Magyars, and half a million Ruthenians (who are Little Russians). Its only outlets are by rail or river, and the transport of goods on the Danube has become much more complicated as the result of the splitting up of the Old Austrian Empire.

There are several Irredenta in this new state. The slopes from Saxony up to the Horsts on the west are relatively gentle. Partly for this reason large numbers of Germans have spilled over into Bohemia, where they form large majorities in a westernmost strip some 30 miles wide. These, of course, have not been affected by the Treaty of Versailles. In the Moravian Gate there are also many Germans, who have spread south from Silesia. Southern Slovakia is very largely Magyar (Fig. 122), but this area was given to Czecho-Slovakia to enable it to control the east-west rail and river traffic. It is stated that much of this area has only recently been settled by the Magyars, but it is to-day a definite centre of Irredentism. However, the conditions of the small farmer, both Slovak and Magyar, in Slovakia are much better now than formerly; and it is probable that with time all the Slovakian region will acquiesce in the present rule. Ruthenia, though its language is that of the Ukraine, voted to join the Czechs and Slovaks in 1918, if it were given autonomy. This extension of the Republic so far to the east permits Czecho-Slovakia to march with Rumania for a considerable distance. This would be a matter of great importance in case of war, since these two nations are equally desirous of maintaining the *status quo*.

Bohemia and Moravia are two of the best endowed states in Europe. They have large coal resources, and notable agricultural lands which are especially important for beet-sugar. Nearly four-fifths of all the industries of the former huge Austrian Empire were contained in Czecho-Slovakia. Topography has led to the main railways *crossing* the elongated region rather than remaining within the state. However, national lines already run from Pilsen to Kosice (close to Ruthenia) and these will no doubt soon be linked up with those entering Rumania.

D. *The Build of Hungary and its Effect on the Nation*

Perhaps no other country in Europe has the uniformity of structure found in present-day Hungary. It is as if the Treaty-Makers in 1919 had consulted a map of the great Carpathian Arc and its enclosed Basin, and had drawn the new limits of Hungary so as to include the completely flat central part of this great basin—the Alfold (Fig. 105). Actually, of course, the ethnical boundary has been in large part determined by the topography. The Magyars carried on in the Alfold the life to which they had become accustomed in the many generations they had spent in the Asiatic and Russian steppes. The Alfold is but another steppe, somewhat less dry, and less extreme in temperature than those of Asia. The Magyars divided the plain into a vast number of freehold areas which were granted to the tribes or families. They left the Slavs more or less undisturbed in the border-lands, except in Transylvania, where a separate tribe, the Szeklers, was settled as a Frontier-Guard by Ladislas about 1220. Originally devoted to cattle and horses the Alfold is now a great grain-growing region, with the roads lined with large acacia trees. Only in the northeast near Debreczin are there large areas of the open waste (*puszta*) remaining.

Almost the sole exception to this uniform infilling by late Tertiary and recent deposits consists in a ridge which extends along the west side of Lake Balaton, and (with gaps) reaches the great bend of the Danube (Fig. 105). This consists of faulted blocks of granite and Mesozoic rocks which have not been covered by the universal flood of silt. This upland is known as the Bakony Forest, and it fairly sharply separates the Plain of the Raab from the Alfold proper. This ridge with its forests helped to hold back the Turks in their numerous onslaughts on the Hungarians and Austrians. Just across the Danube from this ridge are some fairly isolated highlands which still form part of Hungary. They are mainly volcanic areas due to lavas welling up on the inner side of the great Fold Arc. Among these the Matra Mountains rise to 3,000 feet (Fig. 105).

The Great Plain is somewhat diversified by large dunes (Fig. 105), by great marshes and by low hills (especially west of the Danube) cut out of the soft surface deposits by the numerous small streams. These are the natural features accompanying a senile river (p. 60), which wanders irregularly over the plain, often with its bed higher than the adjacent fields, so that it must be held in its course by artificial dykes. Much of the area is subject to floods, which indeed form the wide areas of marsh along the Raab, Danube and Theiss. There are few towns in the lower parts of these rivers for this reason. Since the climate is hot and dry for much of the year there are violent sand-storms, which deposit their load as widespread Dunes east of Pest and in the far northeast (Fig. 105).

The Danube had always been the main line of communication until the railway era. Obviously there are three key positions on the river, two of which are in Hungary, while Vienna is only thirty miles to the west. Buda-Pest has arisen where the Bakony Ridge (Fig. 105) determines the elbow-bend of the Danube. Bratislava has a similar defensive position where the ancient crystalline rocks of the core of the Carpathians is breached by the big river. Vienna is the third example, where the folded rocks of the Flysch approach closely to the ancient rocks of the Bohemian Relic-Block. Vienna is the chief Gate, because the country to the west soon opens out into the wide Downfold of Bavaria. The well-defined plain of the Raab and adjacent Danube, between two of these rock-ribs, is often referred to as the Little Alfold.

E. The Growth of the Magyar Nation

It is supposed that the Boii Keltae were in Western Hungary about B.C. 300. Archeology informs us that they had iron coins and lived in little towns. About the time of Christ there were also Dacians and Jazyges (Sarmatians) living in the Alfold Plain. Octavius built several forts along the River Sava about B.C. 20 to control these barbarians; and later this region west of the Danube became the province of Pannonia. Aquincum near Buda-Pest was the chief Roman town, but Sirmium on the Sava was also quite important. About A.D. 108 Dacia east of the Danube became a Roman Province and Ulpia Trajan replaced Sarmizegetusa (north of the Vulcan Pass) as the chief town. About 276 the Romans abandoned Dacia to the Goths, while the latter also acquired Pannonia about 370. Soon after this the Huns made their first raids, rising to great power under Attila about 443. His capital was near Szeged, but his people founded no settled state. The Gepids conquered Dacia, and the Goths occupied Pannonia until 526, when the Gepids moved into Pannonia also. About 560 the Lombards poured into the Alfold from Moravia and settled Pannonia, while the Avars conquered the lands east of the Danube (Fig. 121). These Avars were Asiatics like their predecessors the Huns and their successors the Magyars.

Bajan the Avar established a wide empire which menaced the Byzantines, and later the Franks. The Avars were over-lords of hosts of Slavs, and in 587 penetrated as far as Adrianople. In 602, however, the Avar nation was nearly overthrown by revolts of the Slavs. It is stated that the Byzantine Emperor Heraclius transferred many of the Serbs and Chrobates from the north to the south about this time (see, however, p. 391). In 799 Charlemagne smashed the last armies of the Avars, and his Empire reached to the Sava at Belgrade. Just a hundred years later the first Magyars reached Hungary.

We hear echoes of the Magyars in early writings. They migrated from the Kama Basin in east Russia towards the Caspian Sea, and about A.D. 880 they had reached Moldavia near the Sereth and Dnieper Rivers, where they lived in tents, but grew a good deal of grain. Here they agreed to obey chiefs elected from the tribe of Arpad, to share all land conquered and to allow the lesser chiefs a share in the control. This early compact was the basis of their later government. For a time they aided the Germans against the Moravians (p. 395), and the Byzantines against the Bulgarians. About 895 the Petchenegs attacked them and they migrated over the Passes of the Carpathians and reached Munkacs early in 896 (Fig. 121).

Arpad soon conquered the Khazars, Wallachs and Bulgars who lived between the Theiss and the Carpathians. He then moved west to Buda-Pest, overcame the Franks in Pannonia and assigned lands to each of his eight tribes. It is rather surprising that the Magyars kept their own language when the Bulgar-Finns of a similar stock soon adopted a Slav language (Bulgarian). No doubt the Magyars were more numerous and more readily acquired a working government. At any rate for many centuries they formed a Protector Zone (Fig. 2) against further attacks by barbarians from the southeast.

The Magyars still despised a settled life. They raided far and wide, to Thuringia, to France and even to Spain. They won battles in Italy and even penetrated to Constantinople. But they suffered reverses from the Saxons in 933, and in 955 while attacking Augsburg they were utterly defeated by Otto the Great (p. 268). In 970 the Byzantines captured many Magyar prisoners including Gyula the ruler of Transylvania. His daughter became a Christian and married Geza the Magyar chief. Through the teachings of Adalbert of Prague Geza and his son Stephen became Christians in 985. It is important to note that Stephen joined with the powerful western Roman Catholic Church, which he thought would help him against the Germans, and not with the eastern Greek Church. Thus the Hungarians became and remain the outposts of the Roman Catholics in the east. In 1003 Gyula was defeated by Stephen, and Transylvania was annexed to Hungary. Conrad the German Emperor invaded Hungary in 1031, but he was driven back and forced to acknowledge the complete independence of Hungary. Stephen changed the tribal tenure to freehold, and thereby created a group of lesser nobles who later did much to preserve the independence of Hungary.

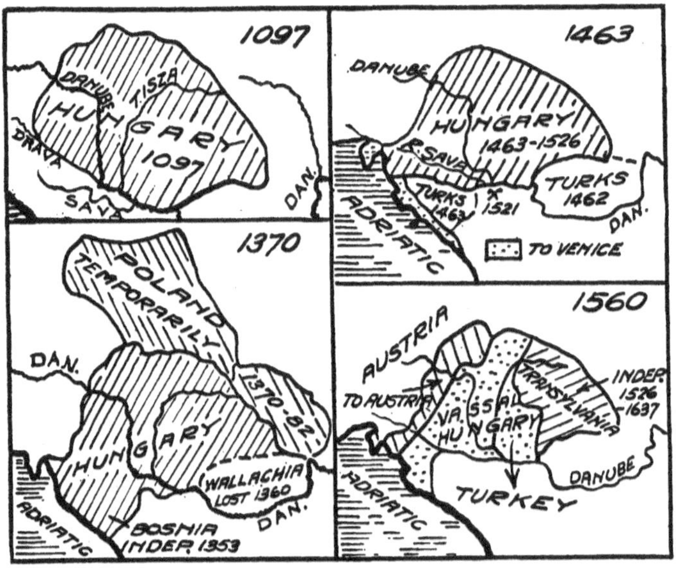

Fig. 113a.—Four Maps showing changes in Hungarian Territory from 1097 to 1560. (See also Fig. 107.)

The early Magyar kingdom occupied the more central part of the Alfold. On the north independent Chrobats and Slovaks still held the southern slopes of the Carpathians, whilst about 1070 Petchenegs and Cumans invaded the Alfold and settled along the lower Theiss (Fig. 121). But in the eleventh century the Magyars pushed their frontiers to the north, and the Slovaks remained their subjects till 1919, having little in common with the independent Bohemians. Early in the twelfth century Croatia and Bosnia in the south were annexed to the Hungarian realm. About 1160 many Saxons were settled in Transylvania and have maintained a somewhat separate national life ever since.

The Byzantine Empire under Manuel Comnenus menaced Hungary during the 12th century. Manuel supported various claimants to the throne, and for a time the Greek Orthodox religion made much headway in Hungary. Soon, however, the Serbian rulers, helped by Hungary, occupied all the attention of the Byzantines on their northern frontier. In 1204 a weak Arpad (Andrew II) came to the throne. He was much like his contemporary John of England in character, and like him was opposed by many strong barons. In 1222 Andrew granted the Golden Bull which opposed the abuse of power by the King to much the same degree as did Magna Charta (1215).

The reign of the wise King Bela IV (1206-70) was marked by the terrible Mongol invasions. He had attempted to unite the adjacent rulers to withstand the barbarians, but quite unsuccessfully. Early in 1241 the Mongols burst through the Carpathian Passes and signally defeated Bela south of the Matra Mountains (Fig. 105). For a year they ravaged Hungary depopulating large areas. Later Bela settled large numbers of Cuman tribes in the southwest. The nobles were encouraged to build strong castles, and their powers increased so greatly that they soon revolted against the Arpad rulers. Finally after much civil war Charles of Naples (a grandson of Stephen V, and allied to the House of Anjou) became King in 1310. He united with Casimir of Poland against the Hapsburgs and the Bohemians, and the Polish King made Louis of Hungary his heir. Under Louis the Great (1342-1382) the Hungarian realm extended almost from the Baltic to the Danube and eastward to the Dnieper (Fig. 113a). But Louis alienated the semi-independent marches of his southern and southeast frontiers, for they preferred the Greek Orthodox religion to the Roman Catholic.

The Wallachians (Rumanians) were rising into prominence about this time. They had settled the land of Etelkos (Moldavia) after the Magyars had abandoned it. They also occupied the march of Wallachia (west of the Aluta River) and about 1350 they also revolted. In Bosnia were many Bogomils (p. 192) whom Louis persecuted, so that in 1353 this region also broke away. Moreover, the marches in Bulgaria and Serbia preferred to become vassals of Moslem Turkey rather than of Catholic Louis. His two daughters reigned over Hungary and Poland. Sigismund of Pomerania married Maria and became King of Hungary in 1387. Hedwig (Jadwiga) was compelled to marry Jagiello, Grand-duke of Lithuania in 1386 (p. 421).

Sigismund was elected head of the Holy Roman Empire in 1411

and King of Bohemia in 1419. It was his destiny to face the Turks during their early onslaughts. He fought in several campaigns, fortified his southern frontiers and built the great fortress at Belgrade. His reign was troubled by revolts of the followers of Huss. Thousands were killed in the southeast, and their lands given to the Rumanians. Sigismund left no sons and his two crowns passed to Albert V (p. 377), who had married his daughter. The troubles arising from the succession in Hungary have been mentioned earlier, and in 1440 Vladislav of Poland was offered the crown.

The Turks had been held back by the efforts of John Hunyadi, Ban of Wallachia. In 1444 the King and the leading nobles were killed at the defeat of Varna. Hunyadi was made regent of Hungary, and by utilising his vast personal resources managed to hold off the Turkish armies. He died in 1456, and his son Matthias in 1458 was elected King of Hungary. Matthias Corvinus was one of the best of European rulers. He founded a strong army, largely composed of Hussites and Rumanians, which also acted as a police force to ensure the peace and security of Hungary. He greatly extended the trade, agriculture and mining in his kingdom, and he much improved many of the towns, especially Buda, Breslau and Debreczin. He made Hungary the greatest power in Central Europe, and he sent embassies as far as Persia (Fig. 113a). Moreover, he maintained peace with the Turks during most of his reign. Unfortunately he antagonised many of the nobles and prelates, and Hungary had no Middle Class which doubtless would have supported his wise government.

F. The Decline of Hungary

After the death of Matthias the nobles acquired all the power, and they chose a puppet King, Vladislaus of Bohemia, in 1490. They actually disbanded the army and ignored the menace in the south. Bathory in Transylvania ruled as a despot, and the great Zapolya family terrorised northern Hungary. Vladislaus was succeeded by his young son Louis, who was placed in the hands of incapable regents. In 1521 the Turks captured Belgrade, and in 1526 they destroyed the new fortress of Peterwardein. Louis thereupon marched south with hastily gathered forces, and practically the whole of the Hungarian army perished at the ill-fated battle of Mohacs. Thereafter the Turks easily overran the great plains, and only in rugged Transylvania did

Hungarian liberty persist, supported here by Saxon towns and Rumanian peasants.

John Zapolya and Ferdinand of Austria struggled for the Hungarian crown, the former being supported by the Sultan. Finally in 1538 Hungary was divided between them, the Austrian gaining Croatia and Western Hungary, while John obtained the crown and ruled over the centre and east of Hungary. Pressburg became the capital of western, *i.e.*, Royal Hungary. When John died the Sultan invaded the country, and in 1547 Hungary was again divided (Fig. 113a). Most of it became part of the Ottoman Empire, but Transylvania was given to the son of John Zapolya as a Principality. The Catholic Church had lost many of its leaders at Mohacs, and Calvinism spread rapidly among the Saxon miners in the north and the allied settlers in Transylvania. In the latter region Francis David made many converts to Unitarianism. A Protestant leader, Bocskay, helped greatly to obtain religious freedom, which persisted here after it had been crushed in Bohemia. Princes of Transylvania, such as Gabor Bethlen, maintained the Hungarian nationality, while the Turks ravaged the central region and the Hapsburgs dominated Royal Hungary. Unfortunately George Rakoczi (1642-1660) allied himself with the Swedes, and opposed the Turks and the Poles. His forces were defeated, and then the Turks overran Transylvania for the first time. However, the time of their expulsion from Hungary was at hand, and the campaigns of 1683 to 1697 drove them from the country (p. 381).

The Hapsburgs, however, settled the depopulated lands with their partisans, producing a new agrarian aristocracy. Hence arose many of those large landed estates which are perhaps unequalled in size in other parts of Europe. Thousands of Serbs also occupied the region south of the Marosh River about 1690. So also the Rumanians migrated *en masse* from the mountains westward to the margins of the Alfold bordering the Theiss, all of which region had before been entirely Magyar. Moreover, from 1720 to 1787 thousands of Germans were settled near the lower Danube, while the Hungarian serfs were restrained from settling in the newly opened lands.[2] After a period of civil war Transylvania was united to the Hungarian crown in 1691. The condition of Hungary improved under the enlightened if absolute governments of Charles, Maria Theresa and Joseph.

[2]F. Eckart, "Short History of the Hungarian People," London, 1931.

The revolution of 1848 saw Kossuth and the Magyars opposed by Jellachich the Croat leader of the Hapsburg armies. For a time the Magyar forces were successful, but in 1849 they were crushed by the Russian allies of the Austrian Emperor. However, in 1867 (p. 385) the status of Hungary was greatly improved by the "Ausgleich," and thereafter it was the Slavs and Rumanians who suffered at the hands of the joint Austria-Hungarian Government.

The present boundary of Hungary includes about eight millions out of the ten million Magyars. About half a million Szeklers live so far away in Transylvania that they had little prospect of being included in the new Hungary, though they remember that Transylvania remained Hungarian almost throughout the Turkish conquest of south-east Europe. The Hungarians have, however, lost large and homogeneous communities of nationals both along the northern boundary and along the lower Theiss (Fig. 122). As regards the former the Czechs claim that the territory is vital as regards communication between Bohemia and Slovakia. There seems, however, no adequate reason for handing over the region south of Szeged to Yugo-Slavia. One can also sympathise with the Hungarians in respect to the lands handed to Rumania, most of which have never been *under the rule* of Rumanians at any time. As the preceding account shows they have been colonised by Rumanians by a process of "peaceful penetration;" and we may be sure that the Magyars never expected that this action would result in the loss of about half their realm.

CHAPTER XXII

POLAND—THE CATHOLIC BUFFER NATION OF THE GREAT NORTHERN PLAIN

A. *The Build of Poland and Some of its Results*

The Build of Poland may be studied in a series of five zones running west-east between the Carpathians and the Baltic Sea. Each of these can be subdivided into smaller units for further description. The Carpathians are built up of the same "elements" as were briefly discussed in the section on Czecho-Slovakia (Fig. 112). The folds were thrust from the south against the softer Secondary and Tertiary rocks of the Plains. Thus we find a series of west-east *Folds* (Fig. 114) of which the core of hard Crystallines just extends into Poland in the High Tatras (8,740 feet). The slopes on the north are mostly built of Flysch or Secondary rocks (5 and 6) as in the Western Beskids. To the east they are rich in petroleum; and the oil-fields extend eastward from Cracow into Rumania. In front of the Carpathians is a shallow *Downfold* like that which is occupied by the Danube in front of the Austrian Alps (Fig. 69). In Poland the corresponding depression contains the Vistula and its tributary the San, and in the far east the upper Dniester, each of which has deposited much alluvial along its bed.

North of this ill-developed Downfold is a third zone of *High-Plains* (or Platforms) which is bisected by the Vistula. They consist mainly of rocks of Cretaceous age or thereabouts, and have an elevation of 1,000 feet above sea level. The western Platform is pierced by the Lysa Gora (Bald Mountain) which is a Permian relic-stump akin to those forming the Harz and similar mountains in Germany. The southwest corner of this Platform contains the Teschen coalfield around Beuthen and Kattowitz, where the older coal measures outcrop beneath the rocks of the Platform. The eastern Platform (to the east of Lwow) also has an elevation of about 1,000 feet.

The fourth Zone consists of the Great Alluvial *Plains* of Poland which are in general about 600 feet above the sea. They are drained by the Warta, the middle Vistula, the Bug, the Narev and the Pripet. Very extensive marshes occupy the basin of the Pripet, which had a vast influence on the early development of the Slav tribes.

Running across these modern rivers, just as we saw in Germany, are a number of Ice-Front Channels. They were eroded by the thaw-waters draining from the front of the ice-lobes, when Poland was covered by ice during the last Ice Age. To-day these channels are

Fig. 114.—Mantle-Map of Poland showing the Flysch Folds in the South, the San-Dniester Downfold, the Cretaceous Platforms and the Morainic Plains of the North. (Partly after De Martonne.)

usually devoid of a large river, though the Narev and Warta occupy them for part of their courses. However, just as in Germany, they have been used in building canals, and three of them (as indicated in Fig. 114) run through Brest, Bromberg and Grodno respectively.

The Fifth Zone consists of tangled heaps of *Moraine* dumped on the plain by the Ice-lobes as they halted for many centuries on their retreat to the north. These moraines have a very indefinite drainage, and contain great numbers of small lakes, the Mazurian Lakes. The little hills are forested and have always been difficult country to traverse. Hence they served as a refuge for centuries for the pagan Prussians and later peoples. They were also the arena where many a fierce battle has been fought since, as witness the two battles of Tannenburg (1410 and 1914) and the battle of Gumbinnen (1914). (Most of this type of country belongs to Prussia, though it is inhabited mainly by Mazurian Poles.)

We may now consider in a little more detail the relation of the environment to the history of Poland. The plains of Poland are intermediate in area between those of Germany and Russia, thus forming a transition, much as does Polish culture, between the German and Russian cultures. The Polish nation originated in the extensive plains between the Warta and the Vistula to which the name of "Great Poland" has been given (Fig. 114). One of its southern towns, Kalisz, is an extremely old settlement which may have been referred to by Ptolemy (A.D. 140) under the name Calisia. This name was later used for the region immediately to the east which is called Galicia. Gnesen near by is the seat of the first Archbishop, and here is the shrine of Saint Adalbert. Posen was the old capital where the early Kings were buried. Lying next to Germany, Great Poland has always been the scene of struggles between the Poles and later German colonisers. The result is that the Western Pole is at once more energetic and more independent than his brother to the east. Moreover, the nobles never acquired so much power in the older provinces as in the new Marches in the east, and so there is a small class of Polish Yeomen in Great Poland.

To the east is the region of Mazovia, somewhat akin to Great Poland, but less fertile and more forested and marshy. Here also a rather democratic type of Pole developed, so much so that Boswell[1] believes that the change from a monarchy to a republic resulted from the union of Mazovia with Poland in the 16th century. Because Warsaw lay between Poland proper and the new regions added in the 14th century it was felt to be a better place for the capital than Posen or Cracow.

The next division of the great plain is Polesia (Fig. 114) to the east

[1]B. Boswell, "Poland and the Poles," 1919, London.

of Brest-Litovsk. Here, as the last name implies, was the beginning of Lithuania, and here Russia placed her military lines of defence against the German menace. They were thus so far west of Germany that she could not be flanked by Prussia on the north or by Austria (from Galicia) on the south. Polesia is a region of marshes and of forests (as the name *Les* implies in Polish). Here is indeed the last great forest left in central Europe, that of Bialowicz, where till lately boar, deer and even bison roamed freely as in prehistoric times. In the adjacent marshes of the Pripet live bear, elk and beaver, and on the islets are to be found the most primitive folk remaining in Europe. These marshes have always been a major obstacle to the invader and a refuge to the helpless. Russia was forced to maintain separate armies on her western front (based on Vilna and on Kiev) because it was so difficult to co-operate across the Polesian Marshes.

Turning now to the zone of Platforms or High-Plains, the most western was that first settled by the Poles. The region of the Oder to the west was soon Germanised, but in Upper Silesia the Poles remain very densely settled near the coalfields. Lesser Poland includes the uplands between the Vistula and Chenstokova. Here a famous monastery was founded by Vladislav Jagiellon, which contains a picture of the Virgin, said to have been painted by Saint Luke. It is visited by nearly half a million Poles each year (Fig. 114). Around the heights of Lysa-Gora are many of the ruined castles of the Slachta (former nobility). Lesser Poland is naturally somewhat less fertile than the alluvial plains of Great Poland. Volhynia, the High-Plain between Lwow and Lublin, is more productive. It was the first region occupied by the Poles as they spread eastward in their conquests of the 12th century. In the southeast corner in Podolia begins that most valued region in Europe, the "Black Earth" of the Ukraine. Always it has attracted the Pole, both lord and peasant, though he had to fight for his living and often for his life against the eastern nomads.

Eastern Galicia is also known as Western Podolia, and is a land of flat uplands dissected by deep, broad ravines. In the valleys are the villages amid many trees, while to-day the uplands are largely devoted to wheat. Lwow (*pron.* Il-vof) is the Polish name for Lemberg, the chief city in this region. It is largely inhabited by Poles and Polish Jews, but the rest of Eastern Galicia is strongly Ruthenian (*i.e.*, Ukrainian or Little Russian). Here was the great medieval route from the east to the west, Lemberg being the meeting place. Two famous sites in Polish history lie to-day just outside the boundary. They are the fort of Kamenets and the battle-field of Chotim on the Dniester in the extreme southeast.

B. The Racial, Linguistic and Religious Groups in Poland

The essential common feature of the Slavs is their language, for there is little uniformity in culture and religion, and there is considerable variety in race. The Slav languages include Russian, Ruthenian, Polish, Czech, Serbish, Bulgarian and a score of dialects. They form a large and important group of the Indo-European Languages, being linked closely with the Baltic group (Lithuanian) and fairly closely with Teutonic and Iranian. They belong to the Satem group as illustrated in the change of k, g, and gh (of the usual Indo-European root) into s, z, z. Examples are *Dekem* (ten) changes to *dest*; *granum* changes to *zrino*; and *hiems* (winter) changes to *zima*. The various Slav languages are closely allied, as the following words will indicate. The Russian word for town is *Gorodu*, and for five is *Penti*; in Polish these are *Grod* and *Pati*; in Czech *Hrod* and *Piat*; in South Slav *Gradu* and *Pet*.

There are five types of people of chief interest to us in this region: the Poles, Lithuanians, White Russians (between Vilna and Minsk), Ukrainians (Little Russians), and Great Russians. What are their differences and resemblances? Linguistically the Lithuanians are farthest removed and indeed are not Slavs. But racially they are indistinguishable from many northern Poles and White Russians. The Polish language differs from those of the three remaining groups, while the White and Great Russian differ very little. Some authorities consider that the White Russian dialect, being most isolated, is likely to be the "purest." If so it is spoken by a group who some ethnologists consider to be largely of Pre-Slav race (see *infra*).

If we examine Ripley's maps of racial characters, we find that the Slavs are all brakeph, though only moderately so in general. In Russia the head-index is almost uniformly 82. In Poland it varies much more, from 80 in the northwest, through 83 in Galicia to 85 in Polesia along the Pripet River. Judging by these averages there seems little room for large numbers of Nordics (77-80) or Mediterraneans (75-79). Considering stature, the Poles are the shortest of the Slavs, being only about 5 feet 4 inches on the average; but these low figures are said to be due to the large numbers of Jews.

The racial composition of Poland is naturally much like that in Germany. The biological boundary runs across Poland so that the southern third is dominantly Alpine (brakeph) and the northern two-thirds is dokeph (Fig. 115). It has usually been assumed that these northern people were either Nordic or mixed Nordic-Alpine. The broadest heads according to Ripley are in the Pripet marshes, but

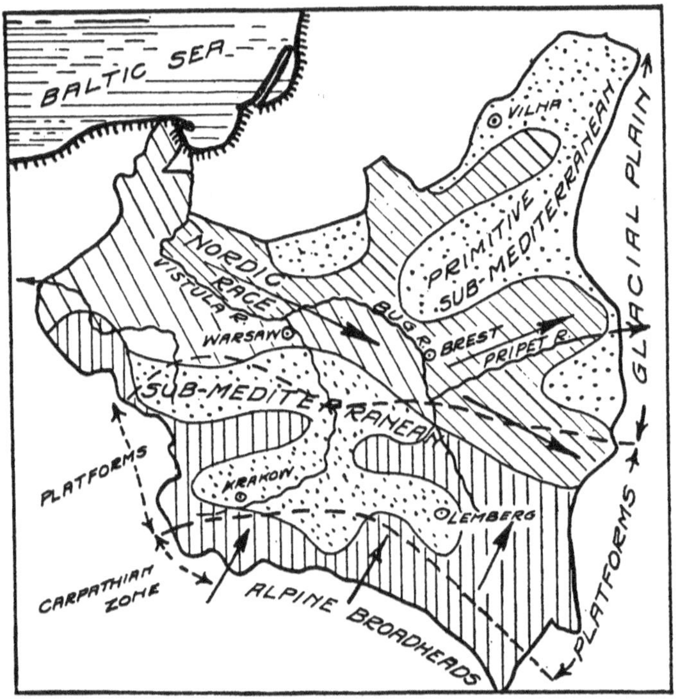

Fig. 115.—The Races in Poland showing Alpine, Nordic and Sub-Mediterranean Groups. (Generalised somewhat from Czekanowski and De Martonne.)

Czekanowski makes these sub-Nordic. Ripley's suggestion that the Western Poles are Slavs "Teutonised away from their original characteristics" seems to the writer as likely as that there is a survival of Pre-Slav Mediterraneans in Poland. However, since the whole problem has a bearing on the origin of all the eastern European peoples the alternative interpretation is given in a later paragraph.

The following table taken from Ripley shows how the general colouring of hair, skin and eyes varies in Russia. Unfortunately only the southeast of Poland is represented; but Ripley states that the Poles are nearly as light in colour as the Lithuanians. The Little

TABLE V

PERCENTAGE OF TYPES IN POLAND

	Lithuanian	White Russian	Podolian	Great Russian	Little Russian	East Galician
Blond.............	67	57	55	40	33	28
Mixed............	28	31	29	40	46	32
Brunet............	5	11	18	20	20	40

Russians still show blue or beer-coloured eyes with light brown hair, but in Galicia we find that the hair is dark and the eye rarely blue, so that these Mountain Ruthenians are sometimes called "Black Russians."

Czekanowski in a recent paper[2] indicates the presence of an early stratum of Pre-Slav tribes, which have been submerged by Alpines from the south and by a great expansion of Nordics from the northwest. In Neolithic times he pictures a merging of Mediterranean and Lapponid (*i.e.*, early Ofnet Alpines) to form his Pre-Slav types, who seem to have settled mainly along the middle Vistula west of Lublin; but nests of these folks are still found along the Warta and Narev. In the Bronze Age the brakephs advanced into Poland; the dark, tall Dinaric type being abundant in the southeast, and the normal Alpine type being more common in Silesia and Little Poland (Fig. 115).

The Nordics expanded into Poland from the northwest. Their chief area to-day is along the Warta and the lower Vistula, which they reached from the Pomeranian area. But in the northeast is a second somewhat isolated area which may have come from Lithuania. A third area includes Volhynia and links the Bug and the Dnieper Rivers. There has, however, been a great mixing all along the margins of all these main racial expansions. It appears as if the Nordics followed the northern water-courses and so isolated the earlier Pre-Slav peoples. Czekanowski thinks that the original Poles were Nordic, since near Halicz (their old capital) this stock is especially strongly represented. The so-called Mediterranean element reached the Dniester basin in very early times as is shown by the "Painted Pottery" found in Podolia. Later, further complications resulted because Rumanian

[2] J. Czekanowski, "Anthropologie von Polen," Petermann Mitt., 1929, pp. 116-117.

shepherds wandered along the Carpathians far to the west. Possibly the Alpines along the Warta are relics of the period when this was the great "Amber route" from the Baltic to the south. These data are charted in Fig. 115, which is much generalised from a map given by De Martonne.

The present writer draws attention to several curious features in the map (Fig. 115) which seem to him to need further explanation. The most obvious "refuge" in central Europe for primitive people would be the Pripet Marshes, yet we do not find the Pre-Slavs here but a sub-Nordic group according to Czekanowski. So also the obvious corridor from the plains to the south was along the Vistula. According to the writer's deductions elsewhere, this corridor should be occupied by the latest and most powerful tribes, whereas from Warsaw to Cracow there seems to be a solid block of Pre-Slav (Sub-Mediterranean). The distribution would indicate that the so-called Pre-Slavs invaded the Alpines rather than the contrary. It seems best, therefore, to await further research before we adopt the theory of the survival of a numerous Pre-Slav, Pre-Nordic people in the upper Vistula basin and elsewhere in Poland.

C. The Problem of the Jews

The Jews in Europe are a group with a marked religious unity, a fairly uniform linguistic unity, but with very little racial unity. To understand the peculiar position of the Jewish people we must study their history far back of modern times. To-day there are about 13 million Jews, of whom four million live in Russia, two million in Poland, two million in New York, one million in or near Hungary, half a million in Rumania, and half a million in Germany. The cruel treatment of the Jews by the Nazis, which seems to be based on an entire ignorance of their history, luckily affects only a small group of this ancient people. Since the Jews are more densely settled in Poland than in any other country of Europe, this seems to be an appropriate place to consider their characteristics in so far as they affect the nations of Europe.

It seems probable that the Jews when they lived in Palestine were essentially of Mediterranean race, like the vast majority of people in that region to-day. They spoke a Semitic tongue, for the most part Aramaic, akin to but not the same as Hebrew. The latter language was, however, taught in the Jewish religious schools as it is to-day.

In modern times we find that the Jews belong to two well-defined classes, often called the Spanish and Polish types. The former (Sephardim) are dark, slender dokephs (long-heads), definitely of the Mediterranean race. The latter type (Ashkenazim) are nine times as numerous and are short, stocky brakephs (broad-heads), with little or nothing in common as regards race with their Spanish co-religionists. The Ashkenazim speak Yiddish, which is essentially a German dialect

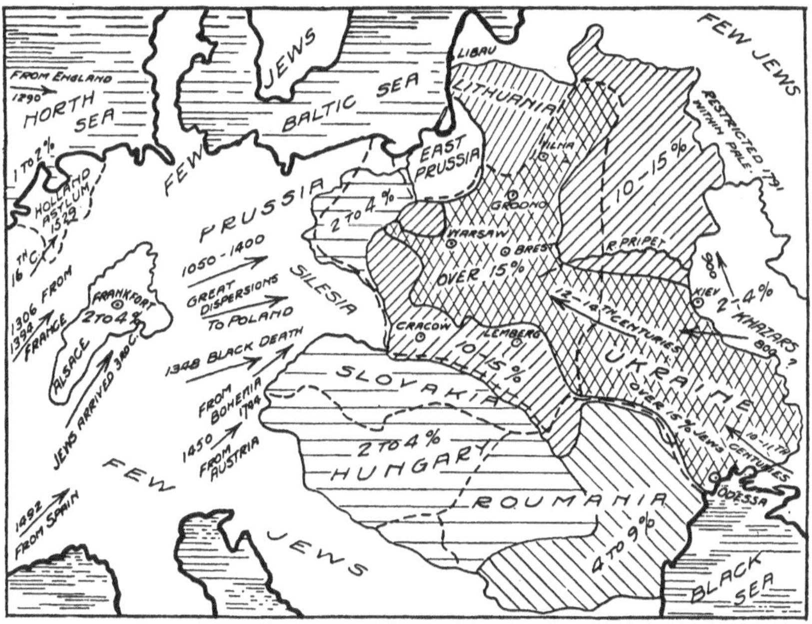

Fig. 116.—The Distribution of the Jews in Europe; mainly in the Ukraine, Poland and Lithuania. Data as to their expulsions from various countries are added. (Somewhat generalised after Bowman and Ripley.)

developed along the Rhine in medieval times. It is written in Hebrew letters and contains a considerable proportion of Hebrew words, but is nevertheless dominantly German. How have these remarkable changes of race, language and habitat come about?

Long before the destruction of Jerusalem (A.D. 70) the Jews had emigrated in large numbers from Palestine. Many had gone to Egypt and thence had travelled into other more western Mediterranean lands. The enforced migrations to Mesopotamia resulted in many Jews staying in the latter country when some returned to the lands of their fathers. When Jerusalem fell a second time in A.D. 135 it is said that half a million Jews perished. For a time they were not allowed to live near Jerusalem, and they naturally joined their brethren in North Africa and Mesopotamia. Already we see two main lines of migration developing, and these have remained somewhat distinct since those early centuries.

After the Moslem conquest the Jews were well treated in Spain for some centuries, and flocked there in large numbers. Their worst persecution began at the end of the 13th century, when England expelled them and many were massacred in Germany. In 1306 they were forced to leave France and in 1348 eighteen hundred were burned in Strasburg. In 1328 six thousand were slain in Spain, and in 1391 fifty thousand more perished in that country. But almost the whole Jewish body was expelled from Spain in 1492, and they fled to Morocco, to Naples, to Amsterdam, and especially to Saloniki and Constantinople, where the Turks gave them an asylum from the ferocity of the Christians. These migrations chiefly concerned the Sephardim, or Jews of the Mediterranean race.

Meanwhile the other wing of the Jews in Mesopotamia had climbed the Great Scarp of Anatolia and had now come in contact with an entirely new type of people, the Alpine Race. Here among the Armenians the so-called "Jewish Nose" is much more common than it ever was among the Jews of Palestine, where indeed it is rare. The Jews spread through Anatolia, ultimately passed the Caucasus and many reached the lands of the Khazars in southeast Russia. There were Jews in Kertch, north of the Black Sea, before the time of Christ, but the numbers increased largely in the next five or six centuries.

The Khazars, also known as the Royal Scythians or White Huns, were a group of people allied to the Turks and Magyars in race, who had largely settled down to a town and trading life during the second century of our era. They lived around the Caspian Sea, which was known to the Arabs as the *Bahr al Khazar*. The capital of the Khazars was first at Samander (north of Baku), but later was removed to Itil at the mouth of the Volga (now Astrakan). During the Golden Age of the Khazars (600 to 950) they were the great traders of southern Russia, and controlled the region between the Dnieper River and the Caspian (Fig. 131). They built a great "White Fort" at Sarket (near Azov) to guard their trade along the north coast of the Black Sea.

About 740 A.D. their ruler, Khan Bulan, was converted to the Jewish religion, as were many of his subjects, and after the time of Khan Obadiah none but a Jew was allowed to reign. Their daughters married neighbouring princes, so that Leo IV, Emperor of the Eastern Roman Empire in 775, was the son of a Khazar Jew.[3] However, in 862 the Nordic Varangians founded a Russian monarchy and this soon came into conflict with the flourishing people to the southeast. In 965 Sviatoslav of Kiev smashed the Khazar kingdom; and though the Khazars survived in the Crimea until 1016, after that date they soon vanished from history. But there is little doubt in the writer's mind that the Polish Jew is largely derived from the Khazar Jews. Indeed it is likely that their keen trading instincts are in part due to their association with the Khazars, just as their facial appearance is possibly due to their Armenian blood. The Khazars undoubtedly traded with the Poles and Russians to their northwest, for Lemberg and Kalisz were on the great Trade Route of early historic times (Fig. 116). It is indeed rather suggestive that the Polish Jew's name for his own people is *Ashkenazim*, which is the same word ("Scythian") as was originally applied to the Khazars.

After the persecution of the Jews in West Europe, there is, of course, no doubt that many of them fled for refuge to Poland, and by the 12th century the two groups had come into contact. During the Crusades many of the Crusaders made little distinction between a Jew in Europe and a Moslem in the East. To his mind they were both better dead. No wonder that Poland appealed to them, for in 1173 the Duke of that country forbad attacks on the Jews. In 1264 Boleslav gave a Charter to the Jews of Poland ensuring them fair treatment. In the 13th century many thousands of folk in Hungary were converted by the Jews. However, the first Jagiellon began to persecute them in 1399, and they were never free from outrage thereafter, though they were better treated in Poland than in most other countries. In 1495 they were expelled from Lithuania, but by the next century there were half a million Jews in Poland. During the civil wars around 1648 Cossacks and others massacred 100,000 Jews. In 1791 the Jews in Russia were confined to the western zone called the Pale (Fig. 116). Many massacres (pogroms) have since occurred, notably in Nishni Novgorod in 1882 and in Kishinev in 1903.

The distribution of the Jews is given in the map (Fig. 116), which

[3]Marx, "History of the Jews," Chicago, 1931. See also the author's paper "Aryan, German, Nordic, Jew," in the "University of Chicago Magazine," Nov., 1935.

also shows some of their main migrations towards their refuge in Poland. The Jewish area in Europe extends from Odessa through the Ukraine and Poland to Lithuania. In this belt over 15 per cent. of the population are Jews, but only rarely and in the cities does the proportion rise above one-third. In Germany the number of Jews is almost negligible among the 63 millions of Gentiles. In all these central countries the Jews are racially the same as their neighbours, *i.e.*, they are a rather mixed variety of the Alpine type. The Khazars were near to the south Russians or Magyars in race, and since then the Jews like the Magyars have become still closer to the general Central European type. Their language Yiddish is essentially an Aryan tongue,[4] though its German origin is hidden by its being written in Hebrew characters. No intelligent person can object strongly to the main philosophy of the Jewish religion, on which indeed Christianity is based. It is, of course, the economic aspects of Jewish culture which excite the hatred of powerful groups of Anti-Semites, but the so-called racial and religious arguments lend themselves better to propaganda. The expulsion of Jews because they are non-Christian is eminently un-Christian. Their persecution because they are of "Non-Aryan Race" is the height of unscientific absurdity. Indeed, if the Nazis were logical most of the South Germans might equally well be expelled from Germany since they are of the same Alpine race as the Jews, and were clearly the originators of the hated Yiddish dialect.

D. The Early Development and Expansion of the Polish Nation

The Danish archeologist Muller has stated that the early Slav graves are marked by the presence of "Ear-chains," wooden pails and specially decorated pots. Using these criteria he believes that in the earliest times the Slavs extended to the Elbe and perhaps even to the Rhine. However, these Proto-Slavs(?) were thrust back to the Oder and here developed the tribes called Poloni by the Romans. The word Pole means plain or field, and the Poles are therefore the "Settlers of the Plain." They lived in little fortified towns (called Grody) for defence against the Norsemen and others. Kruschwitz, Gnesen and Posen, all situated between the Vistula and the Warta, were three of these early towns. To the northwest lived the tribes of Mazurians,

[4]For instance, "*Es is gar all's kein Neues nischt unter der Sunn*" is Yiddish. Yet any beginner in German can read it ("There is no new thing under the sun"). See L. Wiener, "History of Yiddish Literature," London, 1899.

and to the north the Pomerzanie (*i.e.*, "Folk by the Sea"), later known as the Pomeranians. Teuton peoples bounded them on the west, while the Ancient Prussians were a powerful Lithuanian tribe east of the mouth of the Vistula.

Polish history begins with the conversion of Mieszko[5] in 963, at the instance of Otto of Saxony. A bishopric was founded at Posen, and Poland (like Hungary) was thus civilised from the west. The

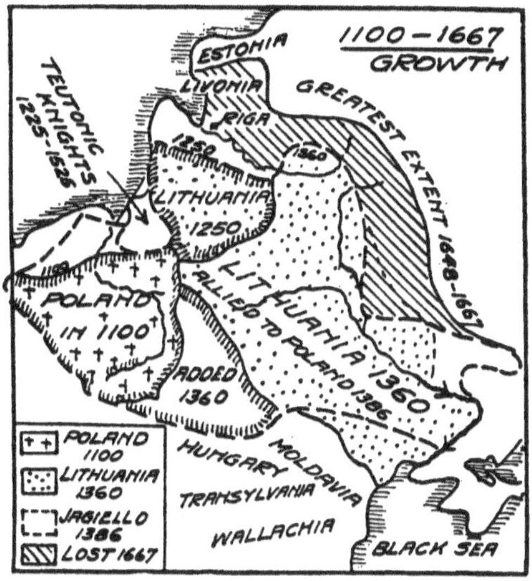

Fig. 117.—The Growth of Poland from 1100 to 1667, chiefly owing to the union with Lithuania.

country thus became a bulwark of the Roman Catholic Church in the northeast. Many of the new monasteries imported German labourers at this early date, who helped to civilise Poland. Boleslav (980-1025) greatly extended Poland, which before his time had included little beyond the basin of the Warta (Fig. 117). He conquered Pomerania, all the basin of the Vistula, and overran Bohemia. Thus Cracow and Kiev became Polish about 999. However, in 1040 Germans and Bohemians united and won back most of the recently annexed lands. Kazimir (1040-1058) was, however, established as a vassal ruler by the German Emperor mainly to curb Bohemia. In 1076 the Poles helped

[5]The following notes will help the reader to pronounce Polish names. *C*=TS; *G* is hard; *J*=Y; *W*=V; *CH*=H; *CZ*=CH; *RZ*=ZH; *SZ*=SH. Accent the Penult in general.

the Pope in his struggles with Henry IV, and in return Boleslav II was crowned an independent King. But civil wars ensued, and Poland once more became vassal to Germany. Boleslav the Wry-mouthed (died 1139) extended his realm to the Baltic and to Cracow and Lublin in the southeast.[6]

Partly as a result of the Mongol invasions, large numbers of Germans colonised the empty lands of Poland during the 13th century, especially in Galicia and Silesia, where ran the main routes from Germany to the Levant. These German communities were given many privileges, including a large measure of self-government. Cracow, Lwow, Posen and Plock also received many Germans in this way, and developed a population which was inevitably unsympathetic to the Polish rural dweller under the Polish nobles. Many Jews, as we have seen, were received hospitably during the 13th century, but they merged with the Poles to a greater degree than did the Germans.

In 1147 the Saxons and Danes succeeded in conquering the land between the Elbe and the Oder. This Duchy they called Brandenburg, from the old Slav town of Branibor. Teutons from the Zuider Zee were settled here, and much of the coast near the Vistula was lost to Poland. In 1194 the Poles in part conquered the "Old Prussians," but were unable to convert them to Christianity. In 1225 the Mazurian (Polish) leader invoked the aid of the German Knights of the Cross against the Prussians. At the end of the war the Knights occupied the conquered lands and defied their Polish allies. After many civil wars Vladislav Lokietek, with the consent of the Pope, was crowned independent King of Poland in 1320. The Emperor, Bohemia and Brandenburg made war on him (in support of a rival, Kasimir), and in 1335 Silesia was ceded to Bohemia by the new King Kasimir. A little later Pomerania was abandoned to the German King.

The relations of Poland to Bohemia and Hungary during the 12th and 13th centuries are of much interest. Bohemia early joined the Empire whose princes were always pressing on Poland's western limits. Poland naturally joined the enemies of the Empire, of whom Hungary was one of the chief. Even when Hapsburgs became the rulers of the Empire this made little difference, since Hungary was hostile to the Hapsburgs until early in the 16th century. Silesia was, of course, a bone of contention between Poland and Bohemia, as it later became

[6]E. H. Lewinski-Corwin, "Political History of Poland," New York, 1917; which has been frequently used in this section.

between Austria and Prussia. In later years the Danube outlet attracted Bohemia and Hungary, while Poland lost touch with Hungary, from which she was, of course, always somewhat isolated by the Carpathians.

Kasimir the Great (1333-1370) won Volhynia in the southeast from the Mongols, and many Poles settled here, and also in the lands recently laid waste by the Knights north of the river Bug. Kasimir encouraged trade, and built many roads. He welcomed the Jews and tried to build up a middle class to balance the power of the Church and the Nobility. In 1364 he founded a University at Cracow, only a few years later than that at Prague. Kasimir was followed by his nephew Louis (Ludwig) of Hungary. The accession of the latter's daughter Jadwiga (Hedwig) was discussed in the last chapter.

In 1386 Vladislav Jagiellon married Jadwiga. This choice by the Poles was mainly due to the fact that the Lithuanians (like the Poles) were bitter enemies of the German Knights. Jagiellon brought into the Polish realm, Lithuania, White Russia, Ukrainia and Volhynia (Fig. 117). Moldavia, Wallachia and Bessarabia became his vassals by 1396, so that Poland controlled almost all eastern Europe between the Baltic and the Black Seas. Indeed Polish was the language of the upper classes in Moldavia until 1800. In 1410 the German Knights met the Polish army at Tannenberg among the Mazurian Lakes. Here the forces of the Knights were crushed, and Poland soon became one of the most Powerful States in Europe.

Let us pause to consider some of the geographical aspects of this period of Poland's great expansion. It is obvious that Poland never made any attempt to subjugate the lands south of the Carpathians. By the time that Cracow was captured their policy consisted of a "Drang nach Osten" almost as marked as that of Austria in the 19th century. The west was soon closed by the growth of powerful German princes, notably Brandenburg. Although there was a slight movement of the boundary to the east after A.D. 1000 the Poles have not since budged from their limits near the Oder as fixed about 1100. On the north the moraines and lakes of Mazuria helped the German Knights to remain fairly independent of the powerful Polish Kings until Brandenburg acquired that territory in 1618. It was, therefore, towards the east and southeast that the lack of any topographic boundary inevitably led to the Poles advancing their territories in that direction. There is, however, in this area an edaphic (soil) and climatic boundary (as shown in Fig. 130), for the steppes reach up to Kiev, which lies on the common boundary of the Forests and the Grasslands. The Poles and Lithuanians always had a firmer grip on the forested lands, the steppes being the abode first of the Petchenegs and Khazars and then of the Cumans and Mongols throughout this early period.

E. The Jagiellions and their Wars with the German Knights

The religious differences in the large domain of Lithuania led to serious trouble. The people of the Ukraine remained true to the Greek Church and were not given the same privileges as the Roman Catholics, and this led to abortive revolts about 1431. Very characteristic of Poland was the caste of nobles, who did not need to own land provided they were possessed of an "escutcheon." Thousands of folk of low economic and social status gradually joined the nobility. The Statutes of Nieszawa in 1454 gave nearly all the power to an oligarchy of these nobles known as the *Slachta*.

Vladislav III became King of Hungary as well as Poland, and in 1443 won a notable victory against the Sultan at Varna, but he was killed in battle with almost all his followers in 1444. Kazimir Jagiellon reigned from 1447 to 1492, striving to reduce the growing power of the Slachta. The cities now began to lose their importance after a period of considerable prosperity. The loss of Constantinople and the advance of the Turk were spoiling the land trade *via* the Ukraine. Moreover, the citizens were largely foreigners such as Germans and Jews, who had little or no patriotic feeling.

Meanwhile Poland was still at war with the German Knights of the Cross, but in 1466, partly owing to aid received from the dissatisfied people of Prussia, Poland defeated the Order. Pomerania and Western Prussia went to Poland, while the Grand Master agreed to hold the rest as a fief of the Polish King. Dantzig now became a very valuable port of Poland and the economic importance of the country was much enhanced. Close relations with Hungary obtained at this period, but unfavourable conditions for Poland were developing in the south and east. In Silesia the process of Germanisation was advancing rapidly. In the southeast the Turks conquered Moldavia in 1480; while Russia was expanding in the east, and Lithuania was growing restless.

During the 16th century the advance in power of the nobles was balanced by a decrease in the rights of the peasants. Gradually grain became of more and more importance, and hence a supply of serf labour was vital. The nobles passed many laws tending to anchor the peasant to the land. Fleets on the Vistula carried cargoes of wheat, rye, hemp, tar, honey, wax, fats, lumber, skins and furs to Dantzig and other ports on the Baltic. Special privileges were given to the nobles as to export and import of goods which were not allowed to the merchant, so that in the 17th century most of the cities had declined still further in prosperity.

The Reformation roused the greatest interest in Poland, and the educated people welcomed it everywhere except in Mazovia, the region around Warsaw. Partly owing to religious troubles the King was willing to allow his nephew, the Grand Master of the German Knights, to become a secular Duke, for which privilege he paid homage to the King in 1525. This step later on contributed greatly to the rising power of Prussia.

At the other extreme of Poland is the Ukraine, the most fertile "prairie" in Europe; but owing to the raids by Mongols or Turks it was not possible for farmers to live there in peace. It is stated that the Poles did not attempt to proselytise the Ukrainians, but left them freely to join the Greek Church. Calvinism made rapid progress in Poland, while Unitarians increased greatly in Ruthenia, where the teachings of Socinus had many followers. The lower classes, however, remained true to the Roman Catholic religion, as did the University. The eminent Pole, Copernicus (1473-1543), wrote not only on astronomy but on political geography in a treatise on the rights of Poland in East Prussia.

Struggles between the Livonian Knights of the Sword and the Russians led to the former offering Livonia to Poland. The Grand Master imitated the Hohenzollern in East Prussia, and became secular Duke of Courland and a vassal of Poland in 1561. In 1569 the Jagiellon King gave up his hereditary rights in Lithuania in order to effect a closer union between Poland, Lithuania and Ruthenia. It was arranged that there should be one King of the two autonomous areas of Poland and Lithuania. The result is sometimes referred to as a Republic with an "elective" King. Warsaw became the capital about this time. In 1572 Sigismund the Second died and the Jagiellon Line came to an end.

In the later centuries of Poland's history she is mainly occupied with maintaining her control of the eastern plains. In such an environment there is no refuge for weaker states in the struggle; they must yield or disappear. Here Russia and Poland challenged each other, and neither would yield, as Lithuania had done to the superior culture of Poland. It was in a sense the age-old war between Byzantium and Rome. The religious cleavage was distinct and permitted of no practical compromise. The struggle was prolonged for three centuries and at the end the state of Poland had vanished from the map.

F. Foreign Kings and Religious Conflicts

By the terms of the Constitution the Polish people met at Warsaw in 1573 to cast their votes for the new King of their "Republic." Henry of Valois was elected, but he resigned almost at once to become Henry III of France. Now began that series of foreign princes whose election often enabled foreign powers to intrigue in Polish affairs. However, in 1576 Stefan Batory of Transylvania (who had married a Jagiellon) was elected and made a good ruler of Poland. Batory, however, strongly supported the Catholic Church, especially the Jesuits who had now acquired great power in Poland as in Bohemia. He built up a strong army utilising the Cossacks as cavalry, but he died after a reign of twelve years. After a period of civil war, Sigismund Vasa (a Swedish prince with a Jagiellon mother) was elected King. He also was a strong partisan of the Catholics and a firm friend of the Hapsburgs. During his reign a Polish army attacked Russia and occupied Moscow in 1606. His son Vladyslav was actually elected Tsar of Russia, but owing to Sigismund's jealousy and to the unpopular religion of his son, this favourable opportunity to unite Poland and Russia was wasted.

Turkey agreed to help the Bohemian Reformers in 1620 in their opposition to Hapsburg rule, while the Polish King ranged himself on the Austrian side. Though the Turks defeated the Poles near Jassy, they were prevented from aiding the Bohemians, and this enabled the Catholics to crush the Reform movement. A few years later wars with Gustavus Adolphus resulted in the loss of Livonia and other territory to the Swedes. However, in 1634 the Russian armies invading Poland met with defeat, and the Tsar lost Smolensk and gave up all claims to Livonia. The year 1660 marks the greatest area under Polish control, as appears on Fig. 117.

In 1595 a very important Synod at Brest attempted to reconcile the Greek and Catholic Churches, as had been tried in Florence in 1439. As a result most of the Ruthenian Clergy joined the compromise or "Uniate" Church. In this they were permitted to keep their own Greek ritual and the priests could marry, but they acknowledged the Pope at Rome as their spiritual Head. The nobles by this time had become yet more powerful, the wealthier living in fortified castles. For instance the Radziwill leader owned 16 cities and 583 villages, while Potocki owned three million acres and controlled 130,000 serfs (Lewinski-Corwin). On the other hand the unfortunate peasants had practically no rights. They had to buy and sell through the Lord of the Manor, and could not leave the estate without his permission.

G. The Wars with the Cossacks, Swedes and Russia

From 1635 to 1656 was the period of the terrible Cossack wars. Many factors, some definitely geographical, led to these disasters. The Ukraine is a vast prairie mostly of rich black soil and in general treeless. In climate and in origin it is in fact very like the fertile western prairie in the United States. Here was a debatable land much like that in America in the early days. Mongols and Turks were ready to raid it, Russia was anxious to acquire it, while Poland controlled it by virtue of its alliance with Lithuania. It was the refuge of the Cossacks, who preferred a wild roving life, in which they raided Poland or Turkey, to the monotony of a farm. The western portion was settled by the Ruthenians (a modification of the word Russian) who were out of touch with Poland owing to their religion which was Uniate; and for the same reason not much in favour with the Russians who were Greek Orthodox. To control the Cossacks the Poles in 1635 built a strong fort at Kudak on the Dnieper, which the Cossacks destroyed. A Polish force under the great land-owner Wisniowiecki (*pron.* Vishny-o-veet-ski) opposed reconciliation and tried to subdue the rebels. Indeed the nobles hoped to conquer the Cossacks and use them as serfs on their estates, and so were averse to peaceful measures.

In 1648 the Great Rebellion of the Cossacks under Chmielnicki broke out, but against the "Magnates, Jews and Jesuits" rather than against the King. The Cossacks defeated Wisniowiecki and with an enormous army invaded Poland, but they withdrew when concessions were made to them by the new King, John Kazimir (1648-1668). Several battles followed later in which the King was defeated by the Cossacks and their allies the Tatars, but a peasant revolt in Poland caused the return of the Polish army. Fortunately for Poland Transylvania and Wallachia joined her to curb the growing power of the Cossacks. In 1654 the Cossack leader turned to Russia and offered the Ukraine to the Tsar provided that the Cossacks had a fair measure of autonomy. The last two years of the war (1656-58) were the most terrible. Russians and Cossacks united against Poles and Tatars, and as a result 120 cities were destroyed and possibly 100,000 people killed.

Meanwhile Russia, Sweden and Brandenburg had turned upon Poland while she was engaged in civil war, and they captured the whole country with little trouble. Popular risings under Czarniecki, helped by dissensions among the invaders, resulted in the Swedes leaving Poland in 1660, but north Livonia was ceded to Sweden. The excesses of the Protestant invaders naturally turned to the advantage of the Catholic party, which has been dominant in Poland ever since.

Troubles with Turkey and the Cossacks led to Poland making peace with Russia in 1667, when Smolensk and much of the Ukraine was handed to the Tsar (Fig. 117). In 1642 Turkey was bought off by yielding Podolia and a portion of the Ukraine. In 1674 John Sobieski, who had fought gallantly against the Turks, was elected King; and he immediately moved against the Turks, winning back most of the Ukraine by 1676. In 1683 he raised the siege of Vienna, which was being attacked by an immense force of Turks, and won several decisive victories against them. It is possible that the Poles altered the whole course of European history by these successes. In 1681 Sobieski gave up Kiev and further areas in the Ukraine to the Tsar in order to gain his aid against the Turks. "With the death of Sobieski in 1696 ended the glory of Old Poland."

A Saxon prince was now elected King as August II, but his attack on the Swedish King Charles X failed ignominiously. The Poles distrusted his plans and he was deposed in 1704. Swedes and Russians now overran the country and each put forward candidates for the crown. Charles X nominated a Pole, Leszczynski (*pron.* Lesh-cheen-ski), and he was successful as long as Charles was in power; but August returned after Charles was crushed by Russia at Pultowa in 1709. A civil war of the worst type was waged during 1715-17 between August and the insurgent Poles. "Everything which had survived former wars was destroyed in this civil strife." August died in 1733 and Russia supported his son, August III, who reigned 28 years. Poland gradually declined in military, mental and moral status. In 1764 the Russian Empress arranged that Poniatowski (said to have been her lover) should become King. Thereafter the policy of Poland was in general decided in Russia and carried out under a guard of Russian armies. Social conditions now became so deplorable in Poland that some writers declare there was some excuse for the Partitions.

H. The Partitions of Poland and Napoleon's Duchy

The partitions of Poland have been referred to in several chapters. They were due to the greed of Prussia, Russia and Austria, Frederick of Prussia being perhaps the chief plotter. Russia already had prac-

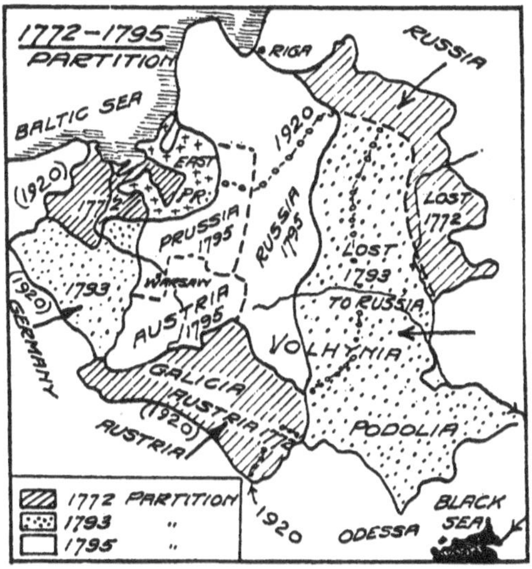

Fig. 118.—The Partitions of Poland from 1772 to 1795. The 1920 boundary is indicated.

tical control of Poland, but when it seemed possible in 1772 that Austria might conquer Moldavia, Catherine of Russia agreed to give up Moldavia to Turkey and recompense herself in Poland. Many heroic defences such as those of Chenstokova and Cracow showed that the Poles could die in the defence of their country. Russia received South Livonia and much of White Russia; Prussia gained West Prussia, and Austria Galicia (Fig. 118).

During the period 1788-1792 the Reform Party in Poland under Malachowski joined Prussia against Russia, and drew up an enlightened constitution. They were opposed by a Pro-Russian Party under Branicki, which was supported by the Russian armies. King Poniatowski and his officers proved timid and the Russians occupied Warsaw. The Second Partition followed in 1793, when Prussia took enough territory from Poland to balance her losses to the French. Russia annexed the rest of White Russia, Ukraine and East Volhynia (Fig. 133). It is to be noted that in this partition, as in the first, Russia claimed no area where the Poles were in the majority. On the other hand, the Second Partition gave Prussia the cradleland of Poland and brought the boundary close to Warsaw and Cracow. Austria took no active part in this Second Partition. At this stage Poland owned much the same area as she does to-day.

In 1794 Kosciusko (*pron.* Kost-see-ush-ko) led the Polish patriots against the Russian armies and at first won many battles. However, the combined Russian and Prussian forces were too powerful for him, and the revolution was ruthlessly crushed. In 1795 the remainder of Poland was divided between the three Empires (Fig. 118). Germany received Warsaw and the area near to East Prussia, Russia obtained Vilna and pushed her frontier west to Brest; Cracow and the Middle Vistula region went to Austria. No European nation except the backward Turks seem to have been much perturbed by Poland's destruction. We read, for instance, that "Revolutionary France did not exhibit any particular enthusiasm for a country of Nobles." On the whole the Russian section was not particularly badly treated, and the new outlets for grain *via* the Black Sea gave increased impetus to industry in Poland.

The hopes of Poland now rested on Bonaparte, who was able to enlist large bodies of Poles by his promises to help their country. After his victories at Eylau and Friedland in 1807 he instituted the Independent Duchy of Poland under a Saxon Prince Frederick August. The new State was created chiefly at the expense of the German gains, and included the basins of the Warta, Narev and Middle Vistula. Napoleon united almost all the Polish-speakers (shown in Fig. 122) in this Duchy. Only near Dantzig was a large homogeneous Polish community omitted, though towards Lemberg in the southeast the population was mostly Polish for some distance outside the boundary. The present 1920 boundary seems much less reasonable, though the White Russians are now the main victims.

In 1809 Austria invaded Poland and occupied Warsaw. Poniatowski thereupon marched into Galicia and captured Cracow, and after Napoleon's victory at Wagram Western Galicia was added to the Duchy. The disastrous French campaign against Moscow in 1812 resulted in Russia occupying the Duchy. At the Congress of Vienna in 1815 the district around Posen was given to Prussia, but the Duchy as a whole was annexed by the Tsar as an independent kingdom with himself as King of Poland. Till 1848 Cracow was made a Free City, but it was then annexed to Austria.

Poland was at first treated fairly by Tsar Alexander, but a revolt broke out against his successor in 1830 with the result that Poland was incorporated in the Russian Empire. Further revolts in 1863 were cruelly punished, 18,000 Poles being exiled to Siberia, and some 60,000 being removed to Russia. The "Vistula Provinces" (*i.e.*, Poland) were now ruled by Military Governors. Every effort was made in Lithuania and Poland to replace Polish culture by Russian; so much so that in 1865 the Pope severed diplomatic relations with the Tsar owing to Russia's attitude to the Polish Catholic Church.

Conditions in Posen were nearly as bad, and in 1886 a Colonisation Commission was established to buy land from the Poles and to replace them by Germans. In Austria, on the other hand, after her defeat in 1866, a considerable measure of Home Rule was given to the Poles in Galicia. Here "Sokol" societies were instituted where military training was given in the guise of athletic exercises, so that Galicia became a rallying-ground for the Polish nation. In Eastern Galicia, however, the Ruthenian population still remains Uniate (Greek Catholic), and there is still some antagonism between the Ruthenian peasant and the Polish artisan or upper-class citizen which time will, one hopes, allay.

In 1914 while the bulk of the Poles longed for a free Poland, there was a strong party in Galicia which favoured union with Austria, and another which believed that more progress could be made as an independent kingdom under Russian Protection. Prussia alone had no political friends. As the war proceeded many promises were made by the three Emperors, but all conditions were changed when the Tsar fell in March, 1917. Pilsudski and his Polish Legions, Paderewski and the help gained in America, Haller and his army directed from Paris, all combined; and Pilsudski was made Head of the Polish Government in November, 1918. In May, 1919, the Poles attacked the Soviets, and in November, 1919, East Galicia was alloted to Poland and not to the Soviets. Further conflict with Russia resulted in the Poles entering Kiev in May, 1920, but a strong Russian counter-attack brought the Russians to Warsaw. With French help Pilsudski repelled the Soviets; and at the Peace of Riga a new frontier, far to the east of the line through Brest suggested in 1919, was agreed upon

(Fig. 122). This new frontier is much like that held by Poland after the second Partition in 1793.

Poland to-day among her population of 29 millions includes about eight millions of "foreigners." These comprise nearly five million Ruthenians and White Russians, two million Jews and one million Germans. Reference to Fig. 122 will show that, as far as areas are concerned, this Ruthenian-Russian Territory (which we may term the "Russian Zone") is the largest single block in Europe governed by a nation speaking a foreign language. Ruthenian in the south and White Russian in the north are both dialects of Russian, and are far removed from Polish. More important still the great culture-cleavage line in Europe between the Roman Catholic and Greek Orthodox (or Greek Uniate) religions passes through Brest, and is therefore to the west of the present boundary. Just across the present boundary (a mere artificial line on a boundless plain) is the White Russian Republic linked closely to the Soviet Union. It seems inevitable that the White Russians in Poland will desire to join their free brothers in the east. An impartial student cannot but see all the elements of a future conflict in this setting.

The Poles, however, appeal to history. They show that this debatable "Russian Zone" in Poland was conquered by the Lithuanians as far back as the 13th century, before Muscovy was an important power. It was civilised from Poland after 1386 and controlled from Poland until the Third Partition in 1795. Its aristocracy spoke Polish for 400 years, and only from 1795 till 1920 (some 125 years out of five centuries) has it been united with Russia. No doubt Poland believes that she can govern this "Russian Zone" as satisfactorily as Switzerland controls its French zone in the west. For the sake of the Peace of Europe it is to be hoped that time and goodwill will reduce the friction between the two nations, for this zone seems to the writer perhaps the most obvious danger-spot in the continent.

PART V

THE EASTERN NATIONS OF EUROPE

CHAPTER XXIII

RUMANIA—A LAND OF HYBRID CULTURE AND MANY INVASIONS

A. The Build of Rumania

Rumania since the Great War consists essentially of the Great Carpathian Arc and its flanking plains. But there is so much diversity in this large area of 122,000 square miles that it is well to divide it into eight smaller homogeneous units, each of which has a rather characteristic environment. Proceeding from west to east we find the following zones:

1. The Eastern Margin of the Alfold, an alluvial plain bordering Transylvania.
2. The Bihar Plateau, a dissected block of ancient rocks.
3. The Transylvanian Basin of about 1,500 feet elevation.
4. The Eastern Carpathians, a relatively low and narrow barrier.
5. The Transylvanian Alps, high peneplains dissected by deep gorges.
6. The Moldavian Platform, of young rocks about 800 feet elevation.
7. The Wallachian Plain, alluvial washed from the Carpathians.
8. The Dobruja, a low Platform of varied rocks.

The general topographic features can be made out from the Block Diagram given in Fig. 119, where the spectator may imagine himself in a balloon over the Black Sea looking westward. Especially noticeable is the striking but imperfect barrier of the Carpathians, in which it is characteristic for the river-divides to be far removed from the highest portion of the mountains. The Danube and the Alt (Aluta)[1] are two large rivers which have kept their pre-uplift or antecedent courses right through the whole series of young folds and plateaux. Moreover, many of the eastern streams if followed upstream, pass through the main heights only to rise in lower divides on the western slopes. Thus the Bistritsa, a tributary of the Sereth, almost reaches the head of the Theiss (Tisa) at the Prislop Pass, well to the *west* of the main mass of the Carpathians. So also the Buzeu nearly reaches the head of the Alt River, while the latter flows through the famous Red Tower Gorge and transects the whole system as stated. The Jiu, further west, rises well north of the Vulcan Pass at Petrosany, whence there is an easy road north to the Marosh (Maros) Basin. Lastly, the

[1]Place-names in Rumania often have Rumanian, Magyar and German variations.

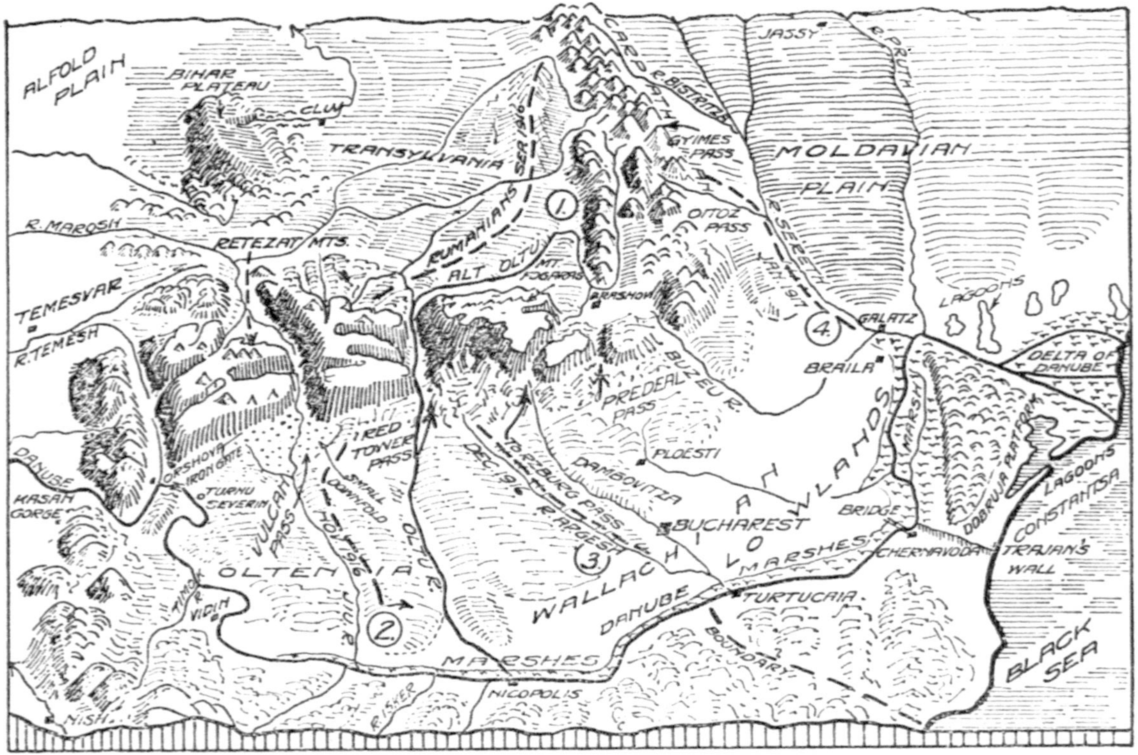

Fig. 119.—A Block Diagram to explain the survival of the Rumanians on the Plateaux of the Carpathians. The main War-Fronts in the German Attack on Rumania in 1916-1917 are shown by numbers. Notice the Antecedent Gorges through the lately formed Mountains. (See also Fig. 105.)

river north of Orshova leads by a low gap (perhaps partly due to a graben) to the head of the Temes River beyond the Carpathians. Often enough these gorges are so narrow that the main roads use higher but wider gaps in preference to the river gorge. Thus the Vulcan Pass is a more important route but much higher than the gorge cut by the Jiu just to the east. The Predéal Pass is also much more used than the lower Oitoz Pass to the east.

It is logical to study the main features of the Carpathian Arc first.[2] This mountain area is as extensive as the Alps and its length is much the same, but it is not nearly so high, nor so great a barrier to communications. Thus the Tatra in Slovakia rises to 8,737 feet, and Mount Negoi in Transylvania is only 8,340 feet, while Mont Blanc in the Alps is 15,871 feet. The Carpathians exhibit the usual arc-like shape of fold-mountains in plan, but the arc is nearly broken in the middle where the ranges are only 60 miles wide, instead of about 180 miles as in the north and south. The passes are low enough for a common vegetation to extend right across the divide, and it was through the low passes of this part of the arc, through Vereczi, Uzsok and Lupkov, that the Magyars reached the Alfold about A.D. 900.

The central part of the Carpathians has apparently been down-warped since the original uplift of the arc, and only the upper "mantle" of Flysch beds is visible here, though elsewhere much older rocks have been exposed by erosion (Fig. 35). Much faulting accompanied by volcanic flows also occurred during the down-warping. These volcanic hills extend from the upper Theiss near Tokay southwards to Brashov (Fig. 112). As usual, the lavas appear only on the inner side of the arc, and the upper Theiss, Upper Maros and Upper Alt flow in the longitudinal valley separating the volcanic ridges from the fold-ridges to the east. In fact the fertile basins of Gyergio and Csik are mainly due to the lavas having blocked these river valleys.

There is not much similarity between the build of the northern Carpathians and the southern Carpathians, which are those crossing Rumania. The latter are inhabited by Rumanians, Magyars appear only in the Transylvania Basin, while Ruthenians occupy the northern corner in Bukovina in the bend of the Upper Prut. In these mountains small villages may be placed as high as 3,300 feet, whereas in the Northern Carpathians we find no villages (unless near mines) above 2,300 feet (De Martonne). Two trunk railways cross these ranges by

[2]E. de Martonne, "The Carpathians", "Geog. Review", New York, June, 1917; which I have used extensively in this section.

the Iron Gates and over the Predeal (Pray-day-al) Pass; and three less-used lines traverse the Red Tower Pass, the Gyimes Pass (between the Alt and the Sereth) and the Jablonitsa (between the Pruth and the Theiss).

The Transylvanian Alps extend from the Danube Gate at Orshova to the Predeal Pass on the Prahova River (Fig. 119). "The most remarkable feature of the Alps is the flatness of most of the ridges, even when they reach an altitude of 6,600 feet. Indeed the steepest climbing in these mountains has to be made as a rule at the beginning while you are ascending from the valleys, which are often wild gorges. After you have reached the edge of the plateau you find comparatively good paths in the midst of a splendid forest. When you have risen above the timber line you travel over extensive pasture grounds, which are known to the Rumanian shepherds as Plaiuri, *i.e.*, paths."[3] In winter the sheep are sometimes taken as far away as the Dobruja. It is the isolation made possible by these elevated mountain-pastures (Fig. 119) which has preserved the Rumanian culture and language in face of the unparalleled number of invasions of Rumania which are discussed in a later section.

These southern plateaux are built of ancient schists, and seem to consist of three levels, probably due to three periods of uplift. The upper level is the Boresco Plateau at 6,600 feet, which is well represented between Orshova and the Upper Jiu River. The second level is the Riu-Ses Plateau somewhat lower and found north of the divide between the Alt and the Marosh; and south of the divide between the Alt and the Predeal Pass. The Vulcan Pass is a sort of shelf (forming part of this level) which lies far above the Surduc Gorge on the Jiu, and was originally one of the shepherd's Plaiuri (Fig. 119). It was much used even in Roman times to connect Transylvania with the Wallachian Plain. Sarmiza-getusa, the old Dacian capital, was just to the north of this Pass. The third level is much lower and is in part represented by the Pliocene valleys. The Surduc Gorge and the Red Tower Gorge of the Alt are cut out at this third level of erosion.

Three of the areas in the Transylvanian Alps rise so high that they were heavily glaciated in the Great Ice Age. They are the Fogaras (8,340), Paringu (east of the Jiu) and Retiezat Plateaux. Here relics of a topography yet older than the Boresco stage remain in the form of crests deeply cut by glacial cirques and at times showing glacial troughs

[3]De Martonne, *loc. cit.*

(Fig. 119). But elsewhere in the Carpathians the erosion has been mainly by water rather than by ice.

There is a Sub-Carpathian Depression in Oltenia analogous to that found in Galicia. Both developed where the folds press upon the younger rocks of the plains. The Oltenian depression is in a drier and warmer region than the adjacent districts and is usually more closely settled than the low hills farther from the great fold ridges (Fig. 119). History tells us that while the peasants in the whole of Wallachia were serfs, most of the villages of the Sub-Carpathian Depression were inhabited by Freemen who owned their own fields and homes. In the Great War the role of the Sub-Carpathian Depression has been quite remarkable. Here the defeated Rumanians rallied near Targu Jiu and Campulung, and stubborn battles were fought in such situations along the upper Jiu River (De Martonne).

The Bihar Block seems to have much the same build as the Transylvanian Alps, but is on a much smaller scale. It has preserved a good deal of the Riu-Ses level, but has been much dissected. It rises to 6,000 feet in Mount Cucurbeta in the northwest, and thence drops rapidly to the west (Fig. 119). It seems to have been a centre of Rumanian culture in very early days, but has been dominated by Magyars throughout historic times.

The Transylvanian Basin is surrounded by mountains, a distribution which results in a dry climate, and an intensification of continental conditions. Thus near Brashov (Kronstadt) the winters are very cold and the summers hot. A layer of loam like the loess of Russia is common as a surface deposit in this dry region, and the flora of the Upper Marosh basin resembles that of the steppes. The Magyars in this area have occupied the broad valleys with their loamy terraces and are also found in the isolated basins. Here are to be seen their large villages and their bighorn cattle. But the Rumanians have greater powers of expansion and have almost assimilated many Magyar and German villages. The margins of the Alfold and the Banat to the south have been discussed in connection with Hungary.

The plain of Wallachia is divided by the Alt River into the western Oltenia or Little Wallachia and the eastern Muntenia or Great Wallachia. The whole region is covered with Pliocene or later material eroded from the Carpathians. These plains slope to the Danube and seem to have caused the great river to flow rather close to the Balkan Highlands, which are not so high or so much dissected as the Carpathians. Wallachia is crossed by a great number of long parallel tributaries, some of which like the Jiu, Alt, Prahova have already been mentioned.

This great plain extending from the Northern Dniester to Bulgaria offers considerable variation in environment. The severe snowy winters of Northern Moldavia are very different from the almost Mediterranean climate of the valleys of Oltenia. Descending from the Carpathians in Wallachia one passes in a few hours from a region of naked rocks, pine forests and streams which race boiling into the gorges, to a hilly region where there are smiling orchards like those of England. Then a little lower one finds oneself under a scorching sun on a plain where corn that broke through the soil only in April bends under the weight of golden ear about the middle of June. Further south is the region of the Danube with forests of knotty willows that seem at first sight impenetrable, though they have clearings where the fisherman mends his nets. In the Dobruja this zone spreads across the river to the right bank passing through a wild masterless country as far as the great lakes, the tortuous delta of the Danube and ultimately the sea.[4]

Moldavia is bounded on the north by the entrenched meanders of the Dniester. Ancient forts at Hotin, Soroca and Bender defend the border. Sandstone hills east of Jassy separate the main southern area of "Black Earth" from small basins of the same rich soil north of Jassy. All along the north bank of the Danube are parallel lagoons (*limans*) which continue along the coast eastward to the Dniester outlet.

Moldavia contains more pasture-land than Wallachia, and here formerly grazed herds of long-horned cattle among the lakes which the Boyars (nobles) made to supply their fish for Lent. Beyond the Pruth is Bessarabia, a thinly populated region with some of the characteristics of a steppe, which on the south borders the *limans* of the coast.

It is interesting to contrast the role of the Danube in Rumanian and in Bulgarian history as Iorga sees it.[4] The Rumanians as soon as they were strong enough hastened to annex and colonise the broad plains to the north of the Danube, their tutelary river. The Bulgarians crossed the river and made it only a stage in their movement ever to the south to reach the Bosporus and Constantinople. Their northern districts remained the almost inaccessible retreats of bands of marauders.

[4]N. Iorga, "A History of Rumania," 1925, London.

The Dobruja like Bessarabia is named after a famous Rumanian chieftain. It consists mainly of a detached block of various rocks which are allied to those forming the Crimea. These rise 1,650 feet above sea level. The Danube makes a wide detour to pass this isolated block, although a depression from Cernavoda to Constantja is little above sea level (Fig. 119). Across this narrow gap Trajan built one of the three main frontier walls which protected the Roman Empire; the second crossed Britain, and the third defended the angle between the Rhine and the Danube (p. 263).

B. Early History and Ethnology of Rumania

Even during Neolithic times it seems likely that the gold ores of Transylvania attracted primitive man. Tribes who made Painted Pottery in Moldavia and others with Incised Pottery from Illyria alike settled in Transylvania before the Bronze Age. Possibly these were the ancestors of the Dacian nation. In the Bronze Age a culture arose in the Carpathians which produced bronze and gold objects unique in Europe, and this may be dated about 1000 B.C. In the Iron Age about 800 B.C. there appeared the Cimmerians (Fig. 28) possibly from the Caucasus, to be followed by the Scythians about 700 B.C., who introduced naturalistic designs of a Siberian type.

The Scythians were nomads, probably speaking Altaic and no doubt of Alpine Race, but dominated by a few Iranian (Indo-Aryan) families. The names of the rivers Danube, Pruth, Tyras (for Dniester) and possibly the Alt are all of Asiatic origin. As we have seen (p. 146), Darius the Persian led an unsuccessful expedition against these people about B.C. 500, which perished in the Wallachian steppes. Later the Scythians were replaced by Sarmatians of somewhat similar origin. Possibly there were Slavs among this confederation, as indicated by certain place-names, which must date earlier than the main Slav migrations of A.D. 500. In the Iron Age the original Dacians seem to have driven the Scythians, *etc.*, to the east and the Cimmerians to the west, and then founded an empire later ruled by Burebista (see below).

In the 4th century B.C. the Greeks founded states along the coasts of the Black Sea. They traded by boats far up the Danube and its tributaries, and Istria on the Delta and Barbusi (near Braila) were two of their towns. They passed up the Sereth, where relics of their citadels have been found, and over the Oitoz Pass; while another route went up the Jalomitsa further to the west. Axiopolis near Cernavoda was a fort to protect their commerce from the Getae, who lived near the Argesh River. This has a Getic name which has not altered in 2,000 years. All along the Dambovitsa River are Greek remains indicating that they used the Predeal Pass to reach the gold of Transylvania.

About B.C. 300 the Keltae arrived on the Lower Danube, and merged to some extent with the Dacians, for they were a similar agricultural people. Probably the Dacians spoke a language akin to that of the Albanians, which is an early type of Indo-Aryan speech rather far removed from allies. At first the Dacians seem to have been a shepherd people, while the Agathyrsi worked the mines in Alexander's time. Their chief centre seems to have been in Transylvania.

The Dacians lived in small square huts built of timber or reeds with straw roofs, and usually there were about 100 dwellings in a village. Large fortresses crowned the hills with stone walls nine feet thick. They were common near the mines of Gradista, where treasures of gold coins have been found. Burebista, whose culture was of this kind, lived about B.C. 50.[5]

In Italy the growth of slave-labour led to many peasants migrating to Illyria and even further east into the Balkans and the Danube valley (Fig. 111). These folk long preceded the conquest of Dacia by Trajan. The first Roman road along the Danube was built about A.D. 34. In 53 Moesia was settled by 100,000 captives brought from north of the Danube, who soon began to export food to Rome. But even earlier there had been trade with the Danube by way of the Pear Tree Pass, and Parvan states that by B.C. 50 Dacia was as full of Roman merchants as was Gaul.

About the time of Christ the Romans occupied a strip of Wallachia north of the Danube, and in A.D. 53 Plautius pacified the whole Moldo-Wallachian Plain, demanding hostages and establishing several stations. The Dobruja was colonised by Romans, but military forces do not seem to have been introduced until much later. Dacia on the north bank was a nation of considerable culture and power when Trajan decided to conquer it. However, the Dacian people were never completely overcome, for many withdrew to the north beyond the reach of Roman armies. After Trajan had overthrown Decebalus in 105 it is worth noting that few Dacians entered the Roman armies. Trajan did not try to conquer the plains, but used Dacia as a mountain

[5]V. Parvan, "Dacia," 1928, Cambridge.

fortress jutting beyond the Danube. The Romans did not love the steppes, and merely placed guards in Dacia and Pannonia, much as they did in the Sahara (Parvan).

Moldavia was never officially Roman, yet Trajan placed a frontier guard at Barbusi where the Sereth joins the Danube. A Roman road led thence over the Oitoz Pass and so reached Bretcu in Transylvania. Another passed through Tyras on the Dniester and along the coast to Olbia on the Bug. When in 270 all the soldiers and officials were withdrawn, the Romanised peasants remained. Indeed the veteran soldiers of Rome had settled on the frontier, while further south there were thousands of foreigners who had been placed there by Rome.

So many ethnic strains have entered into the composition of the Rumanian nation so lately in the history of the country that we cannot use the ethnic map of to-day as a guide to the pre-history of the country. If we glance at the racial map shown in Fig. 120, the outstanding fact is that the people with very broad heads (over 85) are confined to the "refuge areas." For the most part these are the mountains (as the dotted areas show clearly), but Polesia is a marshland, and here also very many brakephs occur, according to Ripley. (Czekanowski, however, gives a somewhat different distribution, see p. 413.) The ethnologist deduces from this distribution that an originally uniform type of brakeph has been displaced by various migrations coming from the lowlands. History tells us of three such migrations.

The Magyars invaded the Alfold about A.D. 900 and seem to have brought a less brakeph type of head to that region. The German migrations down the Danube to Vienna seem to have done the same in Lower Austria and Western Bohemia. Finally the great Finnish migrations of A.D. 600 have entirely altered the racial composition not only of Bulgaria but also of the eastern part of Rumania. Here along the Black Sea is a belt of dokephs as well marked as along the North Sea in Poland. Thus Rumania consists of a very brakeph zone in Transylvania gradually becoming more and more dokeph as it approaches the corridor used by the Bulgar Finns.

Since the Rumanian language is so unusual and is the chief "cement" of the peoples in Rumania, we may well dwell on it for a few minutes. Probably most of the words are Slav, but those in common use are derived from the Latin, and in some cases are nearer to the mother tongue than the equivalents in Italian or Spanish. In many words we find the interesting change from Q to P, which characterised the Keltic languages (p. 108). Thus *quattuor* (four) becomes *patru*;

octo (eight) is *opt*; *aqua* (water) is *apa*. Another common change is for L to become R. One peculiarity which the language shares with Albanian is that the article is post-fixed, and this feature is probably a relic of Dacian speech.

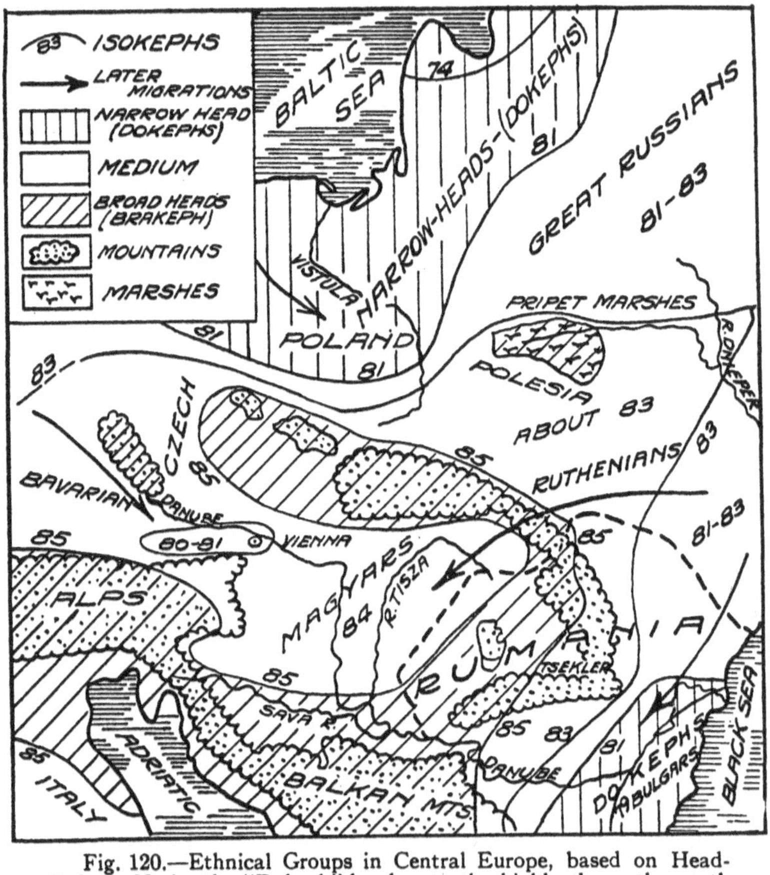

Fig. 120.—Ethnical Groups in Central Europe, based on Head-Index. Notice the "Dokeph" borders to the highlands on the north and south-east. The "Brakeph" folk are found mainly in the mountains and the marshes. (Modified from a map by Ripley.)

C. Upson Clark gives the following interesting commentary on a verse in Rumanian:

Somnoarse pasarele	The sleeping birds
Pe la cuiburi se aduna	At their nests gather
Se ascund in ramurele	Hide themselves in the branches
Noapte buna.	Good Night.

Somnoarse is the feminine of *somnorus* (somnolent). Note R from L.
Pasare-le is the Latin *passer* (sparrow) with the post-fixed article.
Pe la is a double preposition and equals "on to." *La* is from *Illa*.
Cuiburi is from the Vulgar-Latin *Cubium*.
Noapte is the P-Q change from the Latin *Nocte*.

As a guide to the pronunciation of Rumanian names it should be noted that the letters usually have the Italian sounds. Thus *c* before e and i is pronounced ch; g before e and i is j; *j* equals zh; *s cedilla* equals sh; and *t cedilla* is tz; (sometimes the cedilla is omitted). Thus Cetatue is Chetatsoo-ay; Chilia is Kilia; Raduati is Rado-woots.

C. *The Later Period of Barbarian Invasions*

The outstanding feature of the history of Rumania, as emphasised in the title of this chapter is the unceasing series of invasions which devastated the plains on each side of the Carpathians. One could picture a colossal army of barbarians consisting of seven enormous but isolated brigades extending from Rumania to eastern Asia about the year A.D. 400. This concept though somewhat fanciful does enable us to understand the history of Rumania. Each "brigade" marches westward across the steppes, arriving at its terminus in the Alfold about once each century. These seven "brigades" are shown in a very generalised fashion in Figs. 121 and 121a, dated 400 to 1100. They comprise the Huns, Avars, Bulgarians, Magyars, Petchenegs, Cumans and Tatars. Only a nation of tremendous vitality like the Rumanians could survive in the face of such incessant invasions. Of them all only the Magyars founded a permanent kingdom, which dominated the Rumanians almost from their arrival in the Alfold. Only since the Great War have the tables been turned, for Rumania is now by far the more important nation of the two.

The Huns (shown in Fig. 121) were composed of many tribes, and possibly the ancestors of the Bulgarians, at that time settled somewhere near the Urals, contributed the largest quota. The Huns were defeated by the Gepids and Goths about A.D. 455 and soon vanish from history. The Avars have been described in the chapter on Hungary (p. 401), and they made a sojourn of several centuries in the Alfold. They were joined by many other tribes, notably by the Black Bulgars. They subjugated the Slavs of the plains, but probably did not dominate the mountain folk in the southern Carpathians to anything like the same extent. Some of these nomads united with the Khazars about 670 in order to withstand pressure from the Petchenegs, who were thrusting them westward from the rear. Other Khazars took part in the invasion of Moldavia by the Magyars about 840. But the main force of the Khazars did not send their raids beyond the Carpathians from their empire in the Ukraine (Fig. 131).

The Bulgars arrived on the Danube about 482. After many vicissitudes they defeated the Emperor Constantine IV in 679 and then crossed the Danube into their present home. Like the Khazars they affected mainly the people settled east of the Carpathians (Fig. 121 and 121a). During these upheavals the Romano-Dacian culture was preserved in the high-level valleys of the Carpathians by a group of people who called themselves Romani, but who were called Wallach or Vlach (foreigners) by their neighbours. Petchenegs overran the plains from 900 to 1000, and Cumans from 1000 to 1100.

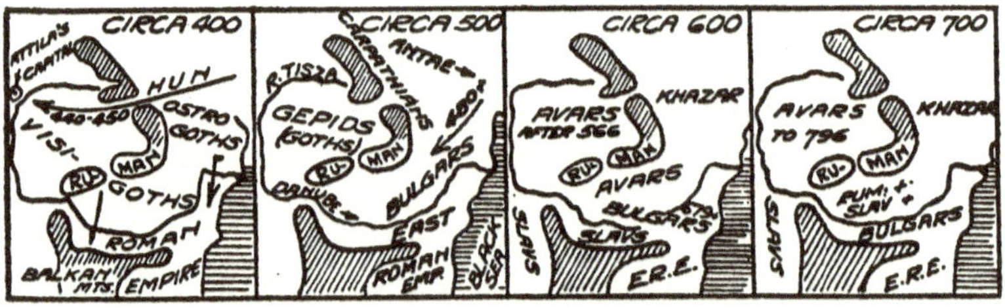

Fig. 121.—Four (out of eight) maps showing the unceasing invasions by Barbarians into the lands now forming Rumania.

History is almost silent as to this early stage of the Rumanian nation. The Magyars declare that they were in possession of Transylvania and the Banat long before the Rumanians. They affirm that the Carpathian Rumanians are derived from the pastoral Vlachs of the Macedonian Mountains, and that only relatively recently did the Wallachs migrate northwards.[6] But it is to-day generally agreed (outside of Hungary) that all through the southern Carpathian area Wallachian shepherds pastured their flocks on the mountain meadows during the summer and drove them down into the plains in winter. Their families lived in scattered farms and villages along the upper water-courses and there preserved their language, costumes and traditions.

The Cuman hordes were of the same general stock as their predecessors. They inhabited the eastern part of modern Rumania from

[6]Charles Upson Clark, "Greater Rumania," 1922, New York.

the 11th to the 14th century, sending raids into the Byzantine and Magyar realms, but making no permanent settlement there at first. Finally many Cumans became incorporated in the Magyar region of the Banat, and in the area between the Theiss and the Danube.

About the dawn of history in Rumania we find that the Slavs of Bulgaria had kept their language but had lost their independence and were controlled by Bulgar (Finn) overlords. Of the Slavs of the Alfold there is neither language, culture nor stock distinguishable among the Magyars with whom they merged. But the Rumanians by virtue of their environment were never obliterated; and even to-day far to the south, the Morlachs preserve their Rumanian culture in the highlands above Ragusa. The Pindus Range between Corfu and Thessaly is still in great part as Wallachian as the Southern Carpa-

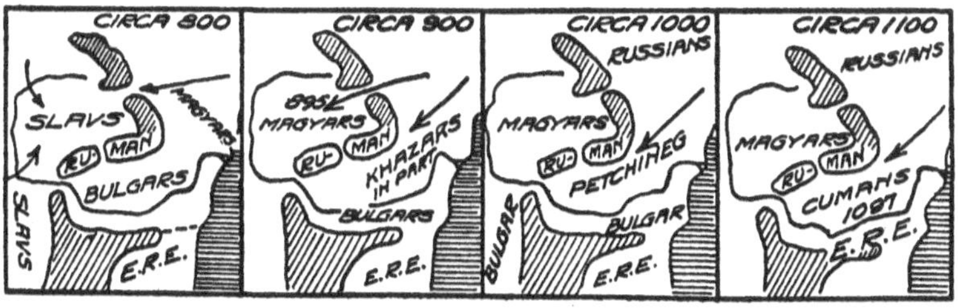

Fig. 121a.—Invasions of Rumania from 800 to 1100 A.D. In most of the period charted (400 to 1100) more Rumanians (or Vlachs) lived east of the Adriatic than in the Carpathians. Diagonal lines show mountains.

thians. Indeed about 1200 these southern tribes fought both the Byzantines and the newly established Latin Empire, and succeeded in capturing Baldwin the Emperor of Constantinople. But they never founded a separate Rumanian state such as developed north of the Danube.

The Carpathian Wallachians seem to have regarded the Emperor of East Rome (Constantinople) as their legitimate ruler. His armies appeared at long intervals to drive away Slav or Magyar invaders. About 1100 the King of Hungary built a castle at Torda to control the gold mines of Transylvania, and a monastery was established at Kertz on the Alt River. There was at this time a great rural Rumania which was ever being pushed into the mountains by the advance of the Catholic Magyars, but there was no definite frontier. Many Saxons were introduced among the Rumanians of Transylvania about 1224, partly to guard the passes from further invasions by the nomads of the east.

The Teutonic Knights founded Kronstadt (Brashov) and the Long Camp (Kampulung) to the southwest about 1211, where their special mission was to fight the pagan Cumans. But as they would not acknowledge the supremacy of the King they were soon expelled, and migrated to the shores of the Baltic (p. 190). Levies of Magyars under German leaders were therefore sent to guard the Passes of Oitoz and Red Tower. These were the Szeklers, whose descendants now number about half a million, and are entirely surrounded by a Rumanian population. The castle of Severin was built near Orshova to guard the Danube Gate about this time.

The flood of Tatars burst through the Carpathians in 1241 and severed the Wallachian people from the Magyar Catholics. They drove out the last of the Cumans, and now the Wallachians in the east paid tribute to Tatars instead of to Magyars or Cumans. From central Asia to the Carpathians there stretched a single state whose peace was guaranteed by the Mongol Emperor. This greatly benefited the Wallachians, through whose lands passed the great Trade Route from Galicia to Akermann on the Black Sea. The last Tatar Prince Demetrius (*i.e.*, Timur) seems to have ruled the Lower Danube about 1330. He was succeeded by a Rumanian Dobrutitch, after whom the Dobruja is named.

D. *The Vassal States of Wallachia and Moldavia*

The history of Rumania during the fourteenth and fifteenth centuries is filled with struggles between the Rumanians, Bulgarians and Magyars. In general Wallachia and Transylvania were dominated by Hungary in this period. Chieftains arose in various parts of the country, such as Basarab, who gained control of Little Wallachia (west of the Alt) about 1290. Later he became Prince of Wallachia, with Hungary as his suzerain. About 1350 Bogdan (or Dragosh) became Prince of Moldavia, and his realm extended from the Dniester to the Carpathians. Thus at the beginning of the 14th century we have the real origin of the two Rumanian states, which existed side by side without really uniting till 1859. The boundary between the two was the little river Milcov, which enters the Sereth above Galatz.

In the following table the chief events of the two kingdoms are arranged in chronological order.

TABLE W

CHIEF DATES IN RUMANIAN HISTORY

Wallachia	*Moldavia*
1247. Seneslav rises on the Argesh.	1335. Bogdan revolts.
1330. Basarab I Voevode of Rumania.	1360. Poland becomes suzerain.
1367. Turkish invasions begin.	
1386. Mircea the Great.	1401. Alexander vassal of Poland.
1400. Wallachia a fief of Turkey.	1455. Turkey gains control.
1442. Hunyadi conquers Wallachia.	1457. Stephen the Great.
1496. Radul the Great.	1513. He defeats Turks, Poles and Hungarians.
1525. Civil Wars.	
1526. Battle of Mohacs (Turkey).	1513. Moldavia vassal of Turkey.
1600. Michael the Bold.	1600. Michael also rules Moldavia.
1618. Phanariots control for Turkey.	
1632. Michael Basarab, Reformer.	
1683. Wallachia helps Turks against Vienna.	1634. Vasili the Wolf.
	1654. Moldavia joins Transylvania.
1688. Brankovan a good ruler.	1699. Treaty of Karlowitz.
1714. Cantacuzenos executed by Turks.	1711. Turko-Russian War.
1716. Phanariots rule till 1821.	1712. Phanariot rule.

It is not easy to discuss the complicated history of Rumania in a few pages. In the first place the development of two separate Rumanian nations side by side for so many centuries introduces special difficulties into our study. Secondly, the Rumanian rulers rarely advanced beyond their national boundaries, which should surely be counted to their merit, but which means that they had no great effect on other countries. Thirdly, Moldavia was at first under the control of Poland, then of Turkey and finally of Russia; while Wallachia was a vassal of Hungary for considerable periods, and then like Moldavia was controlled by Turkey and Russia. It is for this reason that I have dwelt on the early formative years of the Rumanian nation so fully, while the story after the 14th century is much condensed.[7]

It seems probable that Radu (Rudolf) of Fogaras founded a state in the foothills of *Wallachia* at the close of the 13th century with his capital at Argesh or at Kampulung. At times, as in the case of Mircea the Great (1386-1418), the local rulers broke away from Hungary. This King fought against the Turks, but he was badly defeated at Nikopolis in 1396; and thereafter Wallachia had to reckon with

[7]Among the many accounts consulted that of Sir Henry Trotter in the "Encyclopedia Britannica" is the clearest.

Turkish overlords, who alternated with Hungarian Suzerains. All this time the elections to the throne were theoretically in the hands of the Rumanian Boyars (nobles). Vlad the Impaler reigned from 1455 to 1477, though he was temporarily deposed by the Turks around 1460. For a time conditions improved under Basarab (1512-21), but the Turks soon acquired even stronger control, aided by their great victory over the Hungarians at the battle of Mohacs in 1526.

The brightest page in Wallachian history deals with the rule of Michael the Brave (1593-1601), who as usual bought the nomination to the throne from Turkey. He fought bravely against the Turks, whom he signally defeated on the Danube in 1595 and 1596. In 1599 Michael invaded Transylvania, and in 1600 he expelled the voevode (ruler) of Moldavia; so that in the year 1600, for the sole period in history before 1859, the Rumanians were united under one ruler. He was, however, murdered by an Austrian in 1601. Wallachia enjoyed some prosperity under the next rulers, Matthias, Sherban and Brancovan. All of these were vassals of the Sultan but ruled wisely and encouraged progress.

Soon after Brancovan had been cruelly executed by the Sultan, the office of voevode was sold to the highest bidder of *any* nation, and was not as hitherto confined to the Rumanian nobles. The Greek merchants who lived in Phanar (in Constantinople) became rulers of Wallachia from 1716 until 1769, when a Russian army occupied the country. Thereafter its history merged with that of Moldavia, which we may now study briefly.

Moldavia seems to have originated from the migration of Dragosh, a Rumanian chief from the region of Marmoros, which is now the eastern end of Czecho-Slovakia. Probably this was about 1350, but there were many Rumanian folk in Western Moldavia already, though up to this time the Tatars had been in control there. For a century Moldavia was vassal to Poland, but in 1546 the voevode offered tribute to the Sultan. He was deposed and Stephen the Great (1458-1504) began his long and victorious rule of Moldavia. He deposed Vlad the ruler of Wallachia, and tried to enlist Venice and Persia to help him in his attacks on the Turks. When the armies of the latter were too strong he retired to his fastnesses among the Carpathians. In 1487 he invaded Poland and for a time annexed part of Galicia. He also ruled Bukovina and the Danube Delta. His son, however, was compelled to become a vassal of Turkey, which had now occupied Wallachia and the Crimea. Little of note happened until 1600 when

Michael ruled the country jointly with Wallachia. Thereafter the Turkish tax-gatherers administered Moldavia. Vasili the Wolf occupied the throne from 1634 to 1654 and much improved laws and education.

In 1711 the voevode sought the protection of Russia, but a Turkish army reinstated the authority of the Sultan. Jassy was at this time the capital and Galatz the chief port on the Danube, from which timber, grain, honey, wax, etc., were shipped in large quantities. Moldavia was now ruled by the Phanariots as was Wallachia, and some of these like the Mavrocordato and Ypsilanti tried to improve conditions in Moldavia. After the Russo-Turkish war of 1774 Austria was given Bukovina, the cradle of Moldavian rule.

E. Russian Protection Leads to Independent Rumania

After 1774 Russia tried to obtain better conditions for the Rumanian people, but the Turks for many years continued their old policy of bleeding the country by heavy tribute. After a second war in 1805 the Russians acquired Bessarabia from the Turks, but risings by various other Rumanian groups were all crushed. In 1834 Russia compelled the Sultan to agree to the election of Rumanian voevodes in each state, but the Russian representative was given undue power in Rumania. In 1848, the year of European Revolutions, the Rumanians with the aid of France and Austria tried to free themselves from the irksome Russian control. In 1859 the two states were at last united into an almost independent kingdom under Alexander John Cuza. However, he made a poor ruler, and in 1866 he was replaced by Prince Charles of Hohenzollern-Sigmaringen; but Rumania was still nominally under the Sultan. In 1877 Russia again declared war on Turkey, and with the aid of the Rumanians won the battle of Plevna. By the Treaty of Berlin Rumania was declared independent, Bessarabia was awarded to Russia, while the Dobruja was given to Rumania.

The various Balkan wars early in this century in which Rumania took a part cannot be discussed here; but the campaigns in the Great War illustrate so clearly the bearing of the Carpathians and the Danube on the defence of the nation that we can well devote a short time to them. It was not till the war had proceeded for two years that Rumania joined the western allies. At first it seemed likely that her new armies could be poured into Transylvania, which she enveloped on both sides, and so crush the Austrian forces. Moreover,

Rumania was only forty miles from the main railway to the Orient *via* Nish (Fig. 119), and if that were blocked Turkey could be separated from the Central Powers. However, for a number of reasons, Rumania's plans came to naught, and her resources were soon at the mercy of Germany. We may first consider her defences and the roads by which they might be attacked.[8]

The two northern passes are Gyimes with a railway and road, and Oitoz with a wagon road (Fig. 119). In the centre are the Predeal Pass with a railway and road, and the Torzburg Pass with a road. In the west are the Red Tower and Vulcan Passes. The former has a railway and a road, the latter a road only. In every case the slope of the stream-bed is in favour of the attacking army. On the southern boundary the defences of Rumania are superb. The Danube is half a mile wide, and furthermore is fringed by marshes from six to twelve miles wide. There is only one bridge, nine miles long at Cernavoda, along the whole river from Belgrade eastwards to the sea. The Dobruja, however, was Rumanian territory, and on its southern boundary alone there was no natural feature to protect Rumania.

Rumania could therefore adopt one of three plans. She could try to wipe out the Transylvanian Salient, she could attack Nish directly from Orshova, or she could invade Bulgaria and Nish *via* the Dobruja. The latter was the most hopeful plan, since fewer troops were needed. Unfortunately the Rumanians felt compelled to free their nationals first in Transylvania, and so they rapidly advanced their armies beyond the Carpathians. They readily captured Brashov and other towns; but when the German reserves came into action they were quickly driven back to the passes (Fig. 119). Meanwhile the Germans in September attacked Turticaia in the Dobruja, so as to prevent the Rumanians from damaging the Orient Railway. In October, 1916, they captured Constantza, and the Czernavoda bridge was thereupon destroyed. Now the Danube protected the German flank, while the Black Sea prevented the Germans from being attacked on the east.

Persistent attacks were now being made by the Germans on all the passes mentioned previously. If the northern passes were captured the Germans could cut off aid from Russia; while the central passes led directly to Bukharest. The southern passes were better served by railways on the Hungarian side than on the Rumanian, so that in the end the Germans forced their way through the Vulcan Pass and poured

[8]D. W. Johnson, "Topography and Strategy in the War," 1917, New York.

into Rumania. The Rumanians now ran short of munitions, which were to have been supplied by Russia, and their armies were forced to retreat to the northeast. Meanwhile another German army crossed the Danube at Zimnitza and menaced the Rumanian flank. The retreating armies gave battle at the Argesh River on November 30; but unfortunately most of the railways were on the southern side of the

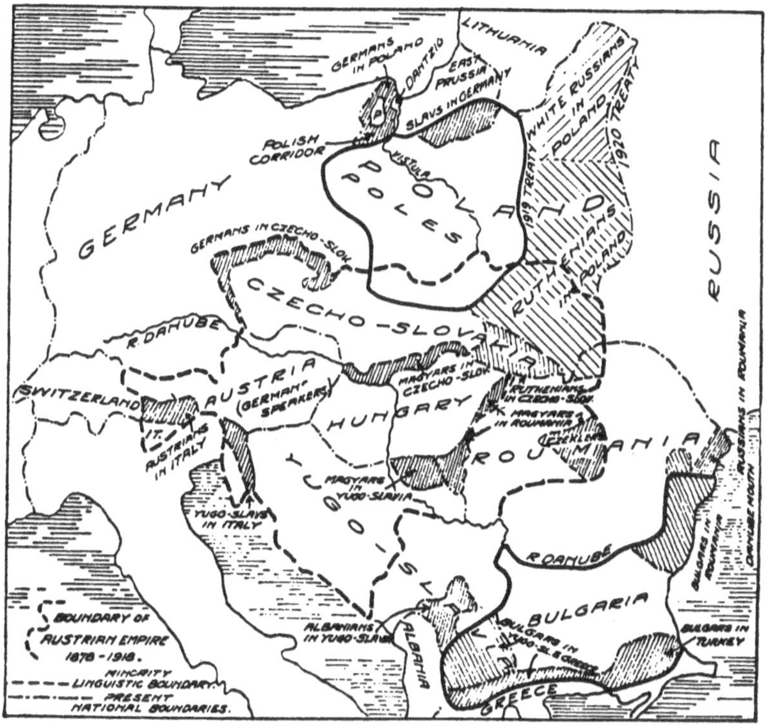

Fig. 122.—The chief "Sore Spots" (shown by rulings) produced in Central Europe by the Peace Treaties about 1919. Notice the great difference between the linguistic areas (shown by heavy lines) and national areas in Poland and Bulgaria.

long parallel rivers of the plain, and so were of little use to the Rumanian armies. Finally along the Putna River, just at the boundary of Moldavia, the Rumanians were able to block the Germans for nearly a year. With the revolution in Russia and the victories of Austria in Galicia, Rumania was compelled to ask for an armistice in December, 1917.

As a result of the various post-war treaties, Rumania has been more fortunate in her new boundaries than any other nation. The Rumanians were entirely unsuccessful in the war. Yet in addition to territory inhabited by their own peoples in Hungary, they were given other lands where nearly four million foreigners dwell. These are on the edge of Hungary, in the Banat, in Bukovina, in southern Bessarabia and in the Dobruja (Fig. 122). Their history shows that Rumania was a vassal state almost continuously until 1829; and she never had a good political claim to the huge area she has obtained in Transylvania. Peaceful Penetration has certainly been a paying proposition for the Rumanian people. (See also the discussion of Fig. 146.)

CHAPTER XXIV

THE BALKAN STATES—BYZANTINE AND TURKISH DOMINATION OF COMPLEX "YOUNG" HIGHLANDS

A. Salient Features in the Build of the Balkans

The key to the topography of Southeast Europe is the little inset map (at A) in Fig. 123. Here the African Shield and Russian Shield are represented as approaching each other, and squeezing the soft young rocks between, as well as the resistant older portions of the crust known as "Median-Masses." The young rocks yield and produce overfolds, falling toward the moving masses, while the two median-masses are not folded. The inset at B shows (in plan) the two series of weak rocks which have yielded to form the Young Mountains. One median-mass is situated in the Hungarian Alfold, and the other in Thrace between the Vardar and Mesta Rivers.

The build in the north of Albania is particularly interesting, and can perhaps be explained as the result of the thrust of the "African Shield" into the gap between the two large median-masses. The crust is folded here into a V (pointing east), so that the Shar Mountains run nearly west-east, and are faced across the Drin Valley by another high range. Between is a complicated lower region which offers a passable route from the great Vardar-Morava Corridor to the Adriatic. (The eleven main regional divisions of the Balkans are indicated in inset B.) In the Balkan Peninsula the two main fold-arcs (the Alpine and Dinaric lines) meet near Belgrade and then turn south (Fig. 123 at B). They run side by side separated only by the Morava Valley; and then the Alpine folds turn to the east again, and cross to Anatolia through Constantinople. Thus the south coast of Thrace is a resistant block which has not yielded to the late Tertiary Folding. The Dinaric folds run parallel to the Adriatic and down through Greece, where they split up to form the "fingers" of Greece and the festoon islands crossing the Aegean Sea (Fig. 123 at B).

According to the Serbian geographer, Cvijic (Svee-yeets), the close of Tertiary times saw great seas invading the Balkans and connecting with the Pannonian Sea in the Alfold. One large gulf extended up the Danube and covered the Rumanian Plain and north Bulgaria. Another lay further south and occupied central Bulgaria. A third extended north from the Aegean into Macedonia and sent an extension right up the Morava Corridor to the Alfold. Later these inland seas were converted into lakes with large rivers draining them to the Black Sea and Aegean ("La Péninsul Balkanique," 1918, Paris).

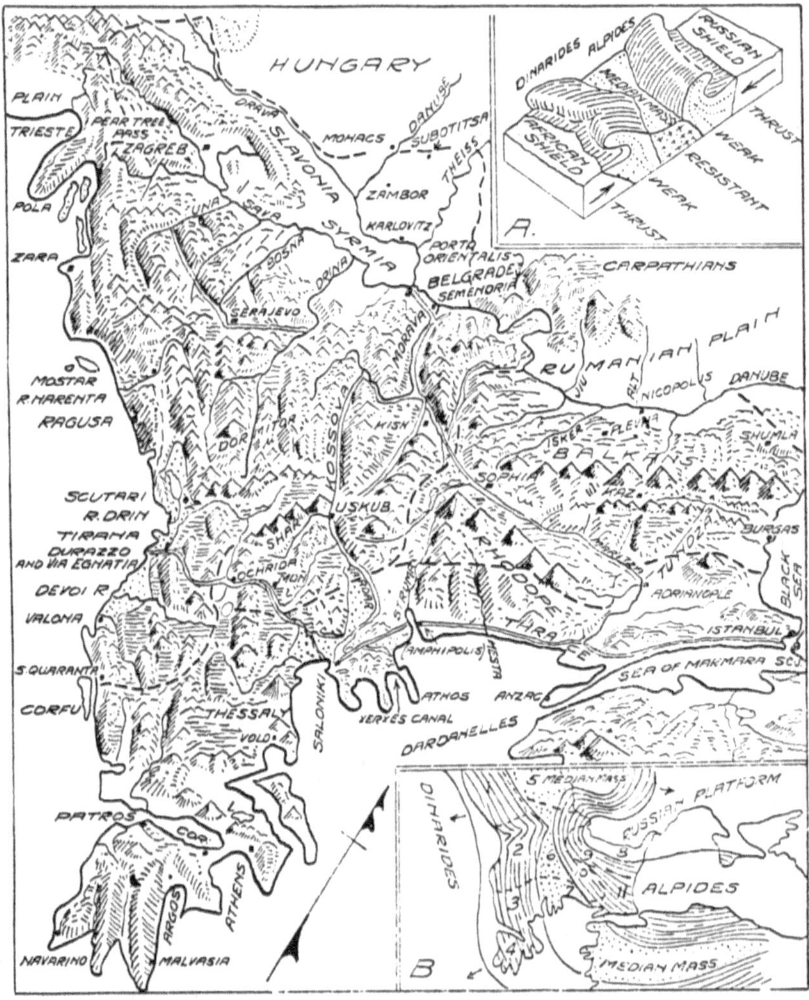

Fig. 123.—A Block Diagram of the Balkans showing the main routes. National Boundaries are shown by broken lines. At Inset A is indicated the production of the Fold-Mountains by the squeezing of the weak strata. Inset B shows the two resulting series of Young Fold-Mountains. The numbers 1 to 11 refer to the Topographic Regions.

Following the major folds there have been widespread subsidences. Graben on a large scale have greatly enlarged the Black Sea and the Aegean Sea; and it seems likely that much of Thrace has subsided from its original elevated position, just as the central part of the Carpathian Arc subsided (p. 435). Hence the Alpine folds in the Maritza Valley (in Bulgaria) do not appear as marked elevations above sea level, but can only be deduced from the geological data. In the adjacent lands enormous dislocations also occurred; so that the crust broke up, especially along the main corridor, into many horsts and graben. Usually these had a north-south direction, but this complexity of build has been to no small degree the cause of the complex history of the Balkans—as the following sections will demonstrate.

The Balkans can be subdivided into a number of topographic units as follows (Fig. 123 at B):

TABLE X

TOPOGRAPHIC UNITS IN THE BALKANS

Major Topographic Class	Topographic Division	Districts
A. Western Dinaric Folds.	1. Northern Karst Highlands	Bosnia and Herzegovina
	2. Albanian Highlands	Albania
	3. Epirus Highlands	North Greece
	4. Peloponnesus Highlands	South Greece
B. Central Corridors......	5. Northern Plains	Slavonia and Syrmia
	6. Morava-Nish-Kossova-Uskub	Southern Serbia
	7. Aegean Plains	Northeast Greece
E. Eastern Balkan Folds..	8. Danube Plain	North Bulgaria
	9. Balkan Ranges	North Bulgaria
	10. Rhodope Plateau	South Bulgaria
	11. Maritza Basin	South Bulgaria and Turkey in Europe

B. *The Dinaric Folds along the Adriatic*

The Dinaric Alps differ from the Swiss Alps in their build, since they have an inner belt of Flysch rocks—which are consolidated sands and marls. These Tertiary rocks weather to form undulating areas of more or less fertile ground. Perhaps we may take the Dinarics as beginning south of a line from Gorizia to Laibach, *i.e.*, at the Pear Tree Pass (Fig. 123). This pass near Laibach is situated where the Sava Valley and the approach to the Alfold come nearest to the sea (at Trieste or Fiume), and so it constitutes the southwest gate to the Hungarian Plain. The Dinaric Ranges rarely rise above 6,000 feet in the north, but offer great obstacles to traffic, owing to their barren limestone surfaces and to the recent dislocations of the crust. They are built up of three parallel belts: (a) The Littoral of pure limestone, (b) the loftier central zone of mixed limestones and older metalliferous rocks, (c) an eastern inner zone of Flysch which slopes to the Sava Valley. The limestone forms a typical Karst country, since the soluble rock allows all the rains to penetrate to underground channels. There is only a very thin soil-cover for the most part, and, indeed, much of the Karst shows naked rock sculptured into queer designs by solution.[1]

The longitudinal folds of limestone have subsided quite lately along the Dalmatian coast, forming numerous elongated limestone islands. These offered safe navigation and good harbours from the earliest times—and were no doubt used by the first traders along the "Amber Route." But inland the littoral is high and barren. There are no cross-valleys except the Narenta, since most of the water sinks into underground channels in the soluble limestone. The powerful Narenta flows down a narrow canyon which has always been difficult to traverse. Nowadays a little rack-railway rises over steep slopes to Serajevo from the narrow plain of Mostar. In the Karst plateau are long boat-shaped hollows—the *polye*—which contain good soils. But these centres of population are isolated by barren rocky expanses.

The inner slopes of the Dinaric highlands in Bosnia, *etc.*, are formed of Flysch—which carry thick forests. Here is a pleasant climate, though at Crknice (Zirknitz) near the Pear Tree Pass is the heaviest rainfall in Europe. Oak and beech are the common trees, and swine in these forests have always been a large factor in the economy of Yugo-Slavia. The highest peak in this region is Dormitor in the north of Montenegro—which rises to 8,290 feet. Owing to the rugged character of this country the Slavs of Montenegro have always managed to retain their independence of the Turk.

The second section consists of Albania and its adjacent highlands. Its chief feature has been referred to, namely, the "cross-corridor" due to the V-shaped reentrant in the folds. Unfortunately, the Drin

[1]M. Newbigin, "Geographical Aspects of Balkan Problems," 1915, London.

offers by no means an easy route to the east. Rather has it cut its way across several "graben," of which four seem to characterise the Balkans in this area. They can be made out—with a little care—in the block diagram (Fig. 123). Nearest the coast is the large lake of Scutari, which shows an early condition of these graben, since the hollow is not yet filled with silts. Then to the east is the graben of the White (or northern) Drin, just on the boundary of Albania. The deep, narrow valley of the lower Drin connects these two. The third graben is the Kossova Valley and the fourth the upper Morava Valley (between Nish and Uskub). These two will be briefly described later on.

In Albania the limestones are not so abundant. The rivers do not sink underground, but cut out zig-zag valleys. These are mainly longitudinal and parallel to the folds—but are connected by narrower transverse valleys (compare p. 296). The coast seems to have been elevated lately—and is swampy and fringed with a level plain. Here malaria is present, and the Albanian prefers a shepherd life on the hills, and has not developed into a seaman like the Dalmatian. The rocks are mainly Flysch which weather to a softer scenery than the limestones. Inland, however, are rugged ranges which offer many obstacles to traffic. The topography and politics have interacted to isolate the primitive Albanians.

Austria has always tried to prevent Serbia from acquiring this natural outlet to the Adriatic, and has preferred a Turkish Albania—or later an independent Albania—rather than that Serbia should control the country. Thus the Albanians have been left to themselves to continue their favourite occupation of "blood feuds between neighbouring clans." As Newbigin writes, it seems hard on the Serbs that they should be shut out of Albania so that the latter may be allowed to preserve their primitive man-hunting proclivities. The Great War, however, produced a few good roads, notably one from Valona up the Vyosa Valley and so across the highlands to Ochrida. Several small railways traverse the lowlands near Tirana, and Durazzo was the western end of the famous Roman Road—the Via Egnatia—which led to Saloniki and Byzantium (Fig. 123).

The southern regions in Greece have already been referred to in Chapter VIII, but their topographic structure may now be briefly described. There is a marked contrast between the western coast with few harbours and with the slopes parallel to the coast, and the eastern and southern coasts cut up into innumerable capes and gulfs

—the latter leading to small enclosed basins. The interior is all highland, which is especially rugged in the west, so that inland routes are rare, and the chief traffic has always been by sea (Fig. 123).

The northeastern portion of modern Greece is a transition region consisting essentially of faulted ancient blocks of the crust, and is described later. Similar faults on the south of the Cyclades Islands are accompanied by the volcanoes of Delos, Santorin and Kos. The mainland consists of "young folds" in the west and late deposits in the east; but both types of structure were shattered and elevated quite recently in geological history. Thus many rivers have an erratic drainage pattern, while irregular basins, sometimes still containing lakes, are a common feature of the topography. All sorts of rocks of different resistance to erosion are present, so that the scenery varies from forms due to hard granites, through karst features due to soluble limestone, to soft pinnacles due to weathered recent rocks.

The district around Athens is the heart of Greece. The main feature is a central east-west basin (Boeotia) draining to Lake Copais, which is flanked on the north by three parallel fold-ranges and on the south by the Pentelikon Range and its extensions. To the south of all these barriers lies Athens on a plain opening to the sea at Piraeus. Hence the famous city was well protected on the landward side, while the long graben of the Gulf of Corinth helped to protect it on the west. The western part of central Greece has a complex topography, as indicated by many small lakes. The parallel valleys do not lend themselves as well to agriculture as do the basins in the east, and communications are also less easy.

Thessaly forms an interesting contrast to the southern areas. It is essentially a tilted crustal block sloping to the west, whose high eastern edge has been carved into Mounts Ossa and Pelion. The inland plain contains the relic of a larger lake, while the southeast corner is drowned to form the Gulf of Volo. Epirus in the west is a region of limestone folds drained by four parallel rivers. The karst surfaces and difficult communications have always made it relatively unimportant in human history.

The Peloponnesus consists of fold ridges more or less meridional in the west, but swinging more to the southeast in the eastern portion. Settlement is confined to the broad valleys—such as Argolis, Laconia and Messenia—between the four "fingers" of Greece. In the interior many of the streams disappear in swallow-holes dissolved in the limestone. The lack of hinterland has caused the decay of many ancient ports, of which Navarino (the site of the naval battle of 1827) and Malvasia (whence "Malmsey" wine was procured by the Venetians) must suffice as examples.

C. The Zone of Corridors

The second major zone in the Balkans consists of the line of corridors connecting the Alfold with the Aegean. In the table (p. 455) there are three divisions numbered 5, 6 and 7 (Fig. 123 at B). As regards the northern division it consists of the southern portions of the Alfold, where the great rivers Drava, Sava and Theiss (or Tisza) unite with the Danube. Here the whole country is flat—the topography is senile (p. 60), and as a result we find a rich agricultural region. Before the war it was the chief wheat granary of Hungary, but now much of the Banat has been given to Yugo-Slavia. The only important upland is the Frushka Gora alongside the Danube at Karlovitz (Fig. 123). Here was a refuge for the Serbs in the early days of Turkish invasion, and here are some of their "holy places." Much of the region between the Theiss and Danube was settled by Serbish refugees, notably by 37,000 families in 1690. As a result of many such migrations this fertile land was allotted to Yugo-Slavia in 1919.

Belgrade is on a spur of the highlands which was the site of a fort in very early days. It is the strategic centre of this part of Europe. The main waterparting of the Balkan Peninsula is so high and so near the sea on the west—within five miles in Montenegro—that all the westward rivers are short and torrential. The three chief rivers are the Morava in Serbia, the Vardar chiefly in Serbia, and the Maritza, chiefly in Bulgaria. The divide between the two former is only 1,300 feet high near Uskub. The divide between the Morava and Maritza (at the Dragoman Pass near Sofia) is about 2,400 feet. It is clear that Serbia has a superior strategic situation since she commands the approaches to both these famous corridors—which indeed meet at Nish and lead to the Danube Corridor close to Belgrade.[2]

The corridor area is, however, not a simple longitudinal valley—but a series of rather disconnected, parallel valleys. Often deeper "basins" (or graben) are arranged along the longitudinal corridors—of which the Kossova (Kosovo) basin is the most famous. Here was fought the disastrous battle in 1389 where Serbia was conquered by

[2]L. W. Lyde, "The Balkan Peninsula," London, 1905.

the Turks; and here the Turks also defeated the Hungarians in 1448. The Morava is fed by a tributary from the Kossova Basin—and this stream rises in a "Valley Divide." In the same swampy area also rises a northern tributary of the Vardar—so that the Kossova Basin is naturally the easiest route from the Alfold to the Aegean.

The main Morava, however, rises south of Nish in a parallel valley where the divide is somewhat lower than at Kossova—though not so easy to traverse. It is this route that is used by the railway to Saloniki, which joins the Kossova corridor at Uskub (Skopli). The projected railway to Scutari would join the corridor a little north of Uskub (at Prishtina), and another ancient road connects Uskub with Sophia. Hence Uskub has been a very important centre throughout Balkan History—and it lies just where the Macedonian Bulgars meet the Serbs of "Old Serbia."

The structure of the Macedonian Plain near Saloniki has been worked out by Ogilvie.[3] At the end of the Tertiary period (about one million years ago) the whole Balkan Peninsula was rent by fractures which culminated in the depression of the Aegean graben. Thus all this region of Macedonia was broken into separate blocks, which were elevated differentially. Water collected on the lowest blocks, and the numerous lakes, such as Beshik, Doiran, etc., around Saloniki originated in this way. The three "fingers" of the Khalkidike Peninsula (one of which carries the twenty monasteries of Mount Athos) are similar parallel horsts, separated by depressed areas. The Plain at the mouth of the Struma is one of the best agricultural areas in the Balkans. The towns are situated on the alluvial fans (below the fault scarps of the Blocks) or on the margins of the marshes. The horst blocks are nearly flat-topped, showing that the original surface, before the dislocations, had been eroded nearly to a peneplain. Saloniki is at the entrance to the main north-south corridor of the Balkans, and was inhabited long before the dawn of history. It is on the famous Via Egnatia, and its history is still illustrated by a Roman arch and a Crusader's Castle. The Jews who fled from Spain in the 15th century now form two-thirds of the townsfolk. They still use the Spanish language almost unchanged.

The Vardar Valley is in part formed of graben as is the Morava. It has built up a large marshy delta which is occupied in winter by the Vlach (Rumanian) shepherds from the Pindus (p. 445). The

[3]A. Ogilvie, "Southern Macedonia," "Geog. Review," New York, 1921.

Khalkidike Peninsula is one of the most backward regions in Europe. There are no roads; but a green strip of grass, fifty yards wide, still shows where Xerxes cut his canal about B.C. 480 (Fig. 123).

D. The Eastern, More Stable, Balkan Structures

Bulgaria consists of four (8, 9, 10 and 11) of the main topographic elements mentioned in the previous table. The "skeleton" consists of the twin ranges of the Balkans together with the range of the Rhodope, which are linked by the Rila (or Western) Rhodope rising to 9,593 feet in Mt. Musa Alla. The Balkans are not very high, the summit (7,800 feet) being due south of Plevna. There are nine fair roads across the Balkans, of which the best known is the Shipka Pass (4,378 feet) near Kazanlik.

The northern slopes of the Balkans consist of a terraced plateau of limestone and sandstone largely covered with loose porous and fertile loess. The rivers flowing to the Danube have cut narrow steepsided canyons in these rocks. Only one river transects the Balkans, the Isker; and just where this river crosses the great Orient Route from Belgrade to Constantinople has developed Sofia (Sophia), the chief city of Bulgaria. At the northern end of the Isker is Nicopolis, always an important crossing place of the Danube, where a battle which ruined the Christian hopes was won by the Turks in 1396. However, the main road in Roman days was from Nish down the Timok Valley. The slope to the south is much steeper than on the north, which has facilitated northern invasions (Lyde, *loc. cit.*).

The axis of these crustal folds runs west-east, and there are in the Balkans (as in the Alps and Carpathians) good examples of longitudinal valleys. In this case the Upper Tundza and its chief tributary occupy a longitudinal depression, the famous Valley of Kazanlik, where Attar of Roses is produced. Then at Jambol the river turns at right angles in a transverse valley to join the Maritza. The southern ridge parallel to the Balkans is often called the Anti-Balkans. It extends with some breaks almost from Sofia to the Black Sea.

The ancient core of the Balkan Peninsula (Region 10 in Fig. 123) is analogous to the massifs of the central plateau in France or Bohemia. It may be termed the Rhodope Massif from the name of its extensive mountain system, and includes the mountain regions of the central and southern part of the peninsula. It begins in the south of the Balkans and sends a branch northwest through the middle of

Serbia to the borders of Hungary; and it extends south to the Aegean and east to the Black Sea.

East of the Vardar Corridor the graben, *etc.*, are nearly absent, and the rocks consist of granites, gneisses and recent eruptive rocks. The main mountain axis is the Rila Range which Cvijic calls the true core. The sections in Western Macedonia and south of the Balkans he terms transition zones.[4]

Within the Rhodope Massif the surface is not invariably formed of crystalline rocks, for beds of younger formation are common. Only the latest Tertiary formation are unfolded but these are often shattered into horsts by fracture. The folding of the Rhodope Massif took place in Tertiary times, and is thus much later than that of the Bohemian Massif which was folded in Paleozoic times.

It seems likely that the Kazanlik longitudinal valley once drained to the Black Sea at Burgas (Fig. 123). So also the Upper Maritza may have proceeded direct to the Black Sea instead of making a right angle bend at Adrianople. Probably these sharp bends in the Tundza and Maritza are deflections due to "block-faulting," when the Aegean graben developed to the south.[5]

Turkey in Europe is now confined to the region east of the lower Maritza. The lowlands occupy the basin between the Tekir Dagh (3,000 feet) and the granitic ranges (3,400 feet) along the Black Sea. They recall the steppes of Russia with their swampy depressions and large numbers of prehistoric tumuli, and in summer the region is similarly enveloped in clouds of dust. Adrianople on the Maritza, however, has a favourable position on the ancient road from Belgrade to Byzantium.

Istanbul is a Turkish name of Constantinople from "*Es-tam-Polin*" (into the city), and lies on a fragment of the Devonian rocks of Asia separated from that continent by the faulted area now occupied by the Bosporus. This famous sea route is twenty miles long and only half a mile wide, and its importance in trade has already been discussed (p. 62). The Golden Horn is the actual harbour of Istanbul (Byzantium), and is a drowned juvenile valley about four miles long just within the southwest end of the Bosporus. Thus the waters form a "Y," with Istanbul in the southwest portion, Galata in the northwest and Scutari to the east.

[4]Peucker on "Balkan Structures," "Geogr. Journal," London, 1902.

[5]M. Newbigin, "Geographical Aspects of Balkan Problems," 1915, London.

E. The Balkans under the Flourishing East Roman Empire[6]

The primitive peoples of the Balkans need not long detain us. At the dawn of history there were Illyrians in the Western highlands and Thracians in the centre and east. Illyrian seems to have been an Indo-Aryan language much like modern Albanian. Some of its dialects such as Messapian have been found in the earliest inscriptions in southern Italy. The Thracians were allied people who probably came from the north as did all the early Indo-Aryan invaders. They sent colonies of Phrygians towards Asia before the Acheans and Dorians arrived from the north during the second millenium before Christ (p. 144). Possibly there were few of the Mediterranean race in the main Balkan area, though we have already seen (p. 144) that a Mykenean civilisation flourished on the south coast of Thrace about 1200 B.C. The main peoples in the Balkans, when the Eastern Roman Empire began, were Thracians and their allies, who were almost certainly of Alpine race. The most notable racial variation resulted from the invasions of Slavs during the 6th and 7th centuries, and these have already been described (p. 391). But this did not alter the physical character of the Balkan people very much, since these Slavs were mostly Alpine people also. Even the Turkish invasions made little difference to the racial composition, for they also were Alpines. Perhaps the only real racial change resulted from the invasions of the Bulgar *Finns* in the 7th century, who certainly brought in a very definite dokeph strain which is clearly indicated in the racial map given in Fig. 120.

The peoples of Bulgaria as in the rest of the Balkans are much mixed. Nowhere does the Bulgarian have a pronounced Slavonic type. The men as a rule are rather below middle height, compactly built and very muscular. The face is broad and rather oval, the nose straight or very slightly curved, the eyes small and narrow, frequently very close together and buried under heavy eyebrows, the hair is fair or more rarely dark. Perhaps the projecting jaws are somewhat characteristic. High cheekbones and slanting eyes indicate the Asiatic strains. The Pomaks number about 26,000 and are Bulgars who

[6]The complexity of the Build of the Balkans is reflected in the complex story of the development of the Balkan Nations. For this reason the present chapter is rather a long one, and I have been compelled to omit many interesting liaisons between topography and history owing to lack of space. With the aid of the maps the reader will be able to supply some of these himself.

adopted Islam. They live in the Rhodope highlands and are perhaps the most genuine Bulgars. In the mountains near Sofia live the Shop tribes who may represent Illyrians mixed with Cumans. In 1861 many thousand Crimean Tatars were settled in Bulgaria, and in 1864 still larger numbers of Circassians. Greeks, Jews and Gypsies are found in large numbers also.[7]

We have already discussed something of the history of the Balkans under the early Greek and Roman rulers (Fig. 38). Constantine in A.D. 328 appreciated the importance of controlling the narrow gap which separated the European and Asiatic cultures, and therefore removed the centre of Roman culture from Rome to this new focus in the East. Soon great roads ran from Rome to the new capital, re-

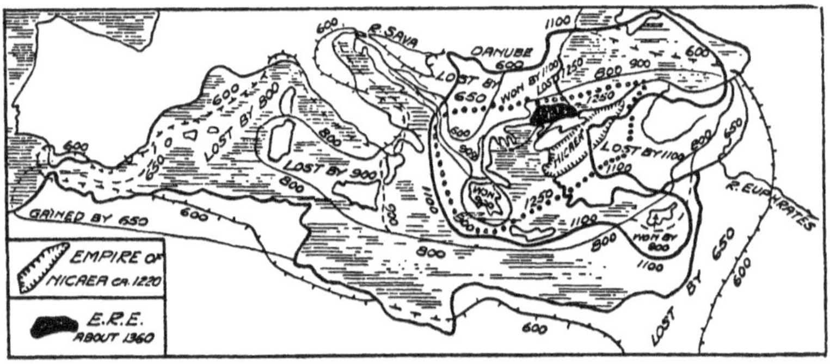

Fig. 124.—The Shrinking of the Byzantine Empire from 600 to 1453. Notice that in the first half of the 13th century, while Latin invaders ruled in Constantinople, the Empire centred about Nikaea.

placing the Via Egnatia which linked Durazzo with Saloniki and Byzantium (Fig. 123).

Even in Diocletian's time Milan—the natural centre of the great agricultural region of Italy—had displaced Rome and become the chief military station in Italy. Hence the importance of the main road from Milan to Aquilea, and thence up the Pear Tree Pass (Fig. 194) to Siscia and Syrmium on the Sava to Singidumum (Belgrade), Naissus (Nish) and Serdica (Sofia), and so along the Maritza Valley to Constantinople. Theodosius in 396 divided the Empire into two parts, and Arcadius was given the Eastern Roman Empire. From this time onward—with rare exceptions such as in the time of Justinian—the two halves of the Empire proceeded almost independently of each

[7]"Handbook of Bulgaria," Naval Staff, London, n.d.

other. We must hurry over the early phases of Byzantine history, although it played a vital part in the culture of the Eastern nations of Europe (p. 170).

The history of this period may be epitomised as a struggle in the main between three rival cultures. The Greeks who dominated Constantinople during most of the era of the Eastern Roman Empire were a cultured people flourishing in a Mediterranean environment, and necessarily familiar with trade rather than agriculture as a means of livelihood. The Slavs were a primitive peasant folk wedded to the soil, and at first almost devoid of any organization beyond tribal custom. The Bulgars were fierce, uncivilised nomads with few desires beyond looting at first, but with an adequate military organisation that enabled them to dominate the Slavs and even to wage war on the powerful Byzantine Empire.

After the fall of Rome and while the Ostrogoths ruled in Italy, Justinian (527-565) controlled the Byzantine Empire (Fig. 41). He was probably of Albanian origin and during his reign his Empire expanded so that for a time it seemed likely to rival the former Roman Empire. Under Belisarius his armies conquered the Vandals in Africa in 533, while in 554 the Gothic Empire was overthrown in Italy. Even the south of Spain was for a time controlled from distant Constantinople. To this city came all the merchandise of the east, as well as great cargoes of wheat from Egypt, and only the ever increasing migrations of Slavs from the north troubled the peace in the Balkans. Justinian adopted the usual Byzantine policy of inviting another barbarian nation to attack his immediate aggressor, when he persuaded the Avars to make war upon the Slavs.

In the east the powerful and civilised Persian nation also waged war against the Emperor. They conquered the vital province of Egypt, and only after years of patient organisation was Heraclius able to take the field against them in 622. The Persians allied themselves with the Avars and attacked Constantinople in 626, but the Avars could do nothing against the great walls of the city, and the Persians, as in an earlier era, were baffled by the Greek fleet. Heraclius pursued the retiring Persians and crushed their King, Chosroes, in 628. He succeeded in extending the Eastern Roman Empire into Mesopotamia, and never again did the Christians control so much of the Near East (Fig. 124).

The Arab invasions of 634 swept back the Byzantine frontier. In

637 Jerusalem fell to the Moslems and has remained in their hands (except for a brief period in the Crusades) until the twentieth century. After the death of Heraclius the Greeks relinquished Syria and maintained their frontier at the barrier of the Cilician Gates in the Taurus scarp (p. 140). The stoutest defenders of the Empire were the Isaurians, a somewhat primitive tribe who lived in the rugged country of southern Anatolia. They produced a successful Emperor in Leo the Isaurian, who ably defended Constantinople against a terrible attack by the Arabs in 717. His reign was marked by the schism between the Image Worshippers (who included most of the monks and were supported by Rome) and the Iconoclasts, headed by Leo. As we have seen elsewhere (p. 182), the former won, and this controversy greatly widened the breach in the Christian Church.

The Byzantine Empire was finally crushed between the hammer of the Turks and the anvil of the Bulgarians (who now appear on the scene); but not until seven centuries after the time of Leo. The Bulgars were a Finnish people, and there seems no reason why they should be called Mongolian, for in all probability they were largely Nordics allied to the Finns of to-day. They crossed the Danube in 679, and speedily compelled most of the Slav and early Romano-Thracian peoples to acknowledge their rule (Fig. 121). Naturally they next extended their realms to the southwest among the primitive Slavs. They marched up the Isker and Timok Valleys, and left the still strong Byzantine region of Thrace for a time in peace. In 811 their Khan, Krum, defeated the Greek army, killed the Emperor Nicephoras, and used his skull as a drinking cup. A year or two later Krum besieged the capital, and from that day to this there has been enmity between the Bulgars and the rulers on the Bosporus.

The greatest hero in Bulgarian history is Simeon (893-927) who had been trained in Constantinople, and who ultimately united Bulgaria, Macedonia and much of Serbia into a powerful realm (Fig.125). The Greeks adopted their usual plan and bribed the Magyars (of Moldavia) to attack the Bulgars on the north. But Simeon was able to defeat the Magyars so drastically that they left Moldavia and migrated to the Alfold (p. 401). Simeon now adopted the title of "Tsar of the Bulgars and Autocrat of the Romans"—thus implying that he was superior to the Greeks of Constantinople, who were universally known as "Romans."[8] His empire fell to pieces on his death

[8]F. Schevill, "History of the Balkan Peninsula," 1922, New York—whom I have used freely in this section.

and was soon conquered by the Byzantines, except for the Macedonian region where the Bulgar rule persisted for many years.

Basil the Macedonian controlled the Empire from 867 to 886 and his dynasty lasted until 1056. Under this family the Empire gained once more something of its early power before its slow decadence from 1200 onwards (Fig. 124). Under Nicephorus and his successor John both the Arabs and Bulgars were repulsed, the latter largely with the aid of new barbarian allies of the Empire, the Russians. In 971 John attacked the Russians and wrested almost all Bulgaria from their hands.

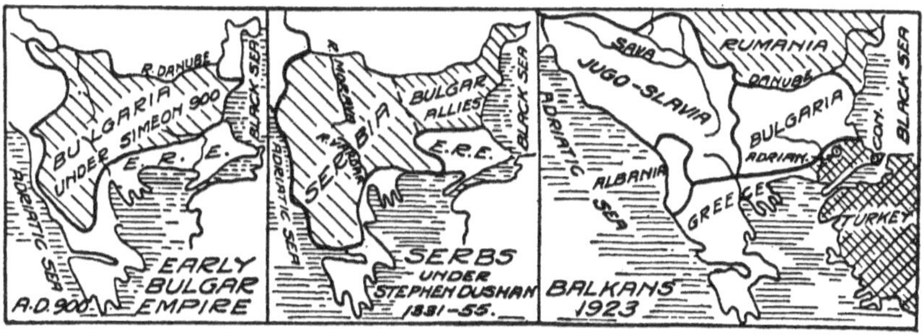

Fig. 125.—Three Maps showing Simeon's Bulgar Empire, Stephen Dushan's Serbian Empire and the modern States resulting from the Great War.

In 976 Basil II became Emperor, and earned the name of "Bulgar-Slayer" by a great victory in 1014 in the Rhodope Mountains. He blinded 15,000 captives and sent the crippled army back to the Bulgarian Tsar in Macedonia. Meanwhile the Arabs had moved eastwards from their old capital of Damascus to Bagdad, and Basil found it possible to push his Empire far into Asia so that he ruled from the Danube to the Euphrates. Already, however, the corrupt nobles were gaining too much power in the Empire, while the middle class was declining in importance as trade suffered from the constant wars and from the rivalry of the merchant peoples of Italy. The attacks by Normans on the west and Turks on the east wrecked the Empire, though it persisted in name and controlled the vicinity of Constantinople for four hundred years longer.

F. The Fall of the Empire and the Rise of the Turks

The last and most dangerous attack upon Europe by Asiatic peoples was now about to begin. A group of Turks from Merv (under a leader called Seljuk) conquered Persia during the eleventh century. They captured Bagdad in 1055 and soon attacked Syria and Anatolia. In 1071 the Byzantine Emperor led an army against them and met the Seljuks at Manzikert in Armenia. The Greeks were completely defeated and the Turks soon overran Anatolia, and, indeed, this once Christian country permeated with Roman culture was then permanently severed from the western community of peoples. A Greek general, Alexius Comnenus, gained the Imperial throne in 1081, and repelled the Norman attacks on Epirus (northwest Greece) in 1085. He encouraged the Crusaders to drive the Turks from Syria, and following their footsteps managed to win back some of the lost Anatolian districts. Alexius hired the ships of Venice to aid him against the Normans, giving the Venetians many trade privileges in return (Fig. 124).

Little of note happened during the twelfth century. The family of Comnenus came to an end in 1185, and the struggle for the throne gave the Venetians an excuse for attacking Constantinople. They persuaded the soldiers of the Fourth Crusade to assist them, and in April, 1204, the great city—the bulwark of Christianity against the Turks—was sacked by an army of Crusaders in the interests of Venice! Baldwin of Flanders became ruler of the short-lived Latin Empire, while across the straits Lascaris ruled the fragment of the Greek Empire surrounding Nikea. By playing off Genoa against Venice, Paleologus, the successor of the Lascaris dynasty, succeeded in regaining Constantinople in 1259. As Schevill points out, the Byzantine Empire—the heir of Rome—perished in 1204. The weakling state ruled by Paleologus and his heirs "dragged out a malingering existence for two hundred years," but exercised little or no influence on European history.

In the northwest corner of the Balkans—separated from the Bulgars by the Morava Corridor—live the Serbs and Croats. The former were converted from Byzantium, the latter from Rome. Hence Yugo-Slavia has always been divided into two by the most significant culture-line in Europe. This distinction in large part results from the fact that the Croats live around the Pear Tree Pass (Fig. 111) with its easy access to North Italy and the influence of the Church of Rome.

Serbia, on the other hand, is the land in the Morava Corridor—which leads directly to Greece and Constantinople. Hence the Serbians have always looked to the south for their culture.

At the end of the eleventh century these Slav tribes won some successes against the Byzantines, under their leader, Stephen Nemania. His dynasty ruled while Constantinople was reeling under the attacks of the Crusaders, so that the expansion both of the Serbish and Bulgarian nations is readily understood. In 1186 the three Asen brothers revolted at Tirnovo in Bulgaria—though it seems likely that they were actually Wallachians. In 1201 the Bulgars under Kaloian Asen obtained their independence from Constantinople. Their realm comprised the natural unit from the Rhodope to the Danube. However, they suffered greatly from Tatar attacks on the north, and in 1330 Stephen Dushan, the most famous of the Serbs, crushed them at Kustendil on the upper Struma, and Bulgaria became a vassal of the Serbish King (Fig. 125). Stephen next made preparations to capture Constantinople, but he died in 1355 just as the Ottoman Turks secured a foothold in Europe.

About 1225 some 50,000 Turks under Soliman migrated from North Persia into Armenia. They were given the region of Bithynia by the Seljuk ruler, and Osman ruled here from 1299 to 1326. His people adopted the name Osmanli, though the corruption "Ottoman" is better known to English readers (Fig. 126). Civil war in Constantinople about 1354 led to the enemies of the Paleologus family calling on the Ottoman Turks for aid. In return the Turks were given a castle near Gallipoli, and soon the Greek Emperor became a vassal of the Moslem. The Sultan Murad (1359-89) conquered Thrace, and in 1366 transferred his capital from Brusa (in Bithynia) to Adrianople. He recruited his armies by training Christian slaves—who became the far-famed Janissaries. In a few years Bulgaria was his, and in 1371 he conquered Macedonia. The main Serbian region was now held in a vice between Nish and Macedonia, and the great battle for supremacy between the Cross and the Crescent was fought in the Kossova Corridor (Fig. 126) in July, 1389. The leaders on both sides perished, but Serbia lay prostrate before the Turk. Seven years later Hungary roused by the Turkish advances lost the battle of Nicopolis on the lower Danube (Fig. 119). However, the Sultan Bayezid now turned his attention to Anatolia, which he soon conquered. But in 1402 he met a greater Asiatic than himself in Timur the Tatar, and at the battle of Ankara the army of Bayezid was overwhelmed. Luckily for the Turks, Timur soon returned to the East, and the Balkan Christians made no attempt to gain their independence.

A later Murad (1421-51) extended the Turkish territories into Greece, but was repulsed by the Albanians under Scanderbeg. Moreover, the Hungarians under Hunyadi repeatedly defeated his armies (p. 404). Mahommed II decided to capture Constantinople, which still remained free by virtue of her stout walls and mobile fleet. In April, 1453, he attacked the city with 400 ships and an army of 150,000

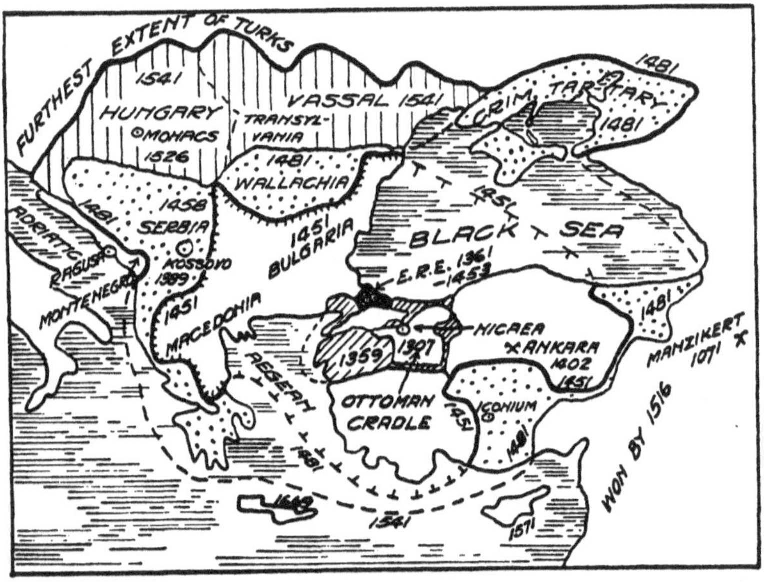

Fig. 126.—Growth of the Turkish Empire from 1300 to 1600. The Magyars in Transylvania were virtually independent from 1526 to 1657.

men. The Emperor Constantine XI had only 8,000 men to man the triple Roman wall built in the fifth century. On May 29 the inevitable happened, and Constantine fell fighting bravely as the Turks stormed the city. Murad now turned his attention to the north, but the bravery of Hunyadi flung him back from Belgrade in 1459. However, Serbia was completely overrun by the Turks, who now attacked Bosnia.

The little Slav Republic of Ragusa, which originated on the Dalmatian coast in the seventh century, deserves a brief mention. It developed into a prosperous trading centre whose ships ("Argosies") carried eastern goods to all the ports of Europe. Though controlled sometimes by Venice, and often paying some tribute to Turkey, Ragusa remained an independent centre of Slav culture when Serbia itself was submerged. It was handed to Austria when Napoleon fell in 1814.

The religious struggles between the Greek and Roman churches had so alienated many Bosnians that they had become followers of Bogumil—who taught an early form of Puritanism (p. 192). These people were persecuted by both the other churches, and when Bosnia was conquered many of them turned Moslem in consequence. Only the region of Cerna Gora (Monte-Negro) now challenged the Turks; and the latter never succeeded in conquering the Serbs in this mountain eyrie. Albania under Scanderbeg also remained free until 1467; but then the tribes united by his genius fell apart and were soon conquered. In 1521 Solyman the Magnificent captured the key fortress of Belgrad. The battle of Mohacs (1526) and its disastrous results have been described already (p. 380). The realm of Solyman reached to the limits of the Alfold, and included the north coasts of the Black Sea. Transylvania, Wallachia and Moldavia were vassal states. In 1676 the Turks acquired Podolia from the Poles, and this marked the zenith of their European expansions. Their failure at Vienna in 1683, and their gradual retreat before the Poles and Austrians have been described elsewhere (p. 381).

In Montenegro we find an environment much like that of the heart of Switzerland, and with somewhat the same results as regards national development. Early in the fifteenth century the Cernojevitch family obtained control of the mountains at the head of the river Zeta, and probably the name Cernagora (Black Mountain or Monte-Negro) is derived from this family name. In 1455 Stephen Cernojevitch allied himself to the Venetians and fought bravely at their side against the Turks. In 1484 Black Ivan, the Montenegrin leader, made Cetinje his capital, and in 1493 he printed the first books in Slavonic. In 1499 the country was annexed by Turkey, but the mountaineers still resisted, and in 1516 the prince-bishop became ruler. In later years the nephew of the bishop succeeded him and this union of church and state continued for three hundred years.[9] In 1690 the Turks sent a large army which captured Cetinje. Again in 1713 they ravaged the country, sending many Montenegrins into slavery. For a time Russia offered the little country some help, but their own

[9] "Yugoslavia," edited by John Buchan, 1923, New York.

prowess enabled them to repel amazing forces of Turkish invaders as in 1768. In 1799 the Sultan recognised their independence with the words, "The Montenegrins have never been subjects of our Sublime Porte."

G. *The Balkan States Become Independent*

There is little of importance to note in Serbian history after Kossova (1389). The dawn of independence owing to the efforts of the Austrians has been described in Chapter XX.

We may picture Serbia about 1800 with a culture relatively unaffected by the Turk rulers, who were only represented by a thousand "spahis," a few judges and a pasha at Belgrad. The "spahis" were landholders who collected a percentage of the peasant's crop, but lived in the towns. In 1804 the Serbs under "Black George" revolted, and while Turkey was at war with Russia met with much success. His revolt collapsed in 1813, but was revived by Milosh Obrenovitch in 1815. The latter was a shrewd diplomat and aided by Russia won the autonomy of Serbia in 1834. These two leading families have alternated as rulers of Serbia ever since.

The Greeks by virtue of their traditions and coastal position had naturally received more attention from western nations in their struggles for freedom. The Turks left the Greek islands relatively free, except for exacting annual tribute. In the Peloponnesus Christians were often empowered to collect the taxes and to keep down the numerous brigands. The latter, like Robin Hood of old, were popular with the peasants, for they kept alive the hope of Greek freedom. Moreover, as the British drove French commerce from the Mediterranean, there was a great expansion of Greek trade to replace that of the French. In 1821 a Greek force under Ypsilanti (a Greek in the Russian army) marched south into Moldavia and strove to arouse the Rumanians against the Sultan. Ypsilanti failed, but the Greeks in Peloponnesus (Morea) revolted and murdered many Turks. After many vicissitudes the British, French and Russians united to compel the Sultan to grant freedom to the Greeks (Fig. 127). Their decision was enforced by the destruction of the Turkish and Egyptian fleets at Navarino (1827). In 1832 the powers agreed to make Greece a kingdom which was offered to Prince Otto of Bavaria. The rise of Rumania through vassalage to independence has been traced in Chapter XXIII.

Bulgaria had vanished from the map after the battle of Kossova. The Bulgars in the Maritza basin were in close contact with the Turkish capital, and unlike the Greeks and Serbs had no chance of preserving their culture. Even their religion was a foreign importation administered in Greek by Greek priests indirectly controlled by the

Fig. 127.—The Shrinking of the Turkish Empire in Europe from 1683 to 1913. Note the Uskub Pass at the head of the Vardar and Morava Rivers.

Sultan. In 1870 the Sultan yielding to Russian pressure granted a separate church to Bulgaria. In 1876 ten thousand Bulgarian peasants were massacred by Turkish irregular troops. This roused a storm of indignation and in 1878 Russia and Rumania forced the Balkan passes. At the Treaty of Berlin modern Bulgaria came into being; but only the part north of the Balkans was autonomous, the southern and main portion of Bulgaria being left in the military occupation of the Turks, while Macedonia was handed back to Turkey (Fig. 127). This curious division was due to the fear of the growing power of Russia, for western Europe preferred Moslem control of the Balkans to a "Big Bulgaria" under the thumb of Russia. Serbia and Rumania were declared entirely independent of Turkey by this Treaty, while Greece was given the grain-growing district of Thessaly. In 1885 the southern Bulgars revolted and unanimously declared for union with the independent northern portion. Serbia was enraged at the great increase of Bulgarian power, and wantonly invaded United Bulgaria only to be completely defeated at Slivnitza. Macedonia still remained outside "Big Bulgaria," but the Bulgarian statesman, Stambuloff, initiated a policy of peaceful penetration by school and church in Macedonia. This region includes the southern part of the Morava-Vardar corridor and the port of Saloniki. The early sections in this chapter have emphasised its importance in the control of the Balkans.

Before the close of the nineteenth century, as Schevill points out, Macedonia was the scene of a triangular struggle conducted chiefly with the tools of church and school for the conquest of the *mind* of the Macedonian peasant. Moreover, there were important minorities of Albanians, wandering Wallachian shepherds, and Spanish Jews, as has been noted earlier. In 1903 an International Committee was appointed to straighten affairs in Macedonia, but was unable to curb the animosity between the Bulgars and Greeks.

In 1908 the reform party in Turkey (the Young Turks) converted many of the army officers to their views and in 1909 the Sultan was deposed. In 1912 Albania, long a rather favoured dependency of Turkey, revolted against the Young Turk government and was accorded home rule, after a short period of warfare. The new boundaries included many districts inhabited by Serbs, Bulgars and Greeks, who accordingly prepared to right affairs by their universal panacea, war. Bulgaria invaded the Maritza Valley, Serbia the upper Vardar Valley, and Greece the Lower Vardar, while little Montenegro attacked Albania. Early in November, 1912, the Bulgars laid siege to the Chataldja lines only twenty miles from Constantinople itself (Fig. 127). At the time of the London Armistice the Turks only held their capital and three invested fortresses!

Quarrels arose among the allies as to the division of the former Turkish lands. Rumania now entered the war on the side of Greece and Serbia, who had united against Bulgaria. The latter country was soon defeated and Macedonia was divided between Serbia and Greece, though Bulgaria received a small portion of Turkish Thrace. This account of the complex struggles in the Balkans must close with a brief reference to the events on the Great War as they affected Balkan territory. Bulgaria promptly joined Germany and attacked Serbia, but on the defeat of the central powers much Bulgarian territory was given to Yugo-Slavia, Greece and Rumania (Fig. 125). Embittered by these losses and jealous of her neighbours the Bulgarian peasant is not likely for many a decade to relinquish the hope of regaining Bulgaria Irredenta, whose area approximately equals that of present-day Bulgaria (Fig. 122).

CHAPTER XXV

RUSSIA—THE LAND OF UNIFORM TOPOGRAPHY AND CULTURE

A. The Topographic Divisions of the Great Plain

Few contrasts could be greater than the subjects of this chapter and of the last. In the Balkans the complexity of the national groups was in no small degree due to the complexity of the topography. In Russia there is one of the most uniform topographic areas in the world. We shall see that it has had an unmistakable effect on the evolution of the Russian Nation.

In earlier sections of this book the Russian Shield has often been referred to; we must now examine this structural unit somewhat closely. Its essential feature is that in general it is buried beneath thick later deposits, and that it has resisted folding and has acted like a buttress (or like the jaw of a vice) against which weaker rocks may be crumpled. It does not, however, follow that the Russian Shield is entirely immobile, for the contrary is the case. It has taken part in the ups and downs of the crust, but not in any paroxysmal fashion. In scientific language the movements have been "epeirogenic," *i.e.*, in broad sweeps, the whole area warping slightly—but never folding sharply. At times this resistant portion of the crust has been cracked, and broad sinks and platforms (*i.e.*, graben and horsts) seem to have developed. Yet all the deposits in Russia—and all the main formations are represented—remain nearly level-bedded. In some cases they have undergone a surprisingly small amount of solidification (from a state of sand, clay, *etc.*) when we consider their great age. This is directly due to the the fact that unlike the Carpathians and Balkans they have not been "kneaded" by Mother Earth; for they have been supported by the underlying Shield.

Since this "Basin and Table" structure is not encountered in Western Europe, it will be well to consider how it developed. Going back to the close of Paleozoic times—some 200 million years ago—we may picture Russia as looking something like the first small map in Fig. 128. Here data by Karpinsky[1] enable the author to reconstruct

[1] "Earth Movements in Russia," "Annales de Geographie," pp. 179-192, 1896, Paris. Also Kayser, "Geology," Stuttgart, 1912.

the sea and land areas at the time of the main coal formation (Carboniferous Period). We see that a broad sea occupied most of Russia—with its chief axis running north-south. In shallow lakes on the west peaty swamps developed, and these ultimately gave rise to the productive coalfields of Tula and Donetz, which are indicated in the map. Shortly after this period the Armorican Folds arose in West Europe (p. 53), and at the same time the great north-south folds of the Urals were developed along the eastern edge of this Carboniferous Sea. These are shown at the right of the map (Fig. 128 at A).

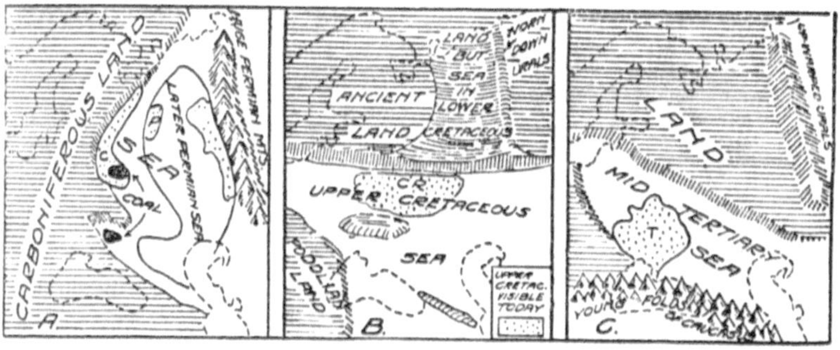

Fig. 128.—Paleo-Geography of Russia. In A (Late Carboniferous) coal is developing in shallow lakes. In Permian times the huge Ural Mountains developed. In Cretaceous times (B) the sea shifts to the south and the Urals are worn down. In late Tertiary (C) the Alpine Folds develop in the south, and the Urals are warped up somewhat. Dotted areas show the respective marine deposits which are visible to-day. (After Karpinsky and Kayser.)

During Mesozoic times this sea became narrower until in the Lower Cretaceous period (some 40 million years ago) it had the shape indicated in the second map (Fig. 128 at B). Sediments of this period were, of course, laid down in this sea on top of the more widely extended beds of Carboniferous, Permian and Triassic times. Then in Upper Cretaceous times the crust began to buckle slightly along a west-east axis, giving rise to the sea shown in the second map (Fig. 128). This in turn contracted somewhat as is apparent by Oligocene times. A little later the Alpine Storm (p. 53) buckled up all the weak deposits in the "Greater Mediterranean Sea," and the eastern extension of these folds crossed the Crimea and built up the high Caucasus to the southeast of Russia. This is indicated in the last map (Fig. 128). In fact, in the Pleistocene the east of Russia yielded again, much as it had done in late Paleozoic times. Long before this period the Ural

Fold Mountains had been worn down to a peneplain, but they were again elevated (*en masse* and without folding) to considerable heights above the sea probably during the Pleistocene. About the same time the northwest half of Russia was covered with an Ice Cap as indicated in Fig. 130.

The retreat of the Ice Cap left the greater part of Russia thickly covered with a tumbled mass of glacial debris—much as we have already observed in Poland and Germany. To the south of these heaps of haphazard silts, gravels and boulders is a belt of fluvio-glacial debris—where the material has been somewhat sorted by the numerous rivers flowing from the melting Ice Cap. This is mixed with a special deposit of loamy material, mainly blown south by the winds from the glacial debris, which is known as *loess*. For climatic reasons the southeast of Russia is a grassland, and the growth of grasses for many thousands of years in these deep soils in the southeast has added a great deal of humus to the loess, *etc.*, and converted it into the famous Black Earth. Its Russian name of "*Cherno-ziom*" is fairly well known (Fig. 130).

It is clear, therefore, that the Build of Russia is very different from anything we have considered in west and central Europe. Perhaps the Alfold is the nearest—but there is no arc of young mountains bounding Russia, except the Caucasus in the southeast corner. We have to deal with enormous areas of level-bedded formations, whose structure has had only a moderate effect on the topography. Indeed, only three small patches in the Russian Plain—the Valdai Hills near Novgorod, a hill near Saratov, and a larger area in western Podilia—exceed 1,000 feet (Fig. 129).

The topography of Russia can best be considered as a series of four undulating uplands separated by three main river valleys (Fig. 129). The former correspond roughly to the Platforms, the latter to the Sinks. These run approximately north-south and are classified in the table on page 481.

The last section in the above table shows us that the geology of Russia is very varied—but it has not so direct a bearing on the Build as in the other countries of Europe, so that we need not here discuss it.

The Build is partly due to the alternation of the structures which we have termed platforms and sinks. The Platforms are usually very broad, flat, crustal blocks, with only a small amount of elevation, while the Sinks are similar broad, flat, downfolds or graben with only a slight depression in the centre. Sometimes these may be bounded by faults, in which case we get a tile-shaped horst, or a very shallow graben; but generally the country is free from obvious faulting. Around the margins of Russia there are, however, more definite structures—which form a sort of broken rim to the vast "platter" of Russia.

In the northwest is one of the most striking features in eastern Europe, the peneplain of Finland. This is a large block of very ancient rocks, which is the best European example of a peneplain—for it is mostly near sea level and shows the characteristic erosion of all kinds of rocks to a *common base-level* (p. 60). On the edge it has been faulted a little, and in these slight graben (or sinks) water has collected to form Lakes Ladoga and Onega, and the White Sea (Fig. 129). The production of this ancient shield goes far back to pre-Cambrian times; and one of the most remarkable journeys in the world that a geologist can take is to travel from Finland to Kazan and thence southwest to Odessa (Fig. 129). He passes across every geological formation, arranged in regular sequence from Cambrian to Recent—which means that as geological time has passed the shorelines have moved eastward to Kazan and thence southwestward to Odessa. (Compare Fig. 128, and see the similar explanation of a journey from Snowdon to London, p. 33.)

In the southwest of Russia is another fragment of the ancient rim in the form of the ancient granites and gneisses of the Podolia Relic-Block (Fig. 129). They also represent a portion of the earth's crust much more ancient than anything in the central Russian area, and are akin to the similar relics in Bohemia and Cevennes. In the southeast is a third such area known as the Ust-Urt Block, while close to the Urals in the east is the Ufa Block. But these two last examples are hidden by later formations and are mapped chiefly by geological deduction.

The Urals appear to-day as a broad belt of elevated land which rises to a height of 5,000 feet at several points in their central and northern sections. They consist of extremely complicated strata—originally folded and first elevated long ago in Permian times. These ancient rocks contain very valuable minerals such as magnetite (iron), gold, platinum and copper—which are especially abundant in the oval area (enclosed in a broken line) shown in Fig. 129. The Caucasus Mountains, belonging to the last great mountain-building era, rise in Elbruz to 18,530 feet, which is indeed the highest peak in Europe. There are only two good roads over the Caucasus—the Darial Pass at 7,805 feet, and the Mamison Pass at 9,270 feet. Both cross the central

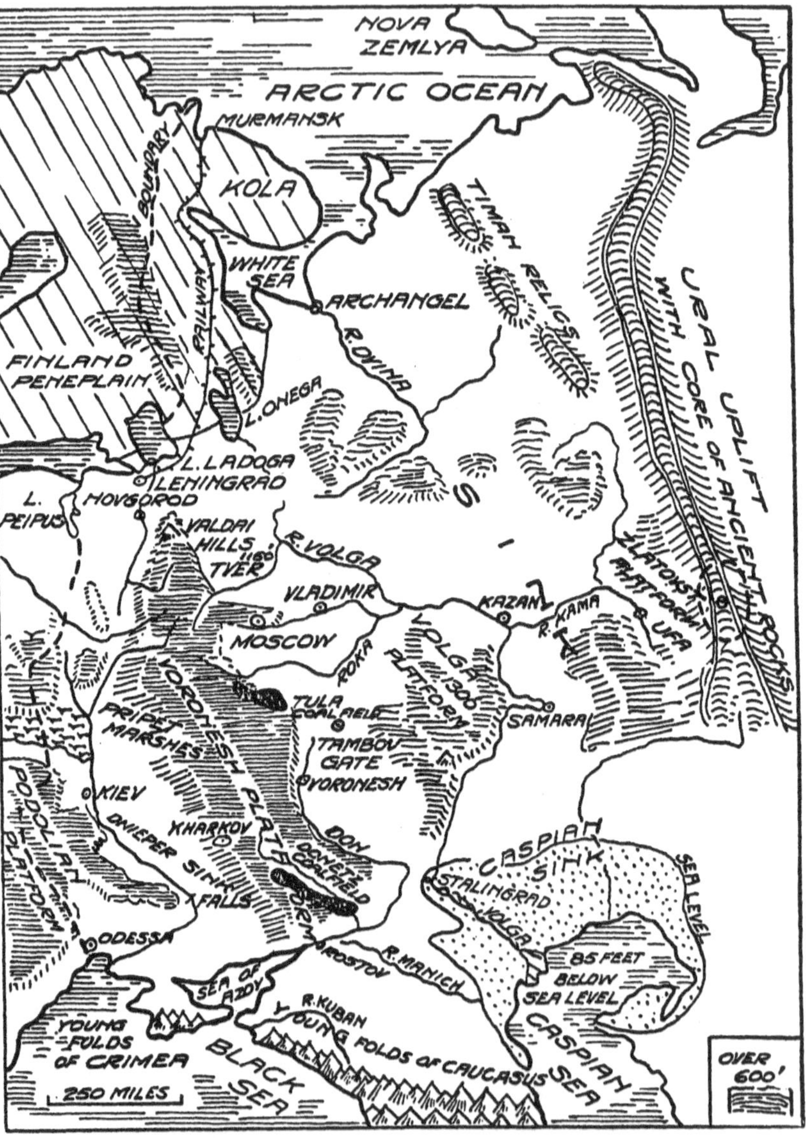

Fig. 129.—Major Physical Features in the Russian Plain. Note the alternate Sinks and Platforms, and the rich mineral area in the core of the Ural Warp at Zlatoust. The black areas are coalfields.

TABLE Y

STRUCTURAL UNITS IN RUSSIA (Fig. 129)

No.	1	2	3	4	5	6	7
Class	Western Borderland	Western Valley	Mid - Russian Upland	Central Valley	Eastern Upland	Eastern Valley	Eastern Borderland
District	West Ukraine	White Russia. Central Ukraine	Moscow to Kharkov	Tambov to Rostov	Kazan to Tsaritsin (Stalingrad)	Kazan to Tsaritsin (Stalingrad)	Tatar A.S.S.R.
Build	Podolia Relic-Block and Galician Platform	Pripet Marshes, Dnieper, South Russian Sink	Valdai Hills, Voronesh Platform (Donetz Coal)	Upper Volga, Oka, and Don Basins. Tambov Gate	Volga Platform	Lower Volga Basin, Caspian Sink	Ural Foreland, Ufa Platform
Geology	Granite, etc.	Eocene, etc.	Carboniferous in North, Chalk in South	Jura, Chalk, Eocene	Trias., Jura, Eocene	Trias., Alluvial	Permian

part of the Caucasus. The dislocations which produced the Black Sea (p. 55) have isolated the Crimean Mountains from the Caucasus. The highest peak in the Crimea is Chatyr-Dagh (5,000 feet).

Turning now to the central part of Russia, there are several of those structural units which I have termed platforms. Most important of all is the Donetz platform, which consists of ancient paleozoic rocks containing good coal seams. This appears at the surface, or is only slightly covered by surface rocks, in the bend of the Donetz River (Fig. 129). It is perhaps the chief source of coal in the Soviet Union. Another relatively isolated fragment of the crust is the Voronesh Block, which lies beneath the surface formations between Kharkov and Tula (Fig. 129). It helps to determine the Mid-Russian Upland, in which so many large rivers have their origin.

In general, however, it may be stated that the relation between these crustal segments (which I here call platforms and sinks) and the present alternation of uplands and basins is only a general one, and must not be pressed too closely. However, the Caspian sink in the southeast is clearly the result of the Pleistocene downfold shown in Fig. 128 at C. The steppes are below sea level here—and in the Caspian Sea the level of the water is 85 feet below mean sea level. Its greatest depth (in the south) is about 3,600 feet. The shore deposits and the biology of the lakes around the Caspian prove that this sea was linked to the Black Sea until after the maximum of the Great Ice Age perhaps 40,000 years ago.

B. *Climate, Soils and Settlement*

If we examine one of the most recent geographies, such as that by Huntington, Williams and Van Valkenburg,[2] we shall find that the distribution of soils is largely a function of climate. The best soils in the world are the Black-Earths, and these have a distribution closely related to the desert lands of the world. In the United States the Black-Earths extend from Western Texas to North Dakota, *i.e.*, just east of the arid region. In Africa they are found south of the Sahara and extend from Senegal to the Sudd on the upper Nile. In Eurasia they form a belt from Orshova, on the Danube, to the Altai Mountains, *i.e.*, on the northern edge of the arid belt. There are four or five other areas elsewhere. Possibly the broadest area of all is found in South Russia, and it is perhaps not too much to say that the prosperity

[2]"Economic and Social Geography" (Plate II at end), 1933, New York.

of the Soviet Union depends more on this factor than on any other. Certainly it is of more importance than Build, or geological strata. Hence we may well devote a little time to the soils which so definitely divide the Russian plain into "natural regions."

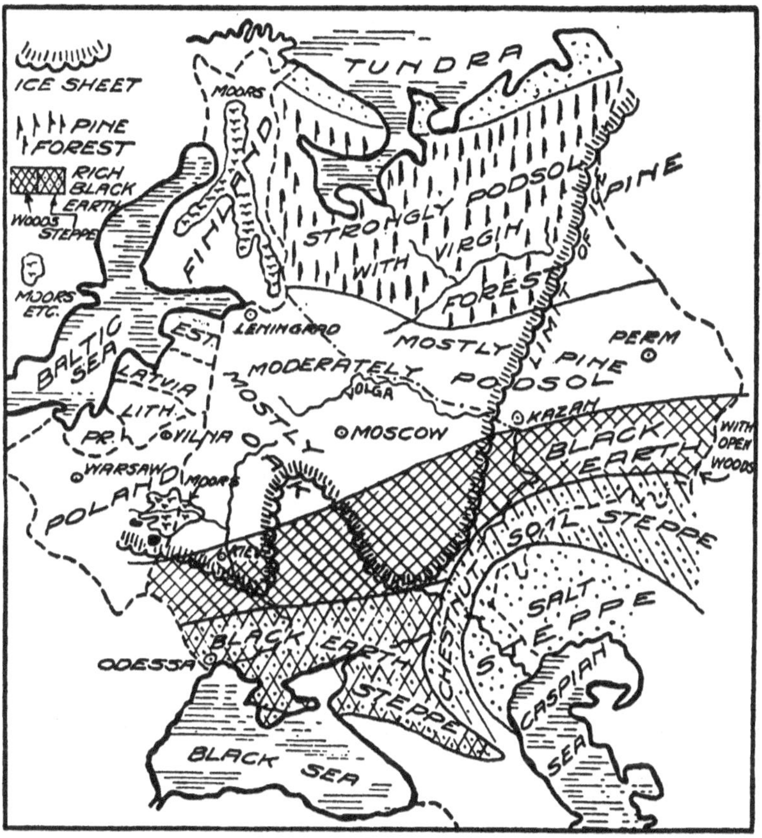

Fig. 130.—The Soils and Forests of Russia. The southern limit of the last Ice-Sheet is shown. "Podsol" is a leached sterile layer often covering Clay. (After Von Seydlitz, Bowman, etc.)

One unfortunate characteristic of many regions with a climate excellently adapted to human comfort is that the regular rains gradually dissolve out (leach) the valuable plant food from the soils. As a result a grayish sandy soil is left which is everywhere called *Podsol.* The Russian name is used because some of the earliest detailed work on soils is due to Russian research. The northern half of Russia in general consists of such Podsols—which agree in distribution with the forest areas (Fig. 130). Indeed, both are due to the fairly *uniform* character of the rains. Kazan is a very interesting district in Russia, for here the three main edaphic (soil and vegetation) controls meet (Fig. 19) as described in an earlier chapter (p. 74). To the north with the long cold winter and podsol soils grow the conifers. To the west with heavier rains but shorter winters grow the broad-leaf forests (oak, etc.) mixed with some conifers. To the south the forests cease—or are represented only by corridors of trees along the rivers, or by little "islands" of trees. This last main boundary runs from Kiev through Kazan to Perm, and is a very important factor in Russian history. Indeed, from the opposition of the Forest and the Steppe has proceeded the historical antagonism, the strife of many centuries, which has divided the two halves of Russia. I refer to the warfare between the Sedentary North and Nomadic South, between Russian and Tatar, and later between the Moscovite of the Forest State and the son of the Steppe, the free Cossack (Leroy-Beaulieu).

The southern portion of the Black-Earth Belt is a steppe land, which becomes drier and less productive towards the southeast, as the rainfall map (Fig. 17) shows. Gradually the black humus (due to the decay of plentiful grass-roots, etc.) becomes less abundant, and a chestnut soil results.[3] This is generally quite fertile if sufficient water can be obtained, but in places it is spoilt by a large content of salt. This in turn merges into the large area of wind-blown sand which covers much of the western littoral of the Caspian Sea (Fig. 130).

There is little need to dwell on the topography of the central area. None of it rises above 1,000 feet, yet the rivers for the most part have cut fairly deep valleys through the glacial debris and also through the level-bedded surface formations. The Volga has cut down so far that it has only a fall of 500 feet in the lower 2,000 miles of its course. A characteristic feature of these vast swelling uplands—such as that south of the Valdai Hills—is that most of the rivers rise in swampy lake-studded regions, where it is easy to pass from one basin to the other. Few divides in the world are so mixed as those south of Novgorod (Fig. 129). Here the headwaters of the Lovat, Dvina, Dnieper, Volga and Oka, all rise close to each other at about the same level, near 800 feet. In the early days, portages with boats were easily

[3]See, however, the Russian chapter by R. M. Fleming in "Studies in Regional Consciousness," Oxford, 1930, for the comparative advantages of the Black-Earth and northern clay soils.

made, and some of the resulting Varangian routes are shown in Fig. 131. Similar undulating uplands build up the other slightly elevated areas. Indeed, the most conspicuous features, in the southern region especially, are the *kurgans* or artificial mounds over the prehistoric graves.

The Dnieper receives the Pripet from Poland, and its vast series of marshes has been described earlier (p. 410). It seems likely that

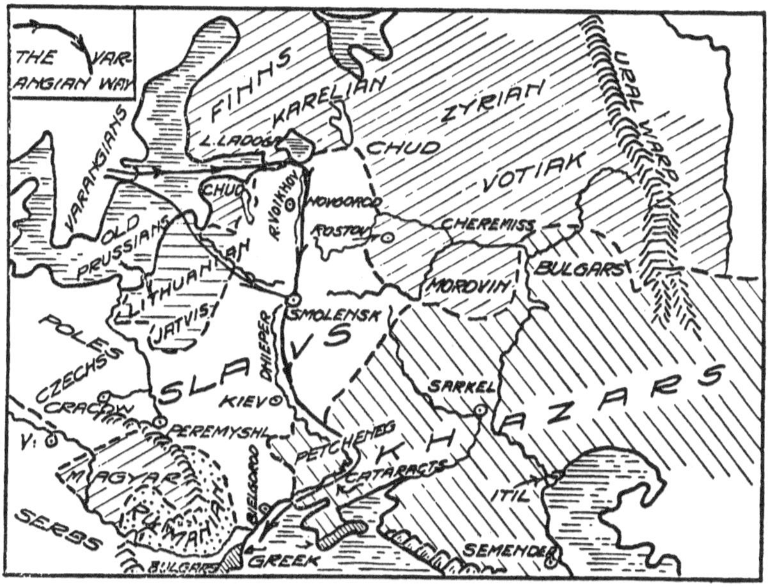

Fig. 131.—Eastern Europe during the 9th and 10th centuries, when much of Russia was controlled by the Finns and Khazars (allied to the Turks). The Varangian Way used by the Swedish Traders is mapped in the west.

the water has been held back mainly by the hard granites of the Podolia Block (Fig. 129), which the river could only erode very slowly. Kiev is situated in the middle of the Dnieper Basin where all its great tributaries converge. The swampy valley is about four miles wide above Kiev but contracts greatly just at the city. Lower down, the river has cut a gorge in the Podolia Block, and below Kremenchug there are many rapids, especially near Dnepropetrovsk (Ekaterinoslav), where it falls 65 feet in ten miles (Fig. 131). Here a famous Power Station has recently been erected. Notice how the rivers

Dnieper, Don and Volga are each eroding their right banks, *i.e.*, the *eastern* edges of the adjacent platform.[4]

The Volga (Finnish—Holy River) rises in one of the swampiest areas of Russia and is only separated by a peat bed from a tributary of the Volkov which enters Lake Ladoga (Reclus). It oozes from bog to bog for 20 miles and then passes through many small lakes, after which it becomes a notable stream and descends down some small rapids to the great plains before it reaches Tver. Its chief tributaries are the Oka—long the boundary between Russian and Tatar—and the Kama.

Moscow, the capital of the Soviet Union, is near the centre of "Economically Developed" Russia, but has not grown up on one of the large rivers. As D'Almeida points out, the Moskva meanders through Moscow, much in the same way as does the Seine through Paris. Both cities grew up where an island filled much of the river—but as the chief danger in Moscow was invasion from the south, the centre of the city is on the north bank (not on the isle, as in Paris). In the 12th century the settlers built their huts on a wooded hill some 120 feet high, where later was placed the Kremlin (or Acropolis). It became the capital of Muscovy, replacing Suzdal and Vladimir. In the 13th century the Prince of Moscow acquired fame by defeating the Lithuanians, and in 1325 the head of the Russian Church established his seat at Moscow. To-day it is the chief railway centre, and contains the chief textile factories of the Union.

Below the junction with the Kama the river hugs the edge of the Volga Block on the west, but the east is flat and much of it was part of the enlarged Caspian Sea of former times. The curious bend at Samara is due to the presence of high limestone hills which are cut through by the river. Further south the west bank rises steeply for 1,000 feet above the river-bed which is here nearly at mean sea level. At Stalingrad the Volga turns to the southeast and wanders as a "braided stream" with many channels across the plains of the Caspian sink. At the angle it is only 40 miles to the bend of the Don which flows at a higher level, and in the middle ages many of the cargoes of the Volga boats (and often the boats also) were dragged overland and placed on the Don, whence the Bosporus and the open ocean could be reached. Sarkel and Sarai—bygone famous towns—

[4]All moving bodies in the Northern Hemisphere (owing to the earth's rotation) tend to swing to the right. This may explain the way these rivers cut into the Blocks on the west banks.

were built near this interesting angle (Fig. 131). During the spring floods all the plain south of Stalingrad is flooded, and in normal times the Volga has fifty regular channels to the Caspian Sea. The Volga is about 2,230 miles long, but its discharge is much less than that of the Danube, since its basin although twice as large is so much drier.

The third great river is the Don, which rises not far from the south bank of the Oka; and Tambov has grown up in the "Gate" on the low divide linking the two basins (Fig. 129). Just to the west of the gap is Kulikovo, where the Russians won their first great victory against the Tatars in 1380. Here also during the middle ages were placed an earth rampart and lines of defence against southern foes. Near Voronesh in the same river basin Peter the Great prepared his fleet for his attacks on the Turks on the Sea of Azov. Along the Lower Don collected the bands of rovers who assumed the name of Cossack, and (like the pioneers in western America) did so much to open up the semi-arid lands. Since the Don has few tributaries and rises in lower lands where the snowfall and rains are lighter, it does not rival the Volga and Dnieper in importance.

C. *The Early Peoples of Russia and their Ethnology*

The effect of the environment on the people before the Soviet Union is well expressed by Maurice Baring.[5] In western Europe the landscape may vary from hour to hour, but you may travel for days in Russia without a change. The life of the people and the cultivation of the land is as monotonous as the landscape. The result is that in Russia the desire for intercourse was never kindled. There was no give and take between mountain and valley as in Western Europe, because there was no mountain and no valley. Since nature, landscape and the manner of life were more or less the same everywhere, the feelings and manner of speech of the people were naturally the same also.

In Russia the absence of natural obstacles was favourable to the expansion, the emigration and spread of the population. A Russian who emigrated from his home to a distant part of the country was able to adapt himself with the greatest of ease wherever he went. This facility of emigration is a cardinal fact in Russian history. In this way the Russian culture is extended, but it has not improved. It is spread over a larger space—but spread thinly. The history of Russia is the

[5] "The Russian People," London, 1911.

history of the colonisation of the Eastern European Plain (Baring). On the other hand, this great plain offers the same advantages to the foreigner as to the native. There is a complete absence of those natural barriers behind which a people may seek shelter from the incursions of others. The only natural protection which the region offers is in its dense forests and swamps. These, however, unlike mountains, offer no variety of conditions or natural products. They afford no stimulation to advance in culture; indeed, they retard civilisation in the act of protecting it. They are better fitted to afford refuge to an exiled people than to encourage progress in a nascent one (Ripley).

We may notice one very striking feature in the political development of Russia. Leningrad (St. Petersburg), Novgorod, Moscow and Kiev have all been capitals at one time or another (Fig. 129). Novgorod flourished in the cold forest region only while the Tatars submerged most of Russia. St. Petersburg was purely the creation of Peter the Great to "open a window" on the northern seas, and had few local resources. Moscow is a large industrial and railway centre; but Kiev (the old capital before 1200) lies on the edge of the rich agricultural belt and is bound to become of increasing importance. Perhaps likely to surpass all is Kharkov, in the centre of the Black-Earth area and close to the Donetz coalfields.

No region in Europe exhibits to-day such ethnological unity as does the great Russian Plain. It is true that there are Ukrainians to the south, White Russians to the east and various Finnish people in the north and northeast (Fig. 135). Moreover, some of these speak entirely different languages, and follow different religions. But they have all developed into a common sub-Alpine race which excludes only a few out of the 120 million peoples concerned. This is particularly interesting when we learn that the skeletons in the innumerable kurgans belong to a rather different stock. The kurgans are mounds from 20 to 50 feet high and date mainly from the late Stone Age. The men of this Kurgan Period betray a notable homogeneity of type, even more uniform than that of the modern living population. The crania are almost invariably of a pure long-headed variety, the head indices ranging as low or lower than that of the purest living Teutonic peoples of to-day. The indices were about 77 or 78, whereas the Russian to-day is 82. The Kurgan folk were much taller than the Slav folk of to-day, and it seems likely that these people migrated right across Central Europe to the Baltic area and Poland (Ripley). There is little doubt in the present writer's mind that the Kurgan people were the ancestors of the Nordics and of the Finns—and that the main bulk of the Slavs are a hybrid between these Proto-Nordics and the abundant Alpine migrations of Central Europe.

There are, however, certain folk in the north of Russia, the Laplanders, and others in the southeast, the later Tatar element, who must be isolated from the main bulk of the Russians (Fig. 136). Both these other types are much more brakeph as well as much more "Mongolian" in appearance. It seems fairly clear to the writer that there have been two very different migrations of these Mongolian-Alpines. The first included the Laplanders—who spread along the coasts of the Arctic Ocean from their earlier Asiatic homes, very early in Russian history. They were followed by the Proto-Nordics who are now represented by many Finnish tribes, Tavasts, Karelians, Ests, Livonians in the west, Zyrians, Votiaks, Cheremiss and Mordvins in the east of Russia (Fig. 135).

Naturally the Finnish people mixed to some extent with the short brakeph Mongoloid Lapps, since they lived close to them for many centuries. The more primitive Lapps learnt the Finnish language—which means nothing as far as descent is concerned—and this has led many writers quite erroneously to link them together racially. At the dawn of history the Finnish people occupied all northern and much of central Russia (Fig. 111). Somewhat similar to the Laplanders in race—but arriving in Russia quite lately—are the Tatar folk, such as the Kazan Tartars, Bashkirs, and their congeners in the Caucasus. These latter tribes have, however, mingled more with Alpine peoples than have the isolated Lapps, and so do not exhibit such a "non-European" appearance. (Further notes on the ethnology appear at the end of this chapter.)

In Chapter XXI (p. 388) the early history of the Slavs is narrated. It seems probable to the writer that the Slavs had originally reached the Vistula from the Siberian Steppes far back in prehistoric times—but that is little more than a speculation at present. Assuming, then, that at the dawn of history the Slavs spread out from the Vistula Basin, we find them displacing the Finns in western Russia about the time of the Gothic migrations (p. 391). According to Peisker[6] they reached the network of rivers in the Valdai region which has already been described. This environment seems to have been created for primitive commerce, and no doubt a much finer civilisation

[6]"Cambridge Medieval History," Vol. II, Chap. XIV, 1913, New York.

would have developed in Muscovy if its southern outlets had been situated elsewhere. But the Dnieper, Don and Volga lead down to the steppes where flourished the mounted nomads—the enemies and stiflers of all growing civilisation.

It is probable that the ancestors of the Ukrainians were the Antae mentioned by early chroniclers in the sixth century. When the Bulgars moved into the Balkans (p. 466) these Slav Antae spread rapidly in the Ukraine and thence migrated to the northeast. They came in contact with the Khazars—a trading people whose capital was near Astrakan (p. 416), and were apparently subjugated by this powerful nation. We hear that the Khazars helped the Slavs to develop a trade along the Dnieper centred at Kiev. This trade route was, of course, linked up with the "Varangian Way" by way of the Dvina to the Baltic.

D. The Varangian Traders and the Princes of Kiev

The Varangians, a group of hardy northern traders, established definite commercial routes very early in Russian history (Fig. 131), and their chief commodities were living Slavs. These they exported to the Roman and Persian Empires and no doubt further, and their payment in part consisted in the thousands of ancient oriental coins which have turned up all along the Russian rivers. About A.D. 850 trading Scandinavian sea-robbers got possession of the Russian network of waterways, and the Dynasty of Rurik founded the first powerful Russian State. (The same thing could be seen only yesterday in Turkestan, where the passive Tajiks languished in servitude to the nomadic Turkomans.) For many centuries the Slav was subjugated by the Swede or the Tatar. Yet he increased greatly in numbers, because the nomads were not always raiding—and were probably always far inferior in number to their serfs, the Slavs. Indeed, the word "slaves" arose from their trade in Slavs.

The cradle of the Russian Empire was the Swedish settlement on Lake Ladoga which may have been ruled by this semi-mythical person, Rurik. Other Swedish districts were Novgorod and Kiev, and in 860 Slavs led by Swedes made a raid as far as Constantinople.

It seems likely that at times the Varangians were mercenaries hired by the Slavs to protect their trading centres. Thus Kiev became a powerful state in the ninth century. Early annals (dated 839) speak of Russian traders to Byzantium, who were unable to return directly to Kiev because of the impending Magyar invasions. No doubt there were native Slav princes in some of these "trading states," but Kiev developed under the leadership of the Varangians. Their chief, Oleg, who lived about 900 A.D., is perhaps the first authentic figure in Russian history. Oleg (or Helgi) conquered Smolensk about 881 and Kiev in 882, and made the latter the centre of his realm. He smashed the Khazars and Bulgarians and founded an Empire which extended from the North Sea to the Black Sea along the whole extent of the Varangian Way. In 907 he led a fleet of 2,000 boats down the Dnieper and levied tribute from Constantinople. His successor Igor (Ingvor) plundered the Persian coast of the Caspian Sea in 914. He besieged Constantinople in 941, but was driven off by Greek fire. His wife, Olga, became a Christian, but his son, Sviatoslav, refused to abandon the old gods. Sviatoslav spent his life in campaigns to extend his realm. By 970 he had conquered the Khazars, Alans and Bulgars, but he was repulsed by the Byzantines and killed by the Petchenegs in 973.

Vladimir was baptised in 988, after his marriage with Anna, a Byzantium princess, and on his death in 1015 Jaroslav became ruler. He supported the people of Novgorod in their struggles with the Varangian overlords and finally banished the latter, so that from this time Russia was controlled by Slavs alone. Jaroslav subdued the Petchenegs, but their place was taken by hordes of Cumans (Polovzes) (Fig. 121a). They defeated the Slavs in an important battle in 1068, and closed the Dnieper waterway, thus ending the trade between Kiev and Constantinople. During Jaroslav's reign Kiev rose to eminence, and he formed connections by marriage with the royal houses of Norway, Poland, Hungary and France. Unfortunately, the complicated law of succession inevitably led to struggles for the throne. Unsuccessful claimants for the Kiev kingdom founded states in the outlying portions of Russia, and so duchies developed in Novgorod, Pskov and in Galicia on the borders of Poland (Fig. 132). About 1200, Roman of Galicia was the most powerful of these Slav princes, but he was killed in 1205 and his kingdom conquered by Hungarians and Poles. The people of Kiev became Christian about 987, and Russian culture naturally developed thereafter from Constantinople and not from Rome. One important consequence of this was that there was little of that constant struggle between Church and Emperor which marked the Medieval period in Western Europe. In Russia the Church was always subordinate to the State—so much so that if the ruler disapproved of a Bishop sent from Constantinople he appointed another.

About 1150 the character of the Russian state changed greatly. Before this the main factor was the trading city—which developed naturally from the Varangian-Russian trade of primitive times. A string of such towns sprang up along the Dnieper-Dvina route and along the adjacent rivers. But after 1150 the local princes acquired much more authority, and the centre of Russia gradually moved from

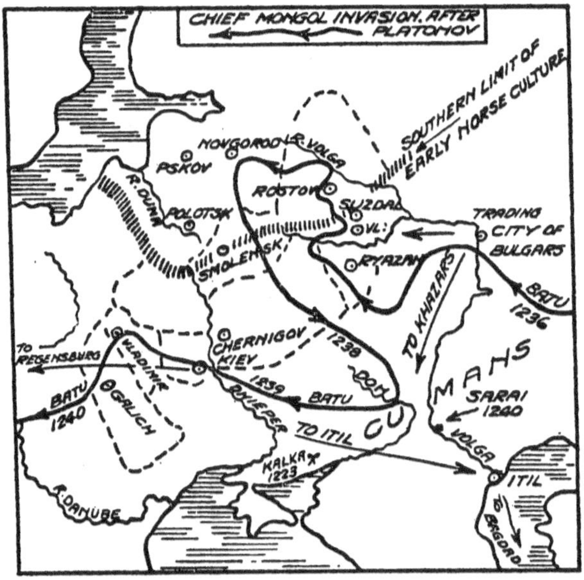

Fig. 132.—Early Trade Routes and the Political Divisions from the 11th to the 13th centuries. The route of the main Mongol Invasion is also shown. (After Mirsky.)

Kiev to Moscow. Rostislav in 1159 ruled from Novgorod to Kiev and thus controlled the whole Dnieper waterway. The wealth of a prince consisted chiefly of the lands he owned and the slaves he could gather to develop his lands. Free land was to be found chiefly to the northeast (Fig. 132), for fierce nomads contested the possession of the Steppes. Moreover, the latter country did not appeal to the Russians, who understood life in forests and swamps, but not in open grasslands. There were already tribes of Finns in the Moscow region, and these blended with the Kiev Russians—who were chiefly Alpine (akin to the bulk of the south and west Slavs to-day)—to form the ancestors of the most abundant people in Europe, the Great Russians. The Bulgars who lived on the middle Volga were a civilised Moslem people who had developed a widespread commerce during the twelfth century (Fig. 132).

In the middle of the twelfth century Prince Andrew of Kiev had acquired control of Vladimir (100 miles east of Moscow); and on the death of his father, who ruled in Kiev, Andrew attacked and conquered the ancient city in 1169. He returned to Vladimir, where he ruled the Suzdal region, and declared himself Grand Duke of Russia, thus making Kiev a vassal of Suzdal (Fig. 132). About this time the common people still had certain political rights and assembled at the "Veche" to decide public questions. In most towns the Dukes soon controlled all the power, but in Novgorod and Pskov the Veche remained the authority and these two towns became republics, which endured until the middle of the sixteenth century.

The large northern realm of Novgorod was of little use for a food supply—and the people obtained grain from the land of Niz (the upper Volga). The north was called Zavoloche (Beyond the Portages), and here the traders obtained furs from the Finns (Zyrians), as well as fish and some minerals. The fur-trade paid so well that Novgorod was the richest city in Russia during the 13th, 14th and 15th centuries.[7]

E. From the Tatar Conquest to Peter the Great

Early in the 13th century the terrible invasions of Jenghiz Khan took place. He was a Mongol, born in central Siberia, and he conquered and united almost all the nomads of north Asia. In 1222 he moved upon Europe and easily subjugated the Alans (Sarmatians) and Cumans—who were of his own Altaic type (Fig. 132). Some of the Cumans fled to their former enemies, the Russians; and in 1223 their combined forces were disastrously defeated at Kalka (near Azov). But the raiders then returned to Asia for some years, until in 1237 Batu, a grandson of Jenghiz, again attacked Russia at the head of 300,000 men. Within the next few years he burnt Moscow and Kiev and invaded Poland, Hungary and Croatia. Only Novgorod of all the large Russian towns was left unconquered—and here the Russian culture was not directly affected by the conquest. Batu established his capital at Sarai on the Volga (Fig. 132), and did not occupy the wooded region north of the steppes. However, the Tatars collected

[7]D. S. Mirsky, "Cambridge Medieval History," Vol. VII, 1932.

tribute from the southern forested belt, where their officer or Bashak represented the Khan.

Gradually the Russian princes returned and obtained control of their special regions—always paying taxes to the Mongol. The Tatars retarded the progress of the country, but they respected the churches on the whole; indeed, some of the later Khans, like Kuyuk Khan and Kublai Khan, occasionally attended Christian services. But the quarrels between the Greek and Roman churches must have alienated the Tatars from the Christian religion. Berkai Khan adopted Islam in the 13th century, and the Tatar domination thus introduced a further cleavage in the religions of the peoples of Eastern Europe. However, the Tatars put an end to most of the civil wars of Russia, for many of the princes and nobles (boyars) had been ruined in the invasion. Thus a way was prepared for the rise of a single Russian leader in the person of the Grand Duke of Muscovy; for Moscow now replaced Vladimir as the centre of Russian culture, and indeed the Grand Duke acted as the representative of the Khan among the Russians.

Early in the 14th century there were lengthy struggles between the rulers at Moscow and Tver, each claiming the title of Grand Duke. Finally Muscovy, under Ivan Kalita, won, and Moscow ranked as the capital after 1328. The expansion of Muscovy in the succeeding centuries is illustrated in Fig. 133 at B.

The Tatar domination destroyed the ancient cumbrous rules of descent, and the duchies now were understood to be the property of the ruler, who could pass the inheritance to his son. However, during the 13th and 14th centuries there were a number of these duchies which tended to develop according to the most obvious geographical divisions, *i.e.*, the river basins. Thus there was one such unit around Vladimir, and others round Riazan, Suzdal, Smolensk, etc. (Fig. 132). Owing to its geographical position Moscow came to the front. It was linked by the Moskva to the Oka River and was the junction of ancient routes from Kiev to Rostov, and from Tver to Riazan. It was also one of the best areas in the new lands taken up by emigrants from Kiev. Its central position among a circle of fairly strong towns ensured that it was spared many of the attacks by foreigners, such as the Lithuanians, Swedes and nomads of the south.

Meanwhile disunion was destroying the power of the Tatars just while the Muscovy princes were learning how to build up a strong state. Dmitri, the son of Ivan II (who reigned from 1362 to 1389)

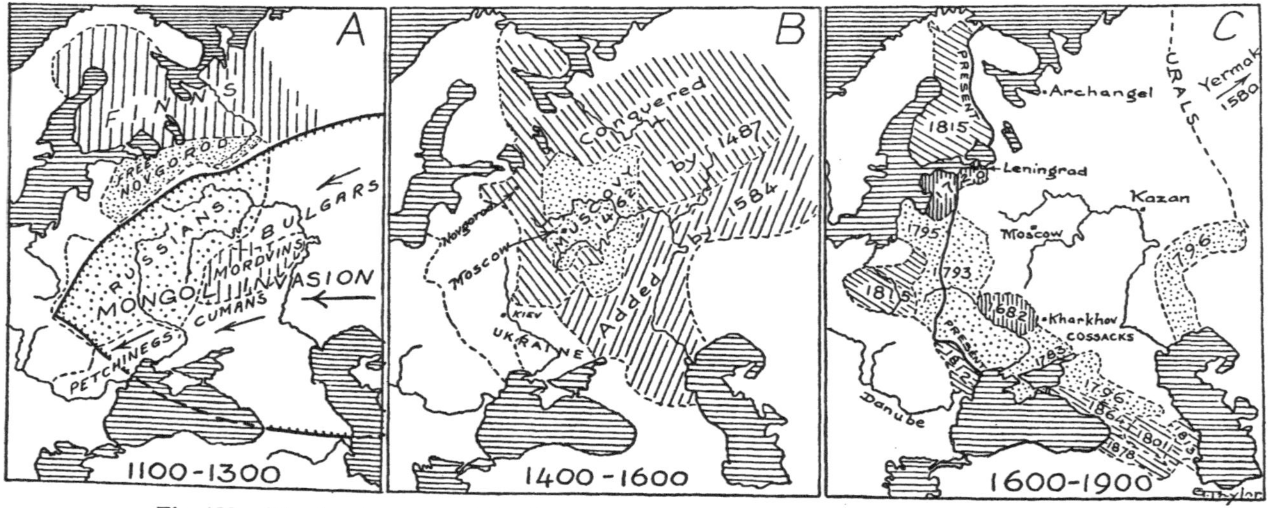

Fig. 133.—The Development of Russia until 1300 (at A). The three main stages in the growth of Muscovy from 1400 to 1600 (at B). The conquests of Russia until 1878 (at C). The present Western Boundary is shown by the heavy line in C.

defied the Tatars and actually made two of them his vassals. In 1380 he felt strong enough to face the armies of the Tatars even though the Lithuanians joined the latter. Moving down the Tambov "Gate" (Fig. 129) at Kulikovo at the source of the Don the Russians decisively defeated the Tatars for the first time since the invasion of 1223 (Fig. 133 at B). This victory naturally greatly strengthened the prestige of the young state of Muscovy, especially as the seat of the Metropolitan (the head of the Church) had been transferred to Moscow in 1325.

By purchase and conquest Moscow in 1462, during the reign of Wassili (Basil), had acquired control of all the lowland to the northeast and southeast as far as Kazan. This may be taken as the *first stage* in the expansion of Muscovy. Much of the region southwest of Muscovy belonged to the Ukrainians as far as Kiev, but Poland (with Lithuania) soon extended her borders until they made contact with those of Muscovy at the end of the 15th century. All the steppe country was still in the hands of the Tatars, but the nomads around Kazan had separated from the main (or Golden) Horde a few decades back. Still opposed to Moscow was the district around Tver, while the only peer of Moscow was the democratic city of Novgorod, which controlled Pskov to the west and the whole of the Russian Plain to the northeast. This latter area of forests was, however, very sparsely inhabited for the most part, chiefly by Finns.

After the accession of Ivan III (1462-1505) the people of Russia themselves realised the advantage of union as never before. In 1463 Yaroslav asked for union with Muscovy. By 1478 Ivan had subjugated Novgorod, and so vastly extended his territories over almost all the whole forested region of Russia. Rostov and Tver declared allegiance in 1474 and 1485. These additions gave Muscovy a realm extending from Pskov to the Urals and from the Arctic Ocean to the edge of the forested lands in the south (Fig. 133 at B). This may be taken as the *second stage* in the expansion of Muscovy.

Ivan III refused to pay tribute to the Tatars, and Russia now regained something of the prestige among the nations which she had lost since the distant days of Jaroslav. But a life and death struggle now ensued between the two Slav powers, Poland and Muscovy, while Ukrainia at times made a bid for recognition and Lithuania was not always willing to adopt the policy of her yoke-mate Poland.

Ivan III married Sofia Paleologa in 1472, just after the sack of Constantinople by the Turks. This Princess claimed to be the chief descendant of the Byzantine Emperors, and hence Ivan adopted the title of Tsar (Caesar) as well as the badge of Byzantium, the "Double Eagle." The reign of Ivan IV (the Terrible) was occupied by crushing the Tatars of Kazan as well as those to the south, and in occupying the new environment of the steppes. In 1552 Kazan was taken and the numerous Finn tribes to the northeast made submission to Muscovy. In 1557 Sarai, the old Tatar capital, yielded, and now the Volga for the first time became a wholly Russian river. 1553 marked the start of important English trade, when Captain Chancellor reached the White Sea.

After 1564 Ivan IV seems to have become insane. He ruthlessly killed all who opposed his wishes, including the Head of the Church who expostulated with him. He gave over large communities to be looted by his special bodyguard and in 1581 he killed his own son. Before his death in 1584 his empire was extended far into Asia by the exploits of a small band of Cossacks under Yermak. This marks the end of the *third stage* of expansion of Russia (Fig. 133 at B).

Feodor, a man of weak character, now became Tsar, and on his death Russia was torn by factions, during which a Polish army controlled Moscow and a Polish Prince almost gained the throne of the Tsars. Sweden under Gustavus Adolphus also compelled the Russians to abandon their hope of territory on the Baltic. Michael, the first of the Romanofs, came to the throne in 1613 and gained some barren lands in the southeast; but it was not until his son, Alexis, began his rule (1645-1676) that Russia continued to advance as she had done under the two Ivans.

The centre of geographical interest in Russia during the 17th century was the Ukraine, which has already been briefly discussed. South Poland was being exploited by the Slachta—the Polish-Lithuanian nobles—who tried to make serfs of all the peasants. The latter fled to the free steppes of the Ukraine. These homeless peasants were known as Cossacks, but later the term was given to mercenary soldiers, who helped to defend the crops against the Tatars. Thus the first homes of the Cossacks in the 15th century were the frontier towns from Riazan southward along the Don to the lower Dnieper (Fig. 133). Many Cossacks lived mainly by raids on the Tatars and Turks, and when the Poles tried to incorporate them as regular guards, they formed a sort of state of their own "Beyond the Waterfalls" of the Dnieper. This is the translation of its name "Zaporozhic."

We have seen earlier how the schism between the Greek Church and the Uniates led to the Cossacks appearing as the defenders of the Orthodox faith. In 1653 the Tsar decided to accept the Cossacks as his subjects, and though Poland won some victories in the wars which followed, Russia gained a large portion of the Ukraine (including Kiev and Poltava) by 1682 (Fig. 133).

By this time the Tatar power had disappeared, but it had placed Russia 500 years behind Western Europe in civilisation. "A despotic government, which treated its subjects like Asiatics; a taxation which emptied the pockets of the people, a brutalization of habits, a growth of servility among the population, and as a consequence a disparagement and even a contempt for culture, an Asiatic arrogance and a tendency to aloofness from the West European world . . . all this was the fruit of the long Tatar thraldom" (Milkowicz).

F. The Development of Modern Russia

Few Kings have deserved fame better than Peter the Great (1689-1725). He soon grasped the necessity for closer contact with the Western culture. Only on the rarely open seas of the Arctic had he a port, that of Archangel, which was now visited by English and Dutch merchants. He first conquered Azov in 1696 so as to have a link with the Black Sea, but was compelled to return it to the Turks in 1711. In 1697 he journeyed through western Europe giving special attention to military and naval matters, including ship-building. On his return he challenged Sweden, and though he was at first often defeated he crushed Charles XII at Poltava in 1709. By 1721 he had conquered Livonia, Esthonia and parts of Finland (Fig. 133 at C). In 1703 he started building St. Petersburg, a city on the swamps at the mouth of the Neva—a large stream draining Lake Ladoga. Here also he constructed a canal along the south of Ladoga linking the Volkov to the Neva. In 1716 he made a second journey to study Art and Science in France, and on his return initiated almost every branch of technical work in Russia. His reforms made many enemies in religious circles and even among his own family, so that his own son died in prison. In area his additions to Russian territory seem small, but they put Russia in contact with the Baltic, which is a relatively open sea (Fig. 133).

The succession after Peter the Great's death is very confused. His wife Catherine, his niece, Anna, and his daughter, Elizabeth, all ruled Russia, while the most powerful monarch of all was Catherine II, the wife of his feeble grandson, Peter III. (She was born Sophia of Anhalt-Zerbst in Germany.) The territorial changes during Anna's reign were not very extensive. She joined with Austria against Turkey and won several victories which resulted in the addition of territory in the lower Dnieper being ceded to Russia.

After this period we may note that Russia became of greater and greater importance in the affairs of Central Europe, while the prestige of Austria began to decline. In 1767 Turkey demanded that Russia should withdraw from the Turkish frontiers, and also that she should evacuate Poland. War followed, and the brilliant victories against the Porte in 1769-71 led indirectly to the partitions of Poland, as has been narrated elsewhere (p. 427). Both Prussia and Austria preferred to see Poland dismembered rather than that Russia should add large territories to her realm in southeast Europe. Again it is to be noted that Russia in the partitions received almost entirely lands where Polish was not the common speech, *i.e.*, White Russia and Lithuania. These, moreover, were lands across which early Russian expansion had taken place. The Poles had no such good claim to them. Indeed, they obtained them at second hand in 1386 by virtue of the Jagiellon marriage with a Lithuanian prince. This marks the *fourth stage* of expansion.

It is worth while to pause a moment to contrast the two nations in 1606, when Polish armies occupied the Kremlin; and again in 1795 when Poland vanished from the map of Europe. They might have worked in unison, Russia as a buffer against the East, Poland—a sister Slav nation—as a teacher of Western ideas to Russia. Russia developed a sense of national unity and emerged from the trials of wars and anarchy as a nation with a definite political aim and a strong executive monarchy. Poland at first far outstripped Russia in progress, culture and political power; so much so that it seemed as if Poland might double her Empire and that Russia might decline to the status of a feeble vassal of Poland. Yet the Poles maintained, in despite of reason, an antiquated feudal aristocratic Constitution which could not help leading to social anarchy and to the political dismemberment of the State. Even when ruin was at their very door the Polish magnates refused to give unanimous support to any one leader. Since they had against them a united kingdom with a woman at the head of it endowed with the very highest political gifts, an unswerving purpose and penetrating foresight, the result was obvious (Baring).

The Wars of Napoleon need not here concern us. The chief

changes in Russia took place in Poland and have been considered elsewhere. However, Alexander's territory was extended in 1801-2 by the incorporation of Georgia—a mountain land fringing the Caucasus. In 1809 he conquered Finland from the Swedes, and in 1812 Bessarabia was yielded to him by Turkey. The Act of Emancipation of the Serfs was signed in 1861, and gave half the land of Russia to the peasants; and the Government recompensed the nobility for their losses. Each peasant received from 5 to 19 acres, depending on the richness of the soil.

Unfortunately many difficulties arose as the result of Emancipation, of which only one or two noted by Baring may be mentioned. The agriculture was controlled by the "Mir," a communal system which tended to put all power into the hands of ignorant peasants. The land was cut up into temporary strips for use by the peasants, and these were often badly tilled, since the peasant only held them for a limited time. Thirdly, a class of Kulaks, richer but uneducated and grasping peasants, gradually acquired much of the power hitherto held by the landlords.

The Crimean War in 1854 had not much effect on Russian development, though she lost her right to place ships of war on the Black Sea. The expansion of Russia in central Asia which occurred between 1859 and 1873 is out of our province. The protection given by Russia to the Southern Slavs and Rumanians has been commented upon in earlier chapters.

The close of the nineteenth century marks the fifth, last and (if we include Asia) greatest period of Russian expansion. As the result of the Great War, and the later struggles in Russia until 1920, various outlying lands have been given up by the Soviet Union (Fig. 133 at C). Their policy of permitting autonomous control by small minorities is something rather new and rather fine in world history. As regards their eastern border, the White Russian Irredenta in Poland has already been described. Little Russians (Ruthenians) also form a solid block in Czecho-Slovakia. There are many Russians in Bessarabia (now Rumania) where, indeed, Russian resentment seems to be most strongly marked. It will be a very convincing argument that the Soviets desire World Peace if these three regions of Irredentism are allowed to develop without bloodshed.

G. The Nations in Russia To-day

Under Peter the Great Russia had not quite 15 million inhabitants, and since that time she has been busy colonising (Fig. 134). The first region occupied was the west, then came the north and centre, and lastly the turn has come to the south and east. The colonial character of this expansion is shown in the dates of the foundation of the well-known cities of Sebastopol, Kherson, Kharkof, Saratof, Samara, Perm

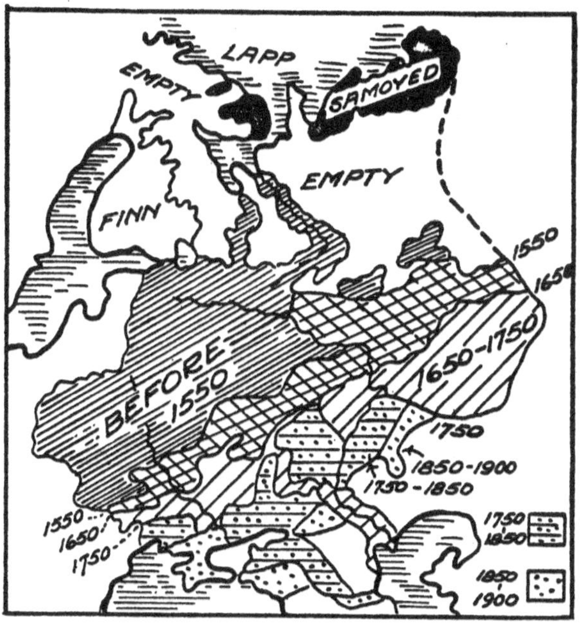

Fig. 134.—Colonisation of the Russian Steppes, 1550 to 1900. (Based on the map by Paul Milyoukov.)

and Orenburg. The greater part of the capitals of the provinces in the south and east are younger than the capitals of the Atlantic States of North America. Indeed, the region round Odessa is called "New Russia" and it is far more recently colonised than the "New England" of the United States.[8]

As LeRoy-Beaulieu observes, the racial composition of Russia resembles a mosaic the ground of which is made up of a single colour—perhaps of three slightly different shades—the border alone showing

[8]A. Leroy-Beaulieu, "The Empire of the Tsars," 1902, New York.

different pieces and colours. The three different "shades" are the Great, White and Little Russians. All three developed in the forest zone (p. 75), but the Great Russians were free to expand to the northeast to Moscow and beyond. The Little Russians (Ukrainians) were hemmed in by the steppes and by the nomads and so kept to Kiev and its immediate vicinity.[9] The people of the Ukraine are generally held to be handsomer and more lively than their brothers to the north. Their leaders became "Polonised" owing to the long period of control by Poland—but the peasant is still definitely a Russian Slav.

It is naturally very difficult to classify the earlier migrants who contributed so much to the Slav population. Among such are the Huns, Avars, Bulgars of the Volga, Alans, Petchenegs, etc. Here our study of past migrations may perhaps help us to differentiate and tabulate the three main types of real Russians. (Some of their physical characteristics were discussed on page 413).

RUSSIANS

WHITE RUSSIANS	GREAT RUSSIANS	UKRAINIANS
Largest proportion of original Slavs, since more isolated. But some Finn strains.	Largest proportion of Finn and some Tatar (Turkish) strains.	Least proportion of original Slav strain. Much Tatar (Turkish) blood, owing to incessant invasions. Less Finn (Nordic) than others. A little of "Mongol" strain.

It is true that the wide flat domain of Russia has been an unrivalled laboratory for the mixing of peoples to form the nearly uniform Russian of to-day, yet while most of Empty Russia has been filled by Russians, many other Europeans entered. Germans and Greek-Orthodox settlers from Austria and the Balkans poured in from 1750 to 1800. Catherine II introduced many Germans who settled in New Russia and along the Lower Volga. There they were exempt from military service (till 1874), and they still keep their German qualities. To-day they constitute an autonomous Republic on the Volga (p. 506) containing over a million people.

The Greco-Slav emigrants of about the same period settled mostly on the northern coasts of the Black Sea, especially in the Crimea. Here they changed places with the Moslem Tatars, many of whom migrated to Turkey to live under the Moslem flag.

[9]See D. S. Mirsky, "Russia, a Social History," 1931, London.

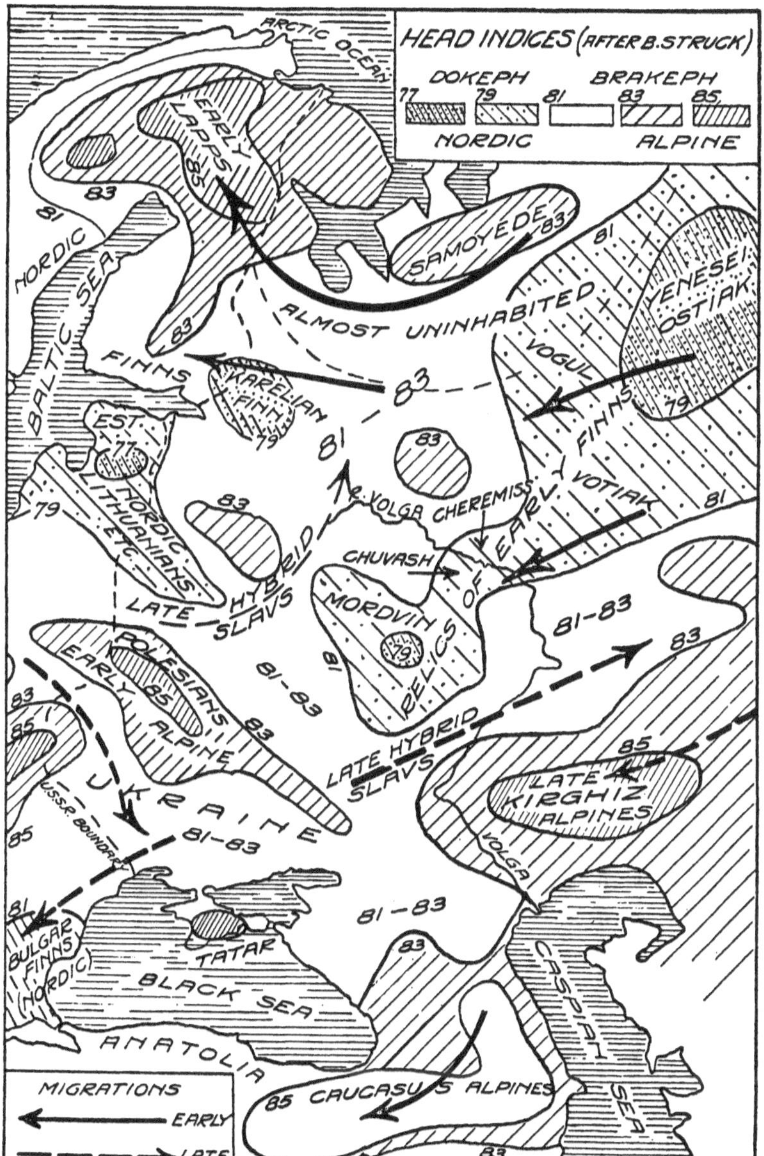

Fig. 135.—A Generalised Map of the Racial Groups now living in Russia. Early Lapps in the north, Finns in the centre and Caucasus Alpines in the south have been disrupted by later hybrid Slav migrations.

The Russian nation still contains numerous Finn "inliers" (*i.e.*, older peoples surrounded by Russians) who still witness to the extent of the country once covered by their people. The main body of the Finns were driven north by the invading Slavs of the early Middle Ages. They settled in Finland, where they came under the control of the Swedes. But along the eastern edge of Finland live the Karelians—who are true Finns, but who have always been influenced by Russian rather than Swedish culture (Fig. 135). Still further from Sweden and naturally less affected by modern civilisation are the Finnish tribes known as the Zyrians, Cheremiss (Mari), Votiaks and those peoples where Finn and Tatar cultures and characters seem to have merged, such as the Chuvashes and Bashkirs (Fig. 136).

The most Asiatic of the stocks present in Russia are the Tatar tribes. They are closely related in race, language and religion to the Osmanli or Turks, and, indeed, the Tatars of Kazan call themselves "Turks." The Turks, of course, are linguistically close to the Finns, though there is little racial relation, one being Alpine, the other mainly Nordic.

Most of the Turks show little of the so-called "Mongolian" type. But the Nogai Tatars of the Crimea and many of the Tatars of the Volga do exhibit the broad face, salient cheekbones, slightly oblique eyes, and yellowish skin which is associated with the word "Mongol." In some Finns—as stated earlier—these features occur and are probably derived from the Lapps, though they may, of course, have mixed with Tatar-Mongol tribes in the dim past. The Kalmyks in the Caspian depression only arrived in the 17th century, and Russia employed them to fight the Tatar tribes—whose ancestral home was far to the north of that of the Kalmyks (Fig. 136). In 1770 many of them returned to their homes near Chinese Turkestan, but a remnant remained, who now constitute the Kalmyk Territory just south of the mouth of the Volga. Until lately most of them were Buddhists—so that these three types of Altaic-speakers each adopted a different religion, for the Turks are Moslem and the Finns became Christian during Medieval times (Fig. 253).

Everyone is familiar with the catch-phrase, "Scratch a Russian and you find a Tatar." Like most popular ideas connected with race there is very little truth in it. It would be a much more accurate phrase to say, "Scratch a Turk and you find a brother European," if race (and not culture) is in question. The Russians had not colonised the steppes when the Tatars invaded them in the 13th century. The contact of the Tatars was much closer with the Finnish aborigines —whose descendants, the Votiaks, Chuvashes, Zyrians, etc. (Fig. 135), still dominate the northeast of Russia.

As the Tatars were gradually conquered in the 15th century they retreated before the Russians and rarely mixed with them. For a time they lived in great numbers in or near the Crimea (Fig. 136), of which the old name was Crim Tartary; but as Russia subjugated this area thousands of them abandoned Europe. Again in 1860-63 two hundred thousand Tatars left the Crimea. In all probability only about one million descendants of the Tatar hordes which terrified Europe in the Middle Ages now live in Russia, mostly near Kazan.

The writer's interpretation of the many racial strains which go to make up the Soviet Union in Europe is given in Fig. 135. We have to deal with three prehistoric groups. In the Far North are the short brakeph Lapps, who are allied to the Samoyedes and like them occupied the tundra. Then in the centre are the tall Finns—who seem to become more Nordic as they are removed farther from chances of Lapp or Tatar mixture. Thus the northeastern Finns are much more dokeph than most of those in Finland to-day. In the south were scattered Alpine peoples like those of the Caucasus and (perhaps) of the Pripet Marshes in Polesia.

This simple pattern was complicated by the surge to the east in early historic times of the Slav peoples—a hybrid between early Kurgan Nordics (Finn?) and the dominant South-European Alpine. They split the earlier Finns into the two groups of to-day, and later colonised the empty southern steppes. Tatars and Kirghiz of the southeast are late invaders from Asia who have no close affinity with the early Lapps, though both groups are brakeph.

H. The Policies of the Soviet Union

In this brief survey of Russia it is impossible to describe the remarkable changes which resulted in the Revolution of November, 1917, and in the later development of the Union of Soviet Socialist Republics (U.S.S.R.). But the political map of Russia to-day is so different from that of the Tsar's régime that some brief reference must be made to it. In Fig. 136 the main divisions of the Soviet Union as far as Europe is concerned are given. Their organisation and population (in millions) are given in the following table:

UNION OF SOVIET SOCIALIST REPUBLICS

- EUROPEAN
  - Ukrainian S.S.R. *29 millions*
  - White Russian S.S. Republic *4 millions*
  - Russian Socialist Federal Soviet Republic *100 millions* linked with seven others,* as below
    - Crimea 0.7
    - Daghestan 0.8
    - Karelia 0.3
    - Chuvash 0.8
    - Bashkir 2.7
    - Tatar 2.6
    - German Volga 0.6
- ASIATIC
  - Transcaucasian
    - Azerbaijan
    - Georgia
    - Armenia
  - Turkoman
  - Uzbek

*Four Asiatic Republics and thirteen small autonomous European areas are omitted. Of the latter only the Votyak and Mari areas have half a million people or more.

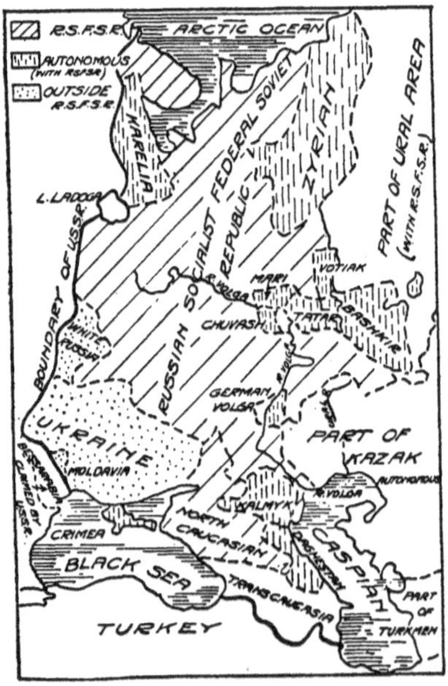

Fig. 136.—Political Divisions of the Soviet Union in Europe in 1929. (After G. Vernadsky.)

The population of U.S.S.R. is composed of 182 different nationalities speaking 149 different languages or dialects. The chief European groups (*i.e.*, those over one million people) at the census of 1926 are as follows:

TABLE Z

CHIEF LINGUISTIC GROUPS IN RUSSIA

	Millions of People	Per Cent.	Language
Great Russians	77.7	53	Slav
Ukrainians	31.2	21	Slav
White Russians	4.7	3	Slav
Tatars	3.0	2	Altaic
Jews	2.6	2	Slav and Yiddish
Mordvinians	1.3	1	Finn (Altaic)
Germans	1.2	1	Teutonic
Chuvashes	1.1	1	Altaic

Since about 85 per cent. of the population of Russia is still dominantly agricultural, a few words on the present position of the peasant may be added. This account puts the case from the Soviet point of view, since it is taken from an official description.

The position and prospects of the farming population were completely transformed as a result of the Revolution. In pre-war days the peasant was virtually land-starved. He worked the land under a three-field system, each field a separate strip usually miles away from the other strips. His methods were most primitive and agricultural education was virtually unknown. To-day, 96 per cent. of the arable land is in the hands of the peasants for their own use. Under the old régime the favoured wealthy peasants, relatively small in numbers, waxed fat at the expense of the mass of their less fortunate brothers. Like the feudal landlords they owned the fields of others and controlled marketing resources. They had become mortgage-holders and middlemen as well as producers. Under the new order the land is worked by individuals or groups under a system of perpetual leasehold, and is held in trust by the State for all the people. Yet the increasing standard of living, the decline in the death rate and the enormous annual increase of population (amounting to three millions a year) mean that the Soviet authorities have an exceedingly difficult problem to feed their people.

The last illustration (Fig. 137) in this chapter may well be devoted to this vital problem of food supply and population. It is based on

Fig. 164 in Bowman's "New World" and shows first of all the northern region where little or no food supply is obtained—the Virgin Forest. The remaining Podsol country in general does not provide enough grain (wheat, rye, barley, maize, etc.) to feed its people, who must import from the south. All the Chernoziom has a surplus in normal years—and the best inter-relation of "food to folk" is found east of the Sea of Azov, where three times (300 per cent.) of the adequate

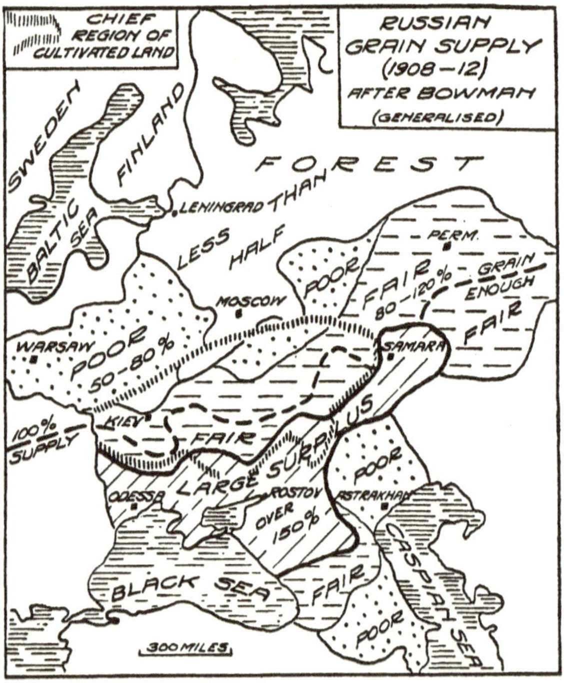

Fig. 137.—The Russian Grain Supply. 100 per cent. means that the region produces sufficient grain for its population. The chief region of cultivation is naturally the region of densest population (approximately 100 people per square mile).

amount is produced. (This does not, of course, mean that the greatest grain-supply is grown here.) In absolute figures this favoured district produced 2,552 lbs. of grain per person, whereas 800 lbs. per person is sufficient. Unfortunately, the large region east of the Volga is subject to bad droughts—in which case it becomes a famine area instead of a land with a food surplus. In the chief croplands (shown in Fig. 137) the grain supply is "Fair," *i.e.*, just sufficient for the farmers' needs.

However, the decline of the birth rate would seem to be becoming world wide, and will soon relieve population pressure in Russia as elsewhere. But during the last few decades for the first time in human history the whole world is beginning to feel crowded. There can never be a century like the twentieth—for the population of the world approximately doubled in the last hundred years. No rate of increase per century approached this before—and the curve is already flattening so rapidly that we may assume it will not occur again.[10]

[10]See Part IV of the author's book "Environment and Race," 1927, Oxford.

CHAPTER XXVI

THE EAST BALTIC REGION—DEPENDENT STATES IN AN ENVIRONMENT OF MORAINES

A. *The Baltic and the Mediterranean Seas Contrasted*

We may profitably commence this study of the Baltic Littoral by discussing the effect of an enclosed sea on the history of the surrounding nations. Obviously one of the historical functions of the Mediterranean is to separate European from African and to a lesser degree from Asiatic cultures. The separation is by no means complete, for as we have seen earlier, the Mediterranean has a characteristic environment of its own based essentially on its winter-rain climate. But once the coasts are left we find African cultures to the south, European to the north and Asiatic to the east. The Mediterranean has not separated *races* at all definitely; indeed, the whole littoral at the dawn of history was inhabited by the dark dokeph race to which the place-name "Mediterranean" has been given. To-day the Alpine race has pushed down to the shores of the Adriatic, of the Aegean and of Anatolia—but elsewhere the Mediterranean still remains essentially a sea surrounded by the primitive "Mediterranean" race. Languages have been limited to a greater degree by the Mediterranean, so that Aryan speech is almost uniformly used on the northern coasts, while the earlier Semitic and Hamitic tongues are general on the south coast, and Altaic (Turk) has thrust its way between the other tongues in the northeast corner. Religion uses the Mediterranean most definitely as a boundary, for we find Christian to the north and Moslem to the south and east.

The commercial aspects have been referred to earlier (p. 61). The control was obtained first by Phoenicians, then by Greeks and lastly by Rome. The latter power was the only one which ever conquered all the coasts, though the Byzantine Emperors nearly managed it later. During the Middle Ages the largest fleets developed through the initiative of trading towns such as Venice and Genoa rather than of powerful Emperors, but somewhat later the Turks came near complete control of the Mediterranean during the height of their power. During the Napoleonic Wars France, to a very limited extent, inherited Rome's Mediterranean Empire. But Britain usually controlled the sea-ways, and while this was the case Napoleon's Empire had no real chance of becoming permanent, or even of extending greatly.

We may now compare the Baltic Sea with the Mediterranean and see if there are any interesting parallels to be observed. The former sea is only one-seventh of the area of the Mediterranean, which should make it easier for one powerful nation to control the whole Baltic Littoral. There has, of course, been no period, however small, when events in the Baltic area so fully determined the path of European History as did those in the Mediterranean for many centuries. But to the regions of northern Europe the Baltic has always been of the greatest significance.

Two geographical factors perhaps determine the chief differences in the two seas. They are climate and aspect. The Baltic is too small to have greatly influenced the climate of its bordering countries. Moreover, the Baltic runs more nearly north-south than does the Mediterranean, and hence it *cuts across* climatic belts; whereas the Mediterranean on the whole lies in one latitudinal belt, and in one climatic zone. For this reason the Mediterranean is more concerned with transport of goods from extra-Mediterranean lands, for its various shores do not differ very greatly in their products. The Baltic, however, links the fur-forest region of the northeast (with a continental climate) to the farming-industrial area of the southwest with a somewhat uniform climate. This promotes local trade and to some extent balances the fact that the Baltic is a "dead-end," with no Suez Canal and no rich lands (like Persia and India) beyond its immediate shores.

Turning now to the cultural aspects of the Baltic, we find that there is still one marked racial stock settled on all the Baltic coasts. There is no good reason why the Nordic Race should not have been named the "Baltic Race," since all its European representatives not long ago lived on its shores. Thus two of the three races of Europe are linked with the two inland seas, and this, of course, but emphasises the fact that coastal plains and protected seas were, as it were, magnets which attracted and held primitive man. In religion also we find that the Baltic Littoral in the 16th century belonged almost entirely to the Protestant Church, though the expansion of the Greek and Roman churches has since destroyed that uniformity in the southeast. In language, however, there is almost as great a variety as around the Mediterranean, for Aryan (Teutonic) languages flourish in the west Baltic lands, Altaic languages have not yet been displaced from Finland and Estonia, while a very peculiar Aryan language, Letto-Lithuanian, is found in the southeast in Latvia, in Lithuania, and was formerly spoken in Prussia. (Slav is not spoken anywhere on the Baltic coasts.)

The following table (from Latham, "Comparative Philology," 1862) illustrates the relations between these varied languages:

TABLE AA

Comparison of Aryan and Altaic Languages

Aryan				Altaic			
English	Latin	Sanskrit	Lithuanian	Finn	Est.	Magyar	Turk
Man	homo, vir	manusha	—	mios	mees	fery	—
Eye	oculus	akshi	akis	silma	silm	szem	gos
Sun	sol	surya	sauli	pouva	paw	nap	gyun
Water	aqua	apa	vandu	wesi	wesi	viz	su

It is rather surprising that no one of the nations around the Baltic ever acquired the whole littoral in historic times. It is probable that the prehistoric Swedish tribes controlled the seas—for their carvings on stone are filled with drawings of boats and galleys of some size. But during historic times Dane, Swede, German, Lithuanian and Finn have all fought for its mastery. As we have seen in previous chapters, the Norwegian, Canute, controlled all the western lands early in the eleventh century (Fig. 67). Then Denmark and Germany took up the challenge, the latter with her numerous Hansa ships (Fig. 74). Although Germany did not conquer the northern coasts her trading towns definitely controlled the seas when Denmark had suffered eclipse in 1223 after the conquest of Livonia.

The rise of the Teutonic Knights brought in a new power which blocked Poland from the sea for several centuries. Sweden, however, came nearest to the complete domination of the Baltic about 1600 (Fig. 139 at D). But the Russian advance under Peter the Great and his successors was too powerful for Sweden to combat, and Russia gradually absorbed all the Swedish conquests—acquiring Finland in 1809. In the next century Russia and Germany were the two strongest countries on the Baltic, with Germany far in the lead as regards industrial and maritime resources.

The policy of the Soviet Union has resulted in the independence of those small Baltic states which are the particular study of this chapter. One outstanding fact in European history is that none of the four, Finland, Esthonia, Latvia and Lithuania, has had any period of complete self-government except Lithuania. Even Lithuania only stood alone for a little over a century (1247 to 1386), just after it had emerged from barbarism. After 1386 Lithuania was definitely subordinated to Poland, although a Jagiellon Dynasty (from Lithuania) thereafter ruled both lands.

B. *The Topography of the East Baltic States*

B1. Finland

In this chapter four European countries are described, all of which have many characteristic features in common. These lands are all low, flat and covered with morainic debris, and all were settled by tribes who are culturally very different from their neighbours. No doubt the combination of these two features—the lack of protection by the topography and the "foreignness"—account for their common historical fate; for we have seen that none of the four enjoyed national independence for any lengthy period. Even from the economic point of view the four East Baltic states are alike, for they have poor natural resources, and are all rather small if we exclude from Finland the half which can support hardly any inhabitants.

The Baltic Sea forms their common boundary on the west, and their common outlet (Fig. 138). It is quite shallow, only two patches exceeding 500 feet, and filled with rather fresh water—in these particulars differing greatly from the Mediterranean Sea. In comparatively recent times it has varied greatly in shape and size. Thus about 10,000 B.C. it extended across south Finland and was then very much larger—covering much of Sweden also. This stage is called the Yoldia Sea. Again about 5000 B.C. it was slightly smaller than at present and was separated from the Atlantic by an isthmus across the Sounds. This stage is known as the Ancylus Lake. Even at the present time, the north of the Baltic is rising more rapidly than most parts of the earth's crust, the rate being about 40 inches a century.

The Gulf of Finland and the largest Russian lakes are relics of the Yoldia Sea (Fig. 63), and this lower region seems to have been (in part) eroded in the softer rocks which fringe the Finland Shield on the south.

Finland itself consists, as we have seen, of a typical low-level peneplain of very ancient rocks (p. 60) thickly covered with tumbled glacial debris. To the south of the Gulf the rocks dip to the south, and repre-

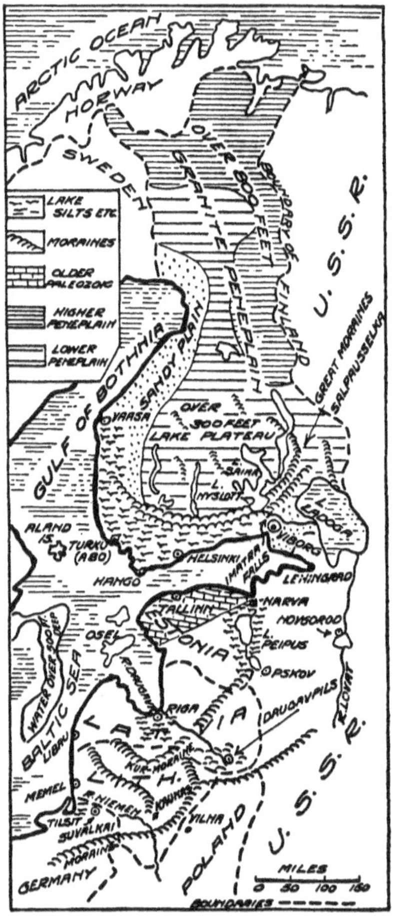

Fig. 138.—The Build of the Low Countries East of the Baltic. The four States are almost covered with moraine, except parts of eastern Finland and northern Estonia. (Data mainly from E. G. Woods.)

sent younger and younger deposits as we move south (Fig. 138). Thus Estonia consists mainly of the deposits laid down on the coasts of the Ancient Continent in a Silurian sea some 400 million years ago. Latvian rocks are largely Devonian, laid down over the others as the sea still retreated southward. Lithuania is so deeply covered with a mantle of quite young deposits that none of the older rocks shows at all. Of course Estonia and Latvia also have great piles of morainic debris, which determine the main features of the topography, but here the underlying ancient sediments are often visible.

Many thousand years after the last ice cap retreated from Germany (p. 260) it still covered northern Finland. About 5000 B.C. the Ancylus Lake (so called from a little shell found in the deposits) covered all southern Finland, though the largest moraine-ridges perhaps projected above the lake. One set of these moraines, the Salpausselka, extends right across Finland for 300 miles, and is by far the largest in Europe. It has helped to dam up many large lakes, of which Lake Saima is the most important. The hard granite of the Peneplain resists erosion, and the waters draining from these lakes are still slowly cutting away the granite, forming a great number of waterfalls in the process. The Imatra Falls near Lake Ladoga are some of the most impressive in Europe (Fig. 138).

The topography of Finland is distinguished by no mountains—but the land rises to the northeast, where the country is mostly over 800 feet above the sea (Fig. 138). This is the Upland Region which is almost uninhabited and may be ignored in our rapid survey. The central part of southern Finland is about 400 feet above the sea and is known as the Lake-Plateau. Thirdly, there is a fairly uniform Coastal Plain—about 50 miles wide—which borders the Baltic Sea (and is clearly shown in Fig. 138). The Lake-Plateau is stated to contain most of the 40,000 lakes of Finland. As the land rose after the phase of the Ancylus Lake, it seems likely that a large central lake was isolated from the Baltic portion. The extremely complicated outlines of the present lakes are due partly to the tumbled masses of moraines, but partly also to rocky ridges corresponding to the Skerries (p. 241) of Scandinavia (E. G. Woods).[1] Most of these ridges run N.N.W. to S.E.E. and this determines the "lie" of the lakes, especially as many curious ridges called "Eskers" (perhaps deposited in subglacial streams) also run in the same direction. The waters of the larger lakes are held back by the Salpausselka—the great double rampart of terminal moraine already referred to.

Settlement took place chiefly on the moraine ridges, which are well-drained and sunny, though the soil is very stony. At Olofsburg near Nyslott is a strong castle built by the Swedes in 1475 to protect their

[1] "The Baltic Region," New York (n.d.). A full study of the physical geography.

conquests against the Russians. Nyslott Isle is now a great tourist centre near the middle of Lake Saima. Another old fortress called Tavastehus dominates the southwest portion of the Lake-Plateau, about 50 miles northeast of Abo. The coastal plain is largely covered with silt and clay (once the old lake-bed), especially on the inland edge. Rounded rocky ridges project from the clay—and are usually striated (scratched) as a result of the great Ice Age. Similar rocky ridges form skerries off the coast—and near Aland are more abundant than in any other part of the Baltic area. The numerous small rivers have cut through the morainic deposits, but have made little impression on the hard granite floor beneath. Many waterfalls have resulted and the possibilities for hydro-electric power are correspondingly great.

Traffic was for long periods along these rivers, but it is very interesting to find that the Stone Age relics are found on the *inner* edge of the coastal plain, indicating that the sea extended so far inland in those early days. Since then the people have spread as lines of little villages outward, as the sea retired to the south and west. This is the opposite of the usual spread of settlement elsewhere, which is *up the valley* from the present coast. This elevation of the coast is still going on, and has greatly reduced the values of harbours at Uleaborg and Vaasa. Meadows have already replaced the harbour of 1606 in the latter town (E. G. Woods). Along the south coast there are deeper deposits of clay which enable a much denser population to flourish there. East of the capital (Helsinki or Helsingfors) the climate becomes much more extreme in character and the plain is much more barren. Hango is the winter port of Finland. Helsingfors is a noble granite city which has expanded greatly since Finland became independent. Viborg was defended by a Swedish castle as early as 1293 to protect the frontier against the Russians. It was a busy Hansa town, but suffered from the growth of St. Petersburg after 1703.

B2. Estonia, Latvia and Lithuania

Estonia has an area of about 20,000 square miles. Its northern zone differs considerably from the rest of these small Baltic states in that the ancient stratified rocks form fine cliffs all along the coast. These are called the Glint. There is only a thin layer of moraine in this Glint area—chiefly in the form of long esker ridges. To the south is a belt of ancient limestone with some karst erosion (p. 456). Most of Estonia is forest-covered, but in the east there are extensive bogs and areas of silt. These are the relics of the vastly extended Lake Peipus which marked the close of the Ice Ages. In the southeast the glacial debris is sometimes over 1,000 feet high. The chief town is Tallinn (Revel) which grew round a castle built by the Danes on one of the cliffs of the Glint. On the plain below are the gabled houses of the Medieval Hanse town of Revel. Tallinn is usually ice-free during the winter. Nearly 88 per cent. of the people are Estonians, Russians (8 per cent.) forming the largest minority.

Latvia is the largest of the three small states south of the Gulf of Finland. Its topography is purely glacial and consists of two horse-shoe-shaped hollows which formerly contained two lobes of the great Ice-Sheet. These are marked to-day by huge terminal moraines, which are clearly shown on the map (Fig. 138). The first lobe covered north Latvia and its relic-lake now surrounds the town of Riga. The second covered southeast Latvia, and now is represented by a similar lake, whose silts surround Dunaburg (Dvinsk or Daugavpils). The chief town is Riga, which was founded about 1200 by the German Bishop of Livonia. It was a flourishing Hanse town, but was most prosperous when it became the great timber-port of the huge Russian Empire. The coast of Latvia was the chief collecting ground in pre-historic times of the fossil pine-gum called Amber. Here also was the northern port of the famous "Varangian Way" which made use of the river Duna (Dvina or Daugava). Inland the forests reign supreme, with some moors, bogs and small lakes in the centre of the Evst Basin (Fig. 138).

In Lithuania we find the same features as in Latvia. It may be divided into Upper (eastern) and Lower (western) Lithuania. The former includes the districts bordering the Suvalkai ridge—and is a confused mass of morainic hills deposited by ice-lobes which moved sometimes from the west and sometimes from the north. Lower Lithuania lies to the northwest and includes the main "End-Moraine" (Fig. 138). It contains some lake-silts near the mouth of the Niemen, but is mostly at a height of three or four hundred feet. The chief river is the Memel (or Niemen), which drains this irregular mass of moraines, and is subject to heavy floods. Formerly a heavily forested country covered with pine and fir, it now has only 17 per cent. of its area under forests. The Lithuanians, when driven west by the Poles, took refuge behind these huge moraines. Even when the Teutonic Knights occupied the coasts the pagan tribes managed to maintain their independence in the forested mazes of the moraines.

C. The Development of the Finnish Nation

The Finns seem to be a somewhat mixed racial group—as we should expect when we realise that they have been driven north by the Alpine (?) Slavs against definitely Alpine-Mongolian Lapps. The maps given by Ripley show a zone of dokephs (80 head index) running from the Baltic to the bend of the Volga and even further east. It includes the western Finns, the Ests, the Karels of Ladoga and Tver, the Cheremiss, Chuvash and Bashkirs (in part). Both north and south other Finnish tribes have broader heads, such as the Tavasts, the Veps, Zyrians, Permiaks, Votiaks to the north of this zone, and the Mordvins to the south (Fig. 136). Especially where the Finns touch the Lapps and Samoyedes have they a high head index. Ripley points out that the broad Mongol face is rather frequently found superimposed on the narrow Finnish head! The original Finn seems to have been a dokeph type (79 or 80), with reddish hair and is distinctly tall. This is perhaps indicated by the Russian name for them which is *Chude*, meaning "giant." There seems little doubt that the Finns, Letto-Lithuanians and Teutons are all offshoots from the same trunk.[2]

At the dawn of history these tribes were in much the same localities as we find them to-day, though it is surmised that the Finns were not abundant in Finland till about A.D. 700. At this time the Danes had already made contacts with the "ancient" Prussians (a Lithuanian tribe). The Swedes had occupied the Duna Valley on their trading journeys to the south, so that they were in close contact with the Letts. Novgorod Russians had conquered the Ingrians (Finns) near Lake Peipus, while Pskov levied tribute on the Letts. It was the Germans, however, who exerted the greatest influence on these backward pagan peoples south of the Gulf; just as the Swedes dominated the Finns north of the Gulf.

The Vikings no doubt made raids on Finland, but it was King Eric of Sweden in 1157 who organised a crusade to convert the heathen Finns and to annex their territory to Sweden. One of the leaders was Bishop Henry of Upsala—who was an Englishman by birth. He became the patron saint of Finland, and the Cathedral at Abo—the old capital—is dedicated to him. Probably the conquest by the Swedes was fortunate for Finland, for the Swedes were never numerous enough to swamp the Finns, but were far in advance of them in culture. Many Swedes settled north and south of Abo—so that even to-day there is a

[2]W. Z. Ripley, "Races of Europe," 1900, London.

majority of Swedes in parts along this coast. The Tavastehus Fort was built in 1249, and the castle at Viborg in 1293. Gustavus Vasa introduced the Protestant religion about 1528, and Gustavus Adolphus founded many churches and schools (Fig. 139).

During the frequent wars between Sweden and Russia, Finland, as a sort of buffer state, suffered greatly. War, famine and pestilence ravaged her—and in the first half of the 18th century many eastern districts were torn from Sweden and annexed to Russia. In 1786 the Finn Sprengporten, disgusted with the feeble rule of the recent Swedish Kings, proposed union with Russia; but his conspiracy came to noth-

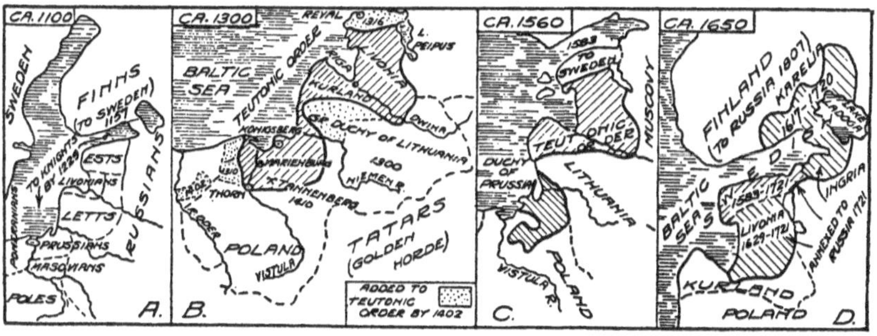

Fig. 139.—The Changes in the Political Control of the East Baltic States from 1100 to 1807. (See also Fig. 117.)

ing. In 1807 Napoleon promised Alexander of Russia that he should be given Finland if he allied himself with France. The Swedish army offered little resistance to the consequent Russian invasion, but the Finlanders rose in defence of their country and resisted bravely if unsuccessfully (Fig. 133). Alexander granted the Finns autonomy under himself as Grand Duke, and Sprengporten was appointed the first civil Governor.[3] During the 19th century Finland was treated justly by the Russians, but in 1901 the process of Russification was much intensified. A general strike in 1905, coupled with domestic trouble, compelled the Russians to reduce the severity of their rule. During the Great War Finland was in a state of blockade and took no active part in the struggle. In March, 1918, the Soviet government granted independence to Finland. Civil war between Whites and Reds broke out in 1918 but was quelled by German intervention. In July, 1919, Finland was declared a Republic.

[3]J. R. Fisher, "Finland and the Tsars," 1901, London.

D. *The Development of Estonia and Latvia*

The earliest significant artefacts in Estonia are pots ornamented with a comb design which are found over much of the area occupied by Finnish tribes to-day. It is stated that the true Estonians arrived in the region during the early Iron Age, a few centuries before Christ. During the later Iron Age the people were skilled in weaving, and iron work and silver ornaments were abundant. History, however, does not go much back of the twelfth century, at which time the Estonians were redoubtable pirates. In 1177 a group of Estonian tribes invaded Pskov, and a few years later crossed the Baltic and destroyed Sigtura, the old capital of Sweden.

In 1186 a German priest converted a number of Livonians and was made Bishop of Uxkull (on the Duna). In 1201 the bishopric was moved to Riga (nearer the coast) by Bishop Albert, and he founded a knightly order on the model of the Templars (Fig. 139). In a few years his "Knights of the Sword" conquered Livonia and Latvia. Estonia to the north was then attacked, but the Novgorodians resented this action and beat off the Knights. With the aid of Waldemar of Denmark Estonia was conquered, but the Danes refused for a time to leave the land in charge of the Knights. However, when Albert died in 1229 the Order controlled all Kurland, Livonia and Estonia. Waldemar built the huge castle of Revel—which now houses the State Assembly—and founded the adjacent city, which was originally called Taani-linn (*i.e.*, Dane's City) by the Estonians.

The Danes sold their Estonian lands to the Knights in 1346, and a period of prosperity ensued marked by the development of Hanseatic trade (p. 271). In the 16th century the people embraced the Protestant religion, causing Poland to intervene on behalf of the Catholics. Russia took part in the resulting troubles and overran the country in 1558. In 1583 North Estonia joined Sweden, while the south was annexed by Poland, and in 1587 Sigismund III of Poland united the whole under his rule. But Gustavus Adolphus conquered Estonia in 1625, and the Swedes soon introduced many social reforms. Charles XII by his challenge to Peter the Great again exposed Estonia to Russian invasion. The people of Narva and Tartu were deported to the interior of Russia and the rights of the peasants much curtailed. In 1721 the Baltic provinces were annexed to Russia (Fig. 139).

During the Great War Germany occupied Estonia (in February, 1918), but abandoned it on their defeat in France in November. The

Russian Bolsheviks invaded the area immediately, but with aid from Finland were soon driven out of Estonia. Warfare with both German and Russian forces occupied 1919, but in February, 1920, the Soviet recognised the independence of Estonia.[4]

The Latvians, who are closely allied to the Lithuanians in race and language, may have arrived in Latvia from the south, since according to Buga the names of some of the rivers in White Russia are Latvian rather than Slav. By the 7th century they were settled in Kurland and soon occupied Livonia as well. In the north of Kurland (Fig. 139 at B) are still found a few Liv people who speak Estonian (Altaic), which is entirely different from the Latvian tongue, an ally of Lithuanian (and Sanskrit). All these Baltic folk are Nordic, the Latvians usually having brown hair, light-coloured eyes, and an average height of 170.4 cm.[5] During the 8th century the Latvians maintained a chain of forts for defence against the Slavs. It is stated that their chief, Viesturs, had a force of cavalry 3,000 strong; but probably not much reliance can be placed on dates and persons before the 13th century. About 1291 German traders entered the Gulf of Riga and first made contact with the Letts.

The German Knights conquered Latvia (as well as Estonia) during the 13th century, and about 1450 the two were united in the republic of Livonia under the control of the Knights and Clergy. In 1561 most of the country was annexed by Sweden and Poland, but Kurland remained practically independent of Poland under Kettler, the last Master of the Order (Fig. 139), who was made Hereditary Duke of Kurland. Livonia was very prosperous during the 17th century under the control of the Swedes, but in 1721 Peter the Great occupied it. The First Partition in 1772 gave the Russians the southeast district and the later partitions added Kurland (Fig. 117). During the 19th century the peasants struggled for land and a measure of political freedom, but with little result till 1861. In 1905 a peasant revolt was quelled by the Russians aided by the Baltic landlords.

During the Great War the Latvians stoutly defended the Riga region against the Germans until the collapse of Russia enabled the Germans to occupy Latvia in 1917. As in Estonia, during 1919 Latvia was menaced by White Russians, Red Russians and a reactionary German force under Von der Goltz. All these were driven from the country early in 1920, and peace was declared both with Germany and

[4]"Estonian Yearbook," 1927, Tallinn.

[5]A. Bihlmans, "Latvia in the Making," 1928, Riga.

Russia in that year. The Latvian territory includes all the Latvians, who, unlike the Lithuanians, have not settled in adjacent countries and so produced very mixed communities (Fig. 140).

E. The Complicated National Problems in Lithuania

The Lithuanians are closely allied to the Letts (of Latvia) in race and culture but differ to-day chiefly in their religion, since they are mostly Catholic while the Latvians are Lutheran. Their language

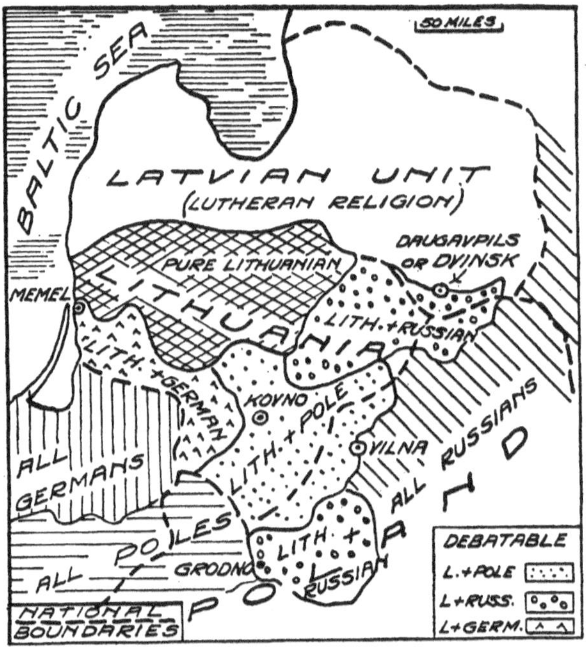

Fig. 140.—The Debatable Areas bordering Lithuania, showing the Catholic wedge of Lithuania between the Lutheran Germans and Latvians and the Soviet Union. (Partly based on Bowman.)

belongs to the Balt (or Letto-Lithuanian) group, which originally included Old Prussian and Jadzwing (or Yatving) now both extinct. They seem to have worshipped three deities, two of whom resemble the Zeus and Poseidon (God of the Sea) of the Greeks. The oak was their sacred tree and they had a complex sacerdotal system. Since they were not completely converted to Christianity until the 15th century, we have an unusually complete knowledge of their pagan customs. They resisted the German Knights much more successfully than did the Prussians or Estonians and indeed were never subjugated by them.

In 1225 Conrad the Polish Prince of Masovia (p. 420) invited the famous southern Teutonic Order to help him to repel the savage Prussians and Lithuanians. This Order had just been expelled from Transylvania and willingly migrated to Prussia. But they accepted the fief from the Emperor Frederic and not from Masovia. In a short time Prussia was overcome, and in 1237 the Knights of Livonia agreed to become a "province" of the Teutonic Order of Prussia. The Knights now became very powerful and their capital at Marienburg rivalled most royal courts in its splendour (Fig. 139 B). When the Lithuanian forests were frozen and traffic was easier the Knights would raid the boggy land and burn many villages. During the 13th century Lithuanian forces took part in the struggles between various Russian princes, and gradually a number of Lithuanian principalities arose in northwest Russia.

Mendog, the first Grand Duke of Lithuania, about 1250 found himself opposed by the German Knights—both north and south—and by the Russians to the east. He was baptised in 1250 and thus made peace with the Livonian Knights of the Sword. In 1316 Gedymin became Grand Duke, and by this time Lithuania was protected not only by its swamps and forests but by castles and walled towns. Russian culture was general among the upper classes, and Lithuania was able to raise armies which could defeat the dreaded Tatars. Even Kiev itself was controlled for a time; and Gedymin, though a pagan, was one of the most powerful rulers in Europe. His six children intermarried with the adjacent princely families and he colonised many desert areas in western Russia. Gedymin was killed by a bullet in 1340—one of the first instances of the use of gunpowder. Olgerd and Keistut, his two sons, greatly extended Lithuanian territory, so that in 1360 it extended from the Baltic to the Black Sea (Fig. 117). On Olgerd's death his son Jagiello became Grand Duke, and on his marriage with Jadwiga of Poland in 1386 Lithuania and Poland became one state (p. 421). As a result, the will of the Polish *Slachta* and of the Catholic priesthood became all important in Lithuania.[6]

It is interesting to read the Lithuanian account of the events of the next century.[7] Here we are told that Jagiello transferred the

[6]H. F. Helmholt, "World's History," Vol. V, London, 1907.

[7]Lithuanian-Polish Dispute, League of Nations Report, 1921, London.

throne of Lithuania to Vitovt in 1392, under whom the Lithuanians beat back the Tatars (at the Blue Waters), and the Teutonic Knights (with the aid of Jagiello) at Tannenberg in 1410. "After Vitovt's death Lithuania continued to elect her own rulers; the Poles, however, adopted the method of repeatedly electing as their king the person chosen as Grand Duke of Lithuania." In 1569 at the Union of Lublin the two countries agreed to form a dual state under one ruler (as did Austria and Hungary in 1866). The Lithuanian nobility, we read, did not become Polish, and tried hard to preserve Lithuania from disintegration. In the Partition of 1795 the whole of Lithuania was attached to Russia (Fig. 133), and the Tsar in 1815 was created Grand Duke of Lithuania. Gradually Russification was intensified, so that in 1840 Russian Law was made compulsory and after 1864 no book or newspaper could be printed in Lithuanian.

During the Great War half a million Lithuanians fought against the Germans, but in 1915 the Germans overran the whole country—which, like Belgium, shared the usual fate of a lowland between two powerful opponents. In February, 1918, Lithuania proclaimed her independence, and when the Germans were beaten the Bolsheviks attacked Vilna. Here the Poles intervened and captured Vilna, and have held the ancient Lithuanian capital almost ever since. The demarcation of the southern border of Lithuania is admittedly very difficult, since Lithuanians seem to be inextricably mingled with Germans near Memel, and with Poles around Vilna (Fig. 140).

The town of Memel with about 25,000 inhabitants has, however, always been almost wholly German; but the country around the town is mainly Lithuanian. Imitating the Poles at Vilna, the Lithuanians seized Memel by a surprise attack on January 15, 1923. The Lithuanian people are almost wholly Roman Catholic, but along the borders in the north and southeast, where the Letts and Germans are numerous, there are many Lutherans. The Lithuanians and Poles are hostile at present, but both belong to the Roman Catholic "outpost," whose northern boundary (*i.e.*, with the Lutherans) agrees with that of Lithuania, while its eastern boundary (with the Greek Church) runs approximately from Dvinsk through Vilna to Grodno (Fig. 140).

Although Vilna is a very notable city in Lithuanian history, being the capital in 1323, it now contains few Lithuanians. The following table gives the percentages in the Russian census for 1897, for the whole Province of Vilna and for the city alone.

TABLE BB

CENSUS FOR VILNA

	City	Province
Jews	40%	13%
Poles	31	8
Russians	25	61
Lithuanians	2	18

Thus this city (like Grodno) is dominantly Jewish, and the importance of the Jews in Lithuania is shown in Fig. 116. However, Polish figures in 1921 for the *Polish Province* of Vilna claim that 57 per cent. are Poles and only 5.6 are Lithuanian. This is partly due to the fact that the Lithuanian language had been abandoned by thousands under the old régime. The Irredentist region as far as Lithuania is concerned is the area between Vilna and Grodno, where about half the people are Lithuanian and half Polish or White Russians. At present this area is still held by Poland (Fig. 140).

CHAPTER XXVII

SOME GENERAL CONCLUSIONS—NATIONAL INDICES AND FUTURE GROUPINGS

A. The Concept of National Indices.

In an earlier chapter (p. 22) some of the complexities underlying the development of a nation have been discussed. These are numerous enough, but one obvious measure of the strength of national life is the *length of time during which it has had relatively independent self-government.* The writer has made several attempts in the past to interpret similar complex problems graphically and quantitatively. For instance, in 1922 he published in the "Geographical Review" (New York) an analogous paper on "The Future White Settlement of the World," where he attempted to show how a very complex problem could be attacked. The result was only a first approximation; but it is encouraging to find it quoted in many texts dealing with population problems. The method of investigation described below was published in the "Journal of Sociology," Chicago, May, 1934.

In the present instance it is clear that one field is pre-eminent as a region for such a study. This field is Europe, whose national development is quite sufficiently complex, but is better understood than that of any other similar part of the world. It was first necessary to decide on the limits of space and time. As in the preceding study, the writer decided to make use of those inevitable units of the geographer, the "natural regions." After a close study of the development of the European nations, Europe was divided into 74 regions, each of which had a measure of homogeneity, whether one considered its environment or its historical development. In general, each of these regions in the south and west has an area of some 24,000 square miles. In the north and east, however, where the development has involved larger and simpler units, the area of each division is about 200,000 square miles (Fig. 141).

As regards the period to be studied, it seemed to the writer that the last thousand years was rather clearly separated from earlier national development. In 873 Charlemagne's Empire was divided into three areas, from two of which developed Germany and France; while Lotharingia gave rise to Holland, Belgium, Switzerland and Italy. About this time also the Scandinavians, Spaniards, Magyars, Poles, Serbs and Bulgarians were each just beginning to unite into nationalities which have persisted more or less successfully ever since. It is true that in Russia and some minor regions *tribal law* still operated. This stage of development is accepted as *preceding* a real national consciousness. Thus A.D. 900 is our starting point; while 1918, when so many new governments came into being, marks the end of my survey. There is, therefore, a period of about ten centuries to be covered in respect of each of the 74 regions.

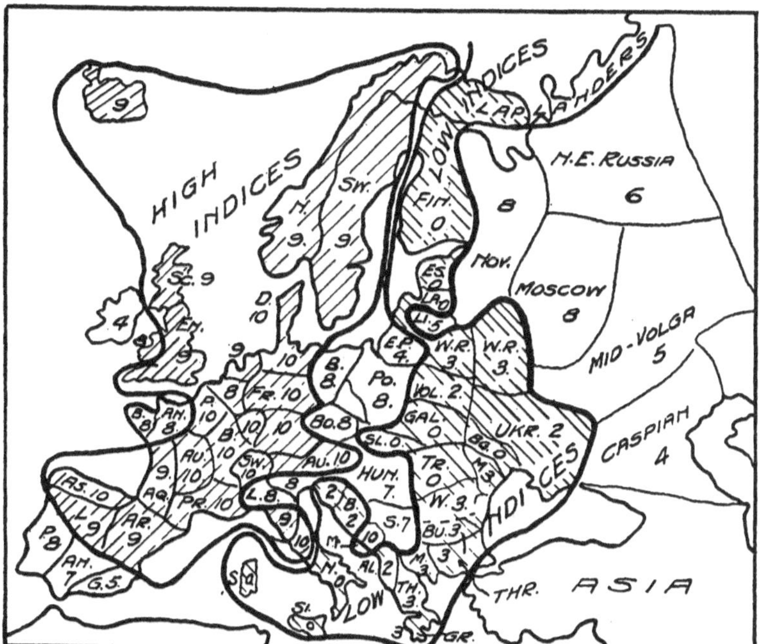

Fig. 141.—National Indices in Europe. The figures represent the number of centuries in each State in which it was Self-governing. High Indices represent over 9 centuries and Low Indices less than 4 centuries.

Since one could only hope to arrive at an approximate conclusion, the method of *typical periods*, at intervals of about a century, has been used. Those dates were chosen which were fairly well charted in such historical atlases as those by W. R. Shepherd (New York, 1921) or Ramsay Muir (London, 1927). In this fashion were collected the data which are given in an abbreviated form in the table (p. 529), "Rulers in Europe, from A.D. 900 to 1914 (Generalised)." In this table the *first* column (headed "900") shows how the 74 regions fared about the year A.D. 900. The column headed "National Index" shows approximately the number of centuries (out of ten) in which the region experienced some form of "home rule."

The data in the column headed 900 may be briefly discussed. *England* in 900 was half Danish—and was suffering from an invasion of tribes not yet assimilated. In *Scotland*, *Wales* and *Ireland* real nationality had not yet developed, for the clans were grouped in antagonised Tribes. This is indicated by TRI in the table. *Anjou* was half Norman. *Aquitaine* was a fief of the French crown, and is shown as FR.F. However, this stage is accepted as an early type of national self-government.[1] *Provence* was independent, but though it later came under Burgundy, and later again was a fief of the Holy Roman Empire (HRE), all these are counted as phases of self-government. *Asturias* was the sole survival of Christian Spain. *Leon* (as limited on the map, Fig. 141) was half Spanish and half Moorish; hence the term "1/2 Moor" in the table. *Holland*, and the region I have labelled *Hanover*, were fiefs of the Holy Roman Empire. *Brandenburg* and *East Prussia* were still uncivilised and controlled by Tribal Law. In my study I have dated their *national birth* later. The remaining regions are not tabulated in this table—but will be found in the original paper.

Proceeding in this fashion at each of the 10 later dates (given at the head of the table), we get a number of typical *cross-sections* of the development of national growth. Let us take the historical development of region 25, *East Prussia*, as an illustration of the method used. Until 1231 this region was inhabited by uncivilised heathen tribes who had no definite national consciousness. The Religious Order of Teutonic Knights then conquered the country, and ruled with little regard for the desires of the peasants, until their order was dispersed, whereupon the Grand Master (allied to the Hohenzollerns) made himself secular ruler. By 1648 it was part of Brandenburg, and later this merged in turn into Prussia, which in 1871 became in turn part of the German Empire. These stages are accordingly entered as "Tribal" (four centuries), "Teutonic Knights" (three centuries), "Brandenburg" and "Prussia." Only during the last three and a half or four centuries has East Prussia had a measure of self-government. Hence the figure 4 at the end of the line which epitomises the history of East Prussia.

[1]As explained earlier, by "Self-Government" is meant that the ruler is of similar nationality to the governed.

TABLE CC

RULERS IN EUROPE, FROM A.D. 900 TO 1914 (GENERALISED)

Regions	900	1000	1100	1190	1360	1477	1519	1648	1740	1815	1914	National Index
1. England	½Dan	Eng	Norm	Eng	Eng	Eng	Eng	Eng	Eng	Eng	Eng	9
2. Scotland	TRI	Norm	Scot	Scot	Scot	Scot	Scot	Eng	Eng	Eng	Eng	9
3. Wales	TRI	Wales	Norm	Eng	Eng	Eng	Eng	Eng	Eng	Eng	Eng	8
4. Ireland	TRI	TRI	TRI	TRI	TRI	Eng	*Eng*	*Eng*	Eng	Eng	Eng	4
5. Anjou	½Norm	FR.F	*Eng*	*Eng*	FR.F	FR.F	FR	FR	FR	FR	FR	8
6. Brittany	TRI	Brit	Brit	*Eng*	*Eng*	Brit	FR	FR	FR	FR	FR	8
7. Aquitaine	FR.F	FR.F	FR.F	*Eng*	*Eng*	FR.F	FR	FR	FR	FR	FR	9
8. Paris	FR	FR	FR	FR	FR	FR	FR	FR	FR	FR	FR	10
9. Auvergne	FR.F	FR.F	FR.F	Eng	FR.F	FR	FR	FR	FR	FR	FR	10
10. Burgundy	FR	Burg	Burg	HRE	HRE	FR	Aus	Spa	FR	FR	FR	10
11. Provence	Prov	Burg	Burg	HRE	HRE	FR	FR	FR	FR	FR	FR	10
12. Asturias	Spa	Spa	Spa	Spa	Spa	Spa	Spa	Spa	Spa	Spa	Spa	10
13. Leon	½*Moor*	Spa	Spa	Spa	Spa	Spa	Spa	Spa	Spa	Spa	Spa	9
14. Aragon	*Moor*	FR	½*Moor*	Spa	Spa	Spa	Spa	Spa	Spa	Spa	Spa	9
15. Andalusia	*Moor*	*Moor*	*Moor*	*Moor*	Spa	Spa	Spa	Spa	Spa	Spa	Spa	7
16. Granada	*Moor*	*Moor*	*Moor*	*Moor*	*Moor*	*Moor*	Spa	Spa	Spa	Spa	Spa	5
17. Portugal	*Moor*	*Moor*	*Moor*	Port	Port	Port	Port	Port	Port	Port	Port	8
18. Holland	HRE	HRE	HRE	HRE	HRE	Burg	*Aus*	Hol	Hol	Hol	Hol	9
19. Belgium	FR	FR.F	FR	FR	HRE	Burg	*Aus*	Spa	Aus	Hol	Bel	8
20. Hanover	HRE	HRE	HRE	HRE	HRE	HRE	HRE	HRE	Eng	Ger	Ger	10
21. Franconia	HRE	HRE	HRE	HRE	HRE	HRE	HRE	HRE	HRE	Ger	Ger	10
22. Suabia	HRE	HRE	HRE	HRE	HRE	Bav	Bav	Bav	Bav	Bav	Ger	10
23. Lorraine	HRE	HRE	HRE	HRE	HRE	HRE	HRE	HRE	FR	FR	FR	10
24. Brandenburg	TRI	POL	POM	POL	BRA	BRA	BRA	BRA	PRUS	PRUS	Ger	8
25. E. Prussia	TRI	TRI	TRI	TRI	Teu	Teu	Teu	BRA	PRUS	PRUS	Ger	4

ABBREVIATIONS are shown by initial letters, as follows:—

Austria, Bavaria, Belgium, Burgundy, Danish, English, French, French Fief, Germany, Holland, Holy Roman Empire, Norman, Poland, Pomerania, Portuguese, Prussia, Spain, Teutonic Knights, Tribal rule.

Italics indicate marked foreign aggression.

34

B. Drawing Isopleths to Indicate National Status

In similar fashion I have obtained the other figures given in the last column. In some cases it has been difficult to decide how to classify a given period, whether Tribal, Foreign Rule, or Self-Government. However, it is hoped that the figures obtained are a fair approximation to reality. It is, of course, the *relative values* of these figures which are of most interest to the geographer. To these figures I have given the name "National Indices."

When these figures in the last column are plotted on the 74 regions, a most interesting set of isopleths[2] is produced (Fig. 141). I have simplified the result by using only three rulings on the map. These regions whose "national index" is *high* (10 or 9) form the first class with *Lengthy National Life.* The second group includes the regions with indices ranging from 8 down to 4. They have had national existences of medium length. Lastly comes the group with *low* indices (from 3 down to 0). These have clearly experienced foreign domination during most of their existence.

There are 24 regions belonging to the first class of *High Index* regions—where national development is pronounced; 23 in the intermediate class; and 27 in the class of *Low Index* regions, where real national development has been prevented during nearly the whole of the 1,000 years under consideration. The map (Fig. 141) shows one isopleth separating indices 9 and 8, and another separated indices 4 and 3.

Clearly there is a *High Index* zone extending from Iceland to Aragon and a *Low Index* zone extending from Lapland to Greece. There are three *Medium Zones,* one from the Baltic to the Adriatic, another in the east of Russia, and a third on the far western shores of Europe. These last three zones are ignored in later comparisons.

C. Frequency Curves for Religion, Race and Health

We may assume that practically all the 74 regions belong to one of three dominant sects, Protestant, Roman Catholic, or Greek Catholic. Their distribution is shown in Fig. 142A. But a simple quantitative method may also be employed. Of the 74 regions, 32 are Roman, 20 are Greek, and 14 are Protestant. If we plot these frequencies in three columns (see Fig. 143) we obtain a characteristic

[2]*Isopleths* are "lines of equal abundance." They include all such lines as contours, isotherms, isohyets, etc.

graph labelled "All Regions" in A. If now we isolate the regions with *high indices* (10 and 9) and plot their frequencies (9, 13, 1), we obtain the graph so named in A; and using the same method with the *low indices* (3 to 0) we obtain a third graph (3, 8, 15). The fourth graph (one-third total) shows what the figures would be if each of the three groups maintained the *same religious proportions* as the total. We can now correlate the High-Index regions as strongly Protestant, and the Low-Index regions as strongly Greek Church. The High-Index regions are a little more strongly Catholic than the Low-Index regions. It is well to be clear that this does not mean that a long national existence is *due* to Protestantism, or inhibited by the Greek Religion.

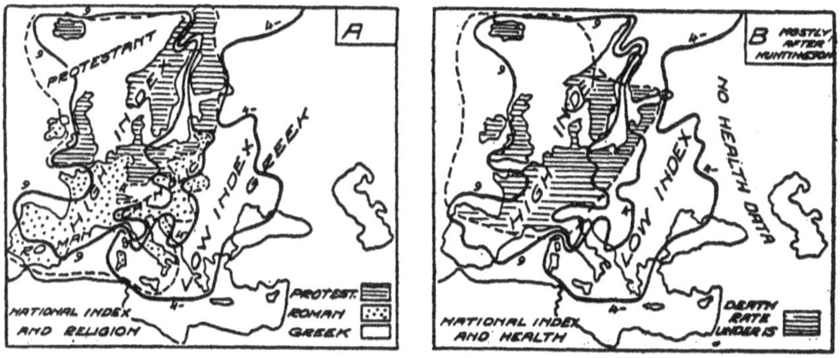

Fig. 142.—The Relation between National Indices and Religion (at A), and Health (at B).

It may well be, however, that the mental independence which predisposed a people to *protest* against orthodox views also determined their resistance to foreign aggression.

Similar graphs are shown for the three dominant *Races* in Europe (Fig. 143B). The conclusion here is that the *High-Index* regions are strongly Nordic, the *Low-Index* regions are strongly Alpine, while the Mediterranean Race does not vary with the national index.

Of course, a mere inspection of isopleths will often show when it is useless to try to correlate two variables. For instance, in Figs. 42-45 appears the result of a study of the spread of Christianity in Europe. It can be seen at a glance that our two critical zones (High-Index and Low-Index regions) run right across the general west-east course of the "Christianity" isopleths. It does not appear, therefore, that *early conversion* had any bearing on *continuity of nationality*, though this might reasonably have been expected.

The map in Fig. 38 illustrates the extent of the Roman Empire—which reached to the isopleth numbered 400 A.D. before it fell to pieces as a result of the Barbarian Invasions. On the whole the High-Index regions are either places like Northern France and Britain which were conquered relatively lately by Rome, or places like Germany and Scandinavia which were never affected by Roman Rule. In effect the latter is correlated with rather a *low* National-Index.

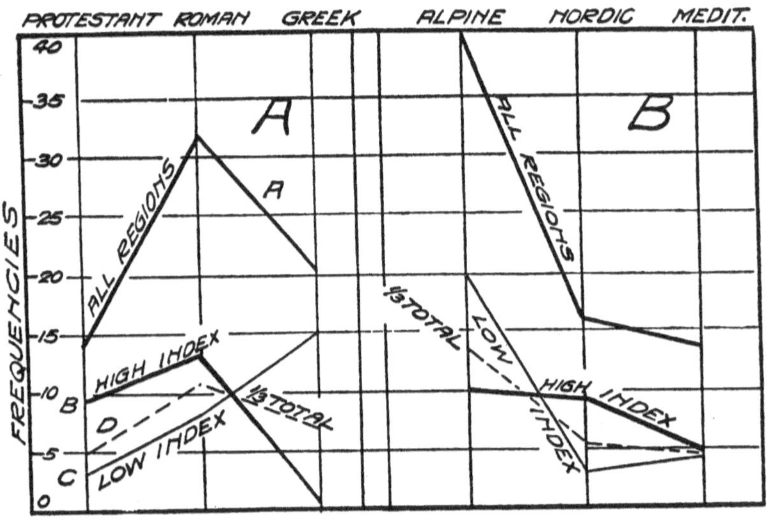

Fig. 143.—Frequency Graphs showing the relation between National Indices and Religion (at A), and Race (at B).

Numerous physical factors, such as Rainfall, Temperature, Elevation, and Distance from the Sea, as well as other non-physical factors, are all discussed in the original paper. However, the relation of Health to national development seems of outstanding importance and may be briefly referred to.

Let us compare the death rates (as given by Huntington and Williams)[3] with our national indices. The most progressive countries, like Holland and Denmark, have death rates below 11. Regions in the south and east of Europe have death rates of more than 20. Let us choose 15 as a mean figure (Fig. 142B). Our High-Index group has only four regions with death rates above 15; *i.e.*, only 17 per cent. have high death rates. When we turn to the regions with low national

[3]"Business Geography," p. 242, 1922, New York.

indices, then we find that 25 out of 27 (or 93 per cent.) have death rates exceeding 15. Of course, we are dealing here with *present* death rates and comparing them with events which occurred during the last thousand years. Possibly we are putting the cart before the horse; for the high correlation between health and national continuity may be explained in at least three ways. (A) The people in these regions were energetic and intelligent, therefore they rebelled not only against evil political conditions but also against evil *social* conditions. (B) The people in these regions early acquired independence, which left them free to pursue an enlightened social programme resulting in great increase in health. (C) The people in these regions were so healthy that this factor helped them in their struggles to remain independent. To the writer reason B seems the most logical. Another interesting map with isopleths agreeing with the National Index Map (Fig. 141) is that of Illiteracy given by A. Hettner on page 56 in his book "Europa," Leipsig, 1925.

The writer does not suggest that this exhausts the major factors controlling the growth of a powerful nation. Many others come to mind. Thus the "Method of Succession" of the rulers is a vital factor. Let us but consider the unsatisfactory elective methods used in the Holy Roman Empire, or in Poland. These inevitably led to rivalry, war and a disturbed realm. Montenegro benefited by a curious rule by warlike bishops, where the descent was from uncle to nephew. Such factors are not easy to examine quantitatively. Then there are various interesting physical factors—such as the site of the capital city. France has grown up around Paris, England around London. Madrid and Leningrad (1709-1917) are not ideal capitals. Boundaries, whether adequate for protection or otherwise, are potent factors. So also is the homogeneity of a nation as regards language, race, religion, *etc.* Lack of space prevents a further discussion of these interesting problems. Any interested reader is referred to the original memoir for further details.

D. *Is Race Negligible in Europe?*

The preceding pages will show that Race is of much less significance in European problems than is usually supposed, primarily because few people understand the racial divisions of Europe and confuse cultural differences with racial differences. The correlation made a few pages back (Fig. 143B) will suggest that Race is still of some importance as a biological factor, and it seems profitable to point out two more examples of the same kind. The first is due to Bertillon, and is charted by Ripley. It shows the frequency of Separation (Divorce) in France (Fig. 144), a country which is especially suited for such racial studies since it is the only European nation "Where three *Races* meet" (Fig. 145). The heavy lines in Fig. 144A are those given also in Fig. 90, and indicate the later "Alpine wedge" in the south separating the primitive Mediterranean types along the Ligurian coast from those in the Dordogne. So also the later Nordic Wedge

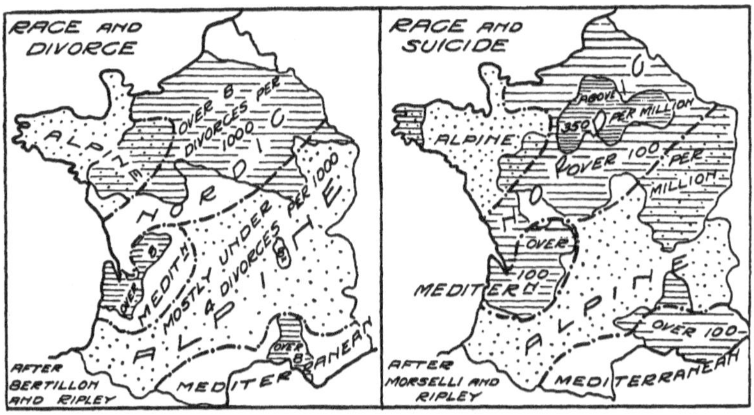

Fig. 144.—The Relation between Race and Divorce in France. Notice that Divorce is infrequent in the regions inhabited by the Alpine Race. At the right is the similar map for Suicide, which again is more prevalent among the Nordic and Mediterranean components in France.

splits the Alpines, isolating a group of brakephs in Brittany. If we plot on such a map the number of divorces per 1,000 marriages we find that all the low figures occur among the Alpine brakephs. The high figures are found with the dokephs, whether they are Mediterranean or Nordic. (In a recent paper on Racial Migration Zones[4] the present writer shows that the blood tests support the belief that the Nordic is only a variety of the Mediterranean Race.)

A second example of social trends apparently closely related to racial differences is illustrated at the right of Fig. 144. Here Morselli's data are charted, and show that the proportion of Suicides agrees closely with racial distribution. Again the Alpine peoples seem to show up best and have a suicide rate below 100 per million, while the

[4]See "Human Biology," February, 1930, Baltimore.

two dokeph races are almost invariably above this figure. I have introduced this brief reference to racial factors because it shows how such data can fairly be used. It is absurd to discuss suicides or divorces or death rates (at any rate in their *racial* aspects) in terms of the whole French *nation* (or the Germans or British, who are also of mixed racial origin). But if we can distribute our social data in the same "compartments" as our racial data (as in Fig. 144), then we can fairly make racial deductions.

Havelock Ellis, using reasonable *provincial* divisions, seems to be able to differentiate the various intellectual endowments of the several ethnical stocks of England. Thus the Angles are prone to compromise, love liberty and produce good Statesmen and Naturalists. The Southwest "Dokephs" are often brilliant Free-thinkers and Inventors. The Welsh Border produces Artists and Technical men. The Danish element seems to be rich in mathematical ability ("Study of British Genius," New York, 1926).

The last illustration of the effect of race on European history is perhaps the most important of all. It deals with the probabilities of racial and national survival in Europe, and is taken from a recent paper by the author entitled "Nordic and Alpine Races."[5] It seems to show quite clearly that the Alpine nations of central Europe have a much better chance of surviving than have the Nordic nations of the northwest.

I have elsewhere[6] given my reasons for believing that the Alpines developed near Turkestan later than the Nordics and from some stock not unlike the latter. But on the present occasion I wish to discuss the population statistics of these two contrasted races in Europe.

A very interesting aspect of the racial rivalry is indicated by the eminent Italian economist, Professor Corrado Gini, in "Population" (University of Chicago Press, 1930). A number of statisticians, including Dublin and Kuczynski, have pointed out that birth and death rates do not tell us very satisfactorily how a nation is progressing. It is much better to consider the "survival-rate," based on the number of mothers in a generation who survive to maintain the population. Thus, among the northwest nations of Europe this survival-rate is not sufficient to maintain the population. For instance, England and Germany are 18 and 10 per cent. short of an adequate number. Estonia, Sweden, Switzerland and Austria also seem to be in a bad posi-

[5]"American Journal of Sociology," July, 1931, Chicago.

[6]"Environment and Race," *passim*.

tion; while Belgium, France (8 per cent.), Finland (4 per cent.) and Norway (6 per cent.) are not so unsatisfactory in this respect. The southeast European nations have populations where the supply of potential mothers is quite adequate. Indeed, Rumania has about 35 per cent. in excess, while Italy (20 per cent.), Bulgaria (17 per cent.), Spain and Russia also belong to this category. Holland is the only northern nation in a satisfactory position, but its survival-rate is falling.

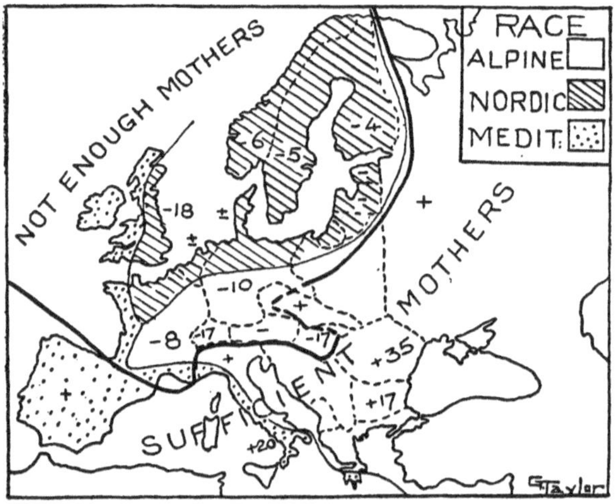

Fig. 145.—The Relation of the three Races of Europe to the "Survival Rate." In the Nordic countries the supply of potential mothers is not enough to maintain the survival of the present population. The figures approximately show (in percentages) excess or deficiency from adequate Survival Rates. (Data from Gini.)

I have shown on Fig. 145 (by the heavy line) the boundary between the declining nations in the northwest and the increasing nations in the southeast. It presages a time when the "Nordic Fetish" will obviously lose its appeal, owing to the dwindling of the Nordic race.

E. *Future European Combinations*

The present is the Apotheosis of National Development and is perhaps the natural reaction to the period of Absolute Monarchies of the 18th and 19th centuries. But everyone who tries to look ahead must feel that it is only a temporary stage in the progress of civilisation. If so, what is the next step? No doubt a series of combinations which will almost inevitably be based largely on economic (*i.e.*, environmental) factors. Commercial intercourse will play a much larger part in the future, and cultural similarities will gradually recede into the background.

Under these circumstances the study of Nationalities[7] published by Arnold Toynbee in February, 1915, a few months after the Great War

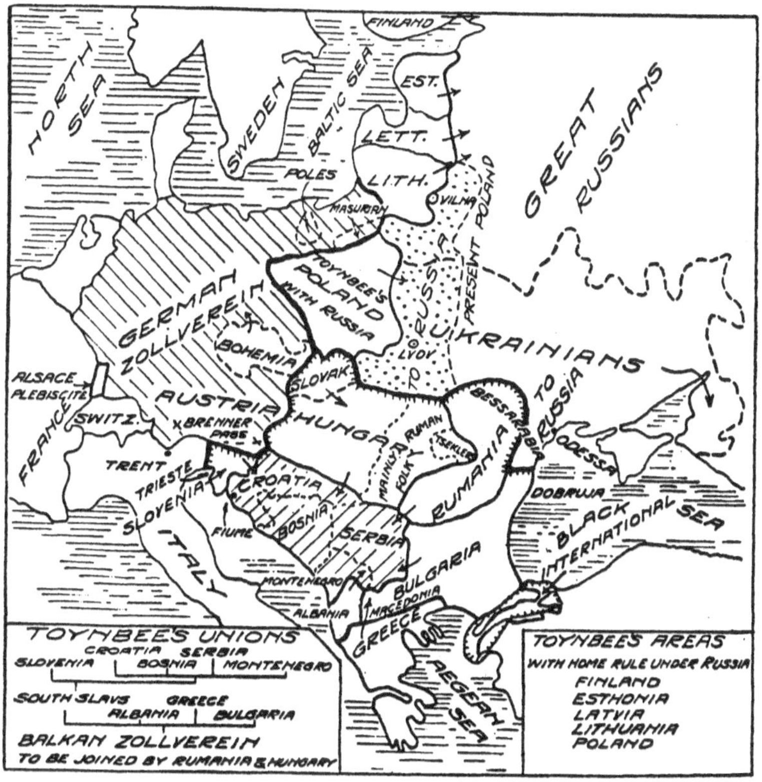

Fig. 146.—Rational Political Boundaries in Europe according to the map published by Toynbee in 1915. He stressed an Economic Union in the Southeast of Europe, and surmised that the Baltic States and Poland would have Home Rule under Russia.

started, is of considerable interest. He was not handicapped by any knowledge as to how the war would turn out! As an Englishman he was less likely to be prejudiced by the innumerable historical and cultural difficulties which necessarily affect most continental publicists. In Fig. 146 some of the main results of his study are indicated in a

[7]A. Toynbee, "Nationality and the War," 1915, London.

somewhat generalised fashion. Many of the changes which he considered advisable have come about. These may be summarised first.

The four Baltic provinces of Russia have obtained not only Home Rule as he suggested but entire independence. Denmark has received slightly less territory than he forecasted. Alsace was restored to France, much as he predicted; though it must be remembered that almost all Alsatian trade is with Germany, whose language she speaks. This will inevitably draw Alsace towards Germany now that the latter country is unable to dominate Alsace officially.

The Polish Corridor in Toynbee's opinion should in general be left in German hands since the "seizure of West Prussia was the most pardonable theft Berlin ever committed." But a strip on the west bank of the Vistula reaching the sea should be added to Poland. His boundary for Poland—which he envisaged as still linked under Home Rule with Russia—certainly never contemplated that the enormous block of White and Little Russians (shown on Fig. 146 by dots) should be added to Poland.

Turning now to Austria, it was his opinion that geographical principles inevitably linked Austria and Bohemia with Germany. Half of the Trentino was to be returned to Italy, since in such cases no argument holds good except the ascertained wish of the living population actually concerned. We may well examine his arguments in regard to the future of Bohemia. It is an industrial state, and, he says, if it separates itself from the Germans they will erect a tariff wall which will cut Bohemia off from the world. All its trade passes through Germany or to Vienna—also in control of German speakers. Few Bohemian products pass through the third outlet by the Moravian Gate, because there is the heart of the Polish industrial district with like goods for sale. His solution is that both Austria and Bohemia should enter the German Empire as individual units with safeguards for the complete preservation of their national cultures.

Finally his solution of the Balkan problems is interesting, and it is not at all impossible that some such federation as he suggests will ultimately develop. He places the major cultural and commercial boundary *between* Austria and Hungary, remarking that "the Balkans begin at the River Leitha." West of the Leitha the nationalities of Europe are mainly grouped in compact blocks, which correspond with considerable accuracy to the physical and economical articulation of the continent. Southeast of the Leitha the nationalities are interlaced in inextricable confusion, where intimate contact has produced mutual exasperation instead of understanding.

One of the most important features is the outlet to the Adriatic by Trieste or Fiume. A railway from Buda-Pest to Fiume was finished in 1873, and was controlled from Buda-Pest and not by the Croatian authorities. A reasonable solution (in Toynbee's opinion) was to erect a separate state of Slovenia in this area which is almost wholly inhabited by that branch of the South Slavs. This small nation, however, Toynbee thinks, would benefit greatly by joining a South Slav Union of Croatia, Serbia and Montenegro. This has, of course, actually happened, save that Italy controls half of Slovenia, and Serbia ruthlessly dominates her Slav sisters.

Furthermore, when Hungary and Austria are separate it is to Hungary's interests to enter into close commercial relations with the South Slavs, who block her outlets, rather than with her former suzerain Austria. A Balkan Zollverein is therefore suggested which includes not only the Balkans proper but Hungary and Rumania. Toynbee in 1915 could only leave the solution of the Transylvanian problem to the war to decide, for he writes, "If Rumania intervenes in the war in favour of the Allies the prize will fall into her grasp." Here, however, it is interesting to note that Slovakia was left in Hungarian hands and not allied with Bohemia, as has actually happened. The entry of Bulgaria into the Zollverein pre-supposed a reasonable boundary of the latter country, which includes Macedonia and the Dobruja. Bessarabia he divided in half—giving Russia the Tatar region around Odessa near the coast, and to Rumania the inland portion where are most of her nationals. The present writer feels that this forecast of 1915 deals with economic and cultural links which still exist. They may temporarily be hidden by post-war animosities, but will ere long once more exert their influence to bring about a wiser and more permanent solution of the national problems in Europe.

An alternative solution to one of Germany's chief geographical problems was offered by Sir Harry Johnston in 1913.[8] He wrote, "Germany is hatefully cramped within her present boundaries. The Baltic Sea, ice-obstructed for five months of the year, is of little help to her commerce. On the North Sea she has only three good ports. In what direction can she find a sea outlet adequate to her desires which should not involve her in a European struggle of the first magnitude?

[8] "Common Sense in Foreign Policy," London, 1913.

She can only do this through the good will of Austria by the acquisition of Trieste." His map shows the Tyrol given to Germany with a narrow outlet to the sea at Trieste (Fig. 146). In compensation he thought that Austria should be allowed to develop a great confederation of South Slavonic States. These possibilities, in spite of the Treaty of Versailles, are engaging the thoughts of Germans to-day, and are therefore worthy of our continued consideration.

During the war Friedrich Naumann (of the Reichstag) developed an elaborate scheme for the Union of Central Europe.[9] This was to be a loosely federated combination of Scandinavia, Germany, Austria, Rumania, the Balkans and Asiatic Turkey. It thus includes much of the Trading Groups II and III, mentioned in the next section (p. 543). In his own words, "Mid Europe is the fruit of war. We have sat together in the war's economic prison, we have fought together, we are determined to live together." The Federation was to have a German nucleus and would voluntarily use the German language, which is already, as he says, the language of intercourse in Central Europe. The following table shows approximately the relative strengths of Naumann's Mid-Europe as compared with the British Empire,[10] European Russia and U.S.A. at the time of his book.

TABLE DD

RESOURCES OF MID-EUROPE AND OTHER AREAS

	Mid-Europe	U.S.A.	British	Russia
Total Population........	166	107	443	169
Total land (mill. hect.)...	116	770	760	516
Agricul. used...........	67	94	146	210
Forest..	220	33	50	108
Wheat (mill. tons).......	11	21	20	23
Coal Prod.: do	307	450	307*	31
Iron Prod.: do	58	90	25	12

*Including Canada and South Africa.

F. *Environmental Factors Tending to Unite the Nations*

Since it seems likely that economic forces will be of such importance in the future history of the nations of Europe, we may well examine this matter a little further. Trade relations are usually developed in

[9]F. Naumann, "Central Europe," p. 316, 1916, London.

[10]Not including Canada and South Africa.

north-south directions rather than east-west directions for obvious reasons of climate. We can divide Europe into four east-west zones (Fig. 19) in accord with the agricultural products, as in the following table:

TABLE EE

AGRICULTURAL ZONES

No.	Position	Type	Countries
1	North Belt.......	Timber supply	Sweden, Finland, N. Russia
2	Cool Agricultural..	Rye, Oats, Potatoes	S. Scandinavia, N. Germany, Poland, Russia
3	Central..........	Wheat and Maize Belt	France, Central Europe, S. Russia
4	Mediterranean....	Olive and Wine Belt	Spain, Italy, S. Balkans

Clearly there will be little trade in agricultural products between nations in an east-west direction in these very well-defined belts.

It is, of course, the interchange between industrial and agricultural regions which is perhaps most characteristic to-day. Omitting the British Isles, the chief coal supplies of Europe are found in the Great Wheat Belt (*i.e.*, the northern half of zone 3); and they extend in a similar east-west zone for reasons explained in an earlier chapter (p. 53). In the coal belt also the trade of the future will tend surely to move from the north or from the south *to the coal* rather than in other directions.

We must not, however, ignore possibilities of other power resources which can be used for manufacturing goods. Hydro-electric power and petroleum luckily occur chiefly in lands without much coal. But so far they are not serious rivals of King Coal for the main development of power, though "White Coal" has ousted the other in certain trades for which electrical treatment is specially advantageous.

The following table gives some of the power resources of the nations of Europe. (Water Power for other nations appears in Fig. 147.)

If we assume that 5 lbs. of coal produces 1 Horse-power per hour, and that there are 3,000 working hours in a year (as A. H. Gibson suggests),[11] then waterfalls generating *one million Horse-power per year* are about equivalent to factories burning 7.5 million tons of coal per

[11] "Natural Sources of Energy," Cambridge, 1913.

TABLE FF

POWER RESOURCES OF CHIEF NATIONS

	Britain	Germany	Russia	Poland	Bohemia	France	Belgium	Italy	Spain
Coal Resources*..................... (thousand million tons)	187	240	55	70†	40	32	11	..	9
Coal mined per year†................. (million tons)	242	317	35	40	36	56	27	..	6
Water Power†...................... (millions of Horsepower)	1	2	8‡	1	1	5	..	4	4
Petroleum Supply†.................. (1928—million barrels)	..	..	88	5	..	..	..	..	..

*Mainly from Dresdner Bank, 1927.
†From Blanchard and Visher.
‡With Caucasus.

year. Of course the former lasts indefinitely, but the latter is cheaper to operate as long as coal is abundant. Hence the water power of Britain (somewhat less than 1 million H.P.) would only supply about one-thirtieth of the power derived from the coal mined there. A nation like Spain, however, with 4 millions of water power, is blessed with the equivalent of 30 million tons of coal, whereas she only mines about 6 million tons a year at present. In the great industrial countries like Britain, Germany, Bohemia and Belgium it is clear that their industrial power will not persist when the coal supplies are exhausted.

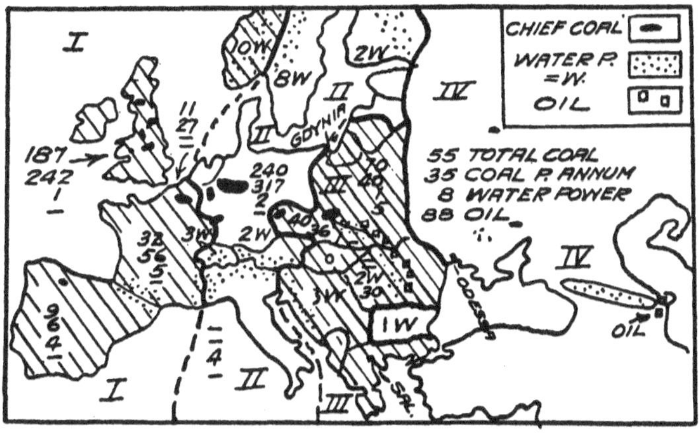

Fig. 147.—The Tentative Division of Europe into four North-South Trade Regions based on the interchange of Power for Crude Supplies. The values of the figures for Coal, Water Power and Oil are explained in the text. Their order is as shown in Russia. (Compare Fig. 14.)

The map in Fig. 147 shows a tentative zoning of Europe along the lines suggested. It does not altogether agree with Toynbee's suggestions. The environmental controls of coal, water transport, *etc.*, here indicated will, of course, be greatly affected by national jealousies and especially by the pet abomination of the present period, Tariff Walls.

Possibly the states of Western Europe (Region I in Fig. 147) will grow closer together as the northern products of Norway and the Mediterranean products of Spain are exchanged for the power concentrated in Britain, North France and Belgium. Region II includes the great German coalfields of which Westphalia alone it is claimed contains half the coal of Europe. It should more and more attract the northern goods from Sweden, Denmark and Finland, and the southern goods from Italy (with Switzerland and Austria).

Region III is based on the Silesian coalfield, which has been split between Germany, Czecho-Slovakia and Poland. Poland, however, has about 90 per cent. of the coal reserves. It should supply power to Hungary and the Balkans—though the Donetz Field may attract Wallachia and Bulgaria via the Black Sea. So also Lithuania and Latvia will probably tend to link up with industrial Poland. The remarkable rise of the Polish port Gdynia at the north end of the corridor strikingly illustrates this urge for north-south trading. From a fishing village of 200 inhabitants it has grown in a decade to a city of 45,000 people (Simonds). Here the Poles are building a harbour of 400 acres which will render them independent of Dantzig, 14 miles away. Region IV in Fig. 147 is the vast country of the Soviets—a little world in itself, which includes all the four agricultural belts, and will develop its industry on the Donetz coalfield to rival similar centres in England, Belgium and the Ruhr. These economic attractions exist all the time despite man-made restrictions.

Man cannot greatly modify his environment. The economic status of a nation depends largely on the wealth with which Nature has endowed her. Not a few of a nation's troubles are due to the fact that her rulers have not learned her economic limitations. Montenegro can never equal Belgium; although judging from past absurdities elsewhere it is not impossible that a time may come when tariff-walls will surround Montenegro to enable her to manufacture Montenegrin tractors (at prohibitive prices) for Montenegrin farmers! It is a platitude to say that the wealthiest nations are not those with the most satisfied people. Man must learn that to solve over-population birth-control is better than big battalions. It is likely that little Denmark, raised from poverty by scientific agriculture, has a larger proportion of happy citizens than England with her wealthy manufacturers and her industrial slums. Or one may cite Switzerland, which seems to the writer to offer a happy augury of what Europe could easily become if man were not so irrational. Difficulties due to differences of Religion and Language have almost disappeared there, though they were very apparent only a century or two ago. Turning to the New World, Canada and the United States fought fiercely in the Revolutionary Period, but to-day the long stretch of their common boundary, a distance of 2,500 miles, is proudly referred to as the "Unguarded Frontier." Europe is not so wide as this, but how much time and energy is wasted on the innumerable European Frontiers because Nations will not trust each other. It is claimed that four-fifths of Europe's income is spent in the unnecessary defence of one section against another! It is one of the ironies of Fate that Woodrow Wilson, the great advocate for moral standards in international politics, should collapse at 64, repudiated by his own people; while "Tiger Clemenceau," the most ruthless of the Treaty makers, should be able at 78 to impose such conditions on the vanquished that Versailles has been cynically termed the "Peace to End Peace."

The writer ventures to close this volume much as he did his earlier book on "Environment and Race." The ideal towards which the most enlightened statesmen are working is surely a World at Peace. The chief obstacles in the way in Europe are cultural prejudices which are entirely man-made. They will surely tend to diminish as each nation realises the place in the world's "Order of Precedence" for which its cultural and economic status equips it. The foregoing study is an attempt to investigate some of the cultural and historical problems of Europe. If it helps in however small a degree to promote the brotherhood of man, the writer's main object will have been accomplished.

INDEX

Geographical terms and Foreign words in *Italics*.
More important references in **bold** type.

www.ingramcontent.com/pod-product-compliance
Lightning Source LLC
LaVergne TN
LVHW100246210926
848480LV00003B/841